D1483046

INTRODUCTION TO
BAYESIAN STATISTICS

INTRODUCTION TO BAYESIAN STATISTICS

Third Edition

WILLIAM M. BOLSTAD
JAMES M. CURRAN

Published by John Wiley & Sons, Inc., Hoboken, New Jersey.
Published simultaneously in Canada.

For general information on our other products and services or for technical support, please contact our Customer Care Department within the United States at (800) 762-2974, outside the United States at (317) 572-3993 or fax (317) 572-4002.

Wiley also publishes its books in a variety of electronic formats. Some content that appears in print may not be available in electronic formats. For more information about Wiley products, visit our web site at www.wiley.com.

Library of Congress Cataloging-in-Publication Data is available.

ISBN 978-1-118-09315-8

Printed in the United States of America

This book is dedicated to

Sylvie,
Ben, Rachel,
Emily, Mary, and
Elizabeth

Contents

Preface   xiii

1 Introduction to Statistical Science   1

1.1   The Scientific Method: A Process for Learning   3
1.2   The Role of Statistics in the Scientific Method   5
1.3   Main Approaches to Statistics   5
1.4   Purpose and Organization of This Text   8

2 Scientific Data Gathering   13

2.1   Sampling from a Real Population   14
2.2   Observational Studies and Designed Experiments   17
      Monte Carlo Exercises   23

3 Displaying and Summarizing Data   31

3.1   Graphically Displaying a Single Variable   32
3.2   Graphically Comparing Two Samples   39
3.3   Measures of Location   41
3.4   Measures of Spread   44

3.5   Displaying Relationships Between Two or More Variables   46
3.6   Measures of Association for Two or More Variables   49
Exercises   52

4   Logic, Probability, and Uncertainty   59

4.1   Deductive Logic and Plausible Reasoning   60
4.2   Probability   62
4.3   Axioms of Probability   64
4.4   Joint Probability and Independent Events   65
4.5   Conditional Probability   66
4.6   Bayes' Theorem   68
4.7   Assigning Probabilities   74
4.8   Odds and Bayes Factor   75
4.9   Beat the Dealer   76
Exercises   80

5   Discrete Random Variables   83

5.1   Discrete Random Variables   84
5.2   Probability Distribution of a Discrete Random Variable   86
5.3   Binomial Distribution   90
5.4   Hypergeometric Distribution   92
5.5   Poisson Distribution   93
5.6   Joint Random Variables   96
5.7   Conditional Probability for Joint Random Variables   100
Exercises   104

6   Bayesian Inference for Discrete Random Variables   109

6.1   Two Equivalent Ways of Using Bayes' Theorem   114
6.2   Bayes' Theorem for Binomial with Discrete Prior   116
6.3   Important Consequences of Bayes' Theorem   119
6.4   Bayes' Theorem for Poisson with Discrete Prior   120
Exercises   122
Computer Exercises   126

7   Continuous Random Variables   129

7.1   Probability Density Function   131
7.2   Some Continuous Distributions   135
7.3   Joint Continuous Random Variables   143

7.4   Joint Continuous and Discrete Random Variables   144
        Exercises   147

8 Bayesian Inference for Binomial Proportion   149

8.1   Using a Uniform Prior   150
8.2   Using a Beta Prior   151
8.3   Choosing Your Prior   154
8.4   Summarizing the Posterior Distribution   158
8.5   Estimating the Proportion   161
8.6   Bayesian Credible Interval   162
        Exercises   164
        Computer Exercises   167

9 Comparing Bayesian and Frequentist Inferences for Proportion   169

9.1   Frequentist Interpretation of Probability and Parameters   170
9.2   Point Estimation   171
9.3   Comparing Estimators for Proportion   174
9.4   Interval Estimation   175
9.5   Hypothesis Testing   178
9.6   Testing a One-Sided Hypothesis   179
9.7   Testing a Two-Sided Hypothesis   182
        Exercises   187
        Monte Carlo Exercises   190

10 Bayesian Inference for Poisson   193

10.1   Some Prior Distributions for Poisson   194
10.2   Inference for Poisson Parameter   200
        Exercises   207
        Computer Exercises   208

11 Bayesian Inference for Normal Mean   211

11.1   Bayes' Theorem for Normal Mean with a Discrete Prior   211
11.2   Bayes' Theorem for Normal Mean with a Continuous Prior   218
11.3   Choosing Your Normal Prior   222
11.4   Bayesian Credible Interval for Normal Mean   224
11.5   Predictive Density for Next Observation   227
        Exercises   230
        Computer Exercises   232

12 Comparing Bayesian and Frequentist Inferences for Mean   **237**

12.1 Comparing Frequentist and Bayesian Point Estimators   238
12.2 Comparing Confidence and Credible Intervals for Mean   241
12.3 Testing a One-Sided Hypothesis about a Normal Mean   243
12.4 Testing a Two-Sided Hypothesis about a Normal Mean   247
Exercises   251

13 Bayesian Inference for Difference Between Means   **255**

13.1 Independent Random Samples from Two Normal Distributions   256
13.2 Case 1: Equal Variances   257
13.3 Case 2: Unequal Variances   262
13.4 Bayesian Inference for Difference Between Two Proportions Using Normal Approximation   265
13.5 Normal Random Samples from Paired Experiments   266
Exercises   272

14 Bayesian Inference for Simple Linear Regression   **283**

14.1 Least Squares Regression   284
14.2 Exponential Growth Model   288
14.3 Simple Linear Regression Assumptions   290
14.4 Bayes' Theorem for the Regression Model   292
14.5 Predictive Distribution for Future Observation   298
Exercises   303
Computer Exercises   312

15 Bayesian Inference for Standard Deviation   **315**

15.1 Bayes' Theorem for Normal Variance with a Continuous Prior   316
15.2 Some Specific Prior Distributions and the Resulting Posteriors   318
15.3 Bayesian Inference for Normal Standard Deviation   326
Exercises   332
Computer Exercises   335

16 Robust Bayesian Methods   **337**

16.1 Effect of Misspecified Prior   338
16.2 Bayes' Theorem with Mixture Priors   340

Exercises   349

Computer Exercises   351

17 Bayesian Inference for Normal
with Unknown Mean and Variance   **355**

17.1   The Joint Likelihood Function   358

17.2   Finding the Posterior when Independent Jeffreys' Priors
       for μ and σ^2 Are Used   359

17.3   Finding the Posterior when a Joint Conjugate Prior for μ
       and σ^2 Is Used   361

17.4   Difference Between Normal Means with Equal Unknown
       Variance   367

17.5   Difference Between Normal Means with Unequal Unknown
       Variances   377

       Computer Exercises   383

       Appendix: Proof that the Exact Marginal Posterior
              Distribution of μ is *Student's t*   385

18 Bayesian Inference for Multivariate Normal Mean Vector   **393**

18.1   Bivariate Normal Density   394

18.2   Multivariate Normal Distribution   397

18.3   The Posterior Distribution of the Multivariate Normal
       Mean Vector when Covariance Matrix Is Known   398

18.4   Credible Region for Multivariate Normal Mean Vector
       when Covariance Matrix Is Known   400

18.5   Multivariate Normal Distribution with Unknown
       Covariance Matrix   402

       Computer Exercises   406

19 Bayesian Inference for the Multiple Linear Regression Model   **411**

19.1   Least Squares Regression for Multiple Linear Regression
       Model   412

19.2   Assumptions of Normal Multiple Linear Regression Model   414

19.3   Bayes' Theorem for Normal Multiple Linear Regression
       Model   415

19.4   Inference in the Multivariate Normal Linear Regression
       Model   419

19.5   The Predictive Distribution for a Future Observation   425

Computer Exercises   428

**20   Computational Bayesian Statistics Including Markov Chain
Monte Carlo   431**

20.1   Direct Methods for Sampling from the Posterior   436
20.2   Sampling–Importance–Resampling   450
20.3   Markov Chain Monte Carlo Methods   454
20.4   Slice Sampling   470
20.5   Inference from a Posterior Random Sample   473
20.6   Where to Next?   475

A   Introduction to Calculus   477

B   Use of Statistical Tables   497

C   Using the Included Minitab Macros   523

D   Using the Included R Functions   543

E   Answers to Selected Exercises   565

References   591

Index   595

PREFACE

Our original goal for this book was to introduce Bayesian statistics at the earliest possible stage to students with a reasonable mathematical background. This entailed coverage of a similar range of topics as an introductory statistics text, but from a Bayesian perspective. The emphasis is on statistical inference. We wanted to show how Bayesian methods can be used for inference and how they compare favorably with the frequentist alternatives. This book is meant to be a good place to start the study of Bayesian statistics. From the many positive comments we have received from many users, we think the book succeeded in its goal. A course based on this goal would include Chapters 1–14.

Our feedback also showed that many users were taking up the book at a more intermediate level instead of the introductory level original envisaged. The topics covered in Chapters 2 and 3 would be old hat for these users, so we would have to include some more advanced material to cater for the needs of that group. The second edition aimed to meet this new goal as well as the original goal. We included more models, mainly with a single parameter. Nuisance parameters were dealt with using approximations. A course based on this goal would include Chapters 4–16.

Changes in the Third Edition

Later feedback showed that some readers with stronger mathematical and statistical background wanted the text to include more details on how to deal with multi-parameter models. The third edition contains four new chapters to satisfy this additional goal, along with some minor rewriting of the existing chapters. Chapter 17 covers Bayesian inference for *Normal* observations where we do not know either the mean or the variance. This chapter extends the ideas in Chapter 11, and also discusses the two sample case, which in turn allows the reader to consider inference on the difference between two means. Chapter 18 introduces the *Multivariate Normal* distribution, which we need in order to discuss multiple linear regression in Chapter 19. Finally, Chapter 20 takes the user beyond the kind of conjugate analysis is considered in most of the book, and into the realm of computational Bayesian inference. The covered topics in Chapter 20 have an intentional light touch, but still give the user valuable information and skills that will allow them to deal with different problems. We have included some new exercises and new computer exercises which use new Minitab macros and *R*-functions. The Minitab macros can be downloaded from the book website: `http://introbayes.ac.nz`. The new R functions have been incorporated in a new and improved version of the R package `Bolstad`, which can either be downloaded from a CRAN mirror or installed directly in R using the internet. Instructions on the use and installation of the Minitab macros and the `Bolstad` package in R are given in Appendices C and D respectively. Both of these appendices have been rewritten to accommodate changes in R and Minitab that have occurred since the second edition.

Our Perspective on Bayesian Statistics

A book can be characterized as much by what is left out as by what is included. This book is our attempt to show a coherent view of Bayesian statistics as a good way to do statistical inference. Details that are outside the scope of the text are included in footnotes. Here are some of our reasons behind our choice of the topics we either included or excluded.

In particular, we did not mention decision theory or loss functions when discussing Bayesian statistics. In many books, Bayesian statistics gets compartmentalized into decision theory while inference is presented in the frequentist manner. While decision theory is a very interesting topic in its own right, we want to present the case for Bayesian statistical inference, and did not want to get side-tracked.

We think that in order to get full benefit of Bayesian statistics, one really has to consider all priors subjective. They are either (1) a summary of what you believe or (2) a summary of all you allow yourself to believe initially. We consider the subjective prior as the relative weights given to each possible parameter value, before looking at the data. Even if we use a flat prior to give all possible values equal prior weight, it is subjective since we chose it. In any case, it gives all values equal weight only in that parameterization, so it can be considered "objective" only in that parameterization. In this book we do not wish to dwell on the problems associated with trying to be objective in Bayesian statistics. We explain why universal objectivity is not possible (in a footnote since we do not want to distract the reader). We want to leave him/her with the "relative weight" idea of the prior in the parameterization in which they have the problem in.

In the first edition we did not mention Jeffreys' prior explicitly, although the $beta(\frac{1}{2}, \frac{1}{2})$ prior for *binomial* and flat prior for *normal* mean are in fact the Jeffreys' prior for those respective observation distributions. In the second edition we do mention Jeffreys' prior for *binomial, Poisson, normal* mean, and *normal* standard deviation. In third edition we mention the independent Jeffreys priors for *normal* mean and standard deviation. In particular, we don't want to get the reader involved with the problems about Jeffreys' prior, such as for mean and variance together, as opposed to independent Jeffreys' priors, or the Jeffreys' prior violating the likelihood principal. These are beyond the level we wish to go. We just want the reader to note the Jeffreys' prior in these cases as possible priors, the relative weights they give, when they may be appropriate, and how to use them. Mathematically, all parameterizations are equally valid; however, usually only the main one is very meaningful. We want the reader to focus on relative weights for their parameterization as the prior. It should be (a) a summary of their prior belief (conjugate prior matching their prior beliefs about moments or median), (b) flat (hence objective) for their parameterization, or (c) some other form that gives reasonable weight over the whole range of possible values. The posteriors will be similar for all priors that have reasonable weight over the whole range of possible values.

The Bayesian inference on the standard deviation of the normal was done where the mean is considered a known parameter. The conjugate prior for the variance is the *inverse chi-squared* distribution. Our intuition is about the standard deviation, yet we are doing Bayes' theorem on the variance. This required introducing the change of variable formula for the prior density.

In the second edition we considered the mean as known. This avoided the mathematically more advanced case where both mean and standard deviation are unknown. In the third edition we now cover this topic in Chapter 17. In earlier editions the *Student's t* is presented as the required adjustment to credible intervals for the mean when the variance is estimated from the data. In the third edition we show in Chapter 17 that in fact this would be the result when the joint posterior found, and the variance marginalized out. Chapter 17 also covers inference on the difference in two means. This problem is made substantially harder when one relaxes the assumption that both populations have the same variance. Chapter 17 derives the Bayesian solution to the well-known Behrens-Fisher problem for the difference in two population means with unequal population variances. The function `bayes.t.test` in the R package for this book actually gives the user a numerical solution using Gibbs sampling. Gibbs sampling is covered in Chapter 20 of this new edition.

Acknowledgments

WMB would like to thank all the readers who have sent him comments and pointed out misprints in the first and second editions. These have been corrected. WMB would like to thank Cathy Akritas and Gonzalo Ovalles at Minitab for help in improving his Minitab macros. WMB and JMC would like to thank Jon Gurstelle, Steve Quigley, Sari Friedman, Allison McGinniss, and the team at John Wiley & Sons for their support.

Finally, last but not least, WMB wishes to thank his wife Sylvie for her constant love and support.

WILLIAM M. "BILL' BOLSTAD

Hamilton, New Zealand

JAMES M. CURRAN

Auckland, New Zealand

CHAPTER 1

INTRODUCTION TO STATISTICAL SCIENCE

Statistics is the science that relates data to specific questions of interest. This includes devising methods to gather data relevant to the question, methods to summarize and display the data to shed light on the question, and methods that enable us to draw answers to the question that are supported by the data. Data almost always contain uncertainty. This uncertainty may arise from selection of the items to be measured, or it may arise from variability of the measurement process. Drawing general conclusions from data is the basis for increasing knowledge about the world, and is the basis for all rational scientific inquiry. *Statistical inference* gives us methods and tools for doing this despite the uncertainty in the data. The methods used for analysis depend on the way the data were gathered. It is vitally important that there is a probability model explaining how the uncertainty gets into the data.

Showing a Causal Relationship from Data

Suppose we have observed two variables X and Y. Variable X appears to have an association with variable Y. If high values of X occur with high values of variable Y and low values of X occur with low values of Y, then we say the

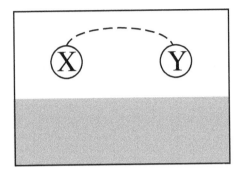

Figure 1.1   Association between two variables.

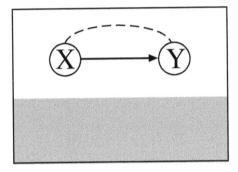

Figure 1.2   Association due to causal relationship.

association is positive. On the other hand, the association could be negative in which high values of variable X occur in with low values of variable Y. Figure 1.1 shows a schematic diagram where the association is indicated by the dashed curve connecting X and Y. The unshaded area indicates that X and Y are observed variables. The shaded area indicates that there may be additional variables that have not been observed.

We would like to determine why the two variables are associated. There are several possible explanations. The association might be a causal one. For example, X might be the cause of Y. This is shown in Figure 1.2, where the causal relationship is indicated by the arrow from X to Y.

On the other hand, there could be an unidentified third variable Z that has a causal effect on both X and Y. They are not related in a direct causal relationship. The association between them is due to the effect of Z. Z is called a *lurking* variable, since it is hiding in the background and it affects the data. This is shown in Figure 1.3.

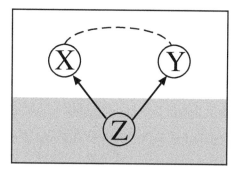

Figure 1.3   Association due to lurking variable.

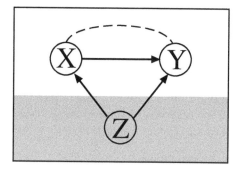

Figure 1.4   Confounded causal and lurking variable effects.

It is possible that both a causal effect and a lurking variable may both be contributing to the association. This is shown in Figure 1.4. We say that the causal effect and the effect of the lurking variable are *confounded*. This means that both effects are included in the association.

Our first goal is to determine which of the possible reasons for the association holds. If we conclude that it is due to a causal effect, then our next goal is to determine the size of the effect. If we conclude that the association is due to causal effect confounded with the effect of a lurking variable, then our next goal becomes determining the sizes of both the effects.

1.1   The Scientific Method: A Process for Learning

In the Middle Ages, science was deduced from principles set down many centuries earlier by authorities such as Aristotle. The idea that scientific theories should be tested against real world data revolutionized thinking. This way of thinking known as the scientific method sparked the Renaissance.

The scientific method rests on the following premises:

- A scientific hypothesis can never be shown to be absolutely true.

- However, it must potentially be disprovable.

- It is a useful model until it is established that it is not true.

- Always go for the simplest hypothesis, unless it can be shown to be false.

This last principle, elaborated by William of Ockham in the 13th century, is now known as *Ockham's razor* and is firmly embedded in science. It keeps science from developing fanciful overly elaborate theories. Thus the scientific method directs us through an improving sequence of models, as previous ones get falsified. The scientific method generally follows the following procedure:

1. Ask a question or pose a problem in terms of the current scientific hypothesis.

2. Gather all the relevant information that is currently available. This includes the current knowledge about parameters of the model.

3. Design an investigation or experiment that addresses the question from step 1. The predicted outcome of the experiment should be one thing if the current hypothesis is true, and something else if the hypothesis is false.

4. Gather data from the experiment.

5. Draw conclusions given the experimental results. Revise the knowledge about the parameters to take the current results into account.

The scientific method searches for cause-and-effect relationships between an experimental variable and an outcome variable. In other words, how changing the experimental variable results in a change to the outcome variable. Scientific modeling develops mathematical models of these relationships. Both of them need to isolate the experiment from outside factors that could affect the experimental results. All outside factors that can be identified as possibly affecting the results must be controlled. It is no coincidence that the earliest successes for the method were in physics and chemistry where the few outside factors could be identified and controlled. Thus there were no *lurking* variables. All other relevant variables could be identified and could then be physically controlled by being held constant. That way they would not affect results of the experiment, and the effect of the experimental variable on the outcome variable could be determined. In biology, medicine, engineering, technology, and the social sciences it is not that easy to identify the relevant factors that must be controlled. In those fields a different way to control outside factors is needed, because they cannot be identified beforehand and physically controlled.

1.2   The Role of Statistics in the Scientific Method

Statistical methods of inference can be used when there is *random* variability in the data. The probability model for the data is justified by the design of the investigation or experiment. This can extend the scientific method into situations where the relevant outside factors cannot even be identified. Since we cannot identify these outside factors, we cannot control them directly. The lack of direct control means the outside factors will be affecting the data. There is a danger that the wrong conclusions could be drawn from the experiment due to these uncontrolled outside factors.

The important statistical idea of *randomization* has been developed to deal with this possibility. The unidentified outside factors can be "averaged out" by randomly assigning each unit to either treatment or control group. This contributes variability to the data. Statistical conclusions always have some uncertainty or error due to variability in the data. We can develop a probability model of the data variability based on the randomization used. Randomization not only reduces this uncertainty due to outside factors, it also allows us to measure the amount of uncertainty that remains using the probability model. Randomization lets us control the outside factors statistically, by averaging out their effects.

Underlying this is the idea of a statistical *population*, consisting of all possible values of the observations that could be made. The data consists of observations taken from a *sample* of the population. For valid inferences about the population *parameters* from the sample *statistics*, the sample must be "representative" of the population. Amazingly, choosing the sample randomly is the most effective way to get representative samples!

1.3   Main Approaches to Statistics

There are two main philosophical approaches to statistics. The first is often referred to as the *frequentist* approach. Sometimes it is called the *classical* approach. Procedures are developed by looking at how they perform over all possible random samples. The probabilities do not relate to the particular random sample that was obtained. In many ways this indirect method places the "cart before the horse."

The alternative approach that we take in this book is the *Bayesian* approach. It applies the laws of probability directly to the problem. This offers many fundamental advantages over the more commonly used frequentist approach. We will show these advantages over the course of the book.

Frequentist Approach to Statistics

Most introductory statistics books take the frequentist approach to statistics, which is based on the following ideas:

- Parameters, the numerical characteristics of the population, are fixed but unknown constants.

- Probabilities are always interpreted as long-run relative frequency.

- Statistical procedures are judged by how well they perform in the long run over an infinite number of hypothetical repetitions of the experiment.

Probability statements are only allowed for random quantities. The unknown parameters are fixed, not random, so probability statements cannot be made about their value. Instead, a sample is drawn from the population, and a sample statistic is calculated. The probability distribution of the statistic over all possible random samples from the population is determined and is known as the *sampling distribution* of the statistic. A parameter of the population will also be a parameter of the sampling distribution. The probability statement that can be made about the statistic based on its sampling distribution is converted to a *confidence* statement about the parameter. The confidence is based on the average behavior of the procedure over all possible samples.

Bayesian Approach to Statistics

The Reverend Thomas Bayes first discovered the theorem that now bears his name. It was written up in a paper *An Essay Towards Solving a Problem in the Doctrine of Chances*. This paper was found after his death by his friend Richard Price, who had it published posthumously in the *Philosophical Transactions of the Royal Society* in 1763 (1763). Bayes showed how *inverse probability* could be used to calculate probability of antecedent events from the occurrence of the consequent event. His methods were adopted by Laplace and other scientists in the 19th century, but had largely fallen from favor by the early 20th century. By the middle of the 20th century, interest in Bayesian methods had been renewed by de Finetti, Jeffreys, Savage, and Lindley, among others. They developed a complete method of statistical inference based on Bayes' theorem.

This book introduces the Bayesian approach to statistics. The ideas that form the basis of the this approach are:

- Since we are uncertain about the true value of the parameters, we will consider them to be random variables.

- The rules of probability are used directly to make inferences about the parameters.

- Probability statements about parameters must be interpreted as "degree of belief." The *prior distribution* must be subjective. Each person can have his/her own prior, which contains the relative weights that person gives to every possible parameter value. It measures how "plausible" the person considers each parameter value to be before observing the data.

- We revise our beliefs about parameters after getting the data by using Bayes' theorem. This gives our *posterior distribution* which gives the relative weights we give to each parameter value after analyzing the data. The posterior distribution comes from two sources: the prior distribution and the observed data.

This has a number of advantages over the conventional frequentist approach. Bayes' theorem is the only consistent way to modify our beliefs about the parameters given the data that actually occurred. This means that the inference is based on the actual occurring data, not all possible data sets that might have occurred but did not! Allowing the parameter to be a random variable lets us make probability statements about it, posterior to the data. This contrasts with the conventional approach where inference probabilities are based on all possible data sets that could have occurred for the fixed parameter value. Given the actual data, there is nothing random left with a fixed parameter value, so one can only make *confidence* statements, based on what could have occurred. Bayesian statistics also has a general way of dealing with a *nuisance parameter*. A nuisance parameter is one which we do not want to make inference about, but we do not want them to interfere with the inferences we are making about the main parameters. Frequentist statistics does not have a general procedure for dealing with them. Bayesian statistics is predictive, unlike conventional frequentist statistics. This means that we can easily find the conditional probability distribution of the next observation given the sample data.

Monte Carlo Studies

In frequentist statistics, the parameter is considered a fixed, but unknown, constant. A statistical procedure such as a particular estimator for the parameter cannot be judged from the value it gives. The parameter is unknown, so we can not know the value the estimator should be giving. If we knew the value of the parameter, we would not be using an estimator.

Instead, statistical procedures are evaluated by looking how they perform in the long run over all possible samples of data, for fixed parameter values over some range. For instance, we fix the parameter at some value. The estimator depends on the random sample, so it is considered a random variable having a probability distribution. This distribution is called the *sampling distribution* of the estimator, since its probability distribution comes from taking all possible random samples. Then we look at how the estimator is distributed around the parameter value. This is called sample space averaging. Essentially it compares the performance of procedures before we take any data.

Bayesian procedures consider the parameter to be a random variable, and its posterior distribution is conditional on the sample data that actually occurred, not all those samples that were possible but did not occur. However, *before* the experiment, we might want to know how well the Bayesian procedure works at some specific parameter values in the range.

To evaluate the Bayesian procedure using sample space averaging, we have to consider the parameter to be both a random variable and a fixed but unknown value at the same time. We can get past the apparent contradiction in the nature of the parameter because the probability distribution we put on the parameter measures our uncertainty about the true value. It shows the relative belief weights we give to the possible values of the unknown parameter! After looking at the data, our belief distribution over the parameter values has changed. This way we can think of the parameter as a fixed, but unknown, value at the same time as we think of it being a random variable. This allows us to evaluate the Bayesian procedure using sample space averaging. This is called *pre-posterior* analysis because it can be done before we obtain the data.

In Chapter 4, we will find out that the laws of probability are the best way to model uncertainty. Because of this, Bayesian procedures will be optimal in the post-data setting, given the data that actually occurred. In Chapters 9 and 11, we will see that Bayesian procedures perform very well in the pre-data setting when evaluated using *pre-posterior* analysis. In fact, it is often the case that Bayesian procedures outperform the usual frequentist procedures even in the pre-data setting.

Monte Carlo studies are a useful way to perform sample space averaging. We draw a large number of samples randomly using the computer and calculate the statistic (frequentist or Bayesian) for each sample. The empirical distribution of the statistic (over the large number of random samples) approximates its sampling distribution (over all possible random samples). We can calculate statistics such as mean and standard deviation on this Monte Carlo sample to approximate the mean and standard deviation of the sampling distribution. Some small-scale Monte Carlo studies are included as exercises.

1.4   Purpose and Organization of This Text

A very large proportion of undergraduates are required to take a service course in statistics. Almost all of these courses are based on frequentist ideas. Most of them do not even mention Bayesian ideas. As a statistician, I know that Bayesian methods have great theoretical advantages. I think we should be introducing our best students to Bayesian ideas, from the beginning. There are not many introductory statistics text books based on the Bayesian ideas. Some other texts include Berry (1996), Press (1989), and Lee (1989).

This book aims to introduce students with a good mathematics background to Bayesian statistics. It covers the same topics as a standard introductory statistics text, only from a Bayesian perspective. Students need reasonable algebra skills to follow this book. Bayesian statistics uses the rules of probability, so competence in manipulating mathematical formulas is required.

Students will find that general knowledge of calculus is helpful in reading this book. Specifically they need to know that area under a curve is found by integrating, and that a maximum or minimum of a continuous differentiable function is found where the derivative of the function equals zero. However, the actual calculus used is minimal. The book is self-contained with a calculus appendix that students can refer to.

Chapter 2 introduces some fundamental principles of scientific data gathering to control the effects of unidentified factors. These include the need for drawing samples randomly, along with some random sampling techniques. The reason why there is a difference between the conclusions we can draw from data arising from an observational study and from data arising from a randomized experiment is shown. Completely randomized designs and randomized block designs are discussed.

Chapter 3 covers elementary methods for graphically displaying and summarizing data. Often a good data display is all that is necessary. The principles of designing displays that are true to the data are emphasized.

Chapter 4 shows the difference between deduction and induction. Plausible reasoning is shown to be an extension of logic where there is uncertainty. It turns out that plausible reasoning must follow the same rules as probability. The axioms of probability are introduced and the rules of probability, including conditional probability and Bayes' theorem are developed.

Chapter 5 covers discrete random variables, including joint and marginal discrete random variables. The *binomial, hypergeometric*, and *Poisson* distributions are introduced, and the situations where they arise are characterized.

Chapter 6 covers Bayes' theorem for discrete random variables using a table. We see that two important consequences of the method are that multiplying the prior by a constant, or that multiplying the likelihood by a constant do not affect the resulting posterior distribution. This gives us the "proportional form" of Bayes' theorem. We show that we get the same results when we analyze the observations sequentially using the posterior after the previous observation as the prior for the next observation, as when we analyze the observations all at once using the joint likelihood and the original prior. We demonstrate Bayes' theorem for binomial observations with a discrete prior and for Poisson observations with a discrete prior.

Chapter 7 covers continuous random variables, including joint, marginal, and conditional random variables. The *beta, gamma*, and *normal* distributions are introduced in this chapter.

Chapter 8 covers Bayes' theorem for the population proportion (*binomial*) with a continuous prior. We show how to find the posterior distribution of the population proportion using either a *uniform* prior or a *beta* prior. We explain how to choose a suitable prior. We look at ways of summarizing the posterior distribution.

Chapter 9 compares the Bayesian inferences with the frequentist inferences. We show that the Bayesian estimator (posterior mean using a uniform prior) has better performance than the frequentist estimator (sample proportion) in terms of mean squared error over most of the range of possible values. This kind of frequentist analysis is useful before we perform our Bayesian analysis. We see the Bayesian credible interval has a much more useful interpretation than the frequentist confidence interval for the population proportion. One-sided and two-sided hypothesis tests using Bayesian methods are introduced.

Chapter 10 covers Bayes' theorem for the *Poisson* observations with a continuous prior. The prior distributions used include the *positive uniform*, the *Jeffreys' prior*, and the *gamma* prior. Bayesian inference for the *Poisson* parameter using the resulting posterior include Bayesian credible intervals and two-sided tests of hypothesis, as well as one-sided tests of hypothesis.

Chapter 11 covers Bayes' theorem for the mean of a *normal* distribution with known variance. We show how to choose a *normal* prior. We discuss dealing with nuisance parameters by marginalization. The predictive density of the next observation is found by considering the population mean a nuisance parameter and marginalizing it out.

Chapter 12 compares Bayesian inferences with the frequentist inferences for the mean of a normal distribution. These comparisons include point and interval estimation and involve hypothesis tests including both the one-sided and the two-sided cases.

Chapter 13 shows how to perform Bayesian inferences for the difference between normal means and how to perform Bayesian inferences for the difference between proportions using the normal approximation.

Chapter 14 introduces the simple linear regression model and shows how to perform Bayesian inferences on the slope of the model. The predictive distribution of the next observation is found by considering both the slope and intercept to be nuisance parameters and marginalizing them out.

Chapter 15 introduces Bayesian inference for the standard deviation σ, when we have a random sample of *normal* observations with known mean μ. This chapter is at a somewhat higher level than the previous chapters and requires the use of the change-of-variable formula for densities. Priors used include *positive uniform for standard deviation*, *positive uniform for variance*, *Jeffreys' prior*, and the *inverse chi-squared* prior. We discuss how to choose an *inverse chi-squared* prior that matches our prior belief about the median. Bayesian inferences from the resulting posterior include point estimates, credible intervals, and hypothesis tests including both the one-sided and two-sided cases.

Chapter 16 shows how we can make Bayesian inference robust against a misspecified prior by using a mixture prior and marginalizing out the mixture parameter. This chapter is also at a somewhat higher level than the others, but it shows how one of the main dangers of Bayesian analysis can be avoided.

Chapter 17 returns to the problem we discussed in Chapter 11 — that is, of making inferences about the mean of a normal distribution. In this chapter, however, we explicitly model the unknown population standard deviation and show how the approximations we suggested in Chapter 11 are exactly true.

We also deal with the two sample cases so that inference can be performed on the difference between two means.

Chapter 18 introduces the multivariate normal distribution and extends the theory from Chapters 11 and 17 to the multivariate case. The multivariate normal distribution is essential for the discussion of linear models and, in particular, multiple regression.

Chapter 19 extends the material from 14 on simple linear regression to the more familiar multiple regression setting. The methodology for making inference about the usefulness of explanatory variables in predicting the response is given, and the posterior predictive distribution for a new observation is derived.

Chapter 20 provides a brief introduction to modern computational Bayesian statistics. Computational Bayesian statistics relies heavily on being able to efficiently sample from potentially complex distributions. This chapter gives an introduction to a number of techniques that are used. Readers might be slightly disappointed that we did not cover popular computer programs such as BUGS and JAGS, which have very efficient general implementations of many computational Bayesian methods and tie in well to R. We felt that these topics require almost an entire book in their own right, and as such we could not do justice to them in such a short space.

Main Points

- An association between two variables does not mean that one causes the other. It may be due to a causal relationship, it may be due to the effect of a third (lurking) variable on both the other variables, or it may be due to a combination of a causal relationship and the effect of a lurking variable.

- Scientific method is a method for searching for cause-and-effect relationships and measuring their strength. It uses controlled experiments, where outside factors that may affect the measurements are controlled. This isolates the relationship between the two variables from the outside factors, so the relationship can be determined.

- Statistical methods extend the scientific method to cases where the outside factors are not identified and hence cannot be controlled. The principle of *randomization* is used to statistically control these unidentified outside factors by averaging out their effects. This contributes to *variability* in the data.

- We can use the probability model (based on the randomization method) to measure the uncertainty.

- The frequentist approach to statistics considers the parameter to be a fixed but unknown constant. The only kind of probability allowed is long-run relative frequency. These probabilities are only for observations and sample statistics, given the unknown parameters. Statistical procedures are judged by how they perform in an infinite number of hypothetical repetitions of the experiment.

- The Bayesian approach to statistics allows the parameter to be considered a random variable. Probabilities can be calculated for parameters as well as observations and sample statistics. Probabilities calculated for parameters are interpreted as "degree of belief" and must be subjective. The rules of probability are used to revise our beliefs about the parameters, given the data.

- A frequentist estimator is evaluated by looking at its sampling distribution for a fixed parameter value and seeing how it is distributed over all possible repetitions of the experiment.

- If we look at the sampling distribution of a Bayesian estimator for a fixed parameter value, it is called pre-posterior analysis since it can be done prior to taking the data.

- A Monte Carlo study is where we perform the experiment a large number of times and calculate the statistic for each experiment. We use the empirical distribution of the statistic over all the samples we took in our study instead of its sampling distribution over all possible repetitions.

CHAPTER 2

SCIENTIFIC DATA GATHERING

Scientists gather data purposefully, in order to find answers to particular questions. Statistical science has shown that data should be relevant to the particular questions, yet be gathered using randomization. The development of methods to gather data purposefully, yet using randomization, is one of the greatest contributions the field of statistics has made to the practice of science.

Variability in data solely due to chance can be averaged out by increasing the sample size. Variability due to other causes cannot be. Statistical methods have been developed for gathering data randomly, yet relevant to a specific question. These methods can be divided into two fields. Sample survey theory is the study of methods for sampling from a finite real population. Experimental design is the study of methods for designing experiments that focus on the desired factors and that are not affected by other possibly unidentified ones.

Inferences always depend on the probability model which we assume generated the observed data being the correct one. When data are not gathered randomly, there is a risk that the observed pattern is due to lurking variables that were not observed, instead of being a true reflection of the underlying

Introduction to Bayesian Statistics, 3rd ed.
By Bolstad, W. M. and Curran, J. M. Copyright © 2016 John Wiley & Sons, Inc.

pattern. In a properly designed experiment, treatments are assigned to subjects in such a way as to reduce the effects of any lurking variables that are present, but unknown to us.

When we make inferences from data gathered according to a properly designed random survey or experiment, the probability model for the observations follows from the design of the survey or experiment, and we can be confident that it is correct. This puts our inferences on a solid foundation. On the other hand, when we make inferences from data gathered from a nonrandom design, we do not have any underlying justification for the probability model, we just assume it is true! There is the possibility the assumed probability model for the observations is not correct, and our inferences will be on shaky ground.

2.1   Sampling from a Real Population

First, we will define some fundamental terms.

- *Population.* The entire group of objects or people the investigator wants information about. For instance, the population might consist of New Zealand residents over the age of eighteen. Usually we want to know some specific attribute about the population. Each member of the population has a number associated with it — for example, his/her annual income. Then we can consider the model population to be the set of numbers for each individual in the real population. Our model population would be the set of incomes of all New Zealand residents over the age of eighteen. We want to learn about the distribution of the population. Specifically, we want information about the population *Parameters*, which are numbers associated with the distribution of the population, such as the population mean, median, and standard deviation. Often it is not feasible to get information about all the units in the population. The population may be too big, or spread over too large an area, or it may cost too much to obtain data for the complete population. So we do not know the parameters because it is infeasible to calculate them.

- *Sample.* A subset of the population. The investigator draws one sample from the population and gets information from the individuals in that sample. Sample *statistics* are calculated from sample data. They are numerical characteristics that summarize the distribution of the sample, such as the sample mean, median, and standard deviation. A statistic has a similar relationship to a sample that a parameter has to a population. However, the sample is known, so the statistic can be calculated.

- *Statistical inference.* Making a statement about population parameters on basis of sample statistics. Good inferences can be made if the sample is representative of the population as a whole! The distribution of the sample must be similar to the distribution of the population from which it came! *Sampling bias*, a systematic tendency to collect a sample which is not representative of the population, must be avoided. It would cause the distribution of the sample to be dissimilar to that of the population, and thus lead to very poor inferences.

Even if we are aware of something about the population and try to represent it in the sample, there is probably some other factors in the population that we are unaware of, and the sample would end up being nonrepresentative in those factors.

▌ EXAMPLE 2.1

Suppose we are interested in estimating the proportion of Hamilton voters who approve the Hamilton City Council's financing a new rugby stadium. We decide to go downtown one lunch break and draw our sample from people passing by. We might decide that our sample should be balanced between males and females the same as the voting age population. We might get a sample evenly balanced between males and females, but not be aware that the people we interview during the day are mainly those on the street during working hours. Office workers would be overrepresented, while factory workers would be underrepresented. There might be other biases inherent in choosing our sample this way, and we might not have a clue as to what these biases are. Some groups would be systematically underrepresented, and others systematically overrepresented. We cannot make our sample representative for classifications we do not know. ▪

Surprisingly, *random samples* give more representative samples than any non-random method such as quota samples or judgment samples. They not only minimize the amount of error in the inference, they also allow a (probabilistic) measurement of the error that remains.

Simple Random Sampling (without Replacement)

Simple random sampling requires a *sampling frame* , which is a list of the population numbered from 1 to N. A sequence of n random numbers are drawn from the numbers 1 to N. Each time a number is drawn it is removed from consideration so that it cannot be drawn again. The items on the list corresponding to the chosen numbers are included in the sample. Thus, at each draw, each item not yet selected has an equal chance of being selected. Every item has equal chance of being in the final sample. Furthermore, every possible sample of the required size is equally likely.

Suppose we are sampling from the population of registered voters in a large city. It is likely that the proportion of males in the sample is close to the proportion of males in the population. Most samples are near the correct proportions; however, we are not certain to get the exact proportion.

All possible samples of size n are equally likely, including those that are not representative with respect to sex.

Stratified Random Sampling

Given that we know what the proportions of males and females are from the list of voters, we should take that information into account in our sampling method. In stratified random sampling, the population is divided into sub-populations called *strata*. In our case this would be males and females. The sampling frame would be divided into separate sampling frames for the two strata. A simple random sample is taken from each *stratum* where the sample size in each stratum is proportional to the stratum size. Every item has equal chance of being selected, and every possible sample that has each stratum represented in the correct proportions is equally likely. This method will give us samples that are exactly representative with respect to sex. Hence inferences from these type of samples will be more accurate than those from simple random sampling when the variable of interest has different distributions over the strata. If the variable of interest is the same for all the strata, then stratified random sampling will be no more (and no less) accurate than simple random sampling. Stratification has no potential downside as far as accuracy of the inference. However, it is more costly, as the sampling frame has to be divided into separate sampling frames for each stratum.

Cluster Random Sampling

Sometimes we do not have a good sampling frame of individuals. In other cases the individuals are scattered across a wide area. In cluster random sampling, we divide that area into neighborhoods called clusters. Then we make a sampling frame for clusters. A random sample of clusters is selected. All items in the chosen clusters are included in the sample. This is very cost effective because the interviewer will not have as much travel time between interviews. The drawback is that items in a cluster tend to be more similar than items in different clusters. For instance, people living in the same neighborhood usually come from the same economic level because the houses were built at the same time and in the same price range. This means that each observation gives less information about the population parameters. It is less efficient in terms of sample size. However, often it is very cost effective, because getting a larger sample is usually cheaper by this method.

Non-sampling Errors in Sample Surveys

Errors can arise in sample surveys or in a complete population census for reasons other than the sampling method used. These non-sampling errors include response bias; the people who respond may be somewhat different than those who do not respond. They may have different views on the matters surveyed.

Since we only get observations from those who respond, this difference would bias the results. A well-planned survey will have callbacks, where those in the sample who have not responded will be contacted again, in order to get responses from as many people in the original sample as possible. This will entail additional costs, but is important as we have no reason to believe that nonrespondents have the same views as the respondents. Errors can also arise from poorly worded questions. Survey questions should be trialed in a pilot study to determine if there is any ambiguity.

Randomized Response Methods

Social science researchers and medical researchers often wish to obtain information about the population as a whole, but the information that they wish to obtain is sensitive to the individuals who are surveyed. For instance, the distribution of the number of sex partners over the whole population would be indicative of the overall population risk for sexually transmitted diseases. Individuals surveyed may not wish to divulge this sensitive personal information. They might refuse to respond or, even worse, they could give an untruthful answer. Either way, this would threaten the validity of the survey results. *Randomized response* methods have been developed to get around this problem. There are two questions, the sensitive question and the dummy question. Both questions have the same set of answers. The respondent uses a randomization that selects which question he or she answers, and also the answer if the dummy question is selected. Some of the answers in the survey data will be to the sensitive question and some will be to the dummy question. The interviewer will not know which is which. However, the incorrect answers are entering the data from known randomization probabilities. This way, information about the population can be obtained without actually knowing the personal information of the individuals surveyed, since only that individual knows which question he or she answered. Bolstad, Hunt, and McWhirter (2001) describe a *Sex, Drugs, and Rock & Roll Survey* that gets sensitive information about a population (Introduction to Statistics class) using randomized response methods.

2.2 Observational Studies and Designed Experiments

The goal of scientific inquiry is to gain new knowledge about the cause-and-effect relationship between a factor and a response variable. We gather data to help us determine these relationships and to develop mathematical models to explain them. The world is complicated. There are many other factors that may affect the response. We may not even know what these other factors are. If we do not know what they are, we cannot control them directly. Unless we can control them, we cannot make inferences about cause-and-effect relationships! Suppose, for example, we want to study a herbal medicine for its

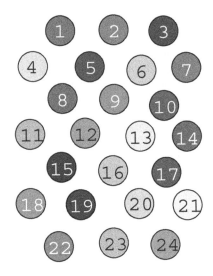

Figure 2.1   Variation among experimental units.

effect on weight loss. Each person in the study is an *experimental unit*. There is great variability between experimental units, because people are all unique individuals with their own hereditary body chemistry and dietary and exercise habits. The variation among experimental units makes it more difficult to detect the effect of a treatment. Figure 2.1 shows a collection of experimental units. The degree of shading shows they are not the same with respect to some unidentified variable. The response variable in the experiment may depend on that unidentified variable, which could be a lurking variable in the experiment.

Observational Study

If we record the data on a group of subjects that decided to take the herbal medicine and compared that with data from a control group who did not, that would be an *observational study*. The treatments have *not* been randomly assigned to treatment and control group. Instead they self-select. Even if we observe a substantial difference between the two groups, we cannot conclude that there is a causal relationship from an observational study. We cannot rule out that the association was due to an unidentified lurking variable. In our study, those who took the treatment may have been more highly motivated to lose weight than those who did not. Or there may be other factors that differed between the two groups. Any inferences we make on an observational study are dependent on the assumption that there are no differences between the distribution of the units assigned to the treatment groups and the control group. We cannot know whether this assumption is actually correct in an observational study.

Designed Experiment

We need to get our data from a designed experiment if we want to be able to make sound inferences about cause-and-effect relationships. The experimenter uses randomization to decide which subjects get into the treatment group(s) and control group respectively. For instance, he/she uses a table of random numbers, or flips a coin.

We are going to divide the experimental units into four treatment groups (one of which may be a control group). We must ensure that each group gets a similar range of units. If we do not, we might end up attributing a difference between treatment groups to the different treatments, when in fact it was due to the lurking variable and a biased assignment of experimental units to treatment groups.

Completely randomized design. We will randomly assign experimental units to groups so that each experimental unit is equally likely to go to any of the groups. Each experimental unit will be assigned (nearly) independently of other experimental units. The only dependence between assignments is that having assigned one unit to treatment group 1 (for example), the probability of the other unit being assigned to group 1 is slightly reduced because there is one less place in group 1. This is known as a completely randomized design. Having a large number of (nearly) independent randomizations ensures that the comparisons between treatment groups and control group are fair since all groups will contain a similar range of experimental units. Units having high values and units having low values of the lurking variable will be in all treatment groups in similar proportions. In Figure 2.2 we see that the four treatment groups have similar range of experimental units with respect to the unidentified lurking variable.

The randomization averages out the differences between experimental units assigned to the groups. The expected value of the lurking variable is the same for all groups, because of the randomization. The average value of the lurking variable for each group will be close to its mean value in the population because there are a large number of independent randomizations. The larger the number of units in the experiment, the closer the average values of the lurking variable in each group will be to its mean value in the population. If we find an association between the treatment and the response, then it will be unlikely that the association was due to any lurking variable. For a large-scale experiment, we can effectively rule out any lurking variable and conclude that the association was due to the effect of different treatments.

Randomized block design. If we identify a variable, then we can control for it directly. It ceases to be a lurking variable. One might think that using

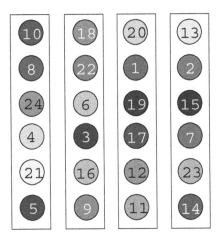

Figure 2.2   Completely randomized design. Units have been randomly assigned to four treatment groups.

judgment about assigning experimental units to the treatment and control groups would lead to similar range of units being assigned to them. The experimenter could get similar groups according to the criterion (identified variable) he/she was using. However, there would be no protection against any other lurking variable that had not been considered. We cannot expect it to be averaged out if we have not done the assignments randomly!

Any prior knowledge we have about the experimental units should be used before the randomization. Units that have similar values of the identified variable should be formed into *blocks*. This is shown in Figure 2.3. The experimental units in each block are similar with respect to that variable. Then the randomization is be done within blocks. One experimental unit in each block is randomly assigned to each treatment group. The blocking controls that particular variable, as we are sure that all units in the block are similar, and one goes to each treatment group. By selecting which one goes to each group randomly, we are protecting against any other lurking variable by randomization. It is unlikely that any of the treatment groups was unduly favored or disadvantaged by the lurking variable. On the average, all groups are treated the same. Figure 2.4 shows the treatment groups found by a randomized block design. We see the four treatment groups are even more similar than those from the completely randomized design.

For example, if we wanted to determine which of four varieties of wheat gave better yield, we would divide the field into blocks of four adjacent plots because plots that are adjacent are more similar in their fertility than plots that are distant from each other. Then within each block, one plot would be

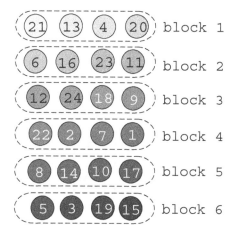

Figure 2.3 Similar units have been put into blocks.

randomly assigned to each variety. This randomized block design ensures that the four varieties each have been assigned to similar groups of plots. It protects against any other lurking variable, by the within-block randomization.

When the response variable is related to the trait we are blocking on, the blocking will be effective, and the randomized block design will lead to more precise inferences about the yields than a completely randomized design with the same number of plots. This can be seen by comparing the treatment groups from the completely randomized design shown in Figure 2.2 with the treatment groups from the randomized block design shown in Figure 2.4. The treatment groups from the randomized block design are more similar than those from the completely randomized design.

Main Points

- *Population.* The entire set of objects or people that the study is about. Each member of the population has a number associated with it, so we often consider the population as a set of numbers. We want to know about the distribution of these numbers.

- *Sample.* The subset of the population from which we obtain the numbers.

- *Parameter.* A number that is a characteristic of the population distribution, such as the mean, median, standard deviation, and interquartile range of the whole population.

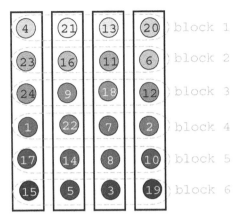

Figure 2.4   Randomized block design. One unit in each block randomly assigned to each treatment group. Randomizations in different blocks are independent of each other.

- *Statistic.* A number that is a characteristic of the sample distribution, such as the mean, median, standard deviation, and interquartile range of the sample.

- *Statistical inference.* Making a statement about population parameters on the basis of sample statistics.

- *Simple random sampling.* At each draw every item that has not already been drawn has an equal chance of being chosen to be included in the sample.

- *Stratified random sampling.* The population is partitioned into subpopulations called strata, and simple random samples are drawn from each stratum where the stratum sample sizes are proportional to the stratum proportions in the population. The stratum samples are combined to form the sample from the population.

- *Cluster random sampling.* The area the population lies in is partitioned into areas called clusters. A random sample of clusters is drawn, and all members of the population in the chosen clusters are included in the sample.

- *Randomized response methods.* These allow the respondent to randomly determine whether to answer a sensitive question or the dummy question, which both have the same range of answers. Thus the respondents personal information is not divulged by the answer, since the interviewer does not know which question it applies to.

- *Observational study.* The researcher collects data from a set of experimental units not chosen randomly, or not allocated to experimental or control group by randomization. There may be lurking variables due to the lack of randomization.

- *Designed experiment.* The researcher allocates experimental units to the treatment group(s) and control group by some form of randomization.

- *Completely randomized design.* The researcher randomly assigns the units into the treatment groups (nearly) independently. The only dependence is the constraint that the treatment groups are the correct size.

- *Randomized block design.* The researcher first groups the units into blocks which contain similar units. Then the units in each block are randomly assigned, one to each group. The randomizations in separate blocks are performed independent of each other.

Monte Carlo Exercises

2.1. **Monte Carlo study comparing methods for random sampling.** We will use a Monte Carlo computer simulation to evaluate the methods of random sampling. Now, if we want to evaluate a method, we need to know how it does in the long run. In a real-life situation, we cannot judge a method by the sample estimate it gives, because if we knew the population parameter, we would not be taking a sample and estimating it with a sample statistic.

One way to evaluate a statistical procedure is to evaluate the *sampling distribution* which summarizes how the estimate based on that procedure varies in the long run (over all possible random samples) for a case when we know the population parameters. Then we can see how closely the sampling distribution is centered around the true parameter. The closer it is, the better the statistical procedure, and the more confidence we will have in it for realistic cases when we do not know the parameter.

If we use computer simulations to run a large number of hypothetical repetitions of the procedure with known parameters, this is known as a Monte Carlo study named after the famous casino. Instead of having the theoretical sampling distribution, we have the empirical distribution of the sample statistic over those simulated repetitions. We judge the statistical procedure by seeing how closely the empirical distribution of the estimator is centered around the known parameter.

The population. Suppose there is a population made up of 100 individuals, and we want to estimate the mean income of the population from a random sample of size 20. The individuals come from three ethnic groups with population proportions of 40%, 40%, and 20%, respectively. There are twenty neighborhoods, and five individuals live in each one. Now, the income distribution may be different for the three ethnic groups. Also, individuals in the same neighborhood tend to be more similar than individuals in different neighborhoods.

[**Minitab:**]   Details about the population are contained in the Minitab worksheet *sscsample.mtw*. Each row contains the information for an individual. Column 1 contains the income, column 2 contains the ethnic group, and column 3 contains the neighborhood. Compute the mean income for the population. That will be the true parameter value that we are trying to estimate.

[**R:**]   Details about the population can be seen by typing

```
help(sscsample.data)
```

In the Monte Carlo study we will approximate the *sampling distribution* of the sample means for three types of random sampling, simple random sampling, stratified random sampling, and cluster random sampling. We do this by drawing a large number (in this case 200) random samples from the population using each method of sampling, calculating the sample mean as our estimate. The empirical distribution of these 200 sample means approximates the sampling distribution of the estimate.

(a) Display the incomes for the three ethnic groups (strata) using boxplots on the same scale. Compute the mean income for the three ethnic groups. Do you see any difference between the income distributions? [**R:**] This may be done in R by typing:

```
boxplot(income~ethnicity, data = sscsample.data)
```

(b) [**Minitab:**] Draw 200 random samples of size 20 from the population using simple random sampling using *sscsample* and put the output in columns c6–c9. Details of how to use this macro are in Appendix C.

[**R:**] Draw 200 random samples of size 20 from the population using simple random sampling using the **sscsample** function.

```
mySamples = list(simple = NULL, strat = NULL,
                 cluster = NULL)
mySamples$simple = sscsample(20, 200)
```

The means and the number of observations sampled from each ethnic group can be seen by typing

```
mySamples$simple
```

More details of how to use this function are in Appendix D.

Answer the following questions from the output:

    i. Does simple random sampling always have the strata repre-sented in the correct proportions?

    ii. On the average, does simple random sampling give the strata in their correct proportions?

    iii. Does the mean of the *sampling distribution* of the sample mean for simple random sampling appear to be close enough to the population mean that we can consider the difference to be due to chance alone? (We only took 200 samples, not all possible samples.)

(c) [**Minitab:**] Draw 200 stratified random samples using the macro and store the output in c11–c14.

[**R:**] Draw 200 stratified random samples using the function and store the output in mySamples$strat.

```
mySamples$strat = sscsample(20, 200, "stratified")
mySamples$strat
```

Answer the following questions from the output:

    i. Does stratified random sampling always have the strata rep-resented in the correct proportions?

    ii. On the average, does stratified random sampling give the strata in their correct proportions?

    iii. Does the mean of the *sampling distribution* of the sample mean for stratified random sampling appear to be close enough to the population mean that we can consider the difference to be due to chance alone? (We only took 200 samples, not all possible samples.)

(d) [**Minitab:**] Draw 200 cluster random samples using the macro and put the output in columns c16–c19.

[**R:**] Draw 200 cluster random samples using the function and store the output in mySamples$cluster.

```
mySamples$cluster = sscsample(20, 200, "cluster")
mySamples$cluster
```

Answer the following questions from the output:

    i. Does cluster random sampling always have the strata repre-sented in the correct proportions?

    ii. On the average, does cluster random sampling give the strata in their correct proportions?

iii. Does the mean of the *sampling distribution* of the sample mean for cluster random sampling appear to be close enough to the population mean that we can consider the difference to be due to chance alone? (We only took 200 samples, not all possible samples.)

(e) Compare the spreads of the sampling distributions (standard deviation and interquartile range). Which method of random sampling seems to be more effective in giving sample means more concentrated about the true mean?

[R:]

```
sapply(mySamples, function(x)sd(x$means))
sapply(mySamples, function(x)IQR(x$means))
```

(f) Give reasons for this.

2.2. **Monte Carlo study comparing completely randomized design and randomized block design.** Often we want to set up an experiment to determine the magnitude of several treatment effects. We have a set of experimental units that we are going to divide into treatment groups. There is variation among the experimental units in the underlying response variable that we are going to measure. We will assume that we have an additive model where each of the treatments has a constant effect. That means the measurement we get for an experimental unit i given treatment j will be the underlying value for unit i plus the effect of the treatment for the treatment it receives:

$$y_{i,j} = u_i + T_j,$$

where u_i is the underlying value for experimental unit i and T_j is the treatment effect for treatment j. The assignment of experimental units to treatment groups is crucial.

There are two things that the assignment of experimental units into treatment groups should deal with. First, there may be a "lurking variable" that is related to the measurement variable, either positively or negatively. If we assign experimental units that have high values of that lurking variable into one treatment group, that group will be either advantaged or disadvantaged, depending if there is a positive or negative relationship. We would be quite likely to conclude that treatment is good or bad relative to the other treatments, when in fact the apparent difference would be due to the effect of the lurking variable. That is clearly a bad thing to occur. We know that to prevent this, the experimental units should be assigned to treatment groups according to some randomization method. On the average, we want all treatment groups to get a similar range of experimental units with respect to the lurking variable. Otherwise, the experimental results may be biased.

Second, the variation in the underlying values of the experimental units may mask the differing effects of the treatments. It certainly makes it harder to detect a small difference in treatment effects. The assignment of experimental units into treatment groups should make the groups as similar as possible. Certainly, we want the group means of the underlying values to be nearly equal.

The *completely randomized design* randomly divides the set of experimental units into treatment groups. Each unit is randomized (almost) independently. We want to ensure that each treatment group contains equal numbers of units. Every assignment that satisfies this criterion is equally likely. This design does not take the values of the other variable into account. It remains a possible lurking variable.

The *randomized block design* takes the other variable value into account. First blocks of experimental units having similar values of the other variable are formed. Then one unit in each block is randomly assigned to each of the treatment groups. In other words, randomization occurs within blocks. The randomizations in different blocks are done independently of each other. This design makes use of the other variable. It ceases to be a lurking variable and becomes the blocking variable.

In this assignment we compare the two methods of randomly assigning experimental units into treatment groups. Each experimental unit has an underlying value of the response variable and a value of another variable associated with it. (If we do not take the other variable in account, it will be a lurking variable.) We will run a small-scale Monte Carlo study to compare the performance of these two designs in two situations.

(a) First we will do a small-scale Monte Carlo study of 500 random assignments using each of the two designs when the response variable is strongly related to the other variable. We let the correlation between them be $\rho = .8$.

[**Minitab:**] The correlation is set by specifying the value of the variable k1 for the Minitab macro *Xdesign*.

[**R:**] The correlation is set by specifying the value of corr in the R function xdesign.

The details of how to use the Minitab macro *Xdesign* or the R function xdesign are in Appendix C and Appendix D, respectively. Look at the boxplots and summary statistics.

  i. Does it appear that, on average, all groups have the same underlying mean value for the other (lurking) variable when we use a completely randomized design?

ii. Does it appear that, on average, all groups have the same underlying mean value for the other (blocking) variable when we use a randomized block design?

iii. Does the distribution of the other variable over the treatment groups appear to be the same for the two designs? Explain any difference.

iv. Which design is controlling for the other variable more effectively? Explain.

v. Does it appear that, on average, all groups have the same underlying mean value for the response variable when we use a completely randomized design?

vi. Does it appear that, on average, all groups have the same underlying mean value for the response variable when we use a randomized block design?

vii. Does the distribution of the response variable over the treatment groups appear to be the same for the two designs? Explain any difference.

viii. Which design will give us a better chance for detecting a small difference in treatment effects? Explain.

ix. Is blocking on the other variable effective when the response variable is strongly related to the other variable?

(b) Next we will do a small-scale Monte Carlo study of 500 random assignments using each of the two designs when the response variable is weakly related to the other variable. We let the correlation between them be $\rho = .4$. Look at the boxplots and summary statistics.

i. Does it appear that, on average, all groups have the same underlying mean value for the other (lurking) variable when we use a completely randomized design?

ii. Does it appear that, on average, all groups have the same underlying mean value for the other (blocking) variable when we use a randomized block design?

iii. Does the distribution of the other variable over the treatment groups appear to be the same for the two designs? Explain any difference.

iv. Which design is controlling for the other variable more effectively? Explain.

v. Does it appear that, on average, all groups have the same underlying mean value for the response variable when we use a completely randomized design?

vi. Does it appear that, on average, all groups have the same underlying mean value for the response variable when we use a randomized block design?

vii. Does the distribution of the response variable over the treatment groups appear to be the same for the two designs? Explain any difference.

viii. Which design will give us a better chance for detecting a small difference in treatment effects? Explain.

ix. Is blocking on the other variable effective when the response variable is strongly related to the other variable?

(c) Next we will do a small-scale Monte Carlo study of 500 random assignments using each of the two designs when the response variable is not related to the other variable. We let the correlation between them be $\rho = 0$. This will make the response variable independent of the other variable. Look at the boxplots for the treatment group means for the other variable.

i. Does it appear that, on average, all groups have the same underlying mean value for the other (lurking) variable when we use a completely randomized design?

ii. Does it appear that, on average, all groups have the same underlying mean value for the other (blocking) variable when we use a randomized block design?

iii. Does the distribution of the other variable over the treatment groups appear to be the same for the two designs? Explain any difference.

iv. Which design is controlling for the other variable more effectively? Explain.

v. Does it appear that, on average, all groups have the same underlying mean value for the response variable when we use a completely randomized design?

vi. Does it appear that, on average, all groups have the same underlying mean value for the response variable when we use a randomized block design?

vii. Does the distribution of the response variable over the treatment groups appear to be the same for the two designs? Explain any difference.

viii. Which design will give us a better chance for detecting a small difference in treatment effects? Explain.

ix. Is blocking on the other variable effective when the response variable is independent from the other variable?

x. Can we lose any effectiveness by blocking on a variable that is not related to the response?

CHAPTER 3

DISPLAYING AND SUMMARIZING DATA

We use statistical methods to extract information from data and gain insight into the underlying process that generated the data. Frequently our data set consists of measurements on one or more variables over the experimental units in one or more samples. The distribution of the numbers in the sample will give us insight into the distribution of the numbers for the whole population.

It is very difficult to gain much understanding by looking at a set of numbers. Our brains were not designed for that. We need to find ways to present the data that allow us to note the important features of the data. The visual processing system in our brain enables us to quickly perceive the overview we want, when the data are represented pictorially in a sensible way. They say a picture is worth a thousand words. That is true, provided that we have the correct picture. If the picture is incorrect, we can mislead ourselves and others very badly!

Introduction to Bayesian Statistics, 3rd ed.
By Bolstad, W. M. and Curran, J. M. Copyright © 2016 John Wiley & Sons, Inc.

3.1   Graphically Displaying a Single Variable

Often our data set consists of a set of measurements on a single variable for a single sample of subjects or experimental units. We want to get some insight into the distribution of the measurements of the whole population. A visual display of the measurements of the sample helps with this.

◼ **EXAMPLE 3.1**

In 1798 the English scientist Cavendish performed a series of 29 measurements on the density of the Earth using a torsion balance. This experiment and the data set are described by Stigler (1977). Table 3.1 contains the 29 measurements.

◼

Table 3.1   Earth density measurements by Cavendish

5.50	5.61	4.88	5.07	5.26	5.55	5.36	5.29	5.58	5.65
5.57	5.53	5.62	5.29	5.44	5.34	5.79	5.10	5.27	5.39
5.42	5.47	5.63	5.34	5.46	5.30	5.75	5.68	5.85	

Dotplot

A dotplot is the simplest data display for a single variable. Each observation is represented by a dot at its value along horizontal axis. This shows the relative positions of all the observation values. It is easy to get a general idea of the distribution of the values. Figure 3.1 shows the dotplot of Cavendish's Earth density measurements.

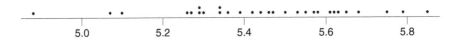

| 5.0 | 5.2 | 5.4 | 5.6 | 5.8 |

Figure 3.1   Dotplot of Earth density measurements by Cavendish.

Boxplot (Box-and-Whisker Plot)

Another simple graphical method to summarize the distribution of the data is to form a boxplot. First we have to sort and summarize the data.

Originally, the sample values are $y_1, \ldots, y_n$. The subscript denotes the order (in time) the observation was taken, y_1 is the first, y_2 is the second, and so on up to y_n which is last. When we order the sample values by size from smallest to largest we get the *order statistics*. They are denoted $y_{[1]}, \cdots, y_{[n]}$, where $y_{[1]}$ is the smallest, $y_{[2]}$ is the second smallest, on up to the largest $y_{[n]}$. We divide the ordered observations into quarters with the quartiles. Q_1, the lower quartile, is the value that 25% of the observations are less than or equal to it, and 75% or more of the observations are greater than or equal to it. Q_2, the middle quartile, is the value that 50% or more of the observations are less than or equal to it, and 50% or more of the observations are greater than or equal to it. Q_2 is also known as the sample median. Similarly Q_3, the upper quartile is the value that 75% of the observations are less than or equal to it, and 25% of the observations are greater than or equal to it. We can find these from the order statistics:

$$Q_1 = y_{[\frac{n+1}{4}]} \,,$$

$$Q_2 = y_{[\frac{n+1}{2}]} \,,$$

$$Q_3 = y_{[\frac{3(n+1)}{4}]} \,.$$

If the subscripts are not integers, we take the weighted average of the two closest order statistics. For example, Cavendish's Earth density data $n = 29$,

$$Q_1 = y_{[\frac{30}{4}]} \,.$$

This is halfway between the 7th- and 8th-order statistics, so

$$Q_1 = \tfrac{1}{2} \times y_{[7]} + \tfrac{1}{2} \times y_{[8]} \,.$$

The five number summary of a data set is $y_{[1]}, Q_1, Q_2, Q_3, y_{[n]}$. This gives the minimum, the three quartiles, and the maximum of the observations. The *boxplot* or *box-and-whisker plot* is a pictorial way of representing the five number summary. The steps are:

- Draw and label an axis.

- Draw a box with ends at the first and third quartiles.

- Draw a line through the box at the second quartile (median).

- Draw a line (whisker) from the lower quartile to the lowest observation, and draw a line (whisker) from the upper quartile to the highest observation.

- **Warning:** Minitab extends the whiskers only to a maximum length of 1.5 × the interquartile range. Any observation further out than that is identified with an asterisk (*) to indicate the observation may be an outlier. This can seriously distort the picture of the sample, because the criterion does not depend on the sample size. A large sample can look very heavy-tailed because the asterisks show that there are many possibly outlying values, when the proportion of outliers is well within the normal range. In Exercise 3.6, we show how this distortion works and how we can control it by editing the outlier symbol in the Minitab boxplot.

The boxplot divides the observations into quarters. It shows you a lot about the shape of the data distribution. Examining the length of the whiskers compared to the box length shows whether the data set has light, normal, or heavy tails. Comparing the lengths of the whiskers show whether the distribution of the data appears to be skewed or symmetric. Figure 3.2 shows the boxplot for Cavendish's Earth density measurements. It shows that the data distribution is fairly symmetric but with a slightly longer lower tail.

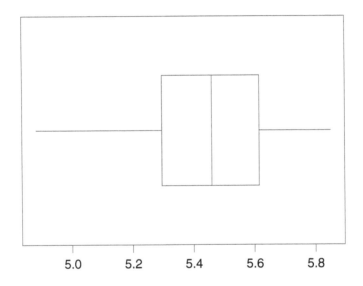

Figure 3.2   Boxplot of Earth density measurements by Cavendish.

Stem-and-Leaf Diagram

The stem-and-leaf diagram is a quick and easy way of extracting information about the distribution of a sample of numbers. The *stem* represents the leading digit(s) to a certain depth (power of 10) of each data item, and the leaf represents the next digit of the data item. A stem-and-leaf diagram can be constructed by hand for a small data set. It is often the first technique used on a set of numbers. The steps are:

- Draw a vertical axis (stem) and scale it for the stem units. Always use a *linear* scale!

- Plot leaf for the next digit. We could round off the leaf digit, but usually we do not bother if we are doing it by hand. In any case, we may have lost some information by rounding off or by truncating.

- Order the leaves with the smallest near stem to the largest farthest away.

- State the leaf unit on your diagram.

The stem-and-leaf plot gives a picture of the distribution of the numbers when we turn it on its side. It retains the actual numbers to within the accuracy of the leaf unit. We can find the order statistics counting up from the lower end. This helps to find the quartiles and the median. Figure 3.3 shows a stem-and-leaf diagram for Cavendish's Earth density measurements. We use a two-digit stem, units and tenths, and a one-digit leaf, hundredths.

	leaf unit .01
48	8
49	
50	7
51	0
52	6799
53	04469
54	2467
55	03578
56	12358
57	59
58	5

Figure 3.3   Stem-and-leaf plot for Cavendish's Earth density measurements.

There are 29 measurements. We can count down to the $X_{\frac{29+1}{2}} = X_{15}$ to find that the median is 5.46. We can count down to $X_{\frac{29+1}{4}} = X_{7\frac{1}{2}}$. Thus the first quartile $Q_1 = \frac{1}{2} \times X_7 + \frac{1}{2} \times X_8$, which is 5.295.

Frequency Table

Another main approach to simplify a set of numbers is to put it in a frequency table. This is sometimes referred to as *binning* the data. The steps are:

- Partition possible values into nonoverlapping groups (bins). Usually we use equal width groups. However, this is not required.

- Put each item into the group it belongs in.

- Count the number of items in each group.

Frequency tables are a useful tool for summarizing data into an understandable form. There is a trade-off between the loss of information in our summary, and the ease of understanding the information that remains. We have lost information when we put a number into a group. We know it lies between the group boundaries, but its exact value is no longer known. The fewer groups we use, the more concise the summary, but the greater loss of information. If we use more groups we lose less information, but our summary is less concise and harder to grasp. Since we no longer have the information about exactly where each value lies in a group, it seems logical that the best assumption we can then make is that each value in the group is equally possible. The Earth density measurements made by Cavendish are shown as a frequency table in Table 3.2.

Table 3.2   Frequency table of Earth density measurements by Cavendish

Boundaries	Frequency
$4.80 < x \leq 5.00$	1
$5.00 < x \leq 5.20$	2
$5.20 < x \leq 5.40$	9
$5.40 < x \leq 5.60$	9
$5.60 < x \leq 5.80$	7
$5.80 < x \leq 6.00$	1

If there are too many groups, some of them may not contain any observations. In that case, it is better to lump two or more adjacent groups into a bigger one to get some observations in every group. There are two ways to show the data in a frequency table pictorially. They are *histograms* and *cumulative frequency polygons*.

Histogram

This is the most common way to show the distribution of data in the frequency table. The steps for constructing a histogram are:

- Put group boundaries on horizontal axis drawn on a *linear* scale.

- Draw a rectangular bar for each group where the *area* of bar is proportional to the frequency of that group. For example, this means that if a group is twice as wide as the others, its height is half that group's frequency. The bar is flat across the top to show our assumption that each value in the group is equally possible.

- Do not put any gaps between the bars if the data are continuous.

- The scale on the vertical axis is density, which is group frequency divided by group width. When the groups have equal width, the scale is proportional to frequency, or relative frequency, and they could be used instead of density. This is not true if unequal width groups are used. It is not necessary to label the vertical axis on the graph. The shape of the graph is the important thing, not its vertical scale.

- **Warning**: [**Minitab:**] If you use unequal group widths in Minitab, you must click on *density* in the *options* dialog box; otherwise, the histogram will have the wrong shape.

The histogram gives us a picture of how the sample data are distributed. We can see the shape of the distribution and relative tail weights. We look at it as representing a picture of the underlying population the sample came from. This underlying population distribution[1] would generally be reasonably smooth. There is always a trade-off between too many and too few groups. If we use too many groups, the histogram has a "saw tooth" appearance and the histogram is not representing the population distribution very well. If we use too few groups, we lose details about the shape. Figure 3.4 shows histogram of the Earth density measurements by Cavendish using 12, 6, and 4 groups, respectively. This illustrates the trade-off between too many and too few groups. We see that the histogram with 12 groups has gaps and a saw-tooth appearance. The histogram with 6 groups gives a better representation of the underlying distribution of Earth density measurements. The histogram with 4 groups has lost too much detail. The last histogram has unequal width groups. The height of the wider bars is shortened to keep the area proportional to frequency.

Cumulative Frequency Polygon

The other way for displaying the data from a frequency table is to construct a *cumulative frequency polygon*, sometimes called an *ogive*. It is particularly useful because you can estimate the median and quartiles from the graph. The steps are:

- Group boundaries on horizontal axis drawn on a *linear* scale.

- Frequency or percentage shown on vertical axis.

- Plot *(lower boundary of lowest class, 0)*.

---

[1]In this case, the *population* is the set of all possible Earth density measurements that Cavendish could have obtained from his experiment. This population is theoretical, as each of its elements was only brought into existence by Cavendish performing the experiment.

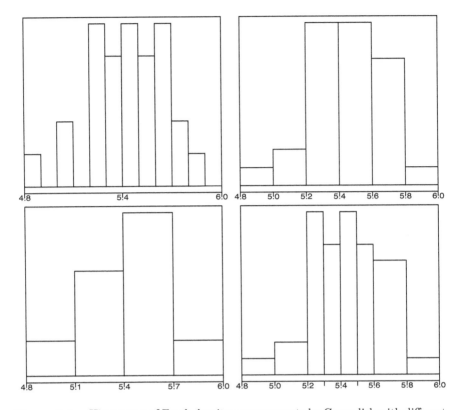

Figure 3.4   Histograms of Earth density measurements by Cavendish with different boundaries. Note that the area is always proportional to frequency.

- For each group, plot *(upper class boundary, cumulative frequency)*. We do not know the exact value of each observation in the group. However, we do know that all the values in a group must be less than or equal to the upper boundary.

- Join the plotted points with a straight line. Joining them with a straight line shows that we consider each value in the group to be equally possible.

We can estimate the median and quartiles easily from the graph. To find the median, go up to 50% on the vertical scale and then draw a horizontal line across to the cumulative frequency polygon, and then a vertical line down to the horizontal axis. The value where it hits the axis is the estimate of the median. Similarly, to find the quartiles, go up to 25% or 75%, go across to cumulative frequency polygon, and go down to horizontal axis to find lower and upper quartile, respectively. The underlying assumption behind these estimates is that all values in a group are evenly spread across the group. Figure 3.5 shows the cumulative frequency polygon for the Earth density measurements by Cavendish.

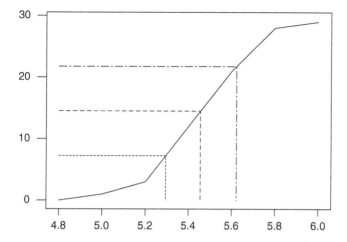

Figure 3.5   Cumulative frequency polygon of Earth density measurements by Cavendish.

3.2   Graphically Comparing Two Samples

Sometimes we have the same variable recorded for two samples. For instance, we may have responses for the treatment group and control group from a randomized experiment. We want to determine whether or not the treatment has been effective.

Often a picture can clearly show us this, and there is no need for any sophisticated statistical inference. The key to making visual comparisons between two data samples is "do not compare apples to oranges." By that, we mean that the pictures for the two samples must be lined up, and with the same scale. Stacked dotplots and stacked boxplots, when they are lined up on the same axis, give a good comparison of the samples. Back-to-back stem-and-leaf diagrams are another good way of comparing two small data sets. The two samples use common stem, and the leaves from one sample are on one side of the stem, and the leaves from the other sample are on the other side of the stem. The leaves of the two sample are ordered, from smallest closest to stem to largest farthest away. We can put histograms back-to-back or stack them. We can plot the cumulative frequency polygons for the two samples on the same axis. If one is always to the left of the other, we can deduce that its distribution is shifted relative to the other.

All of these pictures can show us whether there are any differences between the two distributions. For example, do the distributions seem to have the same location on the number line, or does one appear to be shifted relative to the other? Do the distributions seem to have the same spread, or is one more spread out than the other? Are the shapes similar? If we have more than two samples, we can do any of these pictures that is stacked. Of course, back-to-back ones only work for two samples.

◼ **EXAMPLE 3.2**

Between 1879 and 1882, scientists were devising experiments for determining the speed of light. Table 3.3 contains measurements collected by Michelson in a series of experiments on the speed of light. The first 20 measurements were made in 1879, and the next 23 supplementary measurements were made in 1882. The experiment and the data are described in Stigler (1977).

Table 3.3   Michelson's speed-of-light measurements.[a]

Michelson (1879)		Michelson (1882)	
850	740	883	816
900	1070	778	796
930	850	682	711
950	980	611	599
980	880	1051	781
1000	980	578	796
930	650	774	820
760	810	772	696
1000	1000	573	748
960	960	748	797
		851	809
		723	

[a]Value in table plus 299,000 km/s.

Figure 3.6 shows stacked dotplots for the two data sets. Figure 3.7 shows stacked boxplots for the two data sets. The true value of the speed of light in the air is 2999710. We see from these plots that there was a systematic error (bias) in the first series of measurements that was greatly reduced in the second.

Back-to-back stem-and-leaf diagrams are another good way to show the relationship between two data sets. The stem goes in the middle. We put the leaves for one data set on the right side and put the leaves for the other on the left. The leaves are ascending order moving away from the stem. Back-to-back stem-and-leaf diagrams are shown for Michelson's data in Figure 3.8. The stem is hundreds, and the leaf unit is 10.

◼

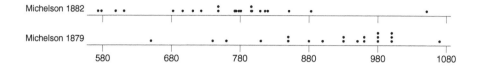

Figure 3.6 Dotplots of Michelson's speed-of-light measurements.

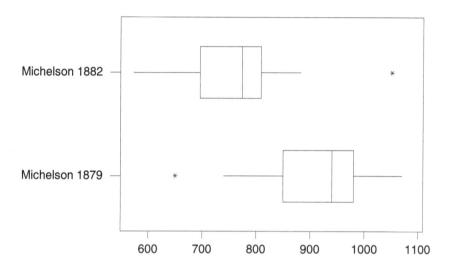

Figure 3.7 Boxplot of Michelson's speed-of-light measurements.

3.3 Measures of Location

Sometimes we want to summarize our data set with numbers. The most important aspect of the data set distribution is determining a value that summarizes its location on the number line. The most commonly used measures of location are the mean and the median. We will look at the advantages and disadvantages of each one.

Both the mean and the median are members of the trimmed mean family, which also includes compromise values between them, depending on the amount of trimming. We do not consider the mode (most common value) to be a suitable measure of location for the following reasons. For continuous data values, each value is unique if we measure it accurately enough. In many cases, the mode is near one end of the distribution, not the central region. The mode may not be unique.

977	5	
1	6	
98	6	5
4412	7	4
9998777	7	6
210	8	1
85	8	558
	9	033
	9	566888
	10	000
5	10	7

leaf unit 10

Figure 3.8   Back-to-back stem-and-leaf plots for Michelson's data.

Mean: Advantages and Disadvantages

The mean is the most commonly used measure of location, because of its simplicity and its good mathematical properties. The mean of a data set $y_1, \cdots, y_n$ is simply the arithmetic average of the numbers.

$$\bar{y} \;=\; \frac{1}{n} \times \sum_{i=1}^{n} y_i \;=\; \frac{1}{n} \times (y_1 + \cdots + y_n).$$

The mean is simple and very easy to calculate.   You just make one pass through the numbers and add them up. Then divide the sum by the size of the sample.

The mean has good mathematical properties.   The mean of a sum is the sum of the means. For example, if y is total income, u is "earned income" (wages and salaries), v is "unearned income" (interest, dividends, rents), and w is "other income" (social security benefits and pensions, etc.). Clearly, a persons total income is the sum of the incomes he or she receives from each source $y_i = u_i + v_i + w_i$. Then

$$\bar{y} = \bar{u} + \bar{v} + \bar{w}.$$

So it doesn't matter if we take the means from each income source and then add them together to find the mean total income, or add the each individuals incomes from all sources to get his/her total income and then take the mean of that. We get the same value either way.

The mean combines well.   The mean of a combined set is the weighted average of the means of the constituent sets, where weights are proportions each constituent set is to the combined set. For example, the data may come from two sources, males and females who had been interviewed separately. The overall mean would be the weighted average of the male mean and the female mean where the weights are the proportions of males and females in the sample.

The mean is the first moment or center of gravity of the numbers. We can think of the mean as the balance point if an equal weight was placed at each of the data points on the (weightless) number line. The mean would be the balance point of the line. This leads to the main disadvantage of the mean. It is strongly influenced by *outliers*. A single observation much bigger than the rest of the observations has a large effect on the mean. That makes using the mean problematic with highly skewed data such as personal income. Figure 3.9 shows how the mean is influenced by an outlier.

Figure 3.9   The mean as the balance point of the data is affected by moving the outlier.

Calculating mean for grouped data.   When the data have been put in a frequency table, we only know between which boundaries each observation lies. We no longer have the actual data values. In that case there are two assumptions we can make about the actual values.

1. All values in a group lie at the group midpoint.

2. All the values in a group are evenly spread across the group.

Fortunately, both these assumptions lead us to the same calculation of the mean value. The total contribution for all the observations in a group is the midpoint times the frequency under both assumptions.

$$\bar{y} = \frac{1}{n} \sum_{j=1}^{J} n_j \times m_j$$

$$= \sum_{j=1}^{J} \frac{n_j}{n} \times m_j \,,$$

where n_j is the number of observations in the jth interval, n is the total number of observations, and m_j is the midpoint of the jth interval.

Median: Advantages and Disadvantages

The median of a set of numbers is the number such that 50% of the numbers are less than or equal to it, and 50% of the numbers are greater than or equal to it. Finding the median requires us to sort the numbers. It is the middle number when the sample size is odd, or it is the average of the two numbers closest to middle when the sample size is even.

$$m = y_{\left[\frac{n+1}{2}\right]} \,.$$

The median is not influenced by outliers at all. This makes it very suitable for highly skewed data like personal income. This is shown in Figure 3.10. However, it does not have same good mathematical properties as mean. The

Figure 3.10   The median as the middle point of the data is not affected by moving the outlier.

median of a sum is not necessarily the sum of the medians. Neither does it have good combining properties similar to those of the mean. The median of the combined sample is not necessarily the weighted average of the medians. For these reasons, the median is not used as often as the mean. It is mainly used for very skewed data such as incomes where there are outliers which would unduly influence the mean, but do not affect the median.

Trimmed mean. We find the trimmed mean with degree of trimming equal to k by first ordering the observations, then trimming the lower k and upper k order statistics, and taking the average of those remaining.

$$\bar{x}_k = \frac{\sum_{i=k+1}^{n-k} x_{[i]}}{n - 2k}.$$

We see that $\bar{x}_0$ (where there is no trimming) is the mean. If n is odd and we let $k = \frac{n}{2}$, then $\bar{x}_k$ is the median. Similarly, if n is even and we let $k = \frac{n-2}{2}$, then $\bar{x}_k$ is the median. If k is small, the trimmed mean will have properties similar to the mean. If k is large, the trimmed mean has properties similar to the median.

3.4   Measures of Spread

After we have determined where the data set is located on the number line, the next important aspect of the data set distribution is determining how spread out the data distribution is. If the data are very variable, the data set will be very spread out. So measuring spread gives a measure of the variability. We will look at some of the common measures of variability.

Range: Advantage and Disadvantage

The range is the largest observation minus the smallest:

$$R = y_{[n]} - y_{[1]}.$$

The range is very easy to find. However, the largest and smallest observation are the observations that are most likely to be outliers. Clearly, the range is extremely influenced by outliers.

Interquartile Range: Advantages and Disadvantages

The interquartile range measures the spread of the middle 50% of the observations. It is the third quartile minus first quartile:

$$IQR = Q_3 - Q_1 \,.$$

The quartiles are not outliers, so the interquartile range is not influenced by outliers. Nevertheless, it is not used very much in inference because like the median it doesn't have good math or combining properties.

Variance: Advantages and Disadvantages

The variance of a data set is the average squared deviation from the mean.[2]

$$\mathrm{Var}[y] = \frac{1}{n} \times \sum_{i=1}^{n} (y_i - \bar{y})^2 \,.$$

In physical terms, it is the second moment of inertia about the mean. Engineers refer to the variance as the *MSD*, mean squared deviation. It has good mathematical properties, although more complicated than those for the mean. The variance of a sum (of independent variables) is the sum of the individual variances.

It has good combining properties, although more complicated than those for the mean. The variance of a combined set is the weighted average of the variances of the constituent sets, plus the weighted average of the squares of the constituent means away from the combined mean, where the weights are the proportions that each constituent set is to the combined set.

Squaring the deviations from the mean emphasizes the observations far from the mean. Those observations have large magnitude in a positive or negative direction already, and squaring them makes them much larger still, and all positive. Thus the variance is very influenced by outliers. The variance is in squared units. Thus its size is not comparable to the mean.

Calculating variance for grouped data.   The variance is the average squared deviation from the mean. When the data have been put in a frequency table, we

---

[2] Note that we are defining the variance of a data set using the divisor n. We aren't making any distinction over whether our data set is the whole population or only a sample from the population. Some books define the variance of a sample data set using divisor $n-1$. One *degree of freedom* has been lost because for a sample, we are using the sample mean instead of the unknown population mean. When we use the divisor $n-1$, we are calculating the *sample estimate of the variance*, not the variance itself.

no longer have the actual data values. In that case there are two assumptions we can make about the actual values.

1. All values in a group lie at the group midpoint.

2. All the values in a group are evenly spread across the group.

Unfortunately, these two assumptions lead us to different calculation of the variance. Under the first assumption we get the approximate formula

$$\text{Var}[y] = \frac{1}{n} \sum_{j=1}^{J} n_j \times (m_j - \bar{y})^2 \,,$$

where n_j is the number of observations in the jth interval, n is the total number of observations, m_j is the midpoint of the jth interval. This formula only contains between-group variation, and ignores the variation for the observations within the same group. Under the second assumption we add in the variation within each group to get the formula

$$\text{Var}[y] = \frac{1}{n} \sum_{j=1}^{J} \left(n_j \times (m_j - \bar{y})^2 + n_j \times \frac{R_j^2}{12} \right) \,,$$

where R_j is the upper boundary minus the lower boundary for the jth group.

Standard Deviation: Advantages and Disadvantages

The standard deviation is the square root of the variance.

$$sd(y) = \sqrt{\frac{1}{n} \times \sum_{i=1}^{n} (y_i - \bar{y})^2} \,.$$

Engineers refer to it as the *RMS*, root mean square. It is not as affected by outliers as the variance is, but it is still quite affected. It inherits good mathematical properties and good combining properties from the variance. The standard deviation is the most widely used measure of spread. It is in the same units as mean, so its size is directly comparable to the mean.

3.5   Displaying Relationships Between Two or More Variables

Sometimes our data are measurements for two variables for each experimental unit. This is called *bivariate* data. We want to investigate the relationship between the two variables.

Scatterplot

The scatterplot is just a two-dimensional dotplot. Mark off the horizontal axis for the first variable, the vertical axis for the second. Each point is plotted on the graph. The shape of the "point cloud" gives us an idea as to whether the two variables are related, and if so, what is the type of relationship.

When we have two samples of bivariate data and want to see if the relationship between the variables is similar in the two samples, we can plot the points for both samples on the same scatterplot using different symbols so we can tell them apart.

▉ EXAMPLE 3.3

The *Bears.mtw* file stored in Minitab contains 143 measurements on wild bears that were anesthetized, measured, tagged, and released. Figure 3.11 shows a scatterplot of head length versus head width for these bears. From this we can observe that head length and head width are related. Bears with large width heads tend to have heads that are long. We can also see that male bears tend to have larger heads than female bears. ▉

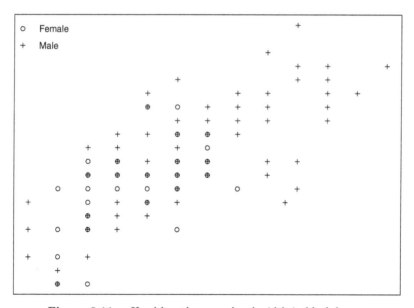

Figure 3.11   Head length versus head width in black bears.

Scatterplot Matrix

Sometimes our data consists of measurements of several variables on each experimental unit. This is called *multivariate* data. To investigate the relationships between the variables, form a *scatterplot matrix*. This means that we construct the scatterplot for each pair of variables, and then display them in an array like a matrix. We look at each scatterplot in turn to investigate the relationship between that pair of the variables. More complicated relationships between three or more of the variables may be hard to see on this plot.

■ EXAMPLE 3.3 (continued)

Figure 3.12 shows a scatterplot matrix showing scatterplots of head length, head width, neck girth, length, chest girth, and weight for the bear measurement data. We see there are strong positive relationships among the variables, and some of them appear to be nonlinear. ■

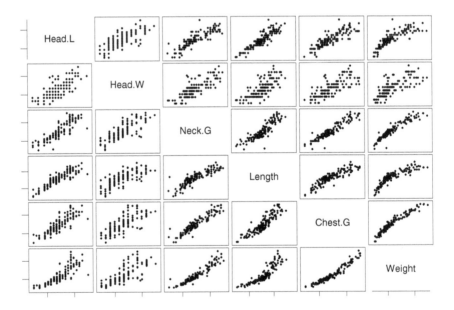

Figure 3.12   Scatterplot matrix of bear data.

3.6   Measures of Association for Two or More Variables

Covariance and Correlation between Two Variables

The covariance of two variables is the average of *first variable minus its mean* times *second variable minus its mean*:

$$\text{Cov}[x, y] = \frac{1}{n} \sum_{i=1}^{n} (x_i - \bar{x})(y_i - \bar{y}).$$

This measures how the variables vary together. Correlation between two variables is the covariance of the two variables divided by product of standard deviations of the two variables. This standardizes the correlation to lie between -1 and $+1$.

$$\text{Cor}[x, y] = \frac{\text{Cov}[x, y]}{\sqrt{\text{Var}[x] \times \text{Var}[y]}}.$$

Correlation measures the strength of the *linear* relationship between two variables. A correlation of $+1$ indicates that the points lie on a straight line with positive slope. A correlation of -1 indicates that the points lie on a straight line with negative slope. A positive correlation that is less than one indicates that the points are scattered, but generally low values of the first variable are associated with low values of the second, and high values of the first are associated with high values of the second. The higher the correlation, the more closely the points are bunched around a line. A negative correlation has low values of the first associated with high values of the second, and high values of the first associated with low values of the second. A correlation of 0 indicates that there is no association of low values or high values of the first with either high or low values of the second. It does not mean the variables are not related, only that they are not linearly related.

When we have more than two variables, we put the correlations in a matrix. The correlation between x and y equals the correlation between y and x, so the correlation matrix is symmetric about the main diagonal. The correlation of any variable with itself equals one.

Table 3.4   Correlation matrix for bear data

	Head.L	Head.W	Neck.G	Length	Chest.G	Weight
Head.L	1.000	.744	.862	.895	.854	.833
Head.W	.744	1.000	.805	.736	.756	.756
Neck.G	.862	.805	1.000	.873	.940	.943
Length	.895	.736	.873	1.000	.889	.875
Chest.G	.854	.756	.940	.889	1.000	.966
Weight	.833	.756	.943	.875	.966	1.000

■   **EXAMPLE 3.3   (continued)**

The correlation matrix for the bear data is given in Table 3.4. We see that all the variables are correlated with each other. Looking at the matrix plot we see that Head.L and Head.W have a correlation of .744, and the scatterplot of those two variables is spread out. We see that the Head.L and Length have a higher correlation of .895, and on the scatterplot of those variables, we see the points lie much closer to a line. We see that Chest.G and Weight are highly correlated at .966. On the scatterplot we see those points lie much closer to a line, although we can also see that actually they seem to lie on a curve that is quite close to a line.   ■

Main Points

- Data should always be looked at in several ways as the first stage in any statistical analysis. Often a good graphical display is enough to show what is going on, and no further analysis is needed. Some elementary data analysis tools are:

  ○ *Order Statistics.* The data when ordered smallest to largest. $y_{[1]}, \ldots, y_{[n]}$.

  ○ *Median.* The value that has 50% of the observations above it and 50% of the observations below it. This is

  $$y_{[\frac{n+1}{2}]} .$$

  It is the middle value of the order statistics when n is odd. When n is even, the median is the weighted average of the two closest order statistics:

  $$y_{[\frac{n+1}{2}]} = \tfrac{1}{2} \times y_{[\frac{n}{2}]} + \tfrac{1}{2} \times y_{[\frac{n}{2}+1]} .$$

  The median is also known as the second quartile.

  ○ *Lower quartile.* The value that 25% of the observations are below it and 75% of the observations are above it. It is also known as the first quartile. It is

  $$Q_1 = y_{[\frac{n+1}{4}]} .$$

  If $\frac{n+1}{4}$ is not an integer, we find it by taking the weighted average of the two closest order statistics.

  ○ *Upper quartile.* The value that 75% of the observations are below it and 25% of the observations are above it. It is also known as the upper quartile. It is

  $$Q_3 = x_{[\frac{3(n+1)}{4}]} .$$

  If $\frac{3(n+1)}{4}$ is not an integer, the quartile is found by taking the weighted average of the two closest order statistics.

- When we are comparing samples graphically, it is important that they be on the same scale. We have to be able to get the correct visual comparison without reading the numbers on the axis. Some elementary graphical data displays are:

  o *Stem-and-leaf diagram.* An quick and easy graphic which allows us to extract information from a sample. A vertical stem is drawn with a numbers up to stem digit along linear scale. Each number is represented using its next digit as a leaf unit at the appropriate place along the stem. The leaves should be ordered away from the stem. It is easy to find (approximately) the quartiles by counting along the graphic. Comparisons are done with back-to-back stem-and-leaf diagrams.

  o *Boxplot.* A graphic along a linear axis where the central box contains the middle 50% of the observation, and a whisker goes out from each end of the box to the lowest and highest observation. There is a line through the box at the median. So it is a visual representation of the five numbers $y_{[1]}, Q_1, Q_2, Q_3, y_{[n]}$ that give a quick summary of the data distribution. Comparisons are done with stacked boxplots.

  o *Histogram.* A graphic where the group boundaries are put on a linear scaled horizontal axis. Each group is represented by a vertical bar where the *area* of the bar is proportional to the frequency in the group.

  o *Cumulative frequency polygon* (ogive). A graphic where the group boundaries are put on a linearly scaled horizontal axis. The point *(lower boundary of lowest group, 0)* and the points *(upper group boundary, cumulative frequency)* are plotted and joined by straight lines. The median and quartiles can be found easily using the graph.

- It is also useful to summarize the data set using a few numerical summary statistics. The most important summary statistic of a variable is a measure of location which indicates where the values lie along the number axis. Some possible measures of location are:

  o *Mean.* The average of the numbers. It is easy to use, has good mathematical properties, and combines well. It is the most widely used measure of location. It is sensitive to outliers, so it is not particularly good for heavy tailed distributions.

  o *Median.* The middle order statistic, or the average of the two closest to the middle. This is harder to find as it requires sorting the data. It is not affected by outliers. The median doesn't have the good mathematical properties or good combining properties of the mean. Because of this, it is not used as often as the mean. Mainly it is used with distributions that have heavy tails or outliers, where it is preferred to the mean.

      o *Trimmed mean.* This is a compromise between the mean and the median. Discard the k largest and the k smallest order statistics and take the average of the rest.

- The second important summary statistic is a measure of spread, which shows how spread out are the numbers. Some commonly used measures of spread are:

      o *Range.* This is the largest order statistic minus the smallest order statistic. Obviously very sensitive to outliers.

      o *Interquartile range (IQR).* This is the upper quartile minus the lower quartile. It measures the spread of the middle 50% of the observations. It is not sensitive to outliers.

      o *Variance.* The average of the squared deviations from the mean. Strongly influenced by outliers. The variance has good mathematical properties, and combines well, but it is in squared units and is not directly comparable to the mean.

      o *Standard deviation.* The square root of the variance. This is less sensitive to outliers than the variance and is directly comparable to the mean since it is in the same units. It inherits good mathematical properties and combining properties from the variance.

- Graphical display for relationship between two or more variables.

      o *Scatterplot.* Look for pattern.

      o *Scatterplot matrix.* An array of scatterplots for all pairs of variables.

- *Correlation* is a numerical measure of the strength of the *linear relationship* between the two variables. It is standardized to always lie between -1 and $+1$. If the points lie on a line with negative slope, the correlation is -1, and if they lie on a line with positive slope, the correlation is $+1$. A correlation of 0 doesn't mean there is no relationship, only that there is no *linear* relationship.

Exercises

3.1. A study on air pollution in a major city measured the concentration of sulfur dioxide on 25 summer days. The measurements were:

3	9	16	23	29
3	11	17	25	35
5	13	18	26	43
7	13	19	27	44
9	14	23	28	46

(a) Form a stem-and-leaf diagram of the sulfur dioxide measurements.

(b) Find the median, lower quartile, and upper quartile of the measurements.

(c) Sketch a boxplot of the measurements.

3.2. Dutch elm disease is spread by bark beetles that breed in the diseased wood. A sample of 100 infected elms was obtained, and the number of bark beetles on each tree was counted. The data are summarized in the following table:

Boundaries	Frequency
$0 < x \leq 50$	8
$50 < x \leq 100$	24
$100 < x \leq 150$	33
$150 < x \leq 200$	21
$200 < x \leq 400$	14

(a) Graph a histogram for the bark beetle data.

(b) Graph a cumulative frequency polygon of the bark beetle data. Show the median and quartiles on your cumulative frequency polygon.

3.3. A manufacturer wants to determine whether the distance between two holes stamped into a metal part is meeting specifications. A sample of 50 parts was taken, and the distance was measured to nearest tenth of a millimeter. The results were:

300.6	299.7	300.2	300.0	300.1
300.0	300 .1	299.9	300.2	300.1
300.5	299.6	300.7	299.9	300.2
299.9	300.4	299.8	300.4	300.4
300.4	300.2	299.4	300.6	299.8
299.7	300.1	299.9	300.0	300.0
300.5	300.1	299.9	299.8	300.2
300.7	300.4	300.0	300.1	300.0
300.2	300.3	300.5	300.0	300.1
300.3	299.9	300.1	300.2	299.5

(a) Form a stem-and-leaf diagram of the measurements.

(b) Find the median, lower quartile, and upper quartile of the measurements.

(c) Sketch a boxplot of the measurements.

(d) Put the measurements in a frequency table with the following classes:

Boundaries	Frequency
$299.2 < x \le 299.6$	
$299.6 < x \le 299.8$	
$299.8 < x \le 300.0$	
$300.0 < x \le 300.2$	
$300.2 < x \le 300.4$	
$300.4 < x \le 300.8$	

(e) Construct a histogram of the measurements.

(f) Construct a cumulative frequency polygon of the measurements. Show the median and quartiles.

3.4. The manager of a government department is concerned about the efficiency in which his department serves the public. Specifically, he is concerned about the delay experienced by members of the public waiting to be served. He takes a sample of 50 arriving customers, and measures the time each waits until service begins. The times (rounded off to the nearest second) are:

98	5	6	39	31
46	129	17	1	64
40	121	88	102	50
123	50	20	37	65
75	191	110	28	44
47	6	43	60	12
150	16	182	32	5
106	32	26	87	137
44	13	18	69	107
5	53	54	173	118

(a) Form a stem-and-leaf diagram of the measurements.

(b) Find the median, lower quartile, and upper quartile of the measurements.

(c) Sketch a boxplot of the measurements.

(d) Put the measurements in a frequency table with the following classes:

Boundaries	Frequency
$0 < x \le 20$	
$20 < x \le 40$	
$40 < x \le 60$	
$60 < x \le 80$	
$80 < x \le 100$	
$100 < x \le 200$	

(e) Construct a histogram of the measurements.

(f) Construct a cumulative frequency polygon of the measurements. Show the median and quartiles.

3.5. A random sample of 50 families reported the dollar amount they had available as a liquid cash reserve. The data have been put in the following frequency table:

Boundaries	Frequency
$0 < x \le 500$	17
$500 < x \le 1,000$	15
$,1000 < x \le 2,000$	7
$2,000 < x \le 4,000$	5
$4,000 < x \le 6,000$	3
$6,000 < x \le 10,000$	3

(a) Construct a histogram of the measurements.

(b) Construct a cumulative frequency polygon of the measurements. Show the median and quartiles.

(c) Calculate the grouped mean for the data.

3.6. In this exercise we see how the default settings in for producing boxplots in Minitab and in R can be misleading because they do not take the sample size into account. We will generate three samples of different sizes from the same distribution, and compare their boxplots.

[**Minitab:**] Generate 250 *normal*(0, 1) observations and put them in column c1 by pulling down the *Calc* menu to the *Random Data* command over to *Normal* and filling in the dialog box. Generate 1,000 *normal*(0, 1) observations the same way and put them in column c2, and generate

4,000 *normal*(0, 1) observations the same way and put them in column c3. Stack these three columns by pulling down the *Data*[3] menu down to *Stack* and over to *Columns* and filling in the dialog box to put the stacked column into c4, with subscripts into c5. Form stacked boxplots by pulling down *Graph* menu to *Boxplot* command and filling in dialog box. The *Graph variable* is c4 and *Categorical variable* is c5.

[**R:**]

```
# We could just use y = rnorm(5250)
# but this the three group sizes clear
y = rnorm(sum(c(250, 1000, 4000)))
x = rep(1:3, c(250, 1000, 4000)))
boxplot(y~x)
```

(a) What do you notice from the resulting boxplot?

(b) Which sample seems to have a heavier tail?

(c) Why is this misleading?

(d) [**Minitab:**] Click on the boxplot. Then pull down the *Editor* menu down to *Select Item* and over to *Outlier Symbols*. Click on *Custom* in the dialog box, and select *Dot*.

[**Minitab version 17.2:**] Left click any one of the outlying points in the boxplot. Then right click to bring up the context menu and select *Edit Outlier Symbols*. Change the symbols to *Custom* and use the dropdown box to select the *Dot* symbol.

[**R:**] In R it is easy to make the box width proportional to the (square root) of the sample size by using the `varwidth` parameter. Simply type:

```
boxplot(y~x, varwidth = TRUE)
```

(e) Is the graph still as misleading as the original?

3.7. Barker and McGhie (1984) collected 100 slugs from the species *Limax maximus* around Hamilton, New Zealand. They were preserved in a relaxed state, and their length in millimeters (mm) and weight in grams (g) were recorded. Thirty of the observations are shown below.

---

[3]Note this used to be labeled the *Manip* menu

Length (mm)	Weight (g)	Length (mm)	Weight (g)	Length (mm)	Weight (g)
73	3.68	21	0.14	75	4.94
78	5.48	26	0.35	78	5.48
75	4.94	26	0.29	22	0.36
69	3.47	36	0.88	61	3.16
60	3.26	16	0.12	59	1.91
74	4.36	35	0.66	78	8.44
85	6.44	36	0.62	90	13.62
86	8.37	22	0.17	93	8.70
82	6.40	24	0.25	71	4.39
85	8.23	42	2.28	94	8.23

[**Minitab:**] The full data are in the Minitab worksheet *slug.mtw*.

[**R:**] The full can be accessed in R by typing

```
data(slug)
```

(a) [**Minitab:**] Plot weight on length using Minitab.

   [**R:**] Plot weight on length using R:

   ```
   plot(weight~length, data = slug)
   ```

   What do you notice about the shape of the relationship?

(b) Often when we have a nonlinear relationship, we can transform the variables by taking logarithms and achieve linearity. In this case, weight is related to volume which is related to length times width times height. Taking logarithms of weight and length should give a more linear relationship.

   [**Minitab:**] Plot log(*weight*) on log(*length*) using Minitab.

   [**R:**] Plot log(*weight*) on log(*length*) using R.

   ```
   plot(log.wt~log.len, data = slug)
   ```

   Does this relationship appear to be linear?

(c) From the scatterplot of log(*weight*) on log(*length*) can you identify any points that do not appear to fit the pattern?

CHAPTER 4

LOGIC, PROBABILITY, AND UNCERTAINTY

Most situations we deal with in everyday life are not completely predictable. If I think about the weather tomorrow at noon, I cannot be certain whether it will or will not be raining. I could contact the Meteorological Service and get the most up-to-date weather forecast possible, which is based on the latest available data from ground stations and satellite images. The forecast could be that it will be a fine day. I decide to take that forecast into account and not take my umbrella. Despite the forecast, it could rain and I could get soaked going to lunch. There is always uncertainty.

In this chapter we will see that deductive logic can only deal with certainty. This is of very limited use in most real situations. We need to develop inductive logic that allows us to deal with uncertainty.

Since we cannot completely eliminate uncertainty, we need to model it. In real life when we are faced with uncertainty, we use plausible reasoning. We adjust our belief about something, based on the occurrence or nonoccurrence of something else. We will see how plausible reasoning should be based on the rules of probability which were originally derived to analyze the outcome of games based on random chance. Thus the rules of probability extend logic to include plausible reasoning where there is uncertainty.

Introduction to Bayesian Statistics, 3rd ed.
By Bolstad, W. M. and Curran, J. M. Copyright © 2016 John Wiley & Sons, Inc.

4.1   Deductive Logic and Plausible Reasoning

Suppose we know "If proposition A is true, then proposition B is true." We are then told "proposition A is true." Therefore, we know that "B is true." It is the only conclusion consistent with the condition. This is a deduction.

Again suppose we know "If proposition A is true, then proposition B is true." Then we are told "B is not true." Therefore, we know that "A is not true." This is also a deduction. When we determine a proposition is true by deduction using the rules of logic, it is certain. Deduction works from the general to the particular.

We can represent propositions using diagrams. Propositions "A is true" and "B is true" are each represented by the interior of a circle. The proposition "*if A is true, then B is true*" is represented by having circle representing A lie completely inside B. This is shown in Figure 4.1. The essence of the first deduction is that if we are in a circle A that lies completely inside circle B, then we must be inside circle B. Similarly, the essence of the second induction is that if we are outside of a circle B that completely contains circle A, then we must be outside circle A.

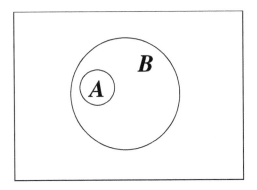

Figure 4.1   "If A is true then B is true." Deduction is possible.

Other propositions can be seen in the diagram. Proposition "*A and B are both true*" is represented by the *intersection*, the region in both the circles simultaneously. In this instance, the intersection equals A by itself. The proposition "A or B is true" is represented by the *union*, region in either one or the other, or both of the circles. In this instance, the union equals B by itself.

On the other hand, suppose we are told "A is not true." What can we now say about B? Traditional logic has nothing to say about this. Both "B is true" and "B is not true" are consistent with the conditions given. Some points outside circle A are inside circle B, and some are outside circle B. No deduction is possible. Intuitively though, we would now believe that it was less plausible that B is true than we previously did before we were told "A

is not true." This is because one of the ways B could be true, namely that
A and B are both true is now no longer a possibility. And the ways that B
could be false have not been affected.

Similarly, when we are told "B is true," traditional logic has nothing to
contribute. Both "A is true" and "A is not true" are consistent with the conditions given. Nevertheless, we see that "B is true" increases the plausibility
of "A is true" because one of the ways A could be false, namely both A and
B are false is no longer possible, and the ways that A are true have not been
affected.

Often propositions are related in such a way that no deduction is possible.
Both "A is true" and "A is false" are consistent with both "B is true" and
"B is false." Figure 4.2 shows this by having the two circles intersect, and
neither is completely inside the other.

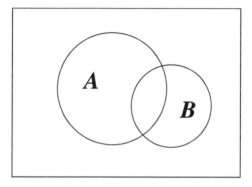

Figure 4.2   Both "A is true" and " A is false" are consistent with both "B is true"
and "B is false." No deduction is possible here.

Suppose we try to use numbers to measure plausibility of propositions.
When we change our plausibility for some proposition on the basis of the
occurrence of some other proposition, we are making an *induction*. Induction
works from the particular to the general.

Desired Properties of Plausibility Measures

1. Degrees of plausibility are represented by nonnegative real numbers.

2. They qualitatively agree with common sense. Larger numbers mean greater
   plausibility.

3. If a proposition can be represented more than one way, then all represen-
   tations must give the same plausibility.

4. We must always take all the relevant evidence into account.

5. Equivalent states of knowledge are always given the same plausibility.

R. T. Cox showed that any set of plausibilities that satisfies the desired properties given above must operate according to the same rules as probability. Thus the sensible way to revise plausibilities is by using the rules of probability. Bayesian statistics uses the rules of probability to revise our belief given the data. Probability is used as an extension of logic to cases where deductions cannot be made. Jaynes and Bretthorst (Editor) gives an excellent discussion on using probability as logic.

4.2   Probability

We start this section with the idea of a random experiment. In a random experiment, though we make the observation under known repeatable conditions, the outcome is uncertain. When we repeat the experiment under identical conditions, we may get a different outcome. We start with the following definitions:

- *Random experiment.* An experiment that has an outcome that is not completely predictable. We can repeat the experiment under the same conditions and not get the same result. Tossing a coin is an example of a random experiment.

- *Outcome.* The result of one single trial of the random experiment.

- *Sample space.* The set of all possible outcomes of one single trial of the random experiment. We denote it Ω. The sample space contains everything we are considering in this analysis of the experiment, so we also can call it the *universe*. In our diagrams we will call it U.

- *Event.* Any set of possible outcomes of a random experiment.

Possible events include the universe, U, and the set containing no outcomes, the empty set ϕ. From any two events E and F we can create other events by the following operations.

- *Union of two events.* The union of two events E and F is the set of outcomes in *either* E <u>or</u> F (inclusive or). Denoted $E \cup F$

- *Intersection of two events.* The intersection of two events E and F is the set of outcomes in both E <u>and</u> F simultaneously. Denoted $E \cap F$.

- *Complement of an event.* The complement of an event E is the set of outcomes not in E. Denoted $\tilde{E}$

We will use Venn diagrams to illustrate the relationship between events. Events are denoted as regions in the universe. The relationship between two events depends on the outcomes they have in common. If all the outcomes in

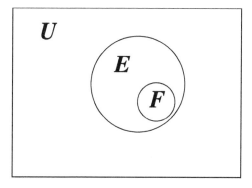

Figure 4.3   Event F is a subset of event E.

one event are also in the other event, the first event is a subset of the other. This is shown in Figure 4.3.

If the events have some outcomes in common, but each has some outcomes that are not in the other, then they are intersecting events. This is shown in Figure 4.4. Neither event is contained in the other.

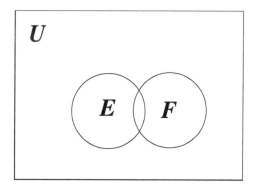

Figure 4.4 E and F are *intersecting* events.

If the two events have no outcomes in common, they are *mutually exclusive* events. In that case the occurrence of one of the events excludes the occurrence of the other, and vice versa. They are also referred to as *disjoint* events. This is shown in Figure 4.5.

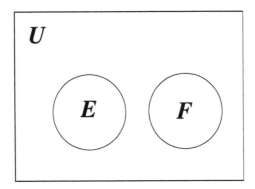

Figure 4.5   Event E and event F are *mutually exclusive* or *disjoint* events.

4.3   Axioms of Probability

The probability assignment for a random experiment is an assignment of probabilities to all possible events the experiment generates. These probabilities are real numbers between 0 and 1. The higher the probability of an event, the more likely it is to occur. A probability that equals 1 means that the event is certain to occur, and a probability of 0 means that the event cannot possibly occur. To be consistent, the assignment of probabilities to events must satisfy the following axioms.

1. $P(A) \geq 0$ for any event A. (Probabilities are nonnegative.)

2. $P(U) = 1$. (Probability of universe $= 1$. Some outcome occurs every time you conduct the experiment.)

3. If A and B are mutually exclusive events, then $P(A \cup B) = P(A) + P(B)$. (Probability is additive over *disjoint* events.)

The other rules of probability can be proved from the axioms.

1. $P(\phi) = 0$. (The empty set has zero probability.)

    - $U = U \cup \phi$ and $U \cap \phi = \phi$. Therefore by axiom 3
    - $1 = 1 + P(\phi)$.
      QED

2. $P(\tilde{A}) = 1 - P(A)$. (The probability of a complement of an event.)

    - $U = A \cup \tilde{A}$ and $A \cap \tilde{A} = \phi$. Therefore by axiom 3
    - $1 = P(A) + P(\tilde{A})$.
      QED

3. $P(A \cup B) = P(A) + P(B) - P(A \cap B)$. (The addition rule of probability.)

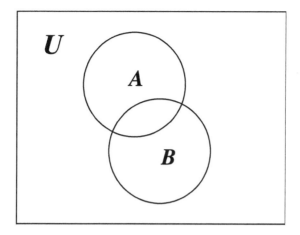

Figure 4.6   Two events A and B in the universe U.

- $A \cup B = A \cup (\tilde{A} \cap B)$ and they are disjoint. Therefore by axiom 3
- $P(A \cup B) = P(A) + P(\tilde{A} \cap B)$.
- $B = (A \cap B) \cup (\tilde{A} \cap B)$, and they are disjoint. Therefore by axiom 3
- $P(B) = P(A \cap B) + P(\tilde{A} \cap B)$. Substituting this in previous equation gives
- $P(A \cup B) = P(A) + P(B) - P(A \cap B)$.
  QED

An easy way to remember this rule is to look at the Venn diagram of the events. The probability of the part $A \cap B$ has been included twice, once in $P(A)$ and once in $P(B)$, so it has to be subtracted out once.

4.4   Joint Probability and Independent Events

Figure 4.6 shows the Venn diagram for two events A and B in the universe U.

The joint probability of events A and B is the probability that both events occur simultaneously, on the same repetition of the random experiment. This would be the probability of the set of outcomes that are in both event A and event B, the intersection $A \cap B$. In other words the joint probability of events A and B is $P(A \cap B)$, the probability of their intersection.

If event A and event B are independent, then $P(A \cap B) = P(A) \times P(B)$. The joint probability is the product of the individual probabilities. If that does not hold, the events are called *dependent* events. Note that whether or not two events A and B are independent or dependent depends on the probabilities assigned.

Distinction between independent events and mutually exclusive events. People often get confused between independent events and mutually exclusive events. This semantic confusion arises because the word *independent* has several meanings. The primary meaning of something being independent of something else is that the second thing has no affect on the first. This is the meaning of the word independent we are using in the definition of independent events. The occurrence of one event does not affect the occurrence or nonoccurrence of the other events.

There is another meaning of the word independent. That is the political meaning of independence. When a colony becomes independent of the mother country, it becomes a distinct separate country. That meaning is covered by the definition of *mutually exclusive* or *disjoint* events.

Independence of two events is not a property of the events themselves, rather it is a property that comes from the probabilities of the events and their intersection. This is in contrast to *mutually exclusive* events, which have the property that they contain no elements in common. Two mutually exclusive events each with non-zero probability cannot be independent. Their intersection is the empty set, so it must have probability zero, which cannot equal the product of the probabilities of the two events!

Marginal probability.   The probability of one of the events A, in the joint event setting is called its marginal probability. It is found by summing $P(A \cap B)$ and $P(A \cap \tilde{B})$ using the axioms of probability.

- $A = (A \cap B) \cup (A \cap \tilde{B})$, and they are disjoint. Therefore by axiom 3

- $P(A) = P(A \cap B) + P(A \cap \tilde{B})$. The marginal probability of event A is found by summing its *disjoint* parts.
  QED

4.5   Conditional Probability

If we know that one event has occurred, does that affect the probability that another event has occurred? To answer this, we need to look at conditional probability.

Suppose we are told that the event A has occurred. Everything outside of A is no longer possible. We only have to consider outcomes inside event A. The *reduced universe* $U_r = A$. The only part of event B that is now relevant is that part which is also in A. This is $B \cap A$. Figure 4.7 shows that, given that event A has occurred, the reduced universe is now the event A, and the only relevant part of event B is $B \cap A$.

Given that event A has occurred, the total probability in the reduced universe must equal 1. The probability of B given A is the unconditional probability of that part of B that is also in A, multiplied by the scale factor $\frac{1}{P(A)}$.

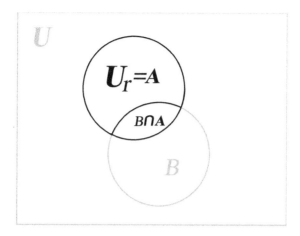

Figure 4.7   The reduced universe, given that event A has occurred.

That gives the conditional probability of event B given event A:

$$P(B|A) = \frac{P(A \cap B)}{P(A)} .$$ (4.1)

We see that the conditional probability $P(B|A)$ is proportional to the joint probability $P(A \cap B)$ but has been rescaled so the probability of the reduced universe equals 1.

Conditional probability for independent events.   Notice that when A and B are independent events we have

$$P(B|A) = P(B) ,$$

since $P(B \cap A) = P(B) \times P(A)$ for independent events, and the factor $P(A)$ will cancel out. Knowledge about A does not affect the probability of B occurring when A and B are independent events! This shows that the definition we used for independent events is a reasonable one.

Multiplication rule.   Formally, we could reverse the roles of the two events A and B. The conditional probability of A given B would be

$$P(A|B) = \frac{P(A \cap B)}{P(B)} .$$

However, we will not consider the two events the same way. B is an unobservable event. That is, the occurrence or nonoccurrence of event B is not observed. A is an observable event that can occur either with event B or with its complement $\tilde{B}$. However, the chances of A occurring may depend on which one of B or $\tilde{B}$ has occurred. In other words, the probability of event A

is conditional on the occurrence or nonoccurrence of event B. When we clear the fractions in the conditional probability formula we get

$$P(A \cap B) = P(B) \times P(A|B). \tag{4.2}$$

This is known as the multiplication rule for probability. It restates the conditional probability relationship of an observable event given an unobservable event in a way that is useful for finding the joint probability $P(A \cap B)$. Similarly,

$$P(A \cap \tilde{B}) = P(\tilde{B}) \times P(A|\tilde{B}).$$

4.6   Bayes' Theorem

From the definition of conditional probability

$$P(B|A) = \frac{P(A \cap B)}{P(A)}.$$

We know that the *marginal* probability of event A is found by summing the probabilities of its *disjoint* parts. Since $A = (A \cap B) \cup (A \cap \tilde{B})$ and clearly $(A \cap B)$ and $(A \cap \tilde{B})$ are disjoint,

$$P(A) = P(A \cap B) + P(A \cap \tilde{B}).$$

We substitute this into the definition of conditional probability to get

$$P(B|A) = \frac{P(A \cap B)}{P(A \cap B) + P(A \cap \tilde{B})}.$$

Now we use the multiplication rule to find each of these joint probabilities. This gives Bayes' theorem for a single event:

$$P(B|A) = \frac{P(A|B) \times P(B)}{P(A|B) \times P(B) + P(A|\tilde{B}) \times P(\tilde{B})}. \tag{4.3}$$

Summarizing, we see Bayes' theorem is a restatement of the conditional probability $P(B|A)$ where:

1. The probability of A is found as the sum of the probabilities of its disjoint parts, $(A \cap B)$ and $(A \cap \tilde{B})$, and

2. Each of the joint probabilities are found using the multiplication rule.

The two important things to note are that the *union* of B and $\tilde{B}$ is the whole universe U, and that they are *disjoint*. We say that events B and $\tilde{B}$ partition the universe.

A set of events partitioning the universe. Often we have a set of more than two events that partition the universe. For example, suppose we have n events $B_1, \cdots, B_n$ such that:

- The union $B_1 \cup B_2 \cup \cdots \cup B_n = U$, the universe, and

- Every distinct pair of the events are disjoint, $B_i \cap B_j = \phi$ for $i = 1, \ldots, n$, $j = 1, \ldots, n$, and $i \neq j$.

Then we say the set of events $B_1, \cdots, B_n$ *partitions* the universe. An observable event A will be partitioned into parts by the partition. $A = (A \cap B_1) \cup (A \cap B_2) \cup \ldots (A \cap B_n)$. $(A \cap B_i)$ and $(A \cap B_j)$ are disjoint since B_i and B_j are disjoint. Hence

$$P(A) = \sum_{j=1}^{n} P(A \cap B_j).$$

This is known as the law of total probability. It just says the probability of an event A is the sum of the probabilities of its disjoint parts. Using the multiplication rule on each joint probability gives

$$P(A) = \sum_{j=1}^{n} P(A|B_j) \times P(B_j).$$

The conditional probability $P(B_i|A)$ for $i = 1, \ldots, n$ is found by dividing each joint probability by the probability of the event A.

$$P(B_i|A) = \frac{P(A \cap B_i)}{P(A)}.$$

Using the multiplication rule to find the joint probability in the numerator, along with the law of total probability in the denominator, gives

$$P(B_i|A) = \frac{P(A|B_i) \times P(B_i)}{\sum_{j=1}^{n} P(A|B_j) \times P(B_j)}. \tag{4.4}$$

This is a result known as Bayes' theorem published posthumously in 1763 after the death of its discoverer, Reverend Thomas Bayes.

■ **EXAMPLE 4.1**

Suppose $n = 4$. Figure 4.8 shows the four unobservable events $B_1, \ldots, B_4$ that partition the universe U, and an observable event A. Now let us look at the conditional probability of B_i given A has occurred. Figure 4.9 shows the reduced universe, given that event A has occurred. The conditional probabilities are the probabilities on the reduced universe, scaled up so they sum to 1. They are given by Equation 4.4.

■

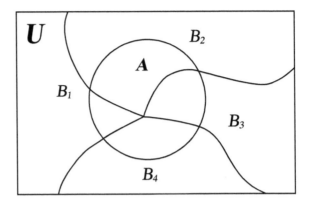

Figure 4.8   Four events B_i for $i = 1, \ldots, 4$ that partition the universe U, along with event A.

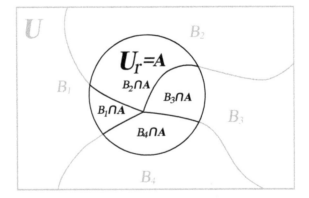

Figure 4.9   The reduced universe given event A has occurred, together with the four events partitioning the universe.

Bayes' theorem is really just a restatement of the conditional probability formula, where the joint probability in the numerator is found by the multiplication rule, and the marginal probability found in the denominator is found using the law of total probability followed by the multiplication rule. Note how the events A and B_i for $i = 1, \ldots, n$ are not treated symmetrically. The events B_i for $i = 1, \ldots, n$ are considered unobservable. We never know which one of them occurred. The event A is an observable event. The marginal probabilities $P(B_i)$ for $i = 1, \ldots, n$ are assumed known before we start and are called our *prior* probabilities.

Bayes' Theorem: The Key to Bayesian Statistics

To see how we can use Bayes' theorem to revise our beliefs on the basis of evidence, we need to look at each part. Let $B_1, \ldots, B_n$ be a set of unobservable events which partition the universe. We start with $P(B_i)$ for $i = 1, \ldots, n$, the *prior* probability for the events B_i, for $i = 1, \ldots, n$. This distribution gives the weight we attach to each of the B_i from our prior belief. Then we find that A has occurred.

The *likelihood* of the unobservable events $B_1, \ldots, B_n$ is the conditional probability that A has occurred given B_i for $i = 1, \ldots, n$. Thus the *likelihood* of event B_i is given by $P(A|B_i)$. We see the *likelihood* is a function defined on the events $B_1, \ldots, B_n$. The *likelihood* is the weight given to each of the B_i events given by the occurrence of A.

$P(B_i|A)$ for $i = 1, \ldots, n$ is the *posterior* probability of event B_i, given that event A has occurred. This distribution contains the weight we attach to each of the events B_i for $i = 1, \ldots n$ after we know event A has occurred. It combines our prior beliefs with the evidence given by the occurrence of event A.

The Bayesian universe.   We can get better insight into Bayes' theorem if we think of the universe as having two dimensions, one observable, and one unobservable. We let the observable dimension be horizontal, and let the unobservable dimension be vertical. The unobservable events no longer partition the universe haphazardly. Instead, they partition the universe as rectangles that cut completely across the universe in a horizontal direction. The whole universe consists of these horizontal rectangles in a vertical stack. Since we do not ever observe which of these events occurred, we never know what vertical position we are in the Bayesian universe.

Observable events are vertical rectangles that cut the universe from top to bottom. We observe that vertical rectangle A has occurred, so we observe the horizontal position in the universe.

Each event $B_i \cap A$ is a rectangle at the intersection of B_i and A. The probability of the event $B_i \cap A$ is found by multiplying the prior probability of B_i times the conditional probability of A given B_i. This is the multiplication rule.

The event A is the union of the disjoint parts $A \cap B_i$ for $i = 1, \ldots, n$. The probability of A is clearly the sum of the probabilities of each of the disjoint parts. The probability of A is found by summing the probabilities of each disjoint part down the vertical column represented by A. This is the *marginal* probability of A.

The posterior probability of any particular B_i given A is the proportion of A that is also in B_i. In other words, the probability of $B_i \cap A$ divided by the sum of $B_j \cap A$ summed over all $j = 1, \ldots, n$.

In Bayes' theorem, each of the joint probabilities is found by multiplying the *prior* probability $P(B_i)$ times the *likelihood* $P(A|B_i)$. In Chapter 5, we will see that the universe set out with two dimensions for two jointly distributed discrete random variables is very similar to that shown in Figures 4.10 and 4.11. One random variable will be observed, and we will determine the conditional probability distribution of the other random variable, given our observed value of the first. In Chapter 6, we will develop Bayes' theorem for two discrete random variables in an analogous manner to our development of Bayes' theorem for events in this chapter.

▣ EXAMPLE 4.1 (continued)

Figure 4.10 shows the four unobservable events B_i for $i = 1, \ldots, 4$ that partition the Bayesian universe, together with event A which is observable. Figure 4.11 shows the reduced universe, given that event A has occurred. These figures will give us better insight than Figures 4.8 and 4.9. We know where in the Bayesian universe we are in the horizontal direction since we know event A occurred. However, we do not know where we are in the vertical direction since we do not know which one of the B_i occurred.

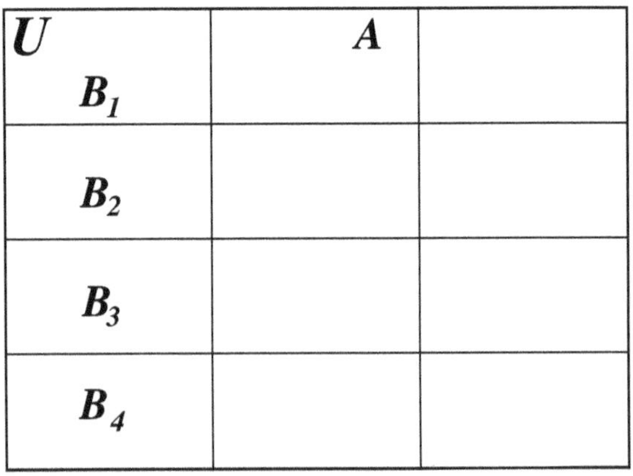

Figure 4.10   The Bayesian universe U with four unobservable events B_i for $i = 1, \ldots, 4$ which partition it shown in the vertical dimension, and the observable event A shown in the horizontal dimension.

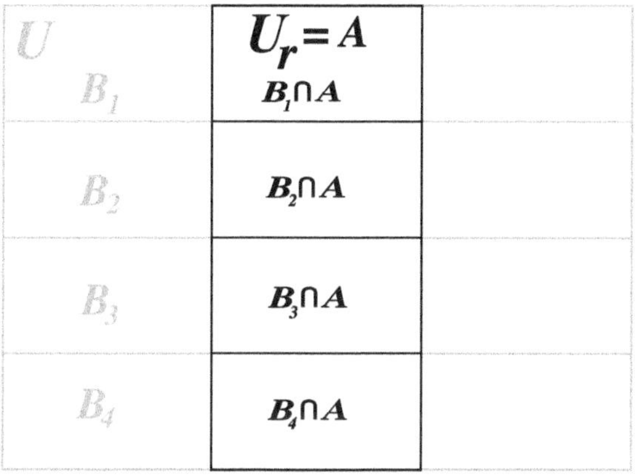

Figure 4.11   The reduced Bayesian universe, given A has occurred, together with the four unobservable events B_i for $i = 1, \ldots, 4$ that partition it.

◼

Multiplying by constant.   The numerator of Bayes' theorem is the prior probability times the likelihood. The denominator is the sum of the prior probabilities times likelihoods over the whole partition. This division of the prior probability times likelihood by the sum of prior probabilities times likelihoods makes the posterior probability sum to 1.

Note that if we multiplied each of the likelihoods by a constant, the denominator would also be multiplied by the same constant. The constant would cancel out in the division, and we would be left with the same posterior probabilities. Because of this, we only need to know the likelihood to within a constant of proportionality. The *relative* weights given to each of the possibilities by the likelihood is all we need. Similarly, we could multiply each prior probability by a constant. The denominator would again be multiplied by the same constant, so we would be left with the same posterior probabilities. The only thing we need in the prior is the *relative* weights we give to each of the possibilities. We often write Bayes' theorem in its proportional form as

$$posterior \propto prior \times likelihood$$

This gives the relative weights for each of the events B_i for $i = 1, \ldots, n$ after we know A has occurred. Dividing by the sum of the relative weights rescales the relative weights so they sum to 1. This makes it a probability distribution.

We can summarize the use of Bayes' theorem for events by the following three steps:

1. Multiply *prior* times *likelihood* for each of the B_i. This finds the probability of $B_i \cap A$ by the multiplication rule.

2. Sum them for $i = 1, \ldots, n$. This finds the probability of A by the law of total probability.

3. Divide each of the prior times likelihood values by their sum. This finds the conditional probability of that particular B_i given A.

4.7   Assigning Probabilities

Any assignment of probabilities to all possible events must satisfy the probability axioms. Of course, to be useful the probabilities assigned to events must correspond to the real world. There are two methods of probability assignment that we will use:

1. *Long-run relative frequency probability assignment*: The probability of an event is considered to be the proportion of times it would occur if the experiment was repeated an infinite number of repetitions. This is the method of assigning probabilities used in frequentist statistics. For example, if I was trying to assign the probability of getting a head on a toss of a coin, I would toss it a large number of times and use the proportion of heads that occurred as an approximation to the probability.

2. *Degree of belief probability assignment*: the probability of an event is what I believe it is from previous experience. This is subjective. Someone else can have a different belief. For example, I could say that I believe the coin is a fair one, so for me, the probability of getting a head equals .5. Someone else might look at the coin and observing a slight asymmetry he/she might decide the probability of getting a head equals .49.

In Bayesian statistics, we will use long-run relative frequency assignments of probabilities for events that are outcomes of the random experiment, given the value of the unobservable variable. We call the unobservable variable the *parameter*. Think about repeating the experiment over and over again an infinite number of times while holding the parameter (unobservable) at a fixed value. The set of all possible observable values of the experiment is called the *sample space* of the experiment. The probability of an event is the long-run relative frequency of the event over all these hypothetical repetitions. We see the *sample space* is the observable (horizontal) dimension of the *Bayesian universe*.

The set of all possible values of the parameter (unobservable) is called the *parameter space*. It is the unobservable (vertical) dimension of the *Bayesian universe*. In Bayesian statistics we also consider the parameter value to be random. The probability I assign to an event "the parameter has a certain value" cannot be assigned by long-run relative frequency. To be consistent with the idea of a fixed but unknown parameter value, I must assign probabilities by degree of belief. This shows the relative plausibility I give to all the possible parameter values before the experiment. Someone else would have different probabilities assigned according to his/her belief.

I am modeling my uncertainty about the parameter value by a single random draw from my prior distribution. I do not consider hypothetical repetitions of this draw. I want to make my inference about the parameter value drawn this particular time, given this particular data. Earlier in the chapter we saw that using the rules of probability is the only consistent way to update our beliefs given the data. So probability statements about the parameter value are always subjective, since they start with subjective prior belief.

4.8   Odds and Bayes Factor

Another way of dealing with uncertain events that we are modeling as random is to form the odds of the event. The *odds* for an event C equals the probability of the event occurring divided by the probability of the event not occurring:

$$odds(C) = \frac{P(C)}{P(\tilde{C})}.$$

Since the probability of the event not occurring equals one minus the probability of the event, there is a one-to-one relationship between the odds of an event and its probability.

$$odds(C) = \frac{P(C)}{(1 - P(C))}.$$

If we are using prior probabilities, we get the *prior* odds — in other words, the ratio before we have analyzed the data. If we are using posterior probabilities, we get the *posterior* odds.

Solving the equation for the probability of event C we get

$$P(C) = \frac{odds(C)}{(1 + odds(C))}.$$

We see that there is a one-to-one correspondence between odds and probabilities.

Bayes Factor (B)

The Bayes factor B contains the evidence in the data D that occurred relevant to the question about C occurring. It is the factor by which the prior odds is changed to the posterior odds:

$$prior\ odds(C) \times B = posterior\ odds(C).$$

We can solve this relationship for the Bayes factor to get

$$B = \frac{posterior\ odds}{prior\ odds}.$$

We can substitute in the ratio of probabilities for both the posterior and prior odds to find

$$B = \frac{P(D|C)}{P(D|\tilde{C})} \, .$$

Thus the Bayes factor is the ratio of the probability of getting the data which occurred given the event, to the probability of getting the data which occurred given the complement of the event. If the Bayes factor is greater than 1, then the data has made us believe that the event is more probable than we thought before. If the Bayes factor is less than 1, then the data has made us believe that the event is less probable than we originally thought.

4.9   Beat the Dealer

In this section we take a diversion into the gambling world. This story recounts the journey of an American mathematician to the Blackjack tables of Las Vegas. Armed with an understanding of the laws of probability, along with early access to a computer, Edward Thorp, a Professor of Mathematics, developed a strategy that could beat the casinos at their own game. It illustrates that observing one event changes the probability of another event, and it also illustrates many other statistical ideas introduced in this chapter.

The game of blackjack, or twenty-one, has the player and the dealer competing to get a score as close as possible to twenty-one, without going bust (over twenty-one). Initially both are dealt two cards, one face up and one face down. Each face card counts ten, and each number card counts its own value, while an ace can be counted either as one or eleven, whichever is advantageous. The player can ask to be dealt a card, face up, as long as he/she has not gone bust. If the player holds before going bust, then the dealer must be dealt a card, face up when the dealer's total is under sixteen, and must hold if the total is seventeen or over.

The casino had set the payoff assuming that the player's probability of winning is calculated starting from a full deck that has just been shuffled. That way they had thought that they were setting a small advantage to the house. The law of averages would ensure that over the long run the house would gain, and the player would lose.

However, as actually played, the deck was not shuffled after every hand. Rather, the cards that had been played (some of which have been observed) were put aside, and the next hand was dealt from the remaining cards. They would continue this way until almost all of the cards had been played before stopping and shuffling. Thorp realized that the real probability the player has of winning a hand depends on the cards that remain in the deck.

- The *conditional* probability of winning given the cards that remain in the deck is what counts, not the unconditional probability calculated assuming a complete shuffled deck.

Although the long-run odds were against the player, sometimes the actual odds would be in favor of the player. If Thorp could identify those times, by making large bets at those times and betting the minimum at other times, overall he would be able to win. This was in the early days of computing, and he had access to IBM 704 computer. He wrote a program that would simulate the playing blackjack with strategies that depend on the cards that had been seen, and he ran the program thousands of times.

- This is a *Monte Carlo* study. He determined that a simple strategy that only depends on the observed ratio of cards over five to those five or under would be effective. This strategy is known as "card counting" and is not illegal.

He went to Las Vegas and proved his strategy by winning lots of money. Of course, the casino was not happy at having to pay out. The reason they were not shuffling the deck between each hand was that they considered shuffling to be dead time during which the casino was not making money. They did not want to shuffle each time, just in case someone was counting cards. One of the first countermeasures they devised was to increase the number of decks of cards used. This would make the ratio of over fives to five and under less variable. However, Thorp continued his Monte Carlo study with more decks and found that it still worked, particularly after many hands had been played and only a few cards remained. He resumed winning, until the casinos banned him. Interested readers can read more about this in Thorp (1962). Card counting continues to be legal, but casinos try to identify those practicing it and ban them from playing. Casinos are private establishments and have the right to ban anyone they wish. Of course, the casinos could just shuffle the deck between each hand. But they have decided that their overall best strategy is to allow card counting but identify successful practitioners and ban them from further play.

One may ask "What about the other cards that have been played, but not observed? Shouldn't they also be taken into account?" Of course, the actual probability of winning depends on all the cards that remain in the deck. The probabilities found by Thorp, which depend only on the cards observed, have averaged the cards that have been played but not seen over all the possible values.

Main Points

- *Deductive logic.* A logical process for determining the truth of a statement from knowing the truth or falsehood of other statements that the first statement is a consequence of. Deduction works from the general to the particular. We can make a deduction from a known population distribution to determine the sampling distribution of a statistic.

- Deductions do not have the possibility of error.

- *Inductive logic.* A process, based on plausible reasoning, for inferring the truth of the statement from knowing the truth or falsehood of other statements which are consequences of the first statement. It works from the particular to the general. Statistical inference is an inductive process for making inferences about the parameter, on the basis of the observed statistic from the sampling distribution given the parameter.

- There is always the possibility of error when making an inference.

- Plausible reasoning should be based on the rules of probability to be consistent. They are:

  o Probability of an event is a nonnegative number.

  o Probability of the sample space (universe) equals 1.

  o The probability is additive over disjoint events.

- A *random experiment* is an experiment where the outcome is not exactly predictable, even when the experiment is repeated under the identical conditions.

- The set of all possible outcomes of a random experiment is called the *sample space* Ω. In frequentist statistics, the sample space is the universe for analyzing events based on the experiment.

- The *union* of two events A and B is the set of outcomes in A or B. This is an inclusive or. The union is denoted $A \cup B$.

- The *intersection* of two events A and B is the set of outcomes in both A and B simultaneously. The intersection is denoted $A \cap B$.

- The *complement* of event A is the set of outcomes not in A. The complement of event A is denoted $\tilde{A}$.

- *Mutually exclusive* events have no elements in common. Their intersection $P(A \cap B)$ equals the empty set, ϕ.

- The conditional probability of event B given event A is given by

$$P(B|A) = \frac{P(A \cap B)}{P(A)}.$$

- The event B is unobservable. The event A is observable. We could nominally write the conditional probability formula for $P(A|B)$, but the relationship is not used in that form. We do not treat the events symmetrically. The multiplication rule is the definition of conditional probability cleared of the fraction.

$$P(A \cap B) = P(B) \times P(A|B).$$

It is used to assign probabilities to compound events.

- The *law of total probability* says that given events $B_1, \ldots, B_n$ that partition the sample space (universe), along with another event A, then

$$P(A) = \sum_{j=1}^{n} P(B_j \cap A)$$

because probability is additive over the disjoint events, $(A \cap B_1) \ldots (A \cap B_n)$. When we find the probability of each of the intersections $A \cap B_j$ by the multiplication rule, we get

$$P(A) = \sum_j P(B_j) \times P(A|B_j).$$

- Bayes' theorem is the key to Bayesian statistics:

$$P(B_i|A) = \frac{P(B_i) \times P(A|B_i)}{\sum_j P(B_j) \times P(A|B_j)}.$$

This comes from the definition of conditional probability. The marginal probability of the event A is found by the law of total probability, and each of the joint probabilities is found from the multiplication rule. $P(B_i)$ is called the prior probability of event B_i, and $P(B_i|A)$ is called the posterior probability of event B_i.

- In the Bayesian universe, the unobservable events $B_1, \ldots, B_n$ which partition the universe are horizontal slices, and the observable event A is a vertical slice. The probability $P(A)$ is found by summing the $P(A \cap B_i)$ down the column. Each of the $P(A \cap B_i)$ is found by multiplying the prior $P(B_i)$ times the likelihood $P(A|B_i)$. So Bayes' theorem can be summarized by saying that the posterior probability is the prior times likelihood divided by the sum of the prior times likelihood.

- The Bayesian universe has two dimensions. The sample space forms the observable (horizontal) dimension of the Bayesian universe. The parameter space is the unobservable (vertical) dimension. In Bayesian statistics, the probabilities are defined on both dimensions of the Bayesian universe.

- The odds of an event A is the ratio of the probability of the event to the probability of its complement:

$$odds(A) = \frac{P(A)}{P(\tilde{A})}.$$

If it is found before analyzing the data, it is the prior odds. If it is found after analyzing the data, it is the posterior odds.

▪ The Bayes factor is the amount of evidence in the data that changes the prior odds to the posterior odds:

$$B \times prior\ odds = posterior\ odds\,.$$

Exercises

4.1. There are two events A and B. $P(A) = .4$ and $P(B) = .5$. The events A and B are independent.

  (a) Find $P(\tilde{A})$.

  (b) Find $P(A \cap B)$.

  (c) Find $P(A \cup B)$.

4.2. There are two events A and B. $P(A) = .5$ and $P(B) = .3$. The events A and B are independent.

  (a) Find $P(\tilde{A})$.

  (b) Find $P(A \cap B)$.

  (c) Find $P(A \cup B)$.

4.3. There are two events A and B. $P(A) = .4$ and $P(B) = .4$. $P(\tilde{A} \cap B) = .24$.

  (a) Are A and B independent events? Explain why or why not.

  (b) Find $P(A \cup B)$.

4.4. There are two events A and B. $P(A) = .7$ and $P(B) = .8$. $P(\tilde{A} \cap \tilde{B}) = .1$.

  (a) Are A and B independent events? Explain why or why not.

  (b) Find $P(A \cup B)$.

4.5. A single fair die is rolled. Let the event A be "the face showing is even." Let the event B be "the face showing is divisible by 3."

  (a) List out the sample space of the experiment.

  (b) List the outcomes in A, and find $P(A)$.

  (c) List the outcomes in B, and find $P(B)$.

  (d) List the outcomes in $A \cap B$, and find $P(A \cap B)$.

  (e) Are the events A and B independent? Explain why or why not.

4.6. Two fair dice, one red and one green, are rolled. Let the event A be "the sum of the faces showing is equal to seven." Let the event B be "the faces showing on the two dice are equal."

   (a) List out the sample space of the experiment.

   (b) List the outcomes in A, and find $P(A)$.

   (c) List the outcomes in B, and find $P(B)$.

   (d) List the outcomes in $A \cap B$, and find $P(A \cap B)$.

   (e) Are the events A and B independent? Explain why or why not.

   (f) How would you describe the relationship between event A and event B?

4.7. Two fair dice, one red and one green, are rolled. Let the event A be "the sum of the faces showing is an even number." Let the event B be "the sum of the faces showing is divisible by 3."

   (a) List the outcomes in A, and find $P(A)$.

   (b) List the outcomes in B, and find $P(B)$.

   (c) List the outcomes in $A \cap B$, and find $P(A \cap B)$.

   (d) Are the events A and B independent? Explain why or why not.

4.8. Two dice are rolled. The red die has been loaded. Its probabilities are $P(1) = P(2) = P(3) = P(4) = \frac{1}{5}$ and $P(5) = P(6) = \frac{1}{10}$. The green die is fair. Let the event A be "the sum of the faces showing is an even number." Let the event B be "the sum of the faces showing is divisible by 3."

   (a) List the outcomes in A, and find $P(A)$.

   (b) List the outcomes in B, and find $P(B)$.

   (c) List the outcomes in $A \cap B$, and find $P(A \cap B)$.

   (d) Are the events A and B independent? Explain why or why not.

4.9. Suppose there is a medical diagnostic test for a disease. The *sensitivity* of the test is .95. This means that if a person has the disease, the probability that the test gives a positive response is .95. The *specificity* of the test is .90. This means that if a person does not have the disease, the probability that the test gives a negative response is .90, or that the *false positive* rate of the test is .10. In the population, 1% of the people have the disease. What is the probability that a person tested has the disease, given the results of the test is positive? Let D be the event "the person has the disease" and let T be the event "the test gives a positive result."

4.10. Suppose there is a medical screening procedure for a specific cancer that has *sensitivity* $= .90$, and *specificity* $= .95$. Suppose the underlying rate of the cancer in the population is .001. Let B be the event "the person has that specific cancer," and let A be the event "the screening procedure gives a positive result."

(a) What is the probability that a person has the disease given the results of the screening is positive?

(b) Does this show that screening is effective in detecting this cancer?

4.11. In the game of blackjack, also known as twenty-one, the player and the dealer are dealt one card face-down and one card face-up. The object is to get as close as possible to the score 21, without exceeding that. Aces count either 1 or 11, face cards count 10, and all other cards count at their face value. The player can ask for more cards to be dealt to him, provided that he has not gone bust (exceeded 21) and lost. Getting 21 on the deal (an ace and a face card or 10) is called a "blackjack." Suppose 4 decks of cards are shuffled together and dealt from. What is the probability the player gets a blackjack?

4.12. After the hand, the cards are discarded, and the next hand continues with the remaining cards in the deck. The player has had an opportunity to see some of the cards in the previous hand, those that were dealt face-up. Suppose he saw a total of 4 cards, and none of them were aces, nor were any of them a face card or a ten. What is the probability the player gets a blackjack on this hand?

CHAPTER 5

DISCRETE
RANDOM VARIABLES

In the previous chapter, we looked at random experiments in terms of events. We also introduced probability defined on events as a tool for understanding random experiments. We showed how conditional probability is the logical way to change our belief about an unobserved event given that we observed another related event. In this chapter we introduce discrete random variables and probability distributions.

A random variable describes the outcome of the experiment in terms of a number. If the only possible outcomes of the experiment are distinct numbers separated from each other (e.g., counts), then we say that the random variable is discrete. There are good reasons why we introduce random variables and their notation:

- It is quicker to describe an outcome as a random variable having a particular value than to describe that outcome in words. Any event can be formed from outcomes described by the random variable using union, intersection, and complements.

- The probability distribution of the discrete random variable is a numerical function. It is easier to deal with a numerical function than with probabilities being a function defined on sets (events). The probability of any possible event can be found from the probability distribution of the random variable using the rules of probability. So instead of having to know the probability of every possible event, we only have to know the probability distribution of the random variable.

- It becomes much easier to deal with compound events made up from repetitions of the experiment.

5.1 Discrete Random Variables

A number that is determined by the outcome of a random experiment is called a random variable. Random variables are denoted by uppercase letters, e.g., Y. The value the random variable takes is denoted by lowercase letters, e.g., y. A discrete random variable, Y, can only take on the distinct values y_k. There can be a finite possible number of values; for example, the random variable defined as "number of heads in n tosses of a coin" has possible values $0, 1, \ldots, n$. Or there can be a countably infinite number of possible values; for example, the random variable defined as "number of tosses until the first head" has possible values $1, 2, \ldots, \infty$. The key thing for discrete random variables is that the possible values are separated by gaps.

Thought Experiment 1: *Roll of a die*
Suppose we have a fair six sided die. Our random experiment is to roll it, and we let the random variable Y be the number on the top face. There are six possible values $1, 2, \ldots, 6$. Since the die is fair, those six values are equally likely. Now, suppose we take independent repetitions of the random variable and record each occurrence of Y. Table 5.1 shows the proportion of times each face has occurred in a typical sequence of rolls of the die, after 10, 100, 1,000, and 10,000 rolls. The last column shows the true probabilities for a fair die.

Table 5.1   Typical results of rolling a fair die

Value	10 Rolls	100 Rolls	1,000 Rolls	10,000 Rolls	...	Probability
			Proportion After			
1	0.1	0.17	0.182	0.1668	...	0.1666
2	0.2	0.13	0.182	0.1739	...	0.1666
3	0.3	0.20	0.176	0.1716	...	0.1666
4	0.1	0.21	0.159	0.1685	...	0.1666
5	0.1	0.09	0.150	0.1592	...	0.1666
6	0.2	0.20	0.151	0.1600	...	0.1666

We note that the proportions taking any value are getting closer and closer to the true probability of that value as n increases to ∞. We could draw graphs of the proportions having each value. These are shown in Figure 5.1. The graphs are at zero for any other y value, and they have a spike at each

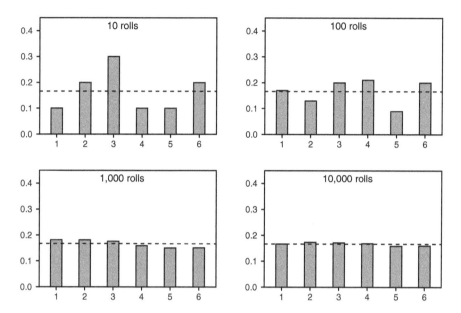

Figure 5.1   Proportions resulting from 10, 100, 1,000, and 10,000 rolls of a fair die.

possible value where the spike height equals the proportion of times that value occurred. The sum of spike heights equals one.

Thought Experiment 2: *Random sampling from a finite population*
Suppose we have a finite population of size N. There can be at most a finite number of possible values, and they must be discrete, since there must be a gap between every pair of two real numbers. Some members of the population have the same value, so there are only K possible values $y_1, \ldots, y_K$. The probability of observing the value y_k is the proportion of population having that value.

We start by randomly drawing from the population with replacement. Each draw is done under identical conditions. If we continue doing the sampling, then eventually we will have seen all possible values. After each draw we update the proportions in the accumulated sample that have each value. We sketch a graph with a spike at each value in the sample equal to the proportion in the sample having that value. The updating of the graph at step n is made by scaling all the existing spikes down by the ratio $\frac{n-1}{n}$ and adding $\frac{1}{n}$ to the spike at the value observed. The scaling changes the proportions after the first $n-1$ observations to the proportions after the first n observations. As the sample size increases, the sample proportions get less variable. In the limit as the sample size n approaches infinity, the spike at each value approaches its probability.

Thought Experiment 3: *Number of tails before first head from independent coin tosses*
Each toss of a coin results in either a head or a tail. The probability of getting a head remains the same on each toss. The outcomes of each toss are independent of each other. This is an example of what we call Bernoulli trials. The outcome of a trial is either a success (head) or failure (tail), the probability of success remains constant over all trials, and we are taking independent trials. We are counting the number of failures before the first success. Every nonnegative integer is a possible value, and there are an infinite number of them. They must be discrete, since there is a gap between every pair of nonnegative integers.

We start by tossing the coin and counting the number of tails until the first head occurs. Then we repeat the whole process. Eventually we reach a state where most of the time we get a value we have gotten before. After each sequence of trials until the first head, we update the proportions that have each value. We sketch a graph with a spike at each value equal to the proportion having that value. As in the previous example, the updating of the graph at step n is made by scaling all the existing spikes down by the ratio $(n-1)/n$ and adding $1/n$ to the spike at the value observed. The sample proportions get less variable as the sample size increases, and in the limit as n approaches infinity, the spike at each value approaches its probability.

5.2   Probability Distribution of a Discrete Random Variable

The proportion functions that we have seen in the three thought experiments are spike functions. They have a spike at each possible value, zero at all other values, and the sum of the spike heights equals one. In the limit as the sample size approaches infinity, the proportion of times a value occurs approaches the probability of that value, and the proportion graphs approach the probability function

$$f(y_k) = P(Y = y_k)$$

for all possible values $y_1, \ldots, y_k$ of the discrete random variable. For any other value y, it equals zero.

Expected Value of a Discrete Random Variable

The expected value of a discrete random variable Y is defined to be the sum over all possible values of each possible value times its probability:

$$E[Y] = \sum_{k=1} y_k \times f(y_k). \tag{5.1}$$

The expected value of a random variable is often called the mean of the random variable and is denoted μ. It is like the sample mean of an infinite sample of independent repetitions of the random variable. The sample mean of a random sample of size n repetitions of the random variable is

$$\bar{y} = \frac{1}{n} \sum_{i=1}^{n} y_i \,.$$

Here y_i is the value that occurs on the ith repetition. We are summing over all repetitions. Grouping together all repetitions that have the same possible value, we get

$$\bar{y} = \sum_{k} \frac{n_k}{n} \times y_k \,,$$

where n_k is the number of observations that have value y_k, and we are now summing over all possible values. Note that each of the y_i (observed values) equals one of the y_k (possible values). But in the limit as n approaches ∞, the relative frequency $\frac{n_k}{n}$ approaches the probability $f(y_k)$, so the sample mean, $\bar{y}$, approaches the expected value, $\mathrm{E}[Y]$. This shows that the expected value of a random variable is like the sample mean of an infinite size random sample of that variable.

The Variance of a Discrete Random Variable

The variance of a random variable is the expected value of square of the variable minus its mean.

$$\begin{aligned}
\mathrm{Var}[Y] &= E[Y - \mathrm{E}[Y]]^2 \\
&= \sum_{k} (y_k - \mu)^2 \times f(y_k) \,.
\end{aligned} \tag{5.2}$$

This is like the sample variance of an infinite size random sample of that variable. We note that if we square the term in brackets, break the sum into three sums, and factor the constant terms out of each sum, we get

$$\begin{aligned}
\mathrm{Var}[Y] &= \sum_{k} y_k^2 \times f(y_k) - 2\mu \times \sum_{k} y_k f(y_k) + \mu^2 \times \sum_{k} f(y_k) \\
&= \mathrm{E}[Y^2] - \mu^2 \,.
\end{aligned}$$

Since $\mu = \mathrm{E}[Y]$, this gives another useful formula for computing the variance.

$$\mathrm{Var}[Y] = \mathrm{E}[Y^2] - [\mathrm{E}[Y]]^2 \,. \tag{5.3}$$

■   **EXAMPLE 5.1**

Let Y be a discrete random variable with probability function given in the following table.

y_i	$f(y_i)$
0	.20
1	.15
2	.25
3	.35
4	.05

To find $E[Y]$ we use Equation 5.1, which gives

$$E[Y] = 0 \times .20 + 1 \times .15 + 2 \times .25 + 3 \times .35 + 4 \times .05$$
$$= 1.90 \, .$$

Note that the expected value does not have to be a possible value of the random variable Y. It represents an average. We will find $Var[Y]$ in two ways and see that they give equivalent results. First, we use the definition of variance given in Equation 5.2.

$$Var[Y] = (0 - 1.90)^2 \times .20 + (1 - 1.90)^2 \times .15 + (2 - 1.90)^2 \times .25$$
$$+ (3 - 1.90)^2 \times .35 + (4 - 1.90)^2 \times .05$$
$$= 1.49 \, .$$

Second, we will use Equation 5.3. We calculate

$$E[Y^2] = 0^2 \times .20 + 1^2 \times .15 + 2^2 \times .25 + 3^2 \times .35 + 4^2 \times .05$$
$$= 5.10 \, .$$

Putting that result in Equation 5.3, we get

$$Var[Y] = 5.10 - 1.90^2$$
$$= 1.49 \, .$$

■

The Mean and Variance of a Linear Function of a Random Variable

Suppose $W = a \times Y + b$, where Y is a discrete random variable. Clearly, W is another number that is the outcome of the same random experiment that

Y came from. Thus W, a linear function of a random variable Y, is another random variable. We wish to find its mean.

$$E[aY + b] = \sum_k (ay_k + b) \times f(y_k)$$

$$= \sum_k ay_k \times f(y_k) + \sum b \times f(y_k)$$

$$= a \sum y_k f(y_k) + b \sum f(y_k).$$

Since $\sum y_k f(y_k) = \mu$ and $\sum f(y_k) = 1$, the mean of the linear function is the linear function of the mean:

$$E[aY + b] = a\,E[Y] + b. \tag{5.4}$$

Similarly, we may wish to know its variance.

$$Var[aY + b] = \sum_k (ay_k + b - E[aY + b])^2 f(y_k)$$

$$= \sum_k [a(y_k - E[Y]) + b - b)]^2 f(y_k)$$

$$= a^2 \sum_k (y_k - E[Y])^2 f(y_k).$$

Thus the variance of a linear function is the square of the multiplicative constant a times the variance :

$$Var[aY + b] = a^2\,Var[Y]. \tag{5.5}$$

The additive constant b does not enter into it.

■ **EXAMPLE 5.1 (continued)**

Suppose $W = -2Y + 3$. Then from Equation 5.4 we have

$$E[W] = -2\,E[Y] + 3$$
$$= -2 \times 1.90 + 3$$
$$= -.80$$

and from Equation 5.5 we have

$$Var[W] = (-2)^2 \times Var[Y]$$
$$= 4 \times 1.49$$
$$= 5.96.$$

■

5.3   Binomial Distribution

Let us look at three situations and see what characteristics they have in common.

Coin tossing.   Suppose we toss the same coin n times, and count the number of heads that occur. We consider that any one toss is not influenced by the outcomes of previous tosses; in other words, the outcome of one toss is independent of the outcomes of previous tosses. Since we are always tossing the same coin, the probability of getting a head on any particular toss remains constant for all tosses. The possible values of the total number of heads observed in the n tosses are $0, \ldots, n$.

Drawing from an urn with replacement.   An urn contains balls of two colors, red and green. The proportion of red balls is π. We draw a ball at random from the urn, record its color, then return it to the urn, and remix the balls before the next random draw. We make a total of n draws and count the number of times we drew a red ball. Since we replace and remix the balls between draws, each draw takes place under identical conditions. The outcome of any particular draw is not influenced by the previous draw outcomes. The probability of getting a red ball on any particular draw remains equal to π, the proportion of red balls in the urn. The possible values of the total number of red balls drawn are $0, \ldots, n$.

Random sampling from a very large population.   Suppose we draw a random sample of size n from a very large population. The proportion of items in the population having some attribute is π. We count the number of items in the sample that have the attribute. Since the population is very large compared to the sample size, removing a few items from the population does not perceptibly change the proportion of remaining items having the attribute. For all intents and purposes it remains π. The random draws are taken under almost identical conditions. The outcome of any draw is not influenced by the previous outcomes. The possible values of the number of items drawn that have the attribute is $0, \ldots, n$.

Characteristics of the Binomial Distribution

These three cases all have the following things in common.

- There are n *independent* trials. Each trial can result either in a "success" or a "failure."

- The probability of "success" is constant over all the trials. Let π be the probability of "success."

- Y is the number of "successes" that occurred in the n trials. Y can take on integer values $0, 1, \ldots, n$.

These are the characteristics of the *binomial*(n, π) distribution. The binomial probability function can be found from these characteristics using the laws of probability. Any sequence having exactly y successes out of the n independent trials has probability equal to $\pi^y(1-\pi)^{n-y}$, no matter in which order they occur. The event $\{Y = y\}$ is the union of all sequences such sequences. The sequences are disjoint, so the probability function of the binomial random variable Y given the parameter value π is written as

$$f(y|\pi) = \binom{n}{y}\pi^y(1-\pi)^{n-y} \qquad (5.6)$$

for $y = 0, 1, \ldots, n$ where the binomial coefficient

$$\binom{n}{y} = \frac{n!}{y! \times (n-y)!}$$

represents the number of sequences having exactly y successes out of n trials and $\pi^y(1-\pi)^{n-y}$ is the probability of any particular sequence having exactly y successes out of n trials.

Mean of binomial.   The mean of the *binomial*(n, π) distribution is the sample size times the probability of success since

$$E[Y|\pi] = \sum_{y=0}^{n} y \times f(y|\pi)$$

$$= \sum_{y=0}^{n} y \times \binom{n}{y}\pi^y(1-\pi)^{n-y} \, .$$

We write this as a conditional mean because it is the mean of Y given the value of the parameter π. The first term in the sum is 0, so we can start the sum at $y = 1$. We cancel y in the remaining terms, and factor out $n\pi$. This gives

$$E[Y|\pi] = \sum_{y=1}^{n} n\pi \binom{n-1}{y-1}\pi^{y-1}(1-\pi)^{n-y} \, .$$

Factoring $n\pi$ out of the sum and substituting $n' = n-1$ and $y' = y-1$, we get

$$E[Y|\pi] = n\pi \sum_{y'=0}^{n'} \binom{n'}{y'}\pi^{y'}(1-\pi)^{n'-y'} \, .$$

We see the sum is a binomial probability function summed over all possible values. Hence it equals one, and the mean of the binomial is

$$E[Y|\pi] = n\pi \, . \qquad (5.7)$$

Variance of binomial. The variance is the sample size times the probability of success times the probability of failure. We write this as a conditional variance since it is the variance of Y given the value of the parameter π. Note that

$$E[Y(Y-1)|\pi] = \sum_{y=0}^{n} y(y-1) \times f(y|\pi)$$

$$= \sum_{y=0}^{n} y(y-1) \times \binom{n}{y} \pi^y (1-\pi)^{n-y}.$$

The first two terms in the sum equal 0, so we can start summing at $y = 2$. We cancel $y(y-1)$ out of the remaining terms and factor out $n(n-1)\pi^2$ to get

$$E[Y(Y-1)|\pi] = \sum_{y=2}^{n} n(n-1)\pi^2 \binom{n-2}{y-2} \pi^{y-2}(1-\pi)^{n-y}.$$

Substituting $y' = y - 2$ and $n' = n - 2$, we get

$$E[Y(Y-1)|\pi] = n(n-1)\pi^2 \sum_{y'=0}^{n-2} \binom{n'}{y'} \pi^{y'} (1-\pi)^{n'}$$

$$= n(n-1)\pi^2$$

since we are summing a binomial distribution over all possible values. The variance can be found by

$$\mathrm{Var}[Y|\pi] = E[Y^2|\pi] - [E[Y|\pi]]^2$$
$$= E[Y(Y-1)|\pi] + E[Y|\pi] - [E[Y|\pi]]^2$$
$$= n(n-1)\pi^2 + n\pi - [n\pi]^2.$$

Hence the variance of the binomial is the sample size times the probability of success times the probability of failure.

$$\mathrm{Var}[Y|\pi] = n\pi(1-\pi). \tag{5.8}$$

5.4 Hypergeometric Distribution

The hypergeometric distribution models sampling from an urn without replacement. There is an urn containing N balls, R of which are red. A sequence of n balls is drawn randomly from the urn *without replacement*. Drawing a red ball is called a "success." The probability of success π does not stay constant over all the draws. At each draw the probability of "success" is the proportion of red balls remaining in the urn, which does depend on the outcomes of previous draws. Y is the number of "successes" in the n trials. Y can take on integer values $0, 1, \ldots, n$.

Probability Function of Hypergeometric

The probability function of the hypergeometric random variable Y given the parameters N, n, R is written as

$$f(y|N, R, n) = \frac{\binom{R}{y} \times \binom{N-R}{n-y}}{\binom{N}{n}}$$

for possible values $y = 0, 1, \ldots, n$.

Mean and variance of hypergeometric. The conditional mean of the hypergeometric distribution is given by

$$E[Y|N, R, n] = n \times \frac{R}{N}.$$

The conditional variance of the hypergeometric distribution is given by

$$\text{Var}[Y|N, R, n] = n \times \frac{R}{N} \times \left(1 - \frac{R}{N}\right) \times \left(\frac{N-n}{N-1}\right)$$

We note that $\frac{R}{N}$ is the proportion of red balls in the urn. The mean and variance of the hypergeometric are similar to that of the binomial, except that the variance is smaller due to the finite population correction factor $\frac{N-n}{N-1}$.

5.5   Poisson Distribution

The Poisson distribution is another distribution for counts.[1] Specifically, the Poisson is a distribution which counts the number of occurrences of rare events over a period of time or space. Unlike the binomial which counts the number of events (successes) in a known number of independent trials, the number of trials in the Poisson is so large that it is not known. Nevertheless, looking at the binomial gives us way to start our investigation of the Poisson. Let Y be a binomial random variable where n is very large, and π is very small. The binomial probability function is

$$P(Y = y|\pi) = \binom{n}{y} \pi^y (1 - \pi)^{n-y}$$

$$= \frac{n!}{(n-y)!y!} \pi^y (1 - \pi)^{n-y}$$

---

[1] First studied by Simeon Poisson (1781–1840).

for $y = 0, \ldots, n$. Since π is small, the only terms that have appreciable probability are those where y is much smaller than n. We will look at the probabilities for those small values of y. Let $\mu = n\pi$. The probability function is

$$P(Y = y|\mu) = \frac{n!}{(n-y)!y!} \left(\frac{\mu}{n}\right)^y \left(1 - \frac{\mu}{n}\right)^{n-y}.$$

Rearranging the terms, we get

$$P(Y = y|\mu) = \frac{n}{n} \times \frac{n-1}{n} \times \cdots \times \frac{n-y+1}{n} \times \frac{\mu^y}{y!} \left(1 - \frac{\mu}{n}\right)^n \left(1 - \frac{\mu}{n}\right)^{-y}.$$

But all the values $\frac{n}{n}, \frac{n-1}{n}, \ldots, \frac{n-y+1}{n}$ are approximately equal to 1 since y is much smaller than n. We let n approach infinity, and π approach 0 in such a way that $\mu = n\pi$ is constant. We know that

$$\lim_{n \to \infty} \left(1 - \frac{\mu}{n}\right)^n = e^{-\mu} \quad \text{and} \quad \lim_{n \to \infty} \left(1 - \frac{\mu}{n}\right)^{-y} = 1,$$

so the Poisson probability function is given by

$$f(y|\mu) = \frac{\mu^y e^{-\mu}}{y!} \tag{5.9}$$

for $y = 0, 1, \ldots$. Thus the *Poisson*(μ) distribution can be used to approximate a *binomial*(n, π) when n is large, π is very small, and $\mu = n\pi$.

Characteristics of the Poisson Distribution

Think of the period of time (or space) divided into n equal parts. The total number of occurrences is the sum of the number of occurrences in all n parts. We see from the Poisson approximation to the binomial that the Poisson distribution is a limiting case of the binomial distribution as $n \to \infty$ and $\pi \to 0$ at such a rate that $n\pi = \mu$ is constant.

- In the binomial, the probability of success remains constant over all the trials. It follows that the instantaneous rate of occurrences per unit time (or space) for the Poisson is constant.

- In the binomial, the trials are independent. Thus the Poisson occurrences in any two non-overlapping intervals will be independent of each other. It follows that the Poisson occurrences are randomly occurring through time at the constant instantaneous rate.

- In the binomial each trial contributes either one success or one failure. It follows that Poisson counts occur one at a time.

- The possible values are $y = 0, 1, \ldots$.

Mean and variance of Poisson.   The mean of the $Poisson(\mu)$ can be found by

$$E[y|\mu] = \sum_{y=0}^{\infty} y \frac{\mu^y e^{-\mu}}{y!}$$

$$= \sum_{y=1}^{\infty} \frac{\mu^y e^{-\mu}}{(y-1)!}$$

We let $y' = y - 1$ and factor out μ:

$$E[y|\mu] = \mu \sum_{y'=0}^{\infty} \frac{\mu^{y'} e^{-\mu}}{y'!} \, .$$

The sum equals one since it is the the sum is over all possible values of a Poisson distribution, so the mean of the $Poisson(\mu)$ is

$$E[y|\mu] = \mu \, .$$

Similarly, we can evaluate

$$E[y \times (y-1)|\mu] = \sum_{y=0}^{\infty} y \times (y-1] \times \frac{\mu^y e^{-\mu}}{y!}$$

$$= \sum_{y=2}^{\infty} \frac{\mu^y e^{-\mu}}{(y-2)!}$$

We let $y' = y - 2$, and factor out μ^2

$$E[y \times (y-1)|\mu] = \mu^2 \sum_{y'=0}^{\infty} \frac{\mu^{y'} e^{-\mu}}{y'!} \, .$$

The sum equals one since it is the the sum is over all possible values of a Poisson distribution, so $E[y \times (y-1)|\mu]$ for a $Poisson(\mu)$ is given by

$$E[y \times (y-1)|\mu] = \mu^2 \, .$$

The Poisson variance is given by

$$\begin{aligned}
Var[y|\mu] &= E[y^2|\mu) - [E[y|\mu]]^2 \\
&= E[y \times (y-1)|\mu] + E[y|\mu] - [E(y|\mu]]^2 \\
&= \mu^2 + \mu - \mu^2 \\
&= \mu \, .
\end{aligned}$$

Thus we see the mean and variance of a $Poisson(\mu)$ are both equal to μ.

Table 5.2   Universe of joint experiment

(x_1, y_1)	.	.	.	(x_1, y_j)	.	.	.	(x_1, y_J)
	.	.	.	.	.	.	.	.
	.	.	.	.	.	.	.	.
	.	.	.	.	.	.	.	.
(x_i, y_1)	.	.	.	(x_i, y_j)	.	.	.	(x_i, y_J)
	.	.	.	.	.	.	.	.
	.	.	.	.	.	.	.	.
	.	.	.	.	.	.	.	.
(x_I, y_1)	.	.	.	(x_I, y_j)	.	.	.	(x_I, y_J)

5.6   Joint Random Variables

When two (or more) numbers are determined from the outcome of a random experiment, we call it a joint experiment. The two numbers are called joint random variables and denoted X, Y. If both the random variables are discrete, they each have separated possible values x_i for $i = 1, \ldots, I$ and y_j for $j = 1, \ldots, J$. The *universe* for the experiment is the set of all possible outcomes of the experiment which are all possible ordered pairs of possible values. The *universe* of the joint experiment is shown in Table 5.2.

The joint probability function of two discrete joint random variables is defined at each point in the universe:

$$f(x_i, y_j) = P(X = x_i, Y = y_j)$$

for $i = 1, \ldots, I$, and $j = 1, \ldots, J$. This is the probability that $X = x_i$ and $Y = y_j$ simultaneously, in other words, the probability of the intersection of the events $X = x_i$ and $Y = y_j$. These joint probabilities can be put in a table.

We might want to consider the probability distribution of just one of the joint random variables, for instance, Y. The event $Y = y_j$ for some fixed value y_j is the union of all events $X = x_i, Y = y_j$, where $i = 1, \ldots, I$, and they are all disjoint. Thus

$$P(Y = y_j) = P(\cup_i (X = x_i, Y = y_j)) = \sum_i P(X = x_i, Y = y_j)$$

for $j = 1, \ldots, J$, since probability is additive over a disjoint union. This probability distribution of Y by itself is called the *marginal* distribution of Y. Putting this relationship in terms of the probability function, we get

$$f(y_j) = \sum_i f(x_i, y_j) \tag{5.10}$$

Table 5.3   Joint and marginal probability distributions

	y_1	.	.	.	y_j	.	.	.	y_J	
x_1	$f(x_1, y_1)$	.	.	.	$f(x_1, y_j)$	.	.	.	$f(x_1, y_J)$	$f(x_1)$
.		.	.	.		.	.	.		.
.		.	.	.		.	.	.		.
.		.	.	.		.	.	.		.
x_i	$f(x_i, y_1)$	.	.	.	$f(x_i, y_j)$	.	.	.	$f(x_i, y_J)$	$f(x_i)$
.		.	.	.		.	.	.		.
.		.	.	.		.	.	.		.
.		.	.	.		.	.	.		.
x_I	$f(x_I, y_1)$	.	.	.	$f(x_I, y_j)$	.	.	.	$f(x_I, y_J)$	$f(x_I)$
	$f(y_1)$	.	.	.	$f(y_j)$	.	.	.	$f(y_J)$	

for $j = 1, \ldots J$. So we see that the individual probabilities of Y is found by summing the joint probabilities down the columns. Similarly the individual probabilities of X can be found by summing the joint probabilities across the rows. We can write them on the margins of the table, hence the names *marginal* probability distribution of Y and X respectively. The joint probability distribution and the marginal probability distributions are shown in Table 5.3. The joint probabilities are in the main body of the table, and the marginal probabilities for X and Y are in the right column and bottom row, respectively.

The expected value of a function of the joint random variables is given by

$$E[h(X, Y)] = \sum_i \sum_j h(x_i, y_j) \times f(x_i, y_j).$$

Often we wish to find the expected value of a sum of random variables. In that case

$$\begin{aligned} E[X + Y] &= \sum_i \sum_j (x_i + y_j) \times f(x_i, y_j] \\ &= \sum_i \sum_j x_i \times f(x_i, y_j) + \sum_i \sum_j y_j \times f(x_i, y_j) \\ &= \sum_i x_i \sum_j f(x_i, y_j) + \sum_j y_j \sum_i f(x_i, y_j) \\ &= \sum_i x_i \times f(x_i) + \sum_j y_j \times f(y_j). \end{aligned}$$

We see the mean of the sum of two random variables is the sum of the means.

$$E[X + Y] = E[X] + E[Y] . \tag{5.11}$$

This equation always holds.

Independent Random Variables

Two (discrete) random variables X and Y are independent of each other if and only if every element in the joint distribution table equals the product of the corresponding marginal distributions. In other words,

$$f(x_i, y_j) = f(x_i) \times f(y_j)$$

for all possible x_i and y_j.

The variance of a sum of random variables is given by

$$\begin{aligned}
\text{Var}[X + Y] &= E(X + Y - E[X + Y])^2 \\
&= \sum_i \sum_j (x_i + y_j - (E[X] + E[Y])^2 \times f(x_i, y_j) \\
&= \sum_i \sum_j [(x_i - E[X]) + (y_j - E[Y])]^2 \times f(x_i, y_j) .
\end{aligned}$$

Multiplying this out and breaking it into three separate sums gives

$$\begin{aligned}
\text{Var}[X + Y] = &\sum_i \sum_j (x_i - E[X])^2 \times f(x_i, y_j) \\
&+ \sum_i \sum_j 2(x_i - E[X])(y_j - E[Y]) f(x_i, y_j) \\
&+ \sum_i \sum_j (y_j - E[Y])^2 \times f(x_i, y_j) .
\end{aligned}$$

The middle term is $2 \times$ the covariance of the random variables. For independent random variables the covariance is given by

$$\begin{aligned}
\text{Cov}[X, Y] &= \sum_i \sum_j (x_i - E[X]) \times (y_j - E[Y]) f(x_i, y_j) \\
&= \sum_i (x_i - E[X]) f(x_i) \times \sum_j (y_j - E[Y]) f(y_j) .
\end{aligned}$$

This is clearly equal to 0. Hence for independent random variables we have

$$\text{Var}[X + Y] = \sum_i (x_i - E[X]))^2 \times f(x_i) + \sum_j (y_j - E[Y])^2 \times f(y_j) .$$

We see the variance of the sum of two independent random variables is the sum of the variances.

$$\text{Var}[X + Y] = \text{Var}[X] + \text{Var}[Y].\tag{5.12}$$

This equation only holds for independent[2] random variables!

◼ EXAMPLE 5.2

Let X and Y be jointly distributed discrete random variables. Their joint probability distribution is given in the following table:

		Y				
		1	2	3	4	$f(x)$
X	1	.02	.04	.06	.08	
	2	.03	.01	.09	.17	
	3	.05	.15	.15	.15	
	$f(y)$					

We find the marginal distributions of X and Y by summing across the rows and summing down the columns, respectively. That gives the table

		Y				
		1	2	3	4	$f(x)$
X	1	.02	.04	.06	.08	.2
	2	.03	.01	.09	.17	.3
	3	.05	.15	.15	.15	.5
	$f(y)$	.1	.2	.3	.4	

We see that the joint probability $f(x_i, y_j)$ is not always equal to the product of the marginal probabilities $f(x_i) \times f(y_j)$. Therefore the two random variables X and Y are not independent.   ◼

Mean and variance of a difference between two independent random variables. When we combine the results of Equations 5.10 and 5.11 with the results of Equations 5.4 and 5.5, we find the that mean of a difference between random variables is

$$\text{E}[X - Y] = \text{E}[X] - \text{E}[Y].\tag{5.13}$$

If the two random variables are independent, we find that the variance of their difference is

$$\text{Var}[X - Y] = \text{Var}[X] + \text{Var}[Y].\tag{5.14}$$

---

[2]In general, the variance of a sum of two random variables is given by $\text{Var}[X + Y] = \text{Var}[X] + 2 \times \text{Cov}[X, Y] + \text{Var}[Y]$.

Variability always adds for independent random variables, regardless of whether we are taking the sum or taking the difference.

5.7   Conditional Probability for Joint Random Variables

If we are given $Y = y_j$, the reduced universe is the set of ordered pairs where the second element is y_j. This is shown in Table 5.4. It is the only part of the universe that remains, given $Y = y_j$. The only part of the event $X = x_i$ that remains is the part in the reduced universe. This is the intersection of the events $X = x_i$ and $Y = y_j$. Table 5.5 shows the original joint probability function in the reduced universe, along with the marginal probability. We see that this is not a probability distribution. The sum of the probabilities in the reduced universe sums to the marginal probability, not to one!

The conditional probability that random variable $X = x_i$, given $Y = y_j$ is the probability of the intersection of the events $X = x_i$ and $Y = y_j$ divided by the probability that $Y = y_j$ from Equation 4.1. Dividing the joint probability by the marginal probability scales it up so the probability of the reduced universe equals 1. The conditional probability is given by

$$f(x_i|y_j) = P(X = x_i|Y = y_j) = \frac{P(X = x_i, Y = y_j)}{P(Y = y_j)} . \qquad (5.15)$$

When we put this in terms of the joint and marginal probability functions, we get

$$f(x_i|y_j) = \frac{f(x_i, y_j)}{f(y_j)} . \qquad (5.16)$$

Table 5.4   Reduced universe given $Y = y_j$

				(x_1, y_j)				
.	.	.	.		.	.	.	.
.	.	.	.		.	.	.	.
.	.	.	.	.	.	.	.	.
.	.	.	.	.	.	.	.	.
.	.	.	.	(x_i, y_j)	.	.	.	.
.	.	.	.	.	.	.	.	.
.	.	.	.	.	.	.	.	.
.	.	.	.	.	.	.	.	.
.	.	.	.	(x_I, y_j)	.	.	.	.

Table 5.5   Joint probability function values in the reduced universe $Y = y_j$. The marginal probability is found by summing down the column.

.	.	.	.	$f(x_1, y_j)$	.	.	.	.
.	.	.	.	.	.	.	.	.
.	.	.	.	.	.	.	.	.
.	.	.	.	.	.	.	.	.
.	.	.	.	$f(x_i, y_j)$	.	.	.	.
.	.	.	.	.	.	.	.	.
.	.	.	.	.	.	.	.	.
.	.	.	.	$f(x_I, y_j)$	.	.	.	.
.	.	.	.	$f(y_j)$	.	.	.	.

The conditional probability distribution.   Letting x_i vary across all possible values of X gives us the conditional probability distribution of $X|Y = y_j$. The conditional probability distribution is defined on the reduced universe given $Y = y_j$. The conditional probability distribution is shown in Table 5.6. Each entry was found by dividing the i,j entry in the joint probability table by j^{th} element in the marginal probability. The marginal probability is $f(y_j) = \sum_i f(x_i, y_j)$ and is found by summing down the jth column of the joint probability table. So the conditional probability of x_i given y_j is the jth column in the joint probability table, divided by the sum of the joint probabilities in the jth column.

◼ **EXAMPLE 5.2   (continued)**

If we want to determine the conditional probability $P(X = 2|Y = 2)$, we plug in the joint and marginal probabilities into Equation 5.15. This gives

$$P(X = 2|Y = 2) = \frac{P(X = 2, Y = 2)}{P(Y = 2)}$$

$$= \frac{.01}{.2}$$

$$= .05 \,.$$

◼

Table 5.6   The conditional probability function defined on the reduced universe
$Y = y_j$

$$
\begin{array}{cccc}
\cdot & \cdot & \cdot & \cdot \\
& & & \\
\end{array}
\boxed{\begin{array}{c}
f(x_1|y_j) \\
\vdots \\
f(x_i|y_j) \\
\vdots \\
f(x_I|y_j)
\end{array}}
\begin{array}{cccc}
\cdot & \cdot & \cdot & \cdot
\end{array}
$$

Conditional probability as multiplication rule. Using similar arguments, we
could find that the conditional probability function of Y given $X = x_i$ is
given by

$$f(y_j|x_i) = \frac{f(x_i, y_j)}{f(x_i)}.$$

However, we will not use the relationship in this form, since we do not consider
the random variables interchangeably. In Bayesian statistics, the random
variable X is the unobservable parameter. The random variable Y is an
observable random variable that has a probability distribution depending on
the parameter. In the next chapter we will use the conditional probability
relationship as the multiplication rule

$$f(x_i, y_j) = f(x_i) \times f(y_j|x_i) \tag{5.17}$$

when we develop Bayes' theorem for discrete random variables.

Main Points

- A random variable Y is a number associated with the outcome of a random experiment.

- If the only possible values of the random variable are a finite set of separated values, $y_1, \ldots, y_K$ the random variable is said to be discrete.

- The probability distribution of the discrete random variable gives the probability associated with each possible value.

- The probability of any event associated with the random experiment can be calculated from the probability function of the random variable using the laws of probability.

- The expected value of a discrete random variable is

$$E[Y] \;=\; \sum_k y_k f(y_k) \,,$$

where the sum is over all possible values of the random variable. It is the mean of the distribution of the random variable.

- The variance of a discrete random variable is the expected value of the squared deviation of the random variable from its mean.

$$\mathrm{Var}[Y] \;=\; E(Y - E[Y])^2 \;=\; \sum_k (y_k - E[Y])^2 f(y_k) \,.$$

Another formula for the variance is

$$\mathrm{Var}[Y] \;=\; E[Y^2] - [E[Y]]^2 \,.$$

- The mean and variance of a linear function of a random variable $aY + b$ are

$$E[aY + b] \;=\; a\,E[Y] + b$$

and

$$\mathrm{Var}[aY + b] \;=\; a^2 \times \mathrm{Var}[Y] \,.$$

- The $binomial(n, \pi)$ distribution models the number of successes in n independent trials where each trial has the same success probability, π.

- The $binomial$ distribution is used for sampling from a finite population with replacement.

- The $hypergeometric$ distribution is used for sampling from a finite population without replacement.

- The $Poisson(\mu)$ distribution counts the number of occurrences of a rare event. Occurrences are occurring randomly through time (or space) at a constant rate and occur one at a time. It is also used to approximate the $binomial(n, \pi)$ where n is large and π is small and we let $\mu = n\pi$.

- The joint probability distribution of two discrete random variables X and Y is written as joint probability function

$$f(x_i, y_j) \;=\; P(X = x_i, Y = y_j) \,.$$

Note: $(X = x_i, Y = y_j)$ is another way of writing the intersection $(X = x_i \cap Y = y_j)$. This joint probability function can be put in a table.

- The marginal probability distribution of one of the random variables can be found by summing the joint probability distribution across rows (for X) or by summing down columns (for Y).

- The mean and variance of a sum of independent random variables are

$$E[X + Y] = E[X] + E[Y]$$

and

$$\mathrm{Var}[X + Y] = \mathrm{Var}[X] + \mathrm{Var}[Y].$$

- The mean and variance of a difference between independent random variables are

$$E[X - Y] = E[X] - E[Y]$$

and

$$\mathrm{Var}[X - Y] = \mathrm{Var}[X] + \mathrm{Var}[Y].$$

- Conditional probability function of X given $Y = y_j$ is found by

$$f(x_i|y_j) = \frac{f(x_i, y_j)}{f(y_j)}.$$

This is the joint probability divided by the marginal probability that $Y = y_j$.

- The joint probabilities on the reduced universe $Y = y_j$ are not a probability distribution. They sum to the marginal probability $f(y_j)$, not to one.

- Dividing the joint probabilities by the marginal probability scales up the probabilities, so the sum of probabilities in the reduced universe is one.

Exercises

5.1. A discrete random variable Y has discrete distribution given in the following table:

y_i	$f(y_i)$
0	.2
1	.3
2	.3
3	.1
4	.1

(a) Calculate $P(1 < Y \leq 3)$.

(b) Calculate $E[Y]$.

(c) Calculate Var[Y].

(d) Let $W = 2Y + 3$. Calculate E[W].

(e) Calculate Var[W].

5.2. A discrete random variable Y has discrete distribution given in the following table:

y_i	$f(y_i)$
0	.1
1	.2
2	.3
5	.4

(a) Calculate $P(0 < Y < 2)$.

(b) Calculate E[Y].

(c) Calculate Var[Y].

(d) Let $W = 3Y - 1$. Calculate E[W].

(e) Calculate Var[W].

5.3. Let Y be *binomial*$(n = 5, \pi = .6)$.

(a) Calculate the mean and variance by filling in the following table:

y_i	$f(y_i)$	$y_i \times f(y_i)$	$y_i^2 \times f(y_i)$
0			
1			
2			
3			
4			
5			
Sum			

      i. E[Y] =

      ii. Var[Y] =

(b) Calculate the mean and variance of Y using Equations 5.7 and 5.8, respectively. Do you get the same results as in part (a)?

5.4. Let Y be *binomial*$(n = 4, \pi = .3)$.

(a) Calculate the mean and variance by filling in the following table:

y_i	$f(y_i)$	$y_i \times f(y_i)$	$y_i^2 \times f(y_i)$
0			
1			
2			
3			
4			
Sum			

    i. $E[Y] =$

    ii. $Var[Y] =$

(b) Calculate the mean and variance of Y using Equations 5.7 and 5.8, respectively. Do you get the same as you got in part (a)?

5.5. Suppose there is an urn containing 20 green balls and 30 red balls. A single trial consists of drawing a ball randomly from the urn, recording its color, and then putting it back in the urn. The experiment consists of 4 independent trials.

(a) List each outcome (sequence of 4 trials) in the sample space together with its probability. What do you notice about the probabilities of outcomes that have the same number of green balls?

(b) Let Y be the number of green balls drawn. List the outcomes that make up each of the following events:
    $Y = 0$ $Y = 1$ $Y = 2$ $Y = 3$ $Y = 4$

(c) What can you say about $P(Y = y)$ in terms of "number of outcomes where $Y = y$, and the probability of any particular sequence of outcomes where $Y = y$.

(d) Explain how this relates to the binomial probability function.

5.6. Suppose there is an urn containing 20 green balls and 30 red balls. A single trial consists of drawing a ball randomly from the urn, recording its color. This time the ball is not returned to the urn. The experiment consists of 4 independent trials.

(a) List each outcome (sequence of 4 trials) in the sample space together with its probability. What do you notice about the probabilities of outcomes that have the same number of green balls.

(b) Let Y be the number of green balls drawn. List the outcomes that make up each of the following events:
    $Y = 0$ $Y = 1$ $Y = 2$ $Y = 3$ $Y = 4$

(c) What can you say about $P(Y = y)$ in terms of "number of outcomes where $Y = y$, and the probability of any particular sequence of outcomes where $Y = y$.

(d) Explain what this means in terms of the hypergeometric distribution. Hint: write this in terms of factorials, then rearrange the terms.

5.7. Let Y have the $Poisson(\mu = 2)$ distribution.

(a) Calculate $P(Y = 2)$.

(b) Calculate $P(Y \le 2)$.

(c) Calculate $P(1 \le Y < 4)$.

5.8. Let Y have the $Poisson(\mu = 3)$ distribution.

(a) Calculate $P(Y = 3)$.

(b) Calculate $P(Y \le 3)$.

(c) Calculate $P(1 \le Y < 5)$.

5.9. Let X and Y be jointly distributed discrete random variables. Their joint probability distribution is given in the following table:

X		Y				
	1	2	3	4	5	$f(x)$
1	.02	.04	.06	.08	.05	
2	.08	.02	.10	.02	.03	
3	.05	.05	.03	.02	.10	
4	.10	.04	.05	.03	.03	
$f(y)$						

(a) Calculate the marginal probability distribution of X.

(b) Calculate the marginal probability distribution of Y.

(c) Are X and Y independent random variables? Explain why or why not.

(d) Calculate the conditional probability $P(X = 3|Y = 1)$.

5.10. Let X and Y be jointly distributed discrete random variables. Their joint probability distribution is given in the following table:

X		Y				$f(x)$
	1	2	3	4	5	
1	.015	.030	.010	.020	.025	
2	.030	.060	.020	.040	.050	
3	.045	.090	.030	.060	.075	
4	.060	.120	.040	.080	.100	
$f(y)$						

(a) Calculate the marginal probability distribution of X.

(b) Calculate the marginal probability distribution of Y.

(c) Are X and Y independent random variables? Explain why or why not.

(d) Calculate the conditional probability $P(X = 2|Y = 3)$.

CHAPTER 6

BAYESIAN INFERENCE
FOR DISCRETE RANDOM VARIABLES

In this chapter we introduce Bayes' theorem for discrete random variables. Then we see how we can use it to revise our beliefs about the parameter, given the sample data that depends on the parameter. This is how we will perform statistical inference in a Bayesian manner.

We will consider the parameter to be random variable X, which has possible values $x_1, \ldots, x_I$. We never observe the parameter random variable. The random variable Y, which depends on the parameter, has possible values $y_1, \ldots, y_J$. We make inferences about the parameter random variable X given the observed value $Y = y_j$ using Bayes' theorem.

The *Bayesian universe* consists of the all possible ordered pairs (x_i, y_j) for $i = 1, \ldots, I$ and $j = 1, \ldots, J$. This is analogous to the universe we used for joint random variables in the last chapter. However, we will not consider the random variables X and Y the same way. The events $(X = x_1), \ldots, (X = x_I)$ partition the universe, but we never observe which one has occurred. The event $Y = y_j$ is observed.

We know that the Bayesian universe has two dimensions, the horizontal dimension which is observable, and the vertical dimension which is unobservable. In the horizontal direction it goes across the sample space which is the set of all possible values, $\{y_1, \ldots, y_J\}$, of the observed random variable Y. In the vertical direction it goes through the parameter space, which is the set of all possible parameter values, $\{x_1, \ldots, x_I\}$. The Bayesian universe for discrete random variables is shown in Table 6.1. This is analogous the Bayesian universe for events described in Chapter 4. The parameter value is unobserved. Probabilities are defined at all points in the Bayesian universe.

Table 6.1   The Bayesian universe

(x_1, y_1)	(x_1, y_2)	.	.	.	(x_1, y_j)	.	.	.	(x_1, y_J)
.	.	.	.	.	.	.	.	.	.
.	.	.	.	.	.	.	.	.	.
.	.	.	.	.	.	.	.	.	.
(x_i, y_1)	(x_i, y_2)	.	.	.	(x_i, y_j)	.	.	.	(x_i, y_J)
.	.	.	.	.	.	.	.	.	.
.	.	.	.	.	.	.	.	.	.
.	.	.	.	.	.	.	.	.	.
(x_I, y_1)	(x_1, y_2)	.	.	.	(x_I, y_j)	.	.	.	(x_I, y_J)

We will change our notation slightly. We will use $f()$ to denote a probability distribution (conditional or unconditional) that contains the observable random variable Y, and $g()$ to denote a probability distribution (conditional or unconditional) that only contains the (unobserved) parameter random variable X. This clarifies the distinction between Y, the random variable that we will observe, and X, the unobserved parameter random variable that we want to make our inference about. Each of the joint probabilities in the Bayesian universe is found using the multiplication rule

$$f(x_i, y_j) = g(x_i) \times f(y_j|x_i) \,.$$

The marginal distribution of Y is found by summing the columns. We show the joint and marginal probability function in Table 6.2. Note that this is similar to how we presented the joint and marginal distribution for two discrete random variables in the previous chapter (Table 5.3). However, now we have moved the marginal probability function of X over to the left-hand side and call it the *prior* probability function of the parameter X to indicate it is known to us at the beginning. We also note the changed notation.

When we observe $Y = y_j$, the reduced Bayesian universe is the set of ordered pairs in the jth column. This is shown in Table 6.3. The posterior probability function of X given $Y = y_j$ is given by

$$g(x_i|y_j) = \frac{g(x_i) \times f(y_j|x_i)}{\sum_{i=1}^{n_i} g(x_i) \times f(y_j|x_i)} \,.$$

Table 6.2   The joint and marginal distributions of X and Y

	prior	y_1				y_j				y_J
x_1	$g(x_1)$	$f(x_1, y_1)$	.	.	.	$f(x_1, y_j)$	.	.	.	$f(x_1, y_J)$
.	.	.		.	.	.	.	.		.
.	.	.	.	.		.	.	.	.	.
.	.	.	.		.	.		.	.	.
x_i	$g(x_i)$	$f(x_i, y_1)$	.	.	.	$f(x_i, y_j)$	.	.	.	$f(x_i, y_J)$
.	.	.	.	.		.	.	.	.	.
.	.	.		.	.	.	.		.	.
.	.	.	.		.	.		.	.	.
x_I	$g(x_I)$	$f(x_I, y_1)$	.	.	.	$f(x_I, y_j)$	.	.	.	$f(x_I, y_J)$
		$f(y_1)$	.	.	.	$f(y_j)$	.	.	.	$f(y_J)$

Table 6.3   The reduced Bayesian universe given $Y = y_j$

.	.	.	.	(x_1, y_j)	.	.	.	.
.	.	.	.		.	.	.	.
.	.	.	.		.	.	.	.
.	.	.	.		.	.	.	.
.	.	.	.	(x_i, y_j)	.	.	.	.
.	.	.	.		.	.	.	.
.	.	.	.		.	.	.	.
.	.	.	.		.	.	.	.
.	.	.	.	(x_I, y_j)	.	.	.	.

Let us look at the parts of the formula.

- The prior distribution of the discrete random variable X is given by the *prior* probability function $g(x_i)$, for $i = 1, \ldots, n$. This is what we believe the probability of each x_i to be before we look at the data. It must come from prior experience, not from the current data.

- Since we observed $Y = y_j$, the likelihood of the discrete parameter random variable is given by the *likelihood function* $f(y_j | x_i)$ for $i = 1, \ldots, n$. This is the conditional probability function of Y given $X = x_i$ evaluated at y_j, the value that actually occurred and where X is allowed to vary over its whole range for $x_i, \ldots, x_n$. We must know the form of the conditional observation distribution, as it shows how the distribution of the observation Y depends on the value of the random variable X, but we see that it only needs to be evaluated at the value that actually occurred, y_j. The likelihood function is the conditional observation distribution evaluated on the reduced universe.

- The posterior probability distribution of the discrete random variable is given by the *posterior* probability function $g(x_i|y_j)$ evaluated at x_i for $i = 1, \ldots, n$, given $Y = y_j$

The formula gives us a method for revising our belief probabilities about the possible values of X given that we observed $Y = y_j$.

■ EXAMPLE 6.1

There is an urn containing a total of 5 balls, some of which may be red and the rest of which are green. We do not know how many of the balls are red. Let the random variable X be the number of red balls in the urn. Possible values of X are $x_i = i$ for $i = 0, \ldots, 5$. Since we do not have any idea about the number of red balls, we will assume all possible values are equally likely. Our prior distribution of X is $g(0) = g(1) = g(2) = g(3) = g(4) = g(5) = 1/6$

We will draw a ball at random from the urn. The random variable Y is equal to 1 if draw is red and 0 otherwise. Conditional observation distribution of $Y|X$ is $P(Y = 1|X = x_i) = i/5$ and $P(Y = 0|X = x_i) = (5 - i)/5$. The joint probabilities are found by multiplying the prior probabilities times the conditional observation probabilities. The marginal probabilities of Y are found by summing the joint probabilities down the columns. These are shown in Table 6.4.

Suppose the selected ball is red, so the reduced universe is in the column labelled $y_j = 1$. The conditional observation probabilities in that column are highlighted. They form the likelihood function. Table 6.5 shows the steps for finding the posterior distribution of X given $Y = 1$.

Notice that the only column that was used to find the posterior probability distribution was the in the reduced universe, the column $Y = 1$. The joint probability came from multiplying the prior probabilities times the likelihood function. The posterior probability equals the prior probability times likelihood divided by the sum of prior probabilities times likelihoods:

$$f(x_i|y_j) = P(X = x_i|Y = y_j) = \frac{g(x_i) \times f(y_j|x_i)}{\sum_{i=1}^{n_i} g(x_i) \times f(y_j|x_i)} .$$

Thus a simpler way of finding the posterior probability is to use only the column in the reduced universe. Its probability is product of the prior times the likelihood. This is shown in Table 6.6.

■

Table 6.4   The joint and marginal probability distributions

x_i	prior probability	$y_j = 0$	$y_j = 1$
0	1/6	$\frac{1}{6} \times \frac{5}{5} = \frac{5}{30}$	$\frac{1}{6} \times \frac{0}{5} = 0$
1	1/6	$\frac{1}{6} \times \frac{4}{5} = \frac{4}{30}$	$\frac{1}{6} \times \frac{1}{5} = \frac{1}{30}$
2	1/6	$\frac{1}{6} \times \frac{3}{5} = \frac{3}{30}$	$\frac{1}{6} \times \frac{2}{5} = \frac{2}{30}$
3	1/6	$\frac{1}{6} \times \frac{2}{5} = \frac{2}{30}$	$\frac{1}{6} \times \frac{3}{5} = \frac{3}{30}$
4	1/6	$\frac{1}{6} \times \frac{1}{5} = \frac{1}{30}$	$\frac{1}{6} \times \frac{4}{5} = \frac{4}{30}$
5	1/6	$\frac{1}{6} \times \frac{0}{5} = \frac{0}{30}$	$\frac{1}{6} \times \frac{5}{5} = \frac{5}{30}$
$f(y_j)$		$\frac{15}{30}$	$\frac{15}{30} = \frac{1}{2}$

Table 6.5   Finding the posterior probabilities of $X|Y = 1$

x_i	prior probability	$y_j = 0$	$y_j = 1$	posterior probability
0	1/6	$\frac{1}{6} \times \frac{5}{5} = \frac{5}{30}$	$\frac{1}{6} \times \frac{0}{5} = 0$	0
1	1/6	$\frac{1}{6} \times \frac{4}{5} = \frac{4}{30}$	$\frac{1}{6} \times \frac{1}{5} = \frac{1}{30}$	$\frac{1}{30} / \frac{1}{2} = \frac{1}{15}$
2	1/6	$\frac{1}{6} \times \frac{3}{5} = \frac{3}{30}$	$\frac{1}{6} \times \frac{2}{5} = \frac{2}{30}$	$\frac{2}{30} / \frac{1}{2} = \frac{2}{15}$
3	1/6	$\frac{1}{6} \times \frac{2}{5} = \frac{2}{30}$	$\frac{1}{6} \times \frac{3}{5} = \frac{3}{30}$	$\frac{3}{30} / \frac{1}{2} = \frac{3}{15}$
4	1/6	$\frac{1}{6} \times \frac{1}{5} = \frac{1}{30}$	$\frac{1}{6} \times \frac{4}{5} = \frac{4}{30}$	$\frac{4}{30} / \frac{1}{2} = \frac{4}{15}$
5	1/6	$\frac{1}{6} \times \frac{0}{5} = \frac{0}{30}$	$\frac{1}{6} \times \frac{5}{5} = \frac{5}{30}$	$\frac{5}{30} / \frac{1}{2} = \frac{5}{15}$
$f(y_j)$		$\frac{15}{30}$	$\frac{15}{30} = \frac{1}{2}$	

Steps for Bayes' Theorem Using a Table

- Set up a table with columns for *parameter value, prior, likelihood, prior × likelihood* and *posterior.*

- Put in the *parameter values*, the *prior*, and the *likelihood* in their respective columns.

- Multiply each element in the *prior* column by the corresponding element in the *likelihood* column and put the results in the *prior × likelihood* column.

- Sum the *prior × likelihood* column.

- Divide each element of *prior × likelihood* column by the sum.

Table 6.6 Simplified table for finding the posterior probabilities of $X|Y = 1$

x_i	prior	likelihood	prior × likelihood	posterior
0	1/6	$\frac{0}{5}$	$\frac{1}{6} \times \frac{0}{5} = 0$	0
1	1/6	$\frac{1}{5}$	$\frac{1}{6} \times \frac{1}{5} = \frac{1}{30}$	$\frac{1}{30} / \frac{1}{2} = \frac{1}{15}$
2	1/6	$\frac{2}{5}$	$\frac{1}{6} \times \frac{2}{5} = \frac{2}{30}$	$\frac{2}{30} / \frac{1}{2} = \frac{2}{15}$
3	1/6	$\frac{3}{5}$	$\frac{1}{6} \times \frac{3}{5} = \frac{3}{30}$	$\frac{3}{30} / \frac{1}{2} = \frac{3}{15}$
4	1/6	$\frac{4}{5}$	$\frac{1}{6} \times \frac{4}{5} = \frac{4}{30}$	$\frac{4}{30} / \frac{1}{2} = \frac{4}{15}$
5	1/6	$\frac{5}{5}$	$\frac{1}{6} \times \frac{5}{5} = \frac{5}{30}$	$\frac{5}{30} / \frac{1}{2} = \frac{5}{15}$
$f(y_j)$			$\frac{15}{30} = \frac{1}{2}$	

▪ Put these posterior probabilities in the *posterior* column!

6.1 Two Equivalent Ways of Using Bayes' Theorem

We may have more than one data set concerning a parameter. They might not even become available at the same time. Should we wait for the second data set, combine it with the first, and then use Bayes' theorem on the combined data set? This would mean that we have to go back to scratch every time more data became available, which would result in a lot of work. Another approach requiring less work would be to use the posterior probabilities given the first data set, as the prior probabilities for analyzing the second data set. We will find that these two approaches lead to the same posterior probabilities. This is a significant advantage to Bayesian methods. In frequentist statistics, we would have to use the first approach, re-analyzing the combined data set when the second one arrives.

Analyzing the observations sequentially one at a time. Suppose that we randomly draw a second ball out of the urn without replacing the first. Suppose the second draw resulted in a green ball, so $Y = 0$. We want to find the posterior probabilities of X given the results of the two observations, red first, green second. We will analyze the observations in sequence using Bayes' theorem each time. We will use the same prior probabilities as before for the first draw. However, we will use the posterior probabilities from the first draw as the prior probabilities for the second draw. The results are shown in Table 6.7.

Analyzing the observations all together in a single step. Alternatively, we could consider both draws together, then revise the probabilities using Bayes' theorem only once. Initially, we are in the same state of knowledge as before. So

Table 6.7   The posterior probability distribution after second observation

x_i	prior	likelihood	prior × likelihood	posterior		
0	0	??	0	$0/\frac{1}{3}$	=	0
1	1/15	$\frac{4}{4}$	$\frac{1}{15}$	$\frac{1}{15}/\frac{1}{3}$	=	$\frac{1}{5}$
2	2/15	$\frac{3}{4}$	$\frac{1}{10}$	$\frac{1}{10}/\frac{1}{3}$	=	$\frac{6}{20}$
3	3/15	$\frac{2}{4}$	$\frac{1}{10}$	$\frac{1}{10}/\frac{1}{3}$	=	$\frac{6}{20}$
4	4/15	$\frac{1}{4}$	$\frac{1}{15}$	$\frac{1}{15}/\frac{1}{3}$	=	$\frac{1}{5}$
5	5/15	$\frac{0}{4}$	0	$0/\frac{1}{3}$	=	0
			$\frac{1}{3}$			1.00

we take the same prior probabilities that we originally used for the first draw when we were analyzing the observations in sequence. All possible values of X are equally likely. The prior probability function is $g(x) = \frac{1}{6}$ for $x = 0, \ldots, 5$.

Let Y_1 and Y_2 be the outcome of the first and second draw, respectively. The probabilities of the second draw depend on the balls left after the first draw. By the multiplication rule, the observation probability conditional on X is

$$f(y_1, y_2|x) = f(y_1|x) \times f(y_2|y_1, x).$$

The joint distribution of X and Y_1, Y_2 is given in Table 6.8. The first ball was red, second was green, so the reduced universe probabilities are in column $y_{j_1}, y_{j_2} = 1, 0$. The likelihood function given by the conditional observation probabilities in that column are highlighted.

Table 6.8   The joint distribution of X, Y_1, Y_2 and marginal distribution of Y_1, Y_2

x_i	prior	y_{j_1}, y_{j_2} 0,0	y_{j_1}, y_{j_2} 0,1	y_{j_1}, y_{j_2} 1,0	y_{j_1}, y_{j_2} 1,1
0	1/6	$\frac{1}{6} \times \frac{5}{5} \times \frac{4}{4}$	$\frac{1}{6} \times \frac{5}{5} \times \frac{0}{4}$	$\frac{1}{6} \times \frac{0}{5} \times \frac{4}{4}$	$\frac{1}{6} \times \frac{0}{5} \times \frac{0}{4}$
1	1/6	$\frac{1}{6} \times \frac{4}{5} \times \frac{3}{4}$	$\frac{1}{6} \times \frac{4}{5} \times \frac{1}{4}$	$\frac{1}{6} \times \frac{1}{5} \times \frac{4}{4}$	$\frac{1}{6} \times \frac{1}{5} \times \frac{0}{4}$
2	1/6	$\frac{1}{6} \times \frac{3}{5} \times \frac{2}{4}$	$\frac{1}{6} \times \frac{3}{5} \times \frac{2}{4}$	$\frac{1}{6} \times \frac{2}{5} \times \frac{3}{4}$	$\frac{1}{6} \times \frac{2}{5} \times \frac{1}{4}$
3	1/6	$\frac{1}{6} \times \frac{2}{5} \times \frac{1}{4}$	$\frac{1}{6} \times \frac{2}{5} \times \frac{3}{4}$	$\frac{1}{6} \times \frac{3}{5} \times \frac{2}{4}$	$\frac{1}{6} \times \frac{3}{5} \times \frac{2}{4}$
4	1/6	$\frac{1}{6} \times \frac{1}{5} \times \frac{0}{4}$	$\frac{1}{6} \times \frac{1}{5} \times \frac{4}{4}$	$\frac{1}{6} \times \frac{4}{5} \times \frac{1}{4}$	$\frac{1}{6} \times \frac{4}{5} \times \frac{3}{4}$
5	1/6	$\frac{1}{6} \times \frac{0}{5} \times \frac{0}{4}$	$\frac{1}{6} \times \frac{0}{5} \times \frac{4}{4}$	$\frac{1}{6} \times \frac{5}{5} \times \frac{0}{4}$	$\frac{1}{6} \times \frac{5}{5} \times \frac{4}{4}$
	$f(y_1, y_2)$	40/120	20/120	20/120	40/120

The first ball was red, second was green, so the reduced universe probabilities are in column $y_{j_1}, y_{j_2} = 1, 0$. The posterior probability of X given $Y_1 = 1$ and $Y_2 = 0$ is found by rescaling the probabilities in the reduced universe so they sum to 1. This is shown in Table 6.9. We see this is the same as the posterior probabilities we found analyzing the observations sequentially, using the posterior after the first as the prior for the second. This shows that it

Table 6.9   The posterior probability distribution given $Y_1 = 1$ and $Y_2 = 0$

x_i	prior	y_{j1}, y_{j2} 0,0	y_{j1}, y_{j2} 0,1	y_{j1}, y_{j2} 1,0	y_{j1}, y_{j2} 1,1	posterior	
0	1/6	$\frac{20}{120}$	0	0	0	0	$=0$
1	1/6	$\frac{12}{120}$	$\frac{4}{120}$	$\frac{4}{120}$	0	$\frac{4}{120}/\frac{20}{120}$	$=\frac{1}{5}$
2	1/6	$\frac{6}{120}$	$\frac{6}{120}$	$\frac{6}{120}$	$\frac{2}{120}$	$\frac{6}{120}/\frac{20}{120}$	$=\frac{3}{10}$
3	1/6	$\frac{2}{120}$	$\frac{6}{120}$	$\frac{6}{120}$	$\frac{6}{120}$	$\frac{6}{120}/\frac{20}{120}$	$=\frac{3}{10}$
4	1/6	0	$\frac{4}{120}$	$\frac{4}{120}$	$\frac{12}{120}$	$\frac{4}{120}/\frac{20}{120}$	$=\frac{1}{5}$
5	1/6	0	0	0	$\frac{20}{120}$	0	$= 0$
	$f(y_1, y_2)$			$20/120$		1.00	

makes no difference whether you analyze the observations one at a time in sequence using the posterior after the previous step as the prior for the next step, or whether you analyze all observations together in a single step starting with your initial prior!

Since we only use the column corresponding to the reduced universe, it is simpler to find the posterior by multiplying prior times likelihood and rescaling to make it a probability distribution. This is shown in Table 6.10.

Table 6.10   The posterior probability distribution after both observations

x_i	prior	likelihood	prior × likelihood	posterior		
0	1/6	$\frac{0}{20}$	$\frac{0}{120}$	$\frac{0}{120}/\frac{1}{6}$	=	0
1	1/6	$\frac{4}{20}$	$\frac{4}{120}$	$\frac{4}{120}/\frac{1}{6}$	=	$\frac{1}{5}$
2	1/6	$\frac{6}{20}$	$\frac{6}{120}$	$\frac{6}{120}/\frac{1}{6}$	=	$\frac{3}{10}$
3	1/6	$\frac{6}{20}$	$\frac{6}{120}$	$\frac{6}{120}/\frac{1}{6}$	=	$\frac{3}{10}$
4	1/6	$\frac{4}{20}$	$\frac{4}{120}$	$\frac{4}{120}/\frac{1}{6}$	=	$\frac{1}{5}$
5	1/6	$\frac{0}{20}$	$\frac{0}{120}$	$\frac{0}{120}/\frac{1}{6}$	=	0
			$\frac{1}{6}$	1.00		

6.2   Bayes' Theorem for Binomial with Discrete Prior

We will look at using Bayes' theorem when the observation comes from the binomial distribution, and there are only a few possible values for the parameter. $Y|\pi$ has the *binomial*(n, π) distribution. (There are n independent trials, each of which can result in "success" or "failure" and the probability of success π remains the same for all trials. Y is the total number of "successes" over the n trials.) There are I discrete possible values of $\pi_1, \ldots, \pi_I$.

Set up a table for the observation distributions. Row i correspond to the *binomial*(n, π_i) probability distribution. Column j corresponds to $Y = j$ (There are $n + 1$ columns corresponding to $0, \ldots, n$.) These binomial probabilities can be found in Table B.1 in Appendix B. The conditional observation probabilities in the reduced universe (column that corresponds to the actual observed value) is called the *likelihood*.

- We decide on our prior probability distribution of the parameter. They give our prior belief about each possible value of the parameter π. If we have no idea beforehand, we can choose the prior distribution that has all values equally likely.

- The joint probability distribution of the parameter π and the observation Y is found by multiplying the conditional probability of $Y|\pi$ by the prior probability of π.

- The marginal distribution of Y is found by summing the joint distribution down the columns.

Now take the observed value of Y. It is the only column that is now relevant. It contains the probabilities of the **reduced universe.** Note that it is the *prior* times the *likelihood*. The posterior probability of each possible value of π is found by dividing that row's element in the relevant column by the marginal probability of Y in that column.

◪ EXAMPLE 6.2

Let $Y|\pi$ be *binomial*$(n = 4, \pi)$. Suppose we consider that there are only three possible values for π, .4,.5, and .6. We will assume they are equally likely. The prior distribution of π and joint distribution of π and Y are given in Table 6.11. The joint probability distribution $f(\pi_i, y_j)$ is found by multiplying the conditional observation distribution $f(y_j|\pi_i)$ times the prior distribution $g(\pi_i)$. In this case, the conditional observation probabilities come from the *binomial*$(n = 4, \pi)$ distribution. These binomial probabilities come from Table B.1 in Appendix B. Suppose $Y = 3$ was observed. The reduced universe is the column for $Y = 3$. The conditional observation probabilities in that column is called the likelihood and is highlighted.

The marginal distribution of Y is found by summing the joint distribution of π and Y down the columns. The prior distribution of π, joint probability distribution of (π, Y), and marginal probability distribution of Y are shown in Table 6.12. Given that $Y = 3$ was observed, only the column labeled 3 is relevant. The prior distribution of π, joint probability distribution of (π, Y), marginal probability distribution of Y, and posterior probability distribution of $\pi|Y = 3$ are shown in Table 6.13.

Table 6.11   The joint probability distribution found by multiplying marginal distribution of π (the *prior*) by the conditional distribution of Y given π (which is binomial). $Y = 3$ was observed, so the binomial probabilities of $Y = 3$ (the *likelihood*) are highlighted.

π	*prior*	0	1	2	3	4
.4	$\frac{1}{3}$	$\frac{1}{3} \times .1296$	$\frac{1}{3} \times .3456$	$\frac{1}{3} \times .3456$	$\frac{1}{3} \times \mathbf{.1536}$	$\frac{1}{3} \times .0256$
.5	$\frac{1}{3}$	$\frac{1}{3} \times .0625$	$\frac{1}{3} \times .2500$	$\frac{1}{3} \times .3750$	$\frac{1}{3} \times \mathbf{.2500}$	$\frac{1}{3} \times .0625$
.6	$\frac{1}{3}$	$\frac{1}{3} \times .0256$	$\frac{1}{3} \times .1536$	$\frac{1}{3} \times .3456$	$\frac{1}{3} \times \mathbf{.3456}$	$\frac{1}{3} \times .1296$

Table 6.12   The joint and marginal probability distributions. $Y = 3$ was observed, so those probabilities are highlighted.

π	*prior*	0	1	2	3	4
.4	$\frac{1}{3}$	.0432	.1152	.1152	**.0512**	.0085
.5	$\frac{1}{3}$	.0208	.0833	.1250	**.0833**	.0208
.6	$\frac{1}{3}$	.0085	.0512	.1152	**.1152**	.0432
marginal		.0725	.2497	.3554	**.2497**	.0725

Table 6.13   The joint, marginal, and posterior probability distribution of π given $Y = 3$. Note the posterior is found by dividing the joint probabilities in the relevant column by their sum.

π	*prior*	0	1	2	3	4	*posterior*		
.4	$\frac{1}{3}$	.0432	.1152	.1152	.0512	.0085	$\frac{.0512}{.2497}$	=	.205
.5	$\frac{1}{3}$	.0208	.0833	.1250	.0833	.0208	$\frac{.0833}{.2497}$	=	.334
.6	$\frac{1}{3}$	.0085	.0512	.1152	.1152	.0432	$\frac{.1152}{.2497}$	=	.461
marginal		.0725	.2497	.3554	.2497	.0725			1.000

Note that the posterior is proportional to prior times likelihood. We did not have to set up the whole joint probability table. It is easier to only look at the reduced universe column. The posterior is equal to prior times likelihood divided by the marginal probability of the observed value. The results are shown in Table 6.14.   ∎

Setting up the Table for Bayes' Theorem on Binomial with Discrete Prior

- Set up a table with columns for *parameter value, prior, likelihood, prior × likelihood,* and *posterior*.

Table 6.14   The simplified table for finding posterior distribution given $Y = 3$

π	prior	likelihood	prior $\times$ likelihood	posterior	
.4	$\frac{1}{3}$	.1536	.0512	$\frac{.0512}{.2497}$ =	.205
.5	$\frac{1}{3}$	.2500	.0833	$\frac{.0833}{.2497}$ =	.334
.6	$\frac{1}{3}$	.3456	.1152	$\frac{.1152}{.2497}$ =	.461
marginal $P(Y = 3)$			.2497		1.000

- Put in the *parameter values*, the *prior*, and the *likelihood* in their respective columns. The *likelihood* values are *binomial(n,π_i)* evaluated at the observed value of y. They can be found in Table B.1, or evaluated from the formula.

- Multiply each element in the *prior* column by the corresponding element in the *likelihood* column and put in the *prior* $\times$ *likelihood* column.

- Sum these *prior* $\times$ *likelihood*.

- Divide each element of *prior* $\times$ *likelihood* column by the sum of *prior* $\times$ *likelihood* column. (This rescales them to sum to 1.)

- Put these in the *posterior* column!

Table 6.15   The simplified table for finding posterior distribution given $Y = 3$. Note we are using the proportional *likelihood* where we have absorbed that part of the binomial distribution that does not depend on π into the constant.

π	prior (proportional)	likelihood (proportional)	prior $\times$ likelihood	posterior	
.4	1	$.4^3 \times .6^1 = .0384$	.0384	$\frac{.0384}{.1873}$ =	.205
.5	1	$.5^3 \times .5^1 = .0625$	.0625	$\frac{.0625}{.1873}$ =	.334
.6	1	$.6^3 \times .4^1 = .0864$	.0864	$\frac{.0864}{.1873}$ =	.461
marginal sum			.1873		1.000

6.3   Important Consequences of Bayes' Theorem

Multiplying all the prior probabilities by a constant does not change the result of Bayes' theorem.   Each of the *prior* $\times$ *likelihood* entries in the table would be multiplied by the constant. The marginal entry found by summing down the column would also be multiplied by the same constant. Thus the posterior probabilities would be the same as before, since the constant would cancel out. The *relative* weights we are giving to each parameter value, not the actual weights, are what counts. If there is a formula for the prior, any part of it that does not contain the parameter can be absorbed into the constant. This may make calculations simpler for us!

Multiplying the likelihood by a constant does not change the result of Bayes' theorem. The *prior* × *likelihood* values would also be multiplied by the same constant, which would cancel out in the posterior probabilities. The likelihood can be considered the weights given to the possible values by the data. Again, it is the *relative* weights that are important, not the actual weights. If there is a formula for the likelihood, any part that does not contain the parameter can be absorbed into the constant, simplifying our calculations!

■ **EXAMPLE 6.2**   **(continued)**

We used a prior that gave each value equal prior probability. In this example there are three possible values, so each has a prior probability equal to $\frac{1}{3}$. Let us multiply each of the 3 prior probabilities by the constant 3 to give prior weights equal to 1. This will simplify our calculations. The observations are *binomial*$(n = 4, \pi)$, and $y = 3$ was observed. The formula for the binomial likelihood is

$$f(y|\pi) = \binom{4}{3}\pi^3(1 - \pi)^1 .$$

The binomial coefficient $\binom{4}{3}$ does not contain the parameter, so it is a constant over the likelihood column. To simplify our calculations, we will absorb it into the constant and use only the part of the likelihood that contains the parameter. In Table 6.15 we see that this gives us the same result we obtained before.   ■

6.4   Bayes' Theorem for Poisson with Discrete Prior

We will see how to apply Bayes' theorem when the observation comes from a *Poisson*(μ) distribution and we have a discrete prior distribution over a few discrete possible values for μ. $Y|mu$ is the number of counts of a process that is occurring randomly through time at a constant rate. The possible values of the parameter are $\mu_1, \ldots, \mu_I$. We decide on the prior probability distribution, $g(\mu_i)$ for $i = 1, \ldots, I$. These give our belief weight for each possible value before we have looked at the data. In Section 6.2 we learned that we do not have to use the full range of possible observations. Instead, we set up a table only using the reduced universe column, i.e., the value that was observed.

Setting up the Table for Bayes' Theorem on Poisson with Discrete Prior

- Set up table with columns for *parameter value, prior, likelihood, prior* × *likelihood,* and *posterior.*

- Put the *parameter value, prior,* and *likelihood* in their respective columns. The likelihood values are $Poisson(\mu)$ probabilities evaluated at the observed value of y. They can be found in Table B.5 in Appendix B, or evaluated from the Poisson formula.

- Multiply each element in the prior column by the corresponding element in the likelihood column, and enter them in the *prior* × *likelihood* column.

- Divide each *prior* × *likelihood* by the sum of the *prior* × *likelihood* column and put them in the *posterior* column.

■ **EXAMPLE 6.3**

Let $Y|\mu$ be Poisson(μ). Suppose that we believe there are only four possible values for μ, 1,1.5,2, and 2.5. Suppose we consider that the two middle values, 1.5 and 2, are twice as likely as the two end values 1 and 2.5. Suppose $y = 2$ was observed. Plug the value $y = 2$ into formula

$$f(y|\mu) = \frac{\mu^y e^{-\mu}}{y!}$$

to give the likelihood. Alternatively, we could find the values for the likelihood from Table B.5 in Appendix B. The results are shown in the Table 6.16. Note: We could use the proportional prior and the proportional likelihood and we would get the same results for the posterior. ■

Table 6.16 The simplified table for finding posterior distribution given $Y = 2$

μ	prior	likelihood	prior × likelihood	posterior	
1.0	$\frac{1}{6}$	$\frac{1.0^2 e^{-1.0}}{2!} = .1839$	.0307	$\frac{.0307}{.2473} =$	.124
1.5	$\frac{1}{3}$	$\frac{1.5^2 e^{-1.5}}{2!} = .2510$	.0837	$\frac{.0837}{.2473} =$	.338
2.0	$\frac{1}{3}$	$\frac{2.0^2 e^{-2.0}}{2!} = .2707$	.0902	$\frac{.0902}{.2473} =$	.365
2.5	$\frac{1}{6}$	$\frac{2.5^2 e^{-2.5}}{2!} = .2565$	.0428	$\frac{.0428}{.2473} =$	.173
marginal $P(Y = 2)$			.2473		1.000

Main Points

- The Bayesian universe has two dimensions. The vertical dimension is the parameter space and is unobservable. The horizontal dimension is the sample space and we observe which value occurs.

- The reduced universe is the column for the observed value.

- For discrete prior and discrete observation, the posterior probabilities are found by multiplying the *prior* × *likelihood* and then dividing by their sum.

- When our data arrives in batches we can use the posterior from the first batch as the prior for the second batch. This is equivalent to combining both batches and using Bayes' theorem only once, using our initial prior.

- Multiplying the *prior* by a constant does not change the result. Only relative weights are important.

- Multiplying the *likelihood* by a constant does not change the result.

- This means we can absorb any part of formula that does not contain the parameter into the constant. This greatly simplifies calculations.

Exercises

6.1. There is an urn containing 9 balls, which can be either green or red. The number of red balls in the urn is not known. One ball is drawn at random from the urn, and its color is observed.

(a) What is the *Bayesian universe* of the experiment.

(b) Let X be the number of red balls in the urn. Assume that all possible values of X from 0 to 9 are equally likely. Let $Y_1 = 1$ if the first ball drawn is red, and $Y_1 = 0$ otherwise. Fill in the joint probability table for X and Y_1 given below:

X	prior	$Y_1 = 0$	$Y_1 = 1$

(c) Find the marginal distribution of Y_1 and put it in the table.

(d) Suppose a red ball was drawn. What is the reduced Bayesian universe?

(e) Calculate the posterior probability distribution of X.

(f) Find the posterior distribution of X by filling in the simplified table:

X	prior	likelihood	prior × likelihood	posterior
marginal $P(Y_1 = 1)$				

6.2. Suppose that a second ball is drawn from the urn, without replacing the first. Let $Y_2 = 1$ if the second ball is red, and let it be 0 otherwise. Use the posterior distribution of X from the previous question as the prior distribution for X. Suppose the second ball is green. Find the posterior distribution of X by filling in the simplified table:

X	prior	likelihood	prior × likelihood	posterior
marginal $P(Y_2 = 0)$				

6.3. Suppose we look at the two draws from the urn (without replacement) as a single experiment. The results were first draw red, second draw green. Find the posterior distribution of X by filling in the simplified table.

X	prior	likelihood	prior × likelihood	posterior
marginal $P(Y_1 = 1, Y_2 = 0)$				

6.4. Let Y_1 be the number of successes in $n = 10$ independent trials where each trial results in a success or failure, and π, the probability of success, remains constant over all trials. Suppose the 4 possible values of π are .20, .40, .60, and .80. We do not wish to favor any value over the others so we make them equally likely. We observe $Y_1 = 7$. Find the posterior distribution by filling in the simplified table.

π	prior	likelihood	prior × likelihood	posterior
marginal $P(Y_1 = 7)$				

6.5. Suppose another 5 independent trials of the experiment are performed and $Y_2 = 2$ successes are observed. Use the posterior distribution for π from Exercise 6.4 as the prior distribution for π. Find the new posterior distribution by filling in the simplified table.

π	prior	likelihood	prior $\times$ likelihood	posterior
marginal $P(Y_2 = 2)$				

6.6. Suppose we combine all the $n = 15$ trials all together and think of them as a single experiment where we observed a total of 9 successes. Start with the initial equally weighted prior from Exercise 6.4 and find the posterior after the single combined experiment. What do the results of Exercises 6.4 – 6.6 show?

π	prior	likelihood	prior $\times$ likelihood	posterior
marginal $P(Y = 9)$				

6.7. Let Y be the number of counts of a Poisson random variable with mean μ. Suppose the 5 possible values of μ are 1, 2, 3, 4, and 5. We do not have any reason to give any possible value more weight than any other value, so we give them equal prior weight. $Y = 2$ was observed. Find the posterior distribution by filling in the simplified table.

μ	prior	likelihood	prior $\times$ likelihood	posterior
marginal $P(Y = 2)$				

Computer Exercises

6.1. The Minitab macro *BinoDP* or the equivalent R function is used to find the posterior distribution of the binomial probability π when the observation distribution of $Y|\pi$ is $binomial(n, \pi)$ and we have a discrete prior for π. Details for invoking *BinoDP* are found in Appendix C, and details for the equivalent R function are found in Appendix D.

Suppose we have 8 independent trials and each has one of two possible either success or failure. The probability of success remains constant for each trial. In that case, $Y|\pi$ is $binomial(n = 8, \pi)$. Suppose we only allow that there are 6 possible values of π, 0, .2, .4, .6, .8, and 1.0. In that case we say that we have a *discrete* distribution for π. Initially we have no reason to favor one possible value over another. In that case we would give all the possible values of π probability equal to $\frac{1}{6}$.

π	$g(\pi)$
0	.166666
.2	.166666
.4	.166666
.6	.166666
.8	.166666
1.0	.166666

Suppose we observe 3 "successes" in the 8 trials.

[**Minitab:**] Use the Minitab macro *BinoDP* to find the posterior distribution $g(\pi|y)$.

[**R:**] Use the R function `binodp` to find the posterior distribution $g(\pi|y)$.

(a) Identify the matrix of conditional probabilities from the output. Relate these conditional probabilities to the binomial probabilities in Table B.1.

(b) What column in the matrix contains the likelihoods?

(c) Identify the matrix of joint probabilities from the output. How are these joint probabilities found?

(d) Identify the marginal probabilities of Y from the output. How are these found?

(e) How are the posterior probabilities found?

6.2. Suppose we take an additional 7 trials and achieve 2 successes.

(a) Let the posterior after the 8 trials and 3 successes in the previous problem be the prior. Use *BinoDP* in Minitab, or `binodp` in R, to find the new posterior distribution for π.

(b) In total, we have taken 15 trials and achieved 5 successes. Go back to the original prior and use *BinoDP* in Minitab, or `binodp` in R, to find the posterior after the 15 trials and 5 successes.

(c) What does this show?

6.3. [**Minitab:**]   The Minitab macro *PoisDP* is used to find the posterior distribution when the observation distribution of $Y|\mu$ is $Poisson(\mu)$ and we have a discrete prior distribution for μ. Details for invoking *PoisDP* are in Appendix C.

[**R:**]   The R function `poisdp` is used to find the posterior distribution when the observation distribution of $Y|\mu$ is $Poisson(\mu)$ and we have a discrete prior distribution for μ. The details for using `poisdp` are in Appendix D.

Suppose there are six possible values $\mu = 1, \ldots, 6$ and the prior probabilities are given by

μ	$g(\mu)$
1	.10
2	.15
3	.25
4	.25
5	.15
6	.10

Suppose the first observation is $Y_1 = 2$. Use *PoisDP* in Minitab, or the R function `poisdp`, to find the posterior distribution $g(\mu|y)$.

(a) Identify the matrix of conditional probabilities from the output. Relate these conditional probabilities to the Poisson probabilities in Table B.5.

(b) What column in the matrix contains the likelihoods?

(c) Identify the matrix of joint probabilities from the output. How are these joint probabilities found?

(d) Identify the marginal probabilities of Y from the output. How are these found?

(e) How are the posterior probabilities found?

6.4. Suppose we take a second observation. We let the posterior after the first observation $Y_1 = 2$ which we found in the previous exercise be the prior for the second observation.

(a) The second observation $Y_2 = 1$. Use *PoisDP* in Minitab, or `poisdp` in R, to find the new posterior distribution for μ.

(b) Identify the matrix of conditional probabilities from the output. Relate these conditional probabilities to the Poisson probabilities in Table B.5.

(c) What column in the matrix contains the likelihoods?

(d) Identify the matrix of joint probabilities from the output. How are these joint probabilities found?

(e) Identify the marginal probabilities of Y from the output. How are these found?

(f) How are the posterior probabilities found?

CHAPTER 7

CONTINUOUS
RANDOM VARIABLES

When we have a continuous random variable, we believe all values over some range are possible if our measurement device is sufficiently accurate. There is an uncountably infinite number of real numbers in an interval, so the probability of getting any particular value must be zero. This makes it impossible to find the probability function of a continuous random variable the same way we did for a discrete random variable. We will have to find a different way to determine its probability distribution. First we consider a thought experiment similar to those done in Chapter 5 for discrete random variables.

Thought Experiment 4 *We start taking a sequence of independent trials of the random variable. We sketch a graph with a spike at each value in the sample equal to the proportion in the sample having that value. After each draw we update the proportions in the accumulated sample that have each value, and update our graph. The updating of the graph at step n is made by scaling all the existing spikes down by the ratio $\frac{n-1}{n}$ and adding $\frac{1}{n}$ to the spike at the value observed at trial n. This keeps the sum of the spike heights equal to 1. Figure 7.1 shows this after 25 draws. Because there are infinitely many possible numbers, it is almost inevitable that we do not draw any of the previous values, so we get a new spike at each draw. After n draws we will* have n spikes, each having height $\frac{1}{n}$. Figure 7.2 shows this after 100 draws. As the sample size, n, approaches infinity, the heights of the spikes shrink to zero. This means the probability of getting any particular value is zero. The output of this thought experiment is not the probability function, which gives the probability of each possible value. This is not like the output of the thought experiments in Chapter 6 where the random variable was discrete.

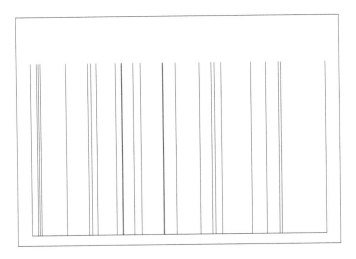

Figure 7.1 Sample probability function after 25 draws.

Figure 7.2 Sample probability function after 100 draws.

What we do notice is that there are some places with many spikes close by, and there are other places with very few spikes close by. In other words, the density of spikes varies. We can think of partitioning the interval into subintervals, and recording the number of observations that fall into each subinterval. We can form a density histogram by dividing the number in each subinterval by the width of the subinterval. This makes the area under the histogram equal to one. Figure 7.3 shows the density histogram for the first 100 observations. Now let n increase, and let the width of the subintervals

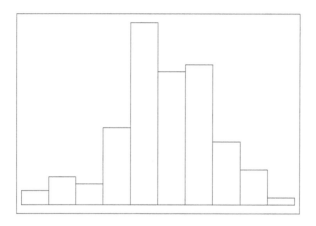

Figure 7.3   Density histogram after 100 draws.

decrease, but at a slower rate than n. Figures 7.4 and 7.5 show the density histogram for the first 1,000 and for the first 10,000 observations, respectively. The proportion of observations in a subinterval approaches the probability of being in the subinterval. As n increases, we get a larger number of shorter subintervals. The histograms get closer and closer to a smooth curve.

7.1   Probability Density Function

The smooth curve is called the probability density function (pdf). It is the limiting shape of the histograms as n goes to infinity, and the width of the bars goes to 0. Its height at a point is not the probability of that point. The thought experiment showed us that probability was equal to zero at every point. Instead, the height of the curve measures how *dense* is the probability at that point.

Since the areas under the histograms all equaled one, the total area under the probability density function must also equal 1:

$$\int_{-\infty}^{\infty} f(y)\, dy = 1\,. \tag{7.1}$$

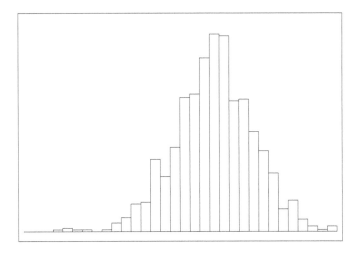

Figure 7.4 Density histogram after 1,000 draws.

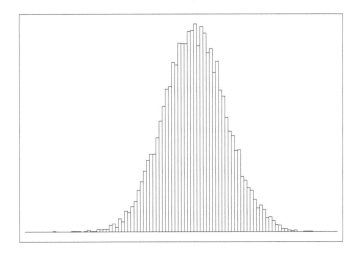

Figure 7.5 Density histogram after 10,000 draws.

The proportion of the observations that lie in an interval (a, b) is given by the area of the histogram bars that lie in the interval. In the limit as n increases to infinity, the histograms become the smooth curve, the probability density function. The area of the bars that lie in the interval becomes the area under the curve over that interval. The proportion of observations that lie in the interval becomes the probability that the random variable lies in the interval.

We know the area under a curve is found by integration, so we can find the probability that the random variable lies in the interval (a, b) by integrating the probability density function over that range:

$$P(a < Y < b) = \int_a^b f(y) \, dy \,. \tag{7.2}$$

Mean of a Continuous Random Variable

In Section 3.3 we defined the mean of the random sample of observations from the random variable to be

$$\bar{y} = \frac{\sum_{i=1}^n y_i}{n} \,.$$

Suppose we put the observations in a density histogram where all groups have equal width. The grouped mean of the data is

$$\bar{y} = \sum_j m_j \frac{n_j}{n} \,,$$

where m_j is the midpoint of the jth bar and $\frac{n_j}{n}$ is its relative frequency. Multiplying and dividing by the width of the bars, we get

$$\bar{y} = \sum_j m_j \times width \times \frac{n_j}{n \times width} \,,$$

where the relative frequency density $\frac{n_j}{n \times width}$ gives the height of bar j. Multiplying it by $width$ gives the area of the bar. Thus the sample mean is the midpoint of each bar times the area of that bar summed over all bars.

Suppose we let n increase without bound, and let the number of bars increase, but at a slower rate. For example, as n increases by a factor of 4, we let the number of bars increase by a factor of 2 so the width of each bar is divided by 2. As n increases without bound, each observation in a group becomes quite close to the midpoint of the group, the number of bars increase without bound, and the width of each bar goes to zero. In the limit, the midpoint of the bar containing the point y approaches y, and the height of the bar containing point y (which is the relative frequency density) approaches $f(y)$. So, in the limit, the relative frequency density approaches the probability density and the sample mean reaches its limit

$$E[Y] = \int_{-\infty}^{\infty} y f(y) \, dy \,, \tag{7.3}$$

which is called the *expected value* of the random variable. The expected value is like the mean of all possible values of the random variable. Sometimes it is referred to as the mean of the random variable Y and denoted μ.

Variance of a Continuous Random Variable

The expected value $E[(Y - E[Y])^2]$ is called the variance of the random variable. We can look at the variance of a random sample of numbers and let the sample size increase.

$$\text{Var}[y] = \frac{1}{n} \times \sum_{i=1}^{n} (y_i - \bar{y})^2 .$$

As we let n increase, we decrease the width of the bars. This makes each observation become closer to the midpoint of the bar it is in. Now, when we sum over all groups, the variance becomes

$$\text{Var}[y] = \sum_{j} \frac{n_j}{n} (m_j - \bar{y})^2 .$$

We multiply and divide by the width of the bar to get

$$\text{Var}[y] = \sum_{j} \frac{n_j}{n \times width} \times width \times (m_j - \bar{y})^2 .$$

This is the square of the midpoint minus the mean times the area of the bar summed over all bars. As n increases to ∞, the relative frequency density approaches the probability density, the midpoint of the bar containing the point y approaches y, and the sample mean $\bar{y}$ approaches the expected value $E[Y]$, so in the limit the variance becomes

$$\text{Var}[Y] = E[(Y - E[Y])^2] = \int_{-\infty}^{\infty} (y - \mu)^2 f(y) \, dy . \tag{7.4}$$

The variance of the random variable is denoted σ^2. We can square the term in brackets,

$$\text{Var}[Y] = \int_{-\infty}^{\infty} (y^2 - 2\mu y + \mu^2) f(y) \, dy ,$$

break the integral into three terms,

$$\text{Var}[Y] = \int_{-\infty}^{\infty} y^2 f(y) \, dy - 2\mu \int_{-\infty}^{\infty} y f(y) \, dy + \mu^2 \int_{-\infty}^{\infty} f(y) \, dy ,$$

and simplify to get an alternate form for the variance:

$$\text{Var}[Y] = E[Y^2] - [E[Y]]^2 . \tag{7.5}$$

7.2   Some Continuous Distributions

Uniform Distribution

The random variable has the *uniform* $(0,1)$ distribution if its probability density function is constant over the interval $[0,1]$, and 0 everywhere else.

$$g(x) = \begin{cases} 1 & \text{for } 0 \leq x \leq 1, \\ 0 & \text{for } x \notin [0,1] \end{cases}$$

It is easily shown that the mean and variance of a uniform $(0,1)$ random variable are $\frac{1}{2}$ and $\frac{1}{12}$, respectively.

Beta Family of Distributions

The *beta(a,b)* distribution is another commonly used distribution for a continuous random variable that can only take on values $0 \leq x \leq 1$. It has the probability density function

$$g(x; a, b) = \begin{cases} k \times x^{a-1}(1-x)^{b-1} & \text{for } 0 \leq x \leq 1, \\ 0 & \text{for } x \notin [0,1] \end{cases}$$

The most important thing is that $x^{a-1}(1-x)^{b-1}$ determines the shape of the curve, and k is only the constant needed to make this a probability density function. Figure 7.6 shows the graphs of this for $a = 2$ and $b = 3$ for a number of values of k. We see that the curves all have the same basic shape but have different areas under the curves. The value of $k = 12$ gives area equal to 1, so that is the one that makes a density function. The distribution with shape given by $x^{a-1}(1-x)^{b-1}$ is called the *beta(a,b)* distribution. The constant needed to make the curve a density function is given by the formula

$$k = \frac{\Gamma(a+b)}{\Gamma(a)\Gamma(b)},$$

where $\Gamma(c)$ is the Gamma function, which is a generalization of the factorial function.[1] The probability density function of the *beta(a,b)* distribution is given by

$$g(x; a, b) = \frac{\Gamma(a+b)}{\Gamma(a)\Gamma(b)} x^{a-1}(1-x)^{b-1}. \tag{7.6}$$

All we need remember is that $\frac{\Gamma(a+b)}{\Gamma(a)\Gamma(b)}$ is the constant needed to make the curve with shape given by $x^{a-1}(1-x)^{b-1}$ a density. a equals one plus the power of x, and b equals one plus the power of $(1-x)$.

---

[1] When c is an integer, $\Gamma(c) = (c-1)!$. The Gamma function always satisfies the equation $\Gamma(c) = (c-1) \times \Gamma(c-1)$ whether or not c is an integer.

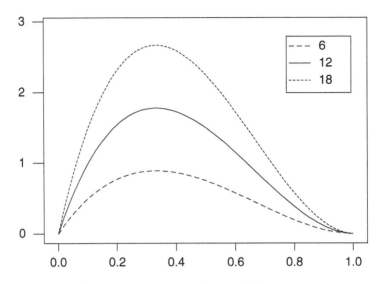

Figure 7.6   The curve $g(x) = kx^1(1 - x)^2$ for several values of k.

This curve can have different shapes depending on the values a and b, so the *beta(a, b)* is actually a family of distributions. The *uniform(0,1)* distribution is a special case of the *beta(a, b)* distribution, where $a = 1$ and $b = 1$.

Mean of a beta distribution.   The expected value of a continuous random variable x is found by integrating the variable times the density function over the whole range of possible values. (Since the *beta(a, b)* density equals 0 for x outside the interval $[0, 1]$, the integration only has to go from 0 to 1, not $-\infty$ to ∞.) For a random variable having the *beta(a, b)* distribution,

$$E[X] = \int_0^1 x \times g(x; a, b)dx = \int_0^1 x \times \frac{\Gamma(a + b)}{\Gamma(a)\Gamma(b)} x^{a-1}(1 - x)^{b-1}\, dx \,.$$

However, by using our understanding of the *beta* distribution, we can evaluate this integral without having to do the integration. First move the constant out in front of the integral, then combine the x terms by adding exponents:

$$E[X] = \frac{\Gamma(a + b)}{\Gamma(a)\Gamma(b)} \int_0^1 x \times x^{a-1}(1 - x)^{b-1}dx = \frac{\Gamma(a + b)}{\Gamma(a)\Gamma(b)} \int_0^1 x^a(1 - x)^{b-1}\, dx \,.$$

We recognize the part under the integral sign as a curve that has the *beta(a + 1, b)* shape. So we must multiply inside the integral by the appropriate constant to make it integrate to 1, and multiply by the reciprocal of the constant outside of the integral to keep the balance:

$$E[X] = \frac{\Gamma(a + b)}{\Gamma(a)\Gamma(b)} \frac{\Gamma(a + 1)\Gamma(b)}{\Gamma(a + b + 1)} \int_0^1 \frac{\Gamma(a + b + 1)}{\Gamma(a + 1)\Gamma(b)} x^a(1 - x)^{b-1}\, dx \,.$$

The integral equals 1, and when we use the fact that $\Gamma(c) = (c-1) \times \Gamma(c-1)$ and do some cancellation, we get the simple formula

$$E[X] = \frac{a}{a+b} \tag{7.7}$$

for the mean of a *beta*(a, b) random variable.

Variance of a beta distribution. The expected value of a *function* of a continuous random variable is found by integrating the *function* times the density function over the whole range of possible values. For a random variable having the *beta*(a, b) distribution,

$$E[X^2] = \int_0^1 x^2 \times \frac{\Gamma(a+b)}{\Gamma(a)\Gamma(b)} x^{a-1}(1-x)^{b-1} \, dx \, .$$

When we evaluate this integral using the properties of the *beta*(a, b) distribution, we get

$$E[X^2] = \frac{a(a+1)}{(a+b+1)(a+b)} \, .$$

When we substitute this formula and the formula for the mean of the *beta*(a, b) into Equation 7.5 and simplify, we find the variance of the random variable having the *beta*(a, b) distribution is given by

$$Var[X] = \frac{ab}{(a+b)^2(a+b+1)} \, . \tag{7.8}$$

Finding beta probabilities. When X has the *beta*(a, b) distribution, we often want to calculate probabilities such as

$$P(X \le x_0) = \int_0^{x_0} g(x; a, b) \, dx \, .$$

[**Minitab:**] This can easily be done in Minitab. Pull down the *Calc* menu to *Probability Distributions* command, over to *Beta...* subcommand, and fill out the dialog box.

Gamma Family of Distributions

The *gamma*(r, v) distribution is used for continuous random variables that can take on nonnegative values $0 \le x < \infty$. Its probability density function is given by

$$g(x; r, v) = k \times x^{r-1} e^{-vx} \text{ for } 0 \le x < \infty \, .$$

The shape of the curve is determined by $x^{r-1}e^{-vx}$, while k is only the constant needed to make this a probability density. Figure 7.7 shows the graphs of this for the case where $r = 4$ and $v = 4$ for several values of k. Clearly the curves

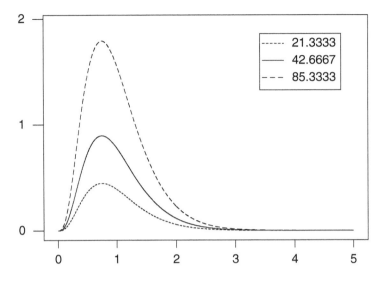

Figure 7.7   The curve $g(x) = kx^3 e^{-4x}$ for several values of k.

have the same basic shape, but have different areas under the curve. The curve with $k = 42.6667$ will have area equal to 1, so it is the exact density.

The distribution having shape given by $x^{r-1}e^{-vx}$ is called the *gamma*(r, v) distribution. The constant needed to make this a probability density function is given by

$$k = \frac{v^r}{\Gamma(r)},$$

where $\Gamma(r)$ is the Gamma function. The probability density of the *gamma*(r, v) distribution is given by

$$g(x; r, v) = \frac{v^r x^{r-1} e^{-vx}}{\Gamma(r)} \tag{7.9}$$

for $0 \leq x < \infty$.

Mean of Gamma distribution.   The expected value of a *gamma*(r, v) random variable x is found by integrating the variable x times its density function over the whole range of possible values. It will be

$$\begin{aligned} \mathrm{E}[X] &= \int_0^\infty x g(x; r, v)\, dx \\ &= \int_0^\infty x \frac{v^r x^{r-1} e^{-vx}}{\Gamma(r)}\, dx \\ &= \frac{v^r}{\Gamma(r)} \int_0^\infty x^r e^{-vx}\, dx. \end{aligned}$$

We recognize the part under the integral to be a curve that has the shape of a *gamma*$(r + 1, v)$ distribution. We multiply inside the integral by the appropriate constant to make it integrate to 1, and outside the integral we multiply by its reciprocal to keep the balance.

$$\mathrm{E}[X] = \frac{v^r}{\Gamma(r)} \times \frac{\Gamma(r+1)}{v^{r+1}} \int_0^\infty \frac{v^{r+1}}{\Gamma(r+1)} x^r e^{-vx} \, dx \,.$$

This simplifies to give

$$\mathrm{E}[X] = \frac{r}{v} \,. \tag{7.10}$$

Variance of a gamma distribution. First we find

$$\mathrm{E}[X^2] = \int_0^\infty x^2 g(x; r, v) \, dx$$

$$= \frac{v^r}{\Gamma(r)} \int_0^\infty x^{r+1} e^{-vx} \, dx \,.$$

We recognize the shape of a *gamma*$(r+2, v)$ under the curve, so this simplifies to

$$\mathrm{E}[X^2] = \frac{(r+1)r}{v^2} \,.$$

When we substitute this, and the formula for the mean of the *gamma*(r, v) into Equation 7.5 and simplify we find the variance of the *gamma*(r, v) distribution to be

$$\mathrm{Var}[X] = \frac{r}{v^2} \,. \tag{7.11}$$

Finding gamma probabilities. When X has the *gamma*(r, v) distribution we often want to calculate probabilities such as

$$P(X \le x_0) = \int_0^{x_0} g(x; r, v) \, dx \,.$$

This can easily be done in Minitab. Pull down the *Calc* menu to *Probability Distributions* command, over to *Gamma...* subcommand, and fill out the dialog box. Note: In Minitab the shape parameter is r and the scale parameter is $\frac{1}{v}$.

Normal Distribution

Very often data appear to have a symmetric bell-shaped distribution. In the early years of statistics, this shape seemed to occur so frequently that it was thought to be normal. The family of distributions with this shape has become known as the *normal* distribution family. It is also known as the *Gaussian* distribution after the mathematician Gauss, who studied its properties. It is the most widely used distribution in statistics. We will see that there is a good reason for its frequent occurrence. However, we must remain aware that the term *normal distribution* is only a name, and distributions with other shapes are not abnormal.

The *normal*(μ, σ^2) distribution is the member of the family having mean μ and variance σ^2. The probability density function of a *normal*(μ, σ^2) distribution is given by

$$g(x|\mu, \sigma^2) = ke^{-\frac{1}{2\sigma^2}(x-\mu)^2}$$

for $-\infty < x < \infty$, where k is the constant value needed to make this a probability density. The shape of the curve is determined by $e^{-\frac{1}{2\sigma^2}(x-\mu)^2}$. Figure 7.8 shows the curve $ke^{-\frac{1}{2\sigma^2}(x-\mu)^2}$ for several values of k. Changing the value of k only changes the area under the curve, not its basic shape. To be a probability density function, the area under the curve must equal 1. The value of k that makes the curve a probability density is $k = \frac{1}{\sqrt{2\pi}\sigma}$.

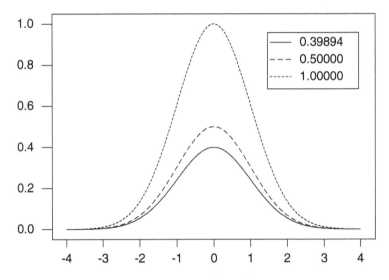

Figure 7.8   The curve $g(x) = ke^{-\frac{1}{2}(x-0)^2}$ for several values of k.

Central limit theorem.   The central limit theorem says that if you take a random sample $y_1, \ldots, y_n$ from any shape distribution having mean μ and variance σ^2, then the limiting distribution of $\frac{\bar{y}-\mu}{\sigma/\sqrt{n}}$ is *normal*$(0, 1)$. The shape of the limiting distribution is *normal* despite the original distribution not necessarily being normal. A linear transformation of a normal distribution is also normal, so the shape of $\bar{y}$ and $\sum y$ are also normal. Amazingly, n does not have to be particularly large for the shape to be approximately normal, $n \geq 25$ is sufficient.

The key factor of the central limit distribution is that when we are averaging a large number of independent effects, each of which is small in relation to the

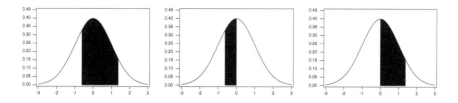

Figure 7.9   The area between $-.62$ and 1.37 split into two parts.

sum, the distribution of the sum approaches the *normal* shape regardless of the shapes of the individual distributions. Thus any random variable that arises as the sum of a large number of independent effects will be approximately normal! This explains why the normal distribution is encountered so frequently.

Finding probabilities using standard normal table.   The standard normal density has mean $\mu = 0$ and variance $\sigma^2 = 1$. Its probability density function is given by

$$f(z) = \frac{1}{\sqrt{2\pi}}e^{-\frac{1}{2}z^2}.$$

We note that this curve is symmetric about $z = 0$. Unfortunately, Equation 7.2, the general form for finding the probability $P(a \leq z \leq b)$, is not of any practical use here. There is no closed form for integrating the standard normal probability density function. Instead, the area between 0 and z for values of z between 0 and 3.99 has been numerically calculated and tabulated in Table B.2 in Appendix B. We use this table to calculate the probability we need.

■ **EXAMPLE 7.1**

Suppose we want to find $P(-.62 \leq Z \leq 1.37)$. In Figure 7.9 we see that the shaded area between $-.62$ and 1.37 is the sum of the two areas between $-.62$ and 0 and between 0 and 1.37, respectively. The area between $-.62$ and 0 is the same as the area between 0 and $+.62$ because the standard normal density is symmetric about 0. In Table B.2 we find this area equals .2324. The area between 0 and 1.37 equals .4147 from the table. So

$$P(-.62 \leq Z \leq 1.37) = .2324 + .4147$$
$$= .6471.$$

■

Any normal distribution can be transformed into a standard normal by subtracting the mean and then dividing by the standard deviation. This lets us find any normal probability using the areas under the standard normal density found in Table B.2.

EXAMPLE 7.2

Suppose we know Y is normal with mean $\mu = 10.8$ and standard deviation $\sigma = 2.1$, and suppose we want to find the probability $P(Y \geq 9.9)$.

$$P(Y \geq 9.9) = P(Y - 10.8 \geq 9.9 - 10.8)$$
$$= P\left(\frac{Y - 10.8}{2.1} \geq \frac{9.9 - 10.8}{2.1}\right).$$

The left side is a standard normal. The right side is a number. We find this probability from the standard normal:

$$P(Y \geq 9.9) = P(Z \geq -.429)$$
$$= .1659 + .5000$$
$$= .6659.$$

Finding beta probabilities using normal approximation.   We can approximate a $beta(a, b)$ distribution by the *normal* distribution having the same mean and variance. This approximation is very effective when both a and b are greater than or equal to ten.

EXAMPLE 7.3

Suppose Y has the $beta(12, 25)$ distribution and we wish to find $P(Y > .4)$. The mean and variance of Y are

$$E[Y] = \frac{12}{37} = .3243 \quad \text{and} \quad \text{Var}[Y] = \frac{12 \times 25}{37^2 \times 38} = .005767,$$

respectively. We approximate the $beta(12, 25)$ distribution with a *normal*$(.3243, .005767)$ distribution. The approximate probability is

$$P(Y > .4) = P\left(\frac{Y - .3243}{\sqrt{.005767}} > \frac{.4 - .3243}{\sqrt{.005767}}\right)$$
$$= P(Z > .997)$$
$$= .1594.$$

Finding gamma probabilities using normal approximation is not recommended. As r approaches infinity the $gamma(r, v)$ distribution approaches the *normal*(m, s^2) distribution where $m = \frac{r}{v}$ and $s^2 = \frac{r}{v^2}$. However, the approach is very slow, and the gamma probabilities calculated using the normal approximation will not be very accurate unless r is quite large (Johnson et al., 1970). Johnson et al. (1970) recommend that the normal approximation to the gamma not be used for this reason, and they give other approximations that are more accurate.

7.3 Joint Continuous Random Variables

We consider two (or more) random variables distributed together. If both X and Y are continuous random variables, they have joint density $f(x, y)$, which measures the probability density at the point (x, y). This would be found by dividing the plane into rectangular regions by partitioning both the x axis and y axis. We look at the proportion of the sample that lie in a region. We increase n, the sample size of the joint random variables without bound, and at the same time decrease the width of the regions (in both dimensions) at a slower rate. In the limit, the proportion of the sample lying in the region centered at (x, y) approaches the joint density $f(x, y)$. Figure 7.10 shows a joint density function.

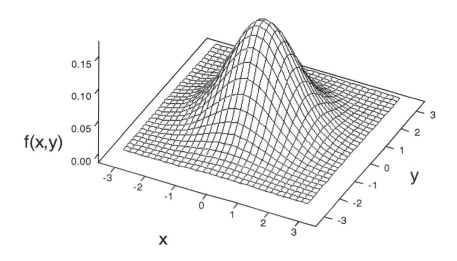

Figure 7.10   A joint density.

We might be interested in determining the density of one of the joint random variables by itself, its *marginal* density. When X and Y are joint random variables that are both continuous, the marginal density of Y is found by integrating the joint density over the whole range of X:

$$f(y) = \int_{-\infty}^{\infty} f(x, y)\, dx \,,$$

and vice versa. (Finding the marginal density by integrating the joint density over the whole range of one variable is analogous to finding the marginal probability distribution by summing the joint probability distribution over all possible values of one variable for jointly distributed discrete random variables.)

Conditional Probability Density

The conditional density of X given $Y = y$ is given by

$$f(x|y) = \frac{f(x,y)}{f(y)} .$$

We see that the conditional density of X given $Y = y$ is proportional to the joint density where $Y = y$ is held fixed. Dividing by the marginal density $f(y)$ makes the integral of the conditional density over the whole range of x equal 1. This makes it a proper density function.

7.4   Joint Continuous and Discrete Random Variables

It may be that one of the variables is continuous, and the other is discrete. For instance, let X be continuous, and let Y be discrete. In that case, $f(x, y_j)$ is a joint probability–probability density function. In the x direction it is continuous, and in the y direction it is discrete. This is shown in Figure 7.11.

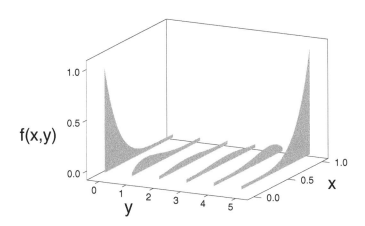

Figure 7.11   A joint continuous and discrete distribution.

In this case, the marginal density of the continuous random variable X is found by

$$f(x) = \sum_j f(x, y_j) ,$$

and the marginal probability function of the discrete random variable Y is found by

$$f(y_j) = \int f(x, y_j) \, dx .$$

The conditional density of X given $Y = y_j$ is given by

$$f(x|y_j) = \frac{f(x, y_j)}{f(y_j)} = \frac{f(x, y_j)}{\int f(x, y_j)\, dx}.$$

We see that this is proportional to the joint probability–probability density function $f(x, y_j)$ where x is allowed to vary over its whole range. Dividing by the marginal probability $f(y_j)$ just scales it to be a proper density function (integrates to 1).

Similarly, the conditional distribution of $Y = y_j$ given x is found by

$$f(y_j|x) = \frac{f(x, y_j)}{f(x)} = \frac{f(x, y_j)}{\sum_j f(x, y_j)}.$$

This is also proportional to the joint probability–probability density function $f(x, y_j)$ where x is fixed, and Y is allowed to take on all the possible values $y_1, \ldots, y_J$.

Main Points

- The probability that a continuous random variable equals any particular value is zero!

- The probability density function of a continuous random variable is a smooth curve that measures the *density* of probability at each value. It is found as the limit of density histograms of random samples of the random variable, where the sample size increases to infinity and the width of the bars goes to zero.

- The probability a continuous random variable lies between two values a and b is given by the area under the probability density function between the two values. This is found by the integral

$$P(a < X < b) = \int_a^b f(x)\, dx.$$

- The expected value of a continuous random variable X is found by integrating x times the density function $f(x)$ over the whole range.

$$\mathrm{E}[X] = \int_{-\infty}^{\infty} x f(x)\, dx.$$

- A *beta(a, b)* random variable has probability density

$$f(x|a, b) = \frac{\Gamma(a + b)}{\Gamma(a)\Gamma(b)} x^{a-1}(1 - x)^{b-1} \quad \text{for} \quad 0 \le x \le 1.$$

- The mean and variance of a *beta(a, b)* random variable are given by

$$E[X] = \frac{a}{a+b} \qquad \text{and} \qquad \text{Var}[X] = \frac{a \times b}{(a+b)^2 \times (a+b+1)}.$$

- A *gamma(r, v)* random variable has probability density

$$g(x; r, v) = \frac{v^r x^{r-1} e^{-vx}}{\Gamma(r)} \quad \text{for } 0 \leq x < \infty.$$

- The mean and variance of a *gamma(r, v)* random variable are given by

$$E[X] = \frac{r}{v} \qquad \text{and} \qquad \text{Var}[X] = \frac{r}{v^2}.$$

- A *normal(μ, σ²)* random variable has probability density

$$g(x|\mu, \sigma^2) = \frac{1}{\sqrt{2\pi}\sigma} e^{-\frac{1}{2\sigma^2}(x-\mu)^2},$$

  where μ is the mean, and σ^2 is the variance.

- The central limit theorem says that for a random sample $y_1, \ldots y_n$ from any distribution $f(y)$ having mean μ and variance σ^2, the distribution of

$$\frac{\bar{y} - \mu}{\sigma/\sqrt{n}}$$

  is approximately *normal(0, 1)* for $n > 25$. This is regardless of the shape of the original density $f(y)$.

- By reasoning similar to that of the central limit theorem, any random variable that is the sum of a large number of independent random variables will be approximately normal. This is the reason why the normal distribution occurs so frequently.

- The marginal distribution of y is found by integrating the joint distribution $f(x, y)$ with respect to x over its whole range.

- The conditional distribution of x given y is proportional to the joint distribution $f(x, y)$ where y fixed and x is allowed to vary over its whole range.

$$f(x|y) = \frac{f(x, y)}{f(y)}.$$

  Dividing by the marginal distribution of $f(y)$ scales it properly so that $f(y|x)$ integrates to 1 and is a probability density function.

Exercises

7.1. Let X have a $beta(3,5)$ distribution.

(a) Find $E[X]$.

(b) Find $Var[X]$.

7.2. Let X have a $beta(12,4)$ distribution.

(a) Find $E[X]$.

(b) Find $Var[X]$.

7.3. Let X have the *uniform* distribution.

(a) Find $E[X]$.

(b) Find $Var[X]$.

(c) Find $P(X \leq .25)$.

(d) Find $P(.33 < X < .75)$.

7.4. Let X be a random variable having probability density function

$$f(x) = 2x \quad \text{for} \quad 0 \leq x \leq 1.$$

(a) Find $P(X \geq .75)$.

(b) Find $P(.25 \leq X \leq .6)$.

7.5. Let Z have the standard normal distribution.

(a) Find $P(0 \leq Z \leq .65)$.

(b) Find $P(Z \geq .54)$.

(c) Find $P(-.35 \leq Z \leq 1.34)$.

7.6. Let Z have the standard normal distribution.

(a) Find $P(0 \leq Z \leq 1.52)$.

(b) Find $P(Z \geq 2.11)$.

(c) Find $P(-1.45 \leq Z \leq 1.74)$.

7.7. Let Y be normally distributed with mean $\mu = 120$ and variance $\sigma^2 = 64$.

(a) Find $P(Y \leq 130)$.

(b) Find $P(Y \geq 135)$.

(c) Find $P(114 \leq Y \leq 127)$.

7.8. Let Y be normally distributed with mean $\mu = 860$ and variance $\sigma^2 = 576$.

(a) Find $P(Y \le 900)$.

(b) Find $P(Y \ge 825)$.

(c) Find $P(840 \le Y \le 890)$.

7.9. Let Y be distributed according to the $beta(10, 12)$ distribution.

(a) Find $E[Y]$.

(b) Find $Var[Y]$.

(c) Find $P(Y > .5)$ using the normal approximation.

7.10. let Y be distributed according to the $beta(15, 10)$ distribution.

(a) Find $E[Y]$.

(b) Find $Var[Y]$.

(c) Find $P(Y < .5)$ using the normal approximation.

7.11. Let Y be distributed according to the $gamma(12, 4)$ distribution.

(a) Find $E[Y]$.

(b) Find $Var[Y]$.

(c) Find $P(Y \le 4)$

7.12. Let Y be distributed according to the $gamma(26, 5)$ distribution.

(a) Find $E[Y]$.

(b) Find $Var[Y]$.

(c) Find $P(Y > 5)$

CHAPTER 8

BAYESIAN INFERENCE FOR BINOMIAL PROPORTION

Frequently there is a large population where π, a proportion of the population, has some attribute. For instance, the population could be registered voters living in a city, and the attribute is "plans to vote for candidate A for mayor." We take a random sample from the population and let Y be the observed number in the sample having the attribute, in this case the number who say they plan to vote candidate A for mayor.

We are counting the total number of "successes" in n independent trials where each trial has two possible outcomes, "success" and "failure." Success on trial i means the item drawn on trial i has the attribute. The probability of success on any single trial is π, the proportion in the population having the attribute. This proportion remains constant over all trials because the population is large.

The conditional distribution of the observation Y, the total number of successes in n trials given the parameter π, is $binomial(n, \pi)$. The conditional probability function for y given π is given by

$$f(y|\pi) = \binom{n}{y}\pi^y(1-\pi)^{n-y} \text{ for } y = 1, \ldots, n.$$

Introduction to Bayesian Statistics, 3rd ed.
By Bolstad, W. M. and Curran, J. M. Copyright © 2016 John Wiley & Sons, Inc.

Here we are holding π fixed and are looking at the probability distribution of y over its possible values.

If we look at this same relationship between π and y, but hold y fixed at the number of successes we observed, and let π vary over its possible values, we have the likelihood function given by

$$f(y|\pi) = \binom{n}{y}\pi^y(1-\pi)^{n-y} \text{ for } 0 \leq \pi \leq 1.$$

We see that we are looking at the same relationship as the distribution of the observation y given the parameter π, but the subject of the formula has changed to the parameter, for the observation held at the value that actually occurred.

To use Bayes' theorem, we need a prior distribution $g(\pi)$ that gives our belief about the possible values of the parameter π before taking the data. It is important to realize that the prior must not be constructed from the data. Bayes' theorem is summarized by *posterior is proportional to the prior times the likelihood*. The multiplication in Bayes' theorem can only be justified when the prior is *independent* of the likelihood![1] This means that the observed data must not have any influence on the choice of prior! The posterior distribution is proportional to prior distribution times likelihood:

$$g(\pi|y) \propto g(\pi) \times f(y|\pi).$$

This gives us the shape of the posterior density, but not the exact posterior density itself. To get the actual posterior, we need to divide this by some constant k to make sure it is a probability distribution, meaning that the area under the posterior integrates to 1. We find k by integrating $g(\pi) \times f(y|\pi)$ over the whole range. So, in general,

$$g(\pi|y) = \frac{g(\pi) \times f(y|\pi)}{\int_0^1 g(\pi) \times f(y|\pi)\, d\pi}, \tag{8.1}$$

which requires an integration. Depending on the prior $g(\pi)$ chosen, there may not necessarily be a closed form for the integral, so it may be necessary to do the integration numerically. We will look at some possible priors.

8.1 Using a Uniform Prior

If we do not have any idea beforehand what the proportion π is, we might like to choose a prior that does not favor any one value over another. Or,

---

[1]We know that for independent events (or random variables) the joint probability (or density) is the product of the marginal probabilities (or density functions). If they are not independent, this does not hold. Likelihoods come from probability functions or probability density functions, so the same pattern holds. They can only be multiplied when they are independent.

we may want to be as objective as possible and not put our personal belief into the inference. In that case we should use the uniform prior that gives equal weight to all possible values of the success probability π. Although this does not achieve universal objectivity (which is impossible to achieve), it is objective for this formulation of the problem[2]:

$$g(\pi) = 1 \quad \text{for} \quad 0 \le \pi \le 1.$$

Clearly, we see that in this case, the posterior density is proportional to the likelihood:

$$g(\pi|y) = \binom{n}{y} \pi^y (1-\pi)^{n-y} \quad \text{for} \quad 0 \le \pi \le 1.$$

We can ignore the part that does not depend on π. It is a constant for all values of π, so it does not affect the shape of the posterior. When we examine that part of the formula that shows the shape of the posterior as a function of π, we recognize that this is a $beta(a, b)$ distribution where $a = y + 1$ and $b = n - y + 1$. So in this case, the posterior distribution of π given y is easily obtained. All that is necessary is to look at the exponents of π and $(1 - \pi)$. We did not have to do the integration.

8.2   Using a Beta Prior

Suppose a $beta(a, b)$ prior density is used for π:

$$g(\pi; a, b) = \frac{\Gamma(a+b)}{\Gamma(a)\Gamma(b)} \pi^{a-1}(1-\pi)^{b-1} \quad \text{for} \quad 0 \le \pi \le 1.$$

The posterior is proportional to prior times likelihood. We can ignore the constants in the prior and likelihood that do not depend on the parameter, since we know that multiplying either the prior or the likelihood by a constant will not affect the results of Bayes' theorem. This gives

$$g(\pi|y) \propto \pi^{a+y-1}(1-\pi)^{b+n-y-1} \quad \text{for} \quad 0 \le \pi \le 1,$$

which is the shape of the posterior as a function of π. We recognize that this is the beta distribution with parameters $a' = a + y$ and $b' = b + n - y$. That is, we add the number of successes to a and add the number of failures to b:

$$g(\pi|y) = \frac{\Gamma(n+a+b)}{\Gamma(y+a)\Gamma(n-y+b)} \pi^{y+a-1}(1-\pi)^{n-y+b-1}$$

---

[2]There are many possible parameterizations of the problem. Any one-to-one function of the parameter would also be a suitable parameter. The prior density for the new parameter could be found from the prior density of the original parameter using the change of variable formula and would not be flat. In other words, it would favor some values of the new parameter over others. You can be objective in a given parameterization, but it would not be objective in the new formulation. *Universal* objectivity is not attainable.

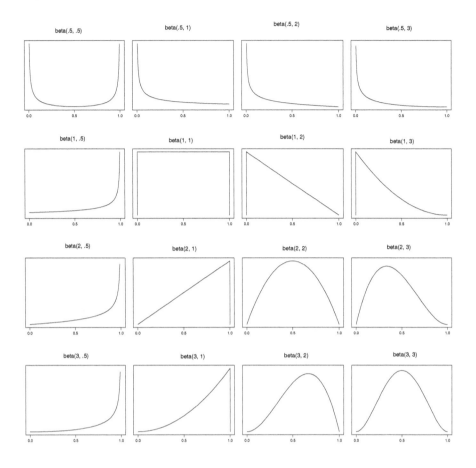

Figure 8.1   Some beta distributions.

for $0 \leq \pi \leq 1$. Again, the posterior density of π has been easily obtained without having to go through the integration.

Figure 8.1 shows the shapes of $beta(a, b)$ densities for values of $a = .5, 1, 2, 3$ and $b = .5, 1, 2, 3$. This shows the variety of shapes that members of the $beta(a, b)$ family can take. When $a < b$, the density has more weight in the lower half. The opposite is true when $a > b$. When $a = b$, the $beta(a, b)$ density is symmetric. When $a = \frac{1}{2}$ much more weight is given to values near 0, and when $b = \frac{1}{2}$ much more weight is given to values near 1. We note that the uniform prior is a special case of the $beta(a, b)$ prior where $a = 1$ and $b = 1$.

Conjugate Family of Priors for Binomial Observation is the Beta Family

When we examine the shape of the binomial likelihood function as a function of π, we see that this is of the same form as the $beta(a, b)$ distribution, a product of π to a power times $(1 - \pi)$ to another power. When we multiply the beta prior times the binomial likelihood, we add the exponents of π and $(1 - \pi)$, respectively. So we start with a beta prior, and we get a beta posterior by the simple rule "add successes to a, add failures to b." This makes using $beta(a, b)$ priors when we have binomial observations particularly easy. Using Bayes' theorem moves us to another member of the same family.

We say that the *beta* distribution is the conjugate[3] family for the *binomial* observation distribution. When we use a prior from the conjugate family, we do not have to do any integration to find the posterior. All we have to do is use the observations to update the parameters of the conjugate family prior to find the conjugate family posterior. This is a big advantage.

Jeffreys' prior for binomial. The $beta(\frac{1}{2}, \frac{1}{2})$ prior is known as the Jeffreys' prior for the binomial. If we think of the parameter as an index of all possible densities the observation could come from, then any continuous function of the parameter would give an equally valid index.[4] Jeffreys' method gives a prior[5] that is invariant under any continuous transformation of the parameter. That means that Jeffreys' prior is objective in the sense that it does not depend on the particular parameterization we used.[6] However, for most parameterizations, the Jeffreys' prior gives more weight to some values than to others so it is usually informative, not noninformative. For further information on Jeffreys' method for finding invariant priors refer to Press (1989), O'Hagan (1994), and Lee (1989). We note that Jeffreys' prior for the binomial is just a particular member of the *beta* family of priors, so the posterior is found using the same updating rules.

---

[3]Conjugate priors only exists when the observation distribution comes from the *exponential* family. In that case the observation distribution can be written $f(y|\theta) = a(\theta)b(y)e^{c(\theta) \times T(y)}$. The conjugate family of priors will then have the same functional form as the likelihood of the observation distribution.

[4]If $\psi = h(\theta)$ is a continuous function of the parameter θ, then $g_\psi(\psi)$, the prior for ψ that corresponds to $g_\theta(\theta)$, the prior for θ is found by the change of variable formula $g_\psi(\psi) = g_\theta(\theta(\psi)) \times \frac{d\theta}{d\psi}$.

[5]Jeffreys' invariant prior for parameter θ is given by $g(\theta) \propto \sqrt{I(\theta|y)}$, where $I(\theta|y)$ is known as Fisher's information and is given by $I(\theta|y) = -E\left(\frac{\partial^2 \log f(y|\theta)}{\partial \theta^2}\right)$.

[6]If we had used another parameterization and found the Jeffreys' prior for that parameterization, then transformed it to our original parameter using the change of variable formula, we would have the Jeffreys' prior for the original parameter.

8.3   Choosing Your Prior

Bayes' theorem gives you a method to revise your (belief) distribution about the parameter, given the data. In order to use it, you must have a distribution that represents your belief about the parameter, before we look at the data.[7] This is your prior distribution. In this section we propose some methods to help you choose your prior, as well as things to consider in prior choice.

Choosing a Conjugate Prior When You Have Vague Prior Knowledge

When you have vague prior knowledge, one of the $beta(a, b)$ prior distributions shown in Figure 8.1 would be a suitable prior. For example, if your prior knowledge about π, is that π is very small, then $beta(.5, 1)$, $beta(.5, 2)$, $beta(.5, 3)$, $beta(1, 2)$, or $beta(1, 3)$ would all be satisfactory priors. All of these conjugate priors offer easy computation of the posterior, together with putting most of the prior probability at small values of π. It does not matter very much which one you chose; the resulting posteriors given the data would be very similar.

Choosing a Conjugate Prior When You Have Real Prior Knowledge by Matching Location and Scale

The $beta(a, b)$ family of distributions is the conjugate family for $binomial(n, \pi)$ observations. We saw in the previous section that priors from this family have significant advantages computationally. The posterior will be a member of the same family, with the parameters updated by simple rules. We can find the posterior without integration. The beta distribution can have a number of shapes. The prior chosen should correspond to your belief. We suggest choosing a $beta(a, b)$ that matches your prior belief about the (location) mean and (scale) standard deviation[8]. Let π_0 be your prior mean for the proportion, and let σ_0 be your prior standard deviation for the proportion.

The mean of $beta(a, b)$ distribution is $\frac{a}{a+b}$. Set this equal to what your prior belief about the mean of the proportion to give

$$\pi_0 = \frac{a}{a+b}.$$

---

[7]This could be elicited from your coherent betting strategy about the parameter value. Having a coherent betting strategy means that if someone started offering you bets about the parameter value, you would not take a worse bet than one you already rejected, nor would you refuse to take a better bet than one you already accepted.

[8]Some people would say that you should not use a conjugate prior just because of these advantages. Instead, you should elicit your prior from your coherent betting strategy. I do not think most people carry around a coherent betting strategy in their head. Their prior belief is less structured. They have a belief about the *location* and *scale* of the parameter distribution. Choosing a prior by finding the conjugate family member that matches these beliefs will give a prior on which a coherent betting strategy could be based!

The standard deviation of beta distribution is $\sqrt{\frac{ab}{(a+b)^2(a+b+1)}}$. Set this equal to what your prior belief about the standard deviation for the proportion. Noting that $\frac{a}{a+b} = \pi_0$ and $\frac{b}{a+b} = 1 - \pi_0$, we see

$$\sigma_0 = \sqrt{\frac{\pi_0(1 - \pi_0)}{a + b + 1}}.$$

Solving these two equations for a and b gives your $beta(a, b)$ prior.

Precautions Before Using Your Conjugate Prior

1. Graph your $beta(a, b)$ prior. If the shape looks reasonably close to what you believe, you will use it. Otherwise, you can adjust π_0 and σ_0 until you find a prior whose graph approximately corresponds to your belief. As long as the prior has reasonable probability over the whole range of values that you think the parameter could possibly be in, it will be a satisfactory prior.

2. Calculate the *equivalent sample size* of the prior. We note that the sample proportion $\hat{\pi} = \frac{y}{n}$ from a $binomial(n,\pi)$ distribution has variance equal to $\frac{\pi(1-\pi)}{n}$. We equate this variance (at π_0, the prior mean) to the prior variance.

$$\frac{\pi_0(1 - \pi_0)}{n_{eq}} = \frac{ab}{(a + b)^2 \times (a + b + 1)}.$$

Since $\pi_0 = \frac{a}{a+b}$ and $(1 - \pi_0) = \frac{b}{a+b}$, the equivalent sample size is $n_{eq} = a+b+1$. It says that the amount of information about the parameter from your prior is equivalent to the amount from a random sample of that size. You should always check if this is unrealistically high. Ask yourself, "Is my prior knowledge about π really equal to the knowledge about π that I would obtain if I checked a random sample of size n_{eq}? If it is not, you should increase your prior standard deviation and recalculate your prior. Otherwise, you would be putting too much prior information about the parameter relative to the amount of information that will come from the data.

Constructing a General Continuous Prior

Your prior shows the *relative* weights you give each possible value before you see the data. The shape of your prior belief may not match the *beta* shape. You can construct a discrete prior that matches your belief weights at several values over the range you believe possible, and then interpolate between them to make the continuous prior. You can ignore the constant needed to make this a density, because when you multiply the prior by a constant, the constant gets cancelled out by Bayes' theorem. However, if you do construct your prior this way, you will have to evaluate the integral of the prior times likelihood numerically to get the posterior. This will be shown in the following example.

Table 8.1   Chris's prior weights. The shape of his continuous prior is found by linearly interpolating between them.

Value	Weight
0	0
.05	1
.1	2
.3	2
.4	1
.5	0

EXAMPLE 8.1

Three students are constructing their prior belief about π, the proportion of Hamilton residents who support building a casino in Hamilton. Anna thinks that her prior mean is .2, and her prior standard deviation is .08. The $beta(a, b)$ prior that satisfies her prior belief is found by

$$\frac{.2 \times .8}{a + b + 1} = .08^2 \,.$$

Therefore her equivalent sample size is $a + b + 1 = 25$. For Anna's prior, $a = 4.8$ and $b = 19.2$.

Bart is a newcomer to Hamilton, so he is not aware of the local feeling for or against the proposed casino. He decides to use a uniform prior. For him, $a = b = 1$. His equivalent sample size is $a + b + 1 = 3$.

Chris cannot fit a $beta(a, b)$ prior to match his belief. He believes his prior probability has a trapezoidal shape. He gives heights of his prior in Table 8.1, and he linearly interpolates between them to get his continuous prior. When we interpolate between these points, we see that Chris's prior is given by

$$g(\pi) = \begin{cases} 20\pi & \text{for } 0 \leq \pi \leq .10 \,, \\ 2 & \text{for } .10 \leq \pi \leq .30 \,, \\ 5 - 10\pi & \text{for } .30 \leq \pi \leq .50 \,. \end{cases}$$

The three priors are shown in the Figure 8.2. Note that Chris's prior is not actually a density since it does not have area equal to one. However, this is not a problem since the relative weights given by the shape of the distribution are all that is needed since the constant will cancel out.   ■

Effect of the Prior

When we have enough data, the effect of the prior we choose will be small compared to the data. In that case we will find that we can get very similar

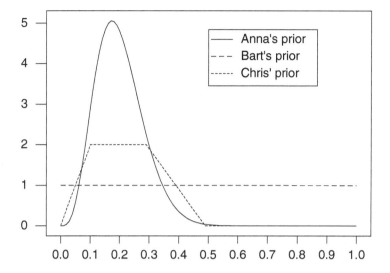

Figure 8.2   Anna's, Bart's, and Chris' prior distribution.

posteriors despite starting from quite different priors. All that is necessary is that they give reasonable weight over the range that is indicated by the likelihood. The exact shape of the prior does not matter very much. The data are said to "swamp the prior."

EXAMPLE 8.1   (continued)

The three students take a random sample of $n = 100$ Hamilton residents and find their views on the casino. Out of the random sample, $y = 26$ said they support building a casino in Hamilton. Anna's posterior is $beta(4.8 + 26, 19.2 + 74)$. Bart's posterior is $beta(1 + 26, 1 + 74)$. Chris' posterior is found using Equation 8.1. We need to evaluate Chris' prior numerically.

[**Minitab:**]   To do this in Minitab, we integrate Chris' $prior \times likelihood$ using the Minitab macro *tintegral.*

[**R:**]   To do this in R, we integrate Chris' $prior \times likelihood$ using the R function `sintegral`.

The three posteriors are shown in Figure 8.3. We see that the three students end up with very similar posteriors, despite starting with priors having quite different shapes. ∎

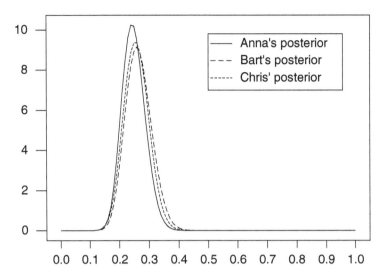

Figure 8.3   Anna's, Bart's, and Chris' posterior distributions.

8.4   Summarizing the Posterior Distribution

The posterior distribution summarizes our belief about the parameter *after* seeing the data. It takes into account our prior belief (the prior distribution) and the data (likelihood). A graph of the posterior shows us all we can know about the parameter, after the data. A distribution is hard to interpret. Often we want to find a few numbers that characterize it. These include measures of location that determine where most of the probability is on the number line, as well as measures of spread that determine how widely the probability is spread. They could also include percentiles of the distribution. We may want to determine an interval that has a high probability of containing the parameter. These are known as *Bayesian credible intervals* and are somewhat analogous to confidence intervals. However, they have the direct probability interpretation that confidence intervals lack.

Measures of Location

First, we want to know where the posterior distribution is located on the number line. There are three possible measures of location we will consider: posterior mode, posterior median, and posterior mean.

Posterior mode.   This is the value that maximizes the posterior distribution. If the posterior distribution is continuous, it can be found by setting the derivative of the posterior density equal to zero. When the posterior $g(\pi|y)$

is $beta(a', b')$, its derivative is given by

$$g'(\pi|y) = (a'-1)\pi^{a'-2} \times (1-\pi)^{b'-1} + \pi^{a'-1} \times (-1)(b'-1)(1-\pi)^{b'-2}.$$

(Note: The prime $'$ has two meanings in this equation; $g'(\pi|y)$ is the derivative of the posterior, while a' and b' are the constants of the *beta* posterior found by the updating rules.) Setting $g'(\pi|y)$ equal to 0 and solving gives the posterior mode

$$mode = \frac{a'-1}{a'+b'-2}.$$

The posterior mode has some potential disadvantages as a measure of location. First, it may lie at or near one end of the distribution, and thus not be representative of the distribution as a whole. Second, there may be multiple local maximums. When we set the derivative function equal to zero and solve, we will find all of them and the local minimums as well.

Posterior median. This is the value that has 50% of posterior distribution below it, 50% above it. If $g(\pi|y)$ is $beta(a', b')$, it is the solution of

$$\int_0^{median} g(\pi|y)\, d\pi = .5.$$

The only disadvantage of the posterior median is that it has to be found numerically. It is an excellent measure of location.

Posterior mean. The posterior mean is a very frequently used measure of location. It is the expected value, or mean, of the posterior distribution.

$$m' = \int_0^1 \pi g(\pi|y)\, d\pi. \tag{8.2}$$

The posterior mean is strongly affected when the distribution has a heavy tail. For a skewed distribution with one heavy tail, the posterior mean may be quite a distance away from most of the probability. When the posterior $g(\pi|y)$ is $beta(a', b')$ the posterior mean equals

$$m' = \frac{a'}{a'+b'}. \tag{8.3}$$

The $beta(a, b)$ distribution is bounded between 0 and 1, so it does not have heavy tails. The posterior mean will be a good measure of location for a *beta* posterior.

Measures of Spread

The second thing we want to know about the posterior distribution is how spread out it is. If it has large spread, then our knowledge about the parameter, even after analyzing the observed data, is still imprecise.

Posterior variance.   This is the variance of posterior distribution.

$$\mathrm{Var}[\pi|y] = \int_0^1 (\pi - m')^2 g(\pi|y)\, d\pi . \tag{8.4}$$

When we have a *beta*(a', b') posterior the posterior variance is

$$\mathrm{Var}[\pi|y] = \frac{a' \times b'}{(a' + b')^2 \times (a' + b' + 1)} . \tag{8.5}$$

The posterior variance is greatly affected for heavy-tailed distributions. For a heavy tailed distribution, the variance will be very large, yet most of the probability is very concentrated quite close the middle of the distribution. It is also in squared units, which makes it hard to interpret its size in relation to the size of the mean. We overcome these disadvantages of the posterior variance by using the posterior standard deviation.

Posterior standard deviation.   This is the square root of posterior variance. It is in terms of units, so its size can be compared to the size of the mean, and it will be less affected by heavy tails.

Percentiles of the posterior distribution.   The kth percentile of the posterior distribution is the value π_k, which has k % of the area below it. It is found numerically by solving

$$k = 100 \times \int_{-\infty}^{\pi_k} g(\pi|y)\, d\pi .$$

Some percentiles are particularly important. The first (or lower) quartile Q_1 is the 25th percentile. The second quartile, Q_2 (or median), is the 50th percentile, and the third (or upper) quartile, Q_3, is the 75th percentile.

The interquartile range.   The interquartile range

$$IQR = Q_3 - Q_1$$

is a useful measure of spread that is not affected by heavy tails.

■   **EXAMPLE 8.1   (continued)**

Anna, Bart, and Chris computed some measures of location and spread for their posterior distributions. Anna and Bart used Equations 8.3 and 8.5 to find their posterior mean and variance, respectively, since they had beta posteriors. Chris used Equations 8.2 and 8.4 to find his posterior mean and variance since his posterior did not have the beta distribution. He evaluated the integrals numerically using the Minitab macro *tintegral*. Their posterior means, medians, standard deviations, and interquartile ranges are shown in Table 8.2. We see clearly that the posterior distributions have similar summary statistics, despite the different priors used.

■

Table 8.2   Measures of location and spread of posterior distributions

Person	Posterior	Mean	Median	Std. Dev.	IQR
Anna	$beta(30.8, 93.2)$	.248	.247	.039	.053
Bart	$beta(27, 75)$	.265	.263	.043	.059
Chris	numerical	.261	.255	.041	.057

8.5   Estimating the Proportion

A point estimate $\hat{\pi}$ is a statistic calculated from the data used as an estimate of the parameter π. Suitable Bayesian point estimates are single values such as measures of location calculated from the posterior distribution. The posterior mean and posterior median are often used as point estimates.

The posterior mean square of an estimate.   The posterior mean square of an estimator $\hat{\pi}$ of the proportion π is

$$\text{PMSE}[\hat{\pi}] = \int_0^1 (\pi - \hat{\pi})^2 g(\pi|y) \, d\pi \,. \tag{8.6}$$

It measures the average squared distance (with respect to the posterior) that the estimate is away from the true value. Adding and subtracting the posterior mean m', we get

$$\text{PMSE}[\hat{\pi}] = \int_0^1 (\pi - m' + m' - \hat{\pi})^2 g(\pi|y) \, d\pi \,.$$

Multiplying out the square we get

$$\text{PMSE}[\hat{\pi}] = \int_0^1 [(\pi - m')^2 + 2(\pi - m')(m' - \hat{\pi}) + (m' - \hat{\pi})^2] g(\pi|y) \, d\pi \,.$$

We split the integral into three integrals. Since both m' and $\hat{\pi}$ are constants with respect to the posterior distribution when we evaluate the integrals, we get

$$\text{PMSE}[\hat{\pi}] = \text{Var}[\pi|y] + 0 + (m' - \hat{\pi})^2 \,. \tag{8.7}$$

This is the posterior variance of π plus the square of the distance $\hat{\pi}$ is away from the posterior mean m'.

The last term is a square and is always greater than or equal to zero. We see that on average, the squared distance the true value is away from the posterior mean m' is less than that for any other possible estimate $\hat{\pi}$, given our prior belief and the observed data. The posterior mean is the optimum estimator *post-data*. That's a good reason to use the posterior mean as the estimate, and it explains why the posterior mean is the most widely used Bayesian estimate. We will use the posterior mean as our estimate for π.

8.6   Bayesian Credible Interval

Often we wish to find a high probability interval for the parameter. A range of values that has a known high *posterior* probability, $(1 - \alpha)$, of containing the parameter is known as a Bayesian credible interval. It is sometimes called *Bayesian confidence interval.* In the next chapter we will see that credible intervals answer a more relevant question than do ordinary frequentist confidence intervals, because of the direct probability interpretation.

There are many possible intervals with same (posterior) probability. The shortest interval with given probability is preferred. It would be found by having the equal heights of the posterior density at the lower and upper endpoints, along with a total tail area of α. The upper and lower tails would not necessarily have equal tail areas. However, it is often easier to split the total tail area into equal parts and find the interval with equal tail areas.

Bayesian Credible Interval for π

If we used a $beta(a, b)$ prior, the posterior distribution of $\pi|y$ is $beta(a', b')$. An equal tail area 95% Bayesian credible interval for π can be found by obtaining the difference between the 97.5$^\text{th}$ and the 2.5$^\text{th}$ percentiles. Using Minitab, pull down *Calc* menu to *Probability Distributions* over to *Beta. . .* and fill in the dialog box. Without Minitab, we approximate the $beta(a', b')$ posterior distribution by the normal distribution having the same mean and variance:

$$(\pi|y) \text{ is approximately } N[m'; (s')^2]$$

where the posterior mean is given by

$$m' = \frac{a'}{a' + b'},$$

and the posterior variance is expressed as

$$(s')^2 = \frac{a'b'}{(a' + b')^2(a' + b' + 1)}.$$

The $(1 - \alpha) \times 100\%$ credible region for π is approximately

$$m' \pm z_{\frac{\alpha}{2}} \times s', \tag{8.8}$$

where $z_{\frac{\alpha}{2}}$ is the value found from the standard normal table. For a 95% credible interval, $z_{.025} = 1.96$. The approximation works very well if we have both $a' \geq 10$ and $b' \geq 10$.

EXAMPLE 8.1   (continued)

Anna, Bart, and Chris calculated 95% credible intervals for π having equal tail areas two ways: using the exact (beta) density function and using the

Table 8.3   Exact and approximate 95% credible intervals

Person	Posterior Distribution	Credible Interval Exact		Credible Interval Normal Approximation	
		Lower	Upper	Lower	Upper
Anna	$beta(30.8, 93.2)$	.177	.328	.173	.324
Bart	$beta(27, 75)$	.184	.354	.180	.350
Chris	$numerical$	.181	.340	.181	.341

normal approximation. These are shown in Table 8.3. Anna, Bart, and Chris have slightly different credible intervals because they started with different prior beliefs. But the effect of the data was much greater than the effect of their priors and they end up with very similar credible intervals. We see that in each case, the 95% credible interval for π calculated using the normal approximation is nearly identical to the corresponding exact 95% credible interval.

∎

Main Points

- The key relationship is *posterior* $\propto$ *prior* $\times$ *likelihood*. This gives us the shape of the posterior density. We must find the constant to divide this by to make it a density, e.g., integrate to 1 over its whole range.

- The constant we need is $k = \int_0^1 g(\pi) \times f(y|\pi) \, d\pi$. In general, this integral does not have a closed form, so we have to evaluate it numerically.

- If the prior is $beta(a, b)$, then the posterior is $beta(a', b')$ where the constants are updated by simple rules $a' = a + y$ (add number of successes to a) and $b' = b + n - y$ (add number of failures to b).

- The *beta* family of priors is called the *conjugate* family for *binomial* observation distribution. This means that the posterior is also a member of the same family, and it can easily be found without the need for any integration.

- It makes sense to choose a prior from the conjugate family, which makes finding the posterior easier. Find the $beta(a, b)$ prior that has mean and standard deviation that correspond to your prior belief. Then graph it to make sure that it looks similar to your belief. If so, use it. If you have no prior knowledge about π at all, you can use the *uniform* prior which gives equal weight to all values. The *uniform* is actually the $beta(1, 1)$ prior.

- If you have some prior knowledge, and you cannot find a member of the conjugate family that matches it, you can construct a discrete prior at several values over the range and interpolate between them to make the prior continuous. Of course, you may ignore the constant needed to make this a density, since any constant gets cancelled out by when you divide by $\int prior \times likelihood$ to find the exact posterior.

- The main thing is that your prior must have reasonable probability over all values that realistically are possible. If that is the case, the actual shape does not matter very much. If there is a reasonable amount of data, different people will get similar posteriors, despite starting from quite different shaped priors.

- The posterior mean is the estimate that has the smallest posterior mean square. This means that, on average (with respect to posterior), it is closer to the parameter than any other estimate. In other words, given our prior belief and the observed data, the posterior mean will be, on average, closer to the parameter than any other estimate. It is the most widely used Bayesian estimate because it is optimal *post-data*.

- A $(1 - \alpha) \times 100\%$ Bayesian credible interval is an interval that has a posterior probability of $1 - \alpha$ of containing the parameter.

- The shortest $(1 - \alpha) \times 100\%$ Bayesian credible interval would have equal posterior density heights at the lower and upper endpoints; however, the areas of the two tails would not necessarily be equal.

- Equal tail area Bayesian credible intervals are often used instead, because they are easier to find.

Exercises

8.1. In order to determine how effective a magazine is at reaching its target audience, a market research company selects a random sample of people from the target audience and interviews them. Out of the 150 people in the sample, 29 had seen the latest issue.

   (a) What is the distribution of y, the number who have seen the latest issue?

   (b) Use a uniform prior for π, the proportion of the target audience that has seen the latest issue. What is the posterior distribution of π?

8.2. A city is considering building a new museum. The local paper wishes to determine the level of support for this project, and is going to conduct a poll of city residents. Out of the sample of 120 people, 74 support the city building the museum.

(a) What is the distribution of y, the number who support the building the museum?

(b) Use a uniform prior for π, the proportion of the target audience that support the museum. What is the posterior distribution of π?

8.3. Sophie, the editor of the student newspaper, is going to conduct a survey of students to determine the level of support for the current president of the students' association. She needs to determine her prior distribution for π, the proportion of students who support the president. She decides her prior mean is .5, and her prior standard deviation is .15.

(a) Determine the $beta(a, b)$ prior that matches her prior belief.

(b) What is the equivalent sample size of her prior?

(c) Out of the 68 students that she polls, $y = 21$ support the current president. Determine her posterior distribution.

8.4. You are going to take a random sample of voters in a city in order to estimate the proportion π who support stopping the fluoridation of the municipal water supply. Before you analyze the data, you need a prior distribution for π. You decide that your prior mean is .4, and your prior standard deviation is .1.

(a) Determine the $beta(a, b)$ prior that matches your prior belief.

(b) What is the equivalent sample size of your prior?

(c) Out of the 100 city voters polled, $y = 21$ support the removal of fluoridation from the municipal water supply. Determine your posterior distribution.

8.5. In a research program on human health risk from recreational contact with water contaminated with pathogenic microbiological material, the National Institute of Water and Atmospheric Research (NIWA) instituted a study to determine the quality of New Zealand stream water at a variety of catchment types. This study is documented in McBride et al. (2002), where $n = 116$ one-liter water samples from sites identified as having a heavy environmental impact from birds (seagulls) and waterfowl. Out of these samples, $y = 17$ samples contained *Giardia* cysts.

(a) What is the distribution of y, the number of samples containing *Giardia* cysts?

(b) Let π be the true probability that a one-liter water sample from this type of site contains *Giardia* cysts. Use a $beta(1, 4)$ prior for π. Find the posterior distribution of π given y.

(c) Summarize the posterior distribution by its first two moments.

(d) Find the *normal* approximation to the posterior distribution $g(\pi|y)$.

(e) Compute a 95% credible interval for π using the normal approximation found in part (d).

8.6. The same study found that $y = 12$ out of $n = 145$ samples identified as having a heavy environmental impact from dairy farms contained *Giardia* cysts.

(a) What is the distribution of y, the number of samples containing *Giardia* cysts?

(b) Let π be the true probability that a one-liter water sample from this type of site contains *Giardia* cysts. Use a $beta(1, 4)$ prior for π. Find the posterior distribution of π given y.

(c) Summarize the posterior distribution by its first two moments.

(d) Find the *normal* approximation to the posterior distribution $g(\pi|y)$.

(e) Compute a 95% credible interval for π using the normal approximation found in part (d).

8.7. The same study found that $y = 10$ out of $n = 174$ samples identified as having a heavy environmental impact from pastoral (sheep) farms contained *Giardia* cysts.

(a) What is the distribution of y, the number of samples containing *Giardia* cysts?

(b) Let π be the true probability that a one-liter water sample from this type of site contains *Giardia* cysts. Use a $beta(1, 4)$ prior for π. Find the posterior distribution of π given y.

(c) Summarize the posterior distribution by its first two moments.

(d) Find the *normal* approximation to the posterior distribution $g(\pi|y)$.

(e) Compute a 95% credible interval for π using the normal approximation found in part (d).

8.8. The same study found that $y = 6$ out of $n = 87$ samples within municipal catchments contained *Giardia* cysts.

(a) What is the distribution of y, the number of samples containing *Giardia* cysts?

(b) Let π be the true probability that a one-liter water sample from a site within a municipal catchment contains *Giardia* cysts. Use a $beta(1, 4)$ prior for π. Find the posterior distribution of π given y.

(c) Summarize the posterior distribution by its first two moments.

(d) Find the *normal* approximation to the posterior distribution $g(\pi|y)$.

(e) Calculate a 95% credible interval for π using the normal approximation found in part (d).

Computer Exercises

8.1. We will use the Minitab macro *BinoBP* or R function `binobp` to find the posterior distribution of the binomial probability π when the observation distribution of $Y|\pi$ is *binomial*(n, π) and we have a *beta*(a, b) prior for π. The *beta* family of priors is the conjugate family for *binomial* observations. That means that if we start with one member of the family as the prior distribution, we will get another member of the family as the posterior distribution. It is especially easy, for when we start with a *beta*(a, b) prior, we get a *beta*(a', b') posterior where $a' = a + y$ and $b' = b + n - y$.

Suppose we have 15 independent trials and each trial results in one of two possible outcomes, success or failure. The probability of success remains constant for each trial. In that case, $Y|\pi$ is *binomial*$(n = 15, \pi)$. Suppose that we observed $y = 6$ successes. Let us start with a *beta*$(1, 1)$ prior.

[**Minitab:**] The details for invoking *BinoBP* are given in Appendix C. Store π, the prior $g(\pi)$, the likelihood $f(y|\pi)$, and the posterior $g(\pi|y)$ in columns c1–c4 respectively.

[**R:**] The details for using `binobp` are given in Appendix D.

(a) What are the posterior mean and standard deviation?

(b) Find a 95% credible interval for π.

8.2. Repeat part (a) with a *beta*$(2, 4)$ prior.

[**Minitab:**] Store the likelihood and posterior in columns c5 and c6 respectively.

8.3. Graph both posteriors on the same graph. What do you notice? What do you notice about the two posterior means and standard deviations? What do you notice about the two credible intervals for π?

8.4. We will use the Minitab macro *BinoGCP* or the R function `binogcp` to find the posterior distribution of the binomial probability π when the observation distribution of $Y|\pi$ is *binomial*(n, π) and we have a general continuous prior for π. Suppose the prior has the shape given by

$$g(\pi) = \begin{cases} \pi & \text{for } \pi \le .2, \\ .2 & \text{for } .2 < \pi \le .3, \\ .5 - \pi & \text{for } .3 < \pi \le .5, \\ 0 & \text{for } .5 < \pi. \end{cases}$$

Store the values of π and prior $g(\pi)$ in columns c1 and c2, respectively. Suppose out of $n = 20$ independent trials, $y = 7$ successes were observed.

(a) [**Minitab:**]   Use *BinoGCP* to determine the posterior distribution $g(\pi|y)$. Details for invoking *BinoGCP* are given in Appendix C.

[**R:**] Use `binogcp` to determine the posterior distribution $g(\pi|y)$. Details for using `binogcp` are given in Appendix D.

(b) Find a 95% credible interval for π by using *tintegral* in Minitab, or the `quantile` function in R upon the results of `binogcp`.

8.5. Repeat the previous question with a *uniform* prior for π.

8.6. Graph the two posterior distributions on the same graph. What do you notice? What do you notice about the two posterior means and standard deviations? What do you notice about the two credible intervals for π?

CHAPTER 9

COMPARING BAYESIAN AND FREQUENTIST INFERENCES FOR PROPORTION

The posterior distribution of the parameter given the data gives the complete inference from the Bayesian point of view. It summarizes our belief about the parameter after we have analyzed the data. However, from the frequentist point of view there are several different types of inference that can be made about the parameter. These include point estimation, interval estimation, and hypothesis testing. These frequentist inferences about the parameter require probabilities calculated from the sampling distribution of the data, given the fixed but unknown parameter. These probabilities are based on all possible random samples that could have occurred. These probabilities are not conditional on the actual sample that did occur!

In this chapter we will see how we can do these types of inferences using the Bayesian viewpoint. These Bayesian inferences will use probabilities calculated from the posterior distribution. That makes them conditional on the sample that actually did occur.

Introduction to Bayesian Statistics, 3rd ed.
By Bolstad, W. M. and Curran, J. M. Copyright © 2016 John Wiley & Sons, Inc.

9.1   Frequentist Interpretation of Probability and Parameters

Most statistical work is done using the frequentist paradigm. A random sample of observations is drawn from a distribution with an unknown parameter. The parameter is assumed to be a fixed but unknown constant. This does not allow any probability distribution to be associated with it. The only probability considered is the probability distribution of the random sample of size n, given the parameter. This explains how the random sample varies over all possible random samples, given the fixed but unknown parameter value. The probability is interpreted as long-run relative frequency.

Sampling Distribution of Statistic

Let $Y_1, \ldots, Y_n$ be a random sample from a distribution that depends on a parameter θ. Suppose a statistic S is calculated from the random sample. This statistic can be interpreted as a random variable, since the random sample can vary over all possible samples. Calculate the statistic for each possible random sample of size n. The distribution of these values is called the *sampling distribution of the statistic*. It explains how the statistic varies over all possible random samples of size n. Of course, the sampling distribution also depends on the unknown value of the parameter θ. We will write this sampling distribution as

$$f(s|\theta) \,.$$

However, we must remember that in frequentist statistics, the parameter θ is a fixed but unknown constant, not a random variable. The sampling distribution measures how the statistic varies over all possible samples, given the unknown fixed parameter value. This distribution does not have anything to do with the actual data that occurred. It is the distribution of values of the statistic that could have occurred, given that specific parameter value. Frequentist statistics uses the sampling distribution of the statistic to perform inference on the parameter. From a Bayesian perspective, this is a backwards form of inference.[1]

This contrasts with Bayesian statistics where the complete inference is the posterior distribution of the parameter given the actual data that occurred:

$$g(\theta|data) \,.$$

Any subsequent Bayesian inference such as a Bayesian estimate or a Bayesian credible interval is calculated from the posterior distribution. Thus the es-

---

[1] Frequentist statistics performs inferences in the parameter space, which is the unobservable dimension of the Bayesian universe, based on a probability distribution in the sample space, which is the observable dimension.

timate or the credible interval depends on the data that actually occurred. Bayesian inference is straightforward.[2]

9.2   Point Estimation

The first type of inference we consider is point estimation, where a single statistic is calculated from the sample data and used to estimate the unknown parameter. The statistic depends on the random sample, so it is a random variable, and its distribution is its sampling distribution. If its sampling distribution is centered close to the true but unknown parameter value θ, and the sampling distribution does not have much spread, the statistic could be used to estimate the parameter. We would call the statistic an *estimator* of the parameter and the value it takes for the actual sample data an *estimate*. There are several theoretical approaches for finding frequentist estimators, such as maximum likelihood estimation (MLE)[3] and uniformly minimum variance unbiased estimation (UMVUE). We will not go into them here. Instead, we will use the sample statistic that corresponds to the population parameter we wish to estimate, such as the sample proportion as the frequentist estimator for the population proportion. This turns out to be the same estimator that would be found using either of the main theoretical approaches (MLE and UMVUE) for estimating the binomial parameter π.

From a Bayesian perspective, point estimation means that we would use a single statistic to summarize the posterior distribution. The most important number summarizing a distribution would be its location. The posterior mean or the posterior median would be good candidates here. We will use the posterior mean as the Bayesian estimate because it minimizes the posterior mean squared error, as we saw in the previous chapter. This means it will be the optimal estimator, given our prior belief and this sample data (i.e., *post-data*).

Frequentist Criteria for Evaluating Estimators

We do not know the true value of the parameter, so we cannot judge an estimator from the value it gives for the random sample. Instead, we will use a criterion based on the sampling distribution of the estimator that is the distribution of the estimator over all possible random samples. We compare possible estimators by looking at how concentrated their sampling distributions are around the parameter value for a range of fixed possible values. When we use the sampling distribution, we are still thinking of the estimator

---

[2]Bayesian statistics performs inference in the parameter space based on a probability distribution in the parameter space.

[3]Maximum likelihood estimation was pioneered by R. A. Fisher.

as a random variable because we have not yet obtained the sample data and calculated the estimate. This is a *pre-data* analysis.

Although this "what if the parameter has this value" type of analysis comes from a frequentist point of view, it can be used to evaluate Bayesian estimators as well. It can be done before we obtain the data, and in Bayesian statistics it is called a *pre-posterior* analysis. The procedure is used to evaluate how the estimator performs over all possible random samples, given that parameter value. We often find that Bayesian estimators perform very well when evaluated this way, sometimes even better than frequentist estimators.

Unbiased Estimators

The expected value of an estimator is a measure of the center of its distribution. This is the average value that the estimator would have averaged over all possible samples. An estimator is said to be *unbiased* if the mean of its sampling distribution is the true parameter value. That is, an estimator $\hat{\theta}$ is unbiased if and only if

$$\mathrm{E}[\hat{\theta}] = \int \hat{\theta} f(\hat{\theta}|\theta) \, d\hat{\theta} = \theta \,,$$

where $f(\hat{\theta}|\theta)$ is the sampling distribution of the estimator $\hat{\theta}$ given the parameter θ. Frequentist statistics emphasizes unbiased estimators because averaged over all possible random samples, an unbiased estimator gives the true value. The bias of an estimator $\hat{\theta}$ is the difference between its expected value and the true parameter value.

$$\mathrm{Bias}[\hat{\theta}, \theta] = \mathrm{E}[\hat{\theta}] - \theta \,. \tag{9.1}$$

Unbiased estimators have bias equal to zero.

In contrast, Bayesian statistics does not place any emphasis on being unbiased. In fact, Bayesian estimators are usually biased.

Minimum Variance Unbiased Estimator

An estimator is said to be a minimum variance unbiased estimator if no other unbiased estimator has a smaller variance. Minimum variance unbiased estimators are often considered the *best* estimators in frequentist statistics. The sampling distribution of a minimum variance unbiased estimator has the smallest spread (as measured by the variance) of all sampling distributions that have mean equal to the parameter value.

However, it is possible that there may be biased estimators that, on average, are closer to the true value than the best unbiased estimator. We need to look at a possible trade-off between bias and variance. Figure 9.1 shows the sampling distributions of three possible estimators of θ. Estimator 1 and estimator 2 are seen to be unbiased estimators. Estimator 1 is the *best unbiased*

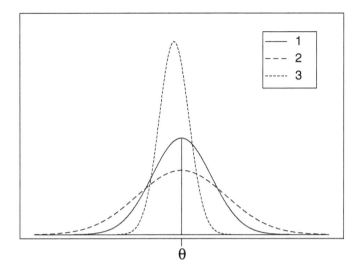

Figure 9.1   Sampling distributions of three estimators.

estimator, since it has the smallest variance among the unbiased estimators. Estimator 3 is seen to be a biased estimator, but it has a smaller variance than estimator 1. We need some way of comparison that includes biased estimators, to find which one will be closest, on average, to the parameter value.

Mean Squared Error of an Estimator

The (frequentist) mean squared error of an estimator $\hat{\theta}$ is the average squared distance the estimator is away from the true value:

$$\mathrm{MSE}[\hat{\theta}] = \mathrm{E}[\hat{\theta} - \theta]^2$$
$$= \int (\hat{\theta} - \theta)^2 \, f(\hat{\theta}|\theta) \, d\hat{\theta}. \tag{9.2}$$

The frequentist mean squared error is calculated from the sampling distribution of the estimator, which means the averaging is over all possible samples given that fixed parameter value. It is *not* the posterior mean square calculated from the posterior distribution that we introduced in the previous chapter. It turns out that the mean squared error of an estimator is the square of the bias plus the variance of the estimator:

$$\mathrm{MSE}[\hat{\theta}] = \mathrm{Bias}[\hat{\theta}, \theta]^2 + \mathrm{Var}[\hat{\theta}]. \tag{9.3}$$

Thus it gives a better frequentist criterion for judging estimators than the bias or the variance alone. An estimator that has a smaller mean squared error is closer to the true value averaged over all possible samples.

9.3   Comparing Estimators for Proportion

Bayesian estimators often have smaller mean squared errors than frequentist estimators. In other words, on average, they are closer to the true value. Thus Bayesian estimators can be better than frequentist estimators, even when judged by the frequentist criterion of mean squared error. The frequentist estimator for π is

$$\hat{\pi}_f = \frac{y}{n},$$

where y, the number of successes in the n trials, has the binomial (n, π) distribution. $\hat{\pi}_f$ is unbiased, and $\text{Var}[\hat{\pi}_f] = \frac{\pi \times (1-\pi)}{n}$. Hence the mean squared error of $\hat{\pi}_f$ equals

$$\text{MSE}[\hat{\pi}_f] = 0^2 + \text{Var}[\hat{\pi}_f]$$
$$= \frac{\pi \times (1-\pi)}{n}.$$

Suppose we use the posterior mean as the Bayesian estimate for π, where we use the Beta(1,1) prior (uniform prior). The estimator is the posterior mean, so

$$\hat{\pi}_B = m' = \frac{a'}{a' + b'},$$

where $a' = 1 + y$ and $b' = 1 + n - y$. We can rewrite this as a linear function of y, the number of successes in the n trials:

$$\hat{\pi}_B = \frac{y + 1}{n + 2} = \frac{y}{n + 2} + \frac{1}{n + 2}.$$

Thus, the mean of its sampling distribution is

$$\frac{n\pi}{n + 2} + \frac{1}{n + 2},$$

and the variance of its sampling distribution is

$$\left[\frac{1}{n + 2} \right]^2 \times n\pi(1 - \pi).$$

Hence from Equation 9.3, the mean squared error is

$$\text{MSE}[\hat{\pi}_B] = \left[\frac{n\pi}{n + 2} + \frac{1}{n + 2} - \pi \right]^2 + \left[\frac{1}{n + 2} \right]^2 \times n\pi(1 - \pi)$$
$$= \left[\frac{1 - 2\pi}{n + 2} \right]^2 + \left[\frac{1}{n + 2} \right]^2 \times n\pi(1 - \pi).$$

For example, suppose $\pi = .4$ and the sample size is $n = 10$. Then

$$\text{MSE}[\hat{\pi}_f] = \frac{.4 \times .6}{10}$$
$$= .024$$

and

$$\text{MSE}[\hat{\pi}_B] = \left[\frac{1 - 2 \times .4}{12}\right]^2 + \left[\frac{1}{12}\right]^2 \times 10 \times .4 \times .6$$
$$= .0169 .$$

Next, suppose $\pi = .5$ and $n = 10$. Then

$$\text{MSE}[\hat{\pi}_f] = \frac{.5 \times .5}{10}$$
$$= .025$$

and

$$\text{MSE}[\hat{\pi}_B] = \left[\frac{1 - 2 \times .5}{12}\right]^2 + \left[\frac{1}{12}\right]^2 \times 10 \times .5 \times .5$$
$$= .01736 .$$

We see that, on average (for these two values of π), the Bayesian posterior estimator is closer to the true value than the frequentist estimator. Figure 9.2 shows the mean squared error for the Bayesian estimator and the frequentist estimator as a function of π. We see that over most (but not all) of the range, the Bayesian estimator (using uniform prior) is better than the frequentist estimator.[4]

9.4 Interval Estimation

The second type of inference we consider is interval estimation. We wish to find an interval (l, u) that has a predetermined probability of containing the parameter. In the frequentist interpretation, the parameter is fixed but unknown; and before the sample is taken, the interval endpoints are random because they depend on the data. After the sample is taken and the endpoints are calculated, there is nothing random, so the interval is said to be a *confidence interval* for the parameter. We know that a predetermined proportion of intervals calculated for random samples using this method will contain the true parameter. But it does not say anything at all about the specific interval we calculate from our data.

In Chapter 8, we found a *Bayesian credible interval* for the parameter π that has the probability that we want. Because it is found from the posterior distribution, it has the coverage probability we want for this specific data.

---

[4]The frequentist estimator, $\hat{\pi}_f = \frac{y}{n}$, would be Bayesian posterior mean if we used the prior $g(\pi) \propto \pi^{-1}(1 - \pi)^{-1}$. This prior is improper since it does not integrate to 1. An estimator is said to be admissible if no other estimator has smaller mean squared error over the whole range of possible values. Wald (1950) showed that Bayesian posterior mean estimators that arose from proper priors are always admissible. Bayesian posterior mean estimators from improper priors sometimes are admissible, as in this case.

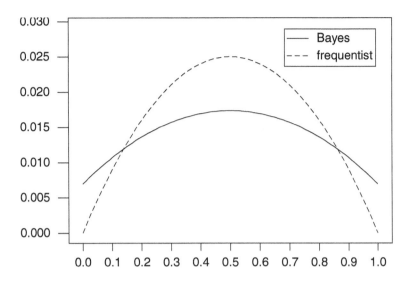

Figure 9.2   Mean squared error for the two estimators.

Confidence Intervals

Confidence intervals are how frequentist statistics tries to find an interval that has a high probability of containing the true value of the parameter θ. A $(1 - \alpha) \times 100\%$ confidence interval for a parameter θ is an interval (l, u) such that

$$P(l \leq \theta \leq u) = 1 - \alpha .$$

This probability is found using the sampling distribution of an estimator for the parameter. There are many possible values of l and u that satisfy this. The most commonly used criteria for choosing them are (1) equal ordinates (heights) on the sampling distribution and (2) equal tail area on the sampling distribution. Equal ordinates will find the shortest confidence interval. However, the equal tail area intervals are often used because they are easier to find. When the sampling distribution of the estimator is symmetric, the two criteria will coincide.

The parameter θ is regarded as a fixed but unknown constant. The endpoints l and u are random variables since they depend on the random sample. When we plug in the actual sample data that occurred for our random sample and calculate the values for l and u, there is nothing left that is random. The interval either contains the unknown fixed parameter or it does not, and we do not know which is true. The interval can no longer be regarded as a probability interval.

Under the frequentist paradigm, the correct interpretation is that $(1 - \alpha) \times 100\%$ of the *random* intervals calculated this way will contain the true value. Therefore we have $(1 - \alpha) \times 100\%$ *confidence* that our interval does. It is a misinterpretation to make a probability statement about the parameter θ from the calculated confidence interval.

Often, the sampling distribution of the estimator used is approximately normal, with mean equal to the true value. In this case, the confidence interval has the form

estimator $\pm$ *critical value* $\times$ *standard deviation of the estimator,*

where the critical value comes from the *standard normal table*. For example, if n is large, then the sample proportion

$$\hat{\pi}_f = \frac{y}{n}$$

is approximately normal with mean π and standard deviation $\sqrt{\frac{\pi(1-\pi)}{n}}$. This gives an approximate $(1-\alpha) \times 100\%$ equal tail area confidence interval for π:

$$\hat{\pi}_f \pm z_{\frac{\alpha}{2}} \times \sqrt{\frac{\hat{\pi}_f(1 - \hat{\pi}_f)}{n}}. \tag{9.4}$$

Comparing Confidence and Credible Intervals for π

The probability calculations for the confidence interval are based on the sampling distribution of the statistic. In other words, how it varies over all possible samples. Hence the probabilities are *pre-data*. They do not depend on the particular sample that occurred. This is in contrast to the Bayesian credible interval calculated from the posterior distribution that has a direct (degree of belief) probability interpretation conditional on the observed sample data. The Bayesian credible interval is more useful to the scientist whose data we are analyzing. It summarizes our beliefs about the parameter values that could credibly be believed given the observed data that occurred. In other words, it is *post-data*. He/she is not concerned about data that could have occurred but did not.

■   **EXAMPLE 9.1   (continued from Chapter 8, p. 162)**

Out of a random sample of $n = 100$ Hamilton residents, $y = 26$ said they support building a casino in Hamilton. A frequentist 95% confidence interval for π is

$$.26 \pm 1.96\sqrt{\frac{.26 \times .74}{100}} = (.174, .346).$$

Compare this with the 95% credible intervals for π calculated by the three students in Chapter 8 and shown in Table 8.3.   ■

9.5   Hypothesis Testing

The third type of inference we consider is hypothesis testing. Scientists do not like to claim the existence of an effect where the discrepancy in the data could be due to chance alone. If they make their claims too quickly, later studies would show their claim was wrong, and their scientific reputation would suffer.

Hypothesis testing, sometimes called significance testing[5], is the frequentist statistical method widely used by scientists to guard against making claims unjustified by the data. The nonexistence of the treatment effect is set up as the *null hypothesis* that "the shift in the parameter value caused by the treatment is zero." The competing hypothesis that there is a nonzero shift in the parameter value caused by the treatment is called the *alternative hypothesis*. Two possible explanations for the discrepancy between the observed data and what would be expected under the null hypothesis are proposed.

(1) The null hypothesis is true, and the discrepancy is due to random chance alone.

(2) The null hypothesis is false. This causes at least part of the discrepancy.

To be consistent with Ockham's razor, we will stick with explanation (1), which has the null hypothesis being true and the discrepancy being due to chance alone, unless the discrepancy is so large that it is very unlikely to be due to chance alone. This means that when we accept the null hypothesis as true, it does not mean that we believe it is literally true. Rather, it means that chance alone remains a reasonable explanation for the observed discrepancy, so we cannot discard chance as the sole explanation.

When the discrepancy is too large, we are forced to discard explanation (1) leaving us with explanation (2), that the null hypothesis is false. This gives us a backward way to establish the existence of an effect. We conclude the effect exists (the null hypothesis is false) whenever the probability of the discrepancy between what occurred and what would be expected under the null hypothesis is too small to be attributed to chance alone.

Because hypothesis testing is very well established in science, we will show how it can be done in a Bayesian manner. There are two situations we will look at. The first is testing a one-sided hypothesis where we are only interested in detecting the effect in one direction. We will see that in this case, Bayesian hypothesis testing works extremely well, without the contradictions required in frequentist tests. The Bayesian test of a one-sided null hypothesis is evaluated from the posterior probability of the null hypothesis.

---

[5]Significance testing was developed by R. A. Fisher as an inferential tool to weigh the evidence against a particular hypothesis. Hypothesis testing was developed by Neyman and Pearson as a method to control the error rate in deciding between two competing hypotheses. These days, the two terms are used almost interchangeably, despite their differing goals and interpretations. This continues to cause confusion.

The second situation is where we want to detect a shift in either direction. This is a two-sided hypothesis test, where we test a point hypothesis (that the effect is zero) against a two-sided alternative. The prior density of a continuous parameter measures probability density, not probability. The prior probability of the null hypothesis (shift equal to zero) must be equal to 0. So its posterior probability must also be zero,[6] and we cannot test a two-sided hypothesis using the posterior probability of the null hypothesis. Rather, we will test the *credibility* of the null hypothesis by seeing if the null value lies in the credible interval. If the null value does lie within the credible interval, we cannot reject the null hypothesis, because the null value remains a credible value.

9.6   Testing a One-Sided Hypothesis

The effect of the treatment is included as a parameter in the model. The hypothesis that the treatment has no effect becomes the *null hypothesis* the parameter representing the treatment effect has the *null* value that corresponds to no effect of the treatment.

Frequentist Test of One-Sided Hypothesis

The probability of the data (or results even more extreme) given that the null hypothesis is true is calculated. If this is below a threshold called the *level of significance*, the results are deemed to be incompatible with the null hypothesis, and the null hypothesis is rejected at that level of significance. This establishes the existence of the treatment effect. This is similar to a "proof by contradiction." However, because of sampling variation, complete contradiction is impossible. Even very unlikely data are possible when there is no treatment effect. So hypothesis tests are actually more like "proof by low probability." The probability is calculated from the sampling distribution, given that the null hypothesis is true. This makes it a *pre-data* probability.

🖳   **EXAMPLE 9.2**

Suppose we wish to determine if a new treatment is better than the standard treatment. If so, π, the proportion of patients who benefit from the new treatment, should be better than π_0, the proportion who benefit from the standard treatment. It is known from historical records that

---

[6]We are also warned that frequentist hypothesis tests of a point null hypothesis never "accept" the null hypothesis; rather, they "cannot reject the null hypothesis."

Table 9.1   Null distribution of Y with a rejection region for a one-sided hypothesis test

Value	$f(y|\pi = .6)$	Region
0	.0001	accept
1	.0016	accept
2	.0106	accept
3	.0425	accept
4	.1115	accept
5	.2007	accept
6	.2508	accept
7	.2150	accept
8	.1209	accept
9	.0403	reject
10	.0060	reject

$\pi_0 = .6$. A random group of 10 patients are given the new treatment. Y, the number who benefit from the treatment will be $binomial(n, \pi)$. We observe $y = 8$ patients that benefit. This is better than we would expect if $\pi = .6$. But, is it enough better for us to conclude that $\pi > .6$ at the 10% level of significance?

The steps are:

1. Set up a null hypothesis about the (fixed but unknown) parameter. For example, $H_0 : \pi \leq .6$. (The proportion who would benefit from the new treatment is less than or equal to the proportion who benefit from the standard treatment.) We include all π values less than the null value .6 in with the null hypothesis because we are trying to determine if the new treatment is better. We have no interest in determining if the new treatment is worse. We will not recommend it unless it is demonstrably better than the standard treatment.

2. The alternative hypothesis is $H_1 : \pi > .6$. (The proportion who would benefit from the new treatment is greater than the proportion who benefit from the standard treatment.)

3. The null distribution of the test statistic is the sampling distribution of the test statistic, given that the null hypothesis is true. In this case, it will be $binomial(n, .6)$ where $n = 10$ is the number of patients given the new treatment.

4. We choose level of significance for the test to be as close as possible to $\alpha = 5\%$. Since y has a discrete distribution, only some values of α are possible, so we will have to choose a value either just above or just below 5%.

5. The rejection region is chosen so that it has a probability of α under the null distribution.[7] If we choose the rejection region $y \geq 9$, then $\alpha = .0463$. The null distribution with the rejection region for the one-sided hypothesis test is shown in Table 9.1.

6. If the value of the test statistic for the given sample lies in the rejection region, then reject the null hypothesis H_0 at level α. Otherwise, we cannot reject H_0. In this case, $y = 8$ was observed. This lies in the acceptance region.

7. The P-value is the probability of getting what we observed, or something even more unlikely, given the null hypothesis is true. The P-value is put forward as measuring the strength of evidence against the null hypothesis.[8] In this case, the P-value $= .1672$.

8. If the P-value $< \alpha$, the test statistic lies in the rejection region, and vice versa. So an equivalent way of testing the hypothesis is to reject if P-value $< \alpha$.[9] Looking at it either way, we cannot reject the null hypothesis $H_0 : \pi \leq .6$. $y = .8$ lies in the acceptance region, and the p-value $> .05$. The evidence is not strong enough to conclude that $\pi > .6$.

∎

There is much confusion about the P-value of a test. It is *not* the posterior probability of the null hypothesis being true given the data. Instead, it is the tail probability calculated using the null distribution. In the binomial case

$$P\text{-value} = \sum_{y_{obs}}^{n} f(y|\pi_0),$$

where y_{obs} is the observed value of y. Frequentist hypothesis tests use a probability calculated on all possible data sets that could have occurred (for the fixed parameter value), but the hypothesis is about the parameter value being in some range of values.

---

[7] This approach is from Neyman and Pearson.

[8] This approach is from R. A. Fisher.

[9] Both α and P-value are tail areas calculated from the null distribution. However, α represents the long-run rate of rejecting a true null hypothesis, and P-value is looked at as the evidence against *this particular null hypothesis by this particular data set*. Using tail areas as simultaneously representing both the long-run and a particular result is inherently contradictory.

Bayesian Tests of a One-Sided Hypothesis

We wish to test

$$H_0 : \pi \leq \pi_0 \quad \text{versus} \quad H_1 : \pi > \pi_0$$

at the level of significance α using Bayesian methods. We can calculate the posterior probability of the null hypothesis being true by integrating the posterior density over the correct region:

$$P(H_0 : \pi \leq \pi_0 | y) = \int_0^{\pi_0} g(\pi | y) \, d\pi \, . \tag{9.5}$$

We reject the null hypothesis if that posterior probability is less than the level of significance α. Thus a Bayesian one-sided hypothesis test is a "test by low probability" using the probability calculated directly from the posterior distribution of π. We are testing a hypothesis about the parameter using the posterior distribution of the parameter. Bayesian one-sided tests use *post-data* probability.

◼ **EXAMPLE 9.2** **(continued)**

Suppose we use a $beta(1,1)$ prior for π. Then given $y = 8$, the posterior density is $beta(9,3)$. The posterior probability of the null hypothesis is

$$P(\pi \leq .6 | y = 8) = \int_0^{.6} \frac{\Gamma(12)}{\Gamma(3)\Gamma(9)} \pi^8 (1 - \pi)^2 d\pi$$
$$= .1189$$

when we evaluate it numerically. This is not less than .05, so we cannot reject the null hypothesis at the 5% level of significance. Figure 9.3 shows the posterior density. The probability of the null hypothesis is the area under the curve to the left of $\pi = .6$.

◼

9.7 Testing a Two-Sided Hypothesis

Sometimes we might want to detect a change in the parameter value in either direction. This is known as a two-sided test since we are wanting to detect any changes from the value π_0. We set this up as testing the point null hypothesis $H_0 : \pi = \pi_0$ against the alternative hypothesis $H_1 : \pi \neq \pi_0$.

Frequentist Test of a Two-Sided Hypothesis

The null distribution is evaluated at π_0, and the rejection region is two-sided, as are *p-values* calculated for this test.

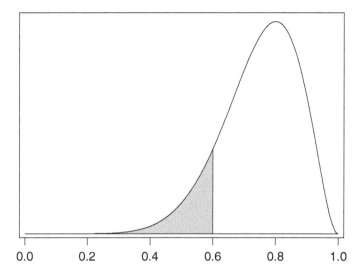

Figure 9.3   Posterior probability of the null hypothesis, $H_0 : \pi \leq .6$ is the shaded area.

■ **EXAMPLE 9.3**

A coin is tossed 15 times, and we observe 10 heads. Are 10 heads out of 15 tosses enough to determine that the coin is not fair? In other words, is π the probability of getting a head different than $\frac{1}{2}$?

The steps are:

1. Set up the null hypothesis about the fixed but unknown parameter π. It is $H_0 : \pi = .5$.

2. The alternative hypothesis is $H_1 : \pi \neq .5$. We are interested in determining a difference in either direction, so we will have a two-sided rejection region.

3. The null distribution is the sampling distribution of Y when the null hypothesis is true. It is $binomial(n = 15, \pi = .5)$.

4. Since Y has a discrete distribution, we choose the level of significance for the test to be as close to 5% as possible.

5. The rejection region is chosen so that it has a probability of α under the null distribution. If we choose rejection region $\{Y \leq 3\} \cup \{Y \geq 12\}$, then $\alpha = .0352$. The null distribution and rejection region for the two-sided hypothesis are shown in Table 9.2.

6. If the value of the test statistic lies in the rejection region, then we reject the null hypothesis H_0 at level α. Otherwise, we cannot reject H_0. In this case, $y = 10$ was observed. This lies in the region where we cannot reject

Table 9.2   Null distribution of Y with the rejection region for two-sided hypothesis test

Value	$f(y|\pi = .5)$	Region
0	.0000	reject
1	.0005	reject
2	.0032	reject
3	.0139	reject
4	.0417	accept
5	.0916	accept
6	.1527	accept
7	.1964	accept
8	.1964	accept
9	.1527	accept
10	.0916	accept
11	.0417	accept
12	.0139	reject
13	.0032	reject
14	.0005	reject
15	.0000	reject

the null hypothesis. We must conclude that chance alone is sufficient to explain the discrepancy, so $\pi = .5$ remains a reasonable possibility.

7. The P-value is the probability of getting what we got (10) or something more unlikely, given the null hypothesis H_0 is true. In this case we have a two-sided alternative, so the p-value is the $P(Y \geq 10) + P(Y \leq 5) = .302$. This is larger than α, so we cannot reject the null hypothesis.

∎

Relationship between two-sided hypothesis tests and confidence intervals. While the null value of the parameter usually comes from the idea of no treatment effect, it is possible to test other parameter values. There is a close relationship between two-sided hypothesis tests and confidence intervals. If you are testing a two-sided hypothesis at level α, there is a corresponding $(1 - \alpha) \times 100\%$ confidence interval for the parameter. If the null hypothesis

$$H_0 : \pi = \pi_0$$

is rejected, then the value π_0 lies outside the confidence interval, and vice versa. If the null hypothesis is accepted (cannot be rejected), then π_0 lies inside the confidence interval, and vice versa. The confidence interval "summarizes" all possible null hypotheses that would be accepted if they were tested.

Bayesian Test of a Two-Sided Hypothesis

From the Bayesian perspective, the posterior distribution of the parameter given the data sums up our entire belief after the data. However, the idea of hypothesis testing as a protector of scientific credibility is well established in science. So we look at using the posterior distribution to test a point null hypothesis versus a two-sided alternative in a Bayesian way.

If we use a continuous prior, we will get a continuous posterior. The probability of the exact value represented by the point null hypothesis will be zero. We cannot use posterior probability to test the hypothesis. Instead, we use a correspondence similar to the one between confidence intervals and hypothesis tests, but with credible interval instead.

Compute a $(1 - \alpha) \times 100\%$ credible interval for π. If π_0 lies inside the credible interval, accept (do not reject) the null hypothesis $H_0 : \pi = \pi_0$; and if π_0 lies outside the credible interval, then reject the null hypothesis.

◨ **EXAMPLE 9.3 (continued)**

If we use a uniform prior distribution, then the posterior is the $beta(10 + 1, 5 + 1)$ distribution. A 95% Bayesian credible interval for π found using the normal approximation is

$$\frac{11}{17} + 1.96 \times \sqrt{\frac{11 \times 6}{((11+6)^2 \times (11+6+1))}} = .647 \pm .221$$

$$= (.426, .868) \, .$$

The null value $\pi = .5$ lies within the credible interval, so we cannot reject the null hypothesis. It remains a credible value. ∎

Main Points

- The posterior distribution of the parameter given the data is the entire inference from a Bayesian perspective. Probabilities calculated from the posterior distribution are *post-data* because the posterior distribution is found after the observed data has been taken into the analysis.

- Under the frequentist perspective there are specific inferences about the parameter: point estimation, confidence intervals, and hypothesis tests.

- Frequentist statistics considers the parameter a fixed but unknown constant. The only kind of probability allowed is long-run relative frequency.

- The sampling distribution of a statistic is its distribution over all possible random samples given the fixed parameter value. Frequentist statistics is based on the sampling distribution.

- Probabilities calculated using the sampling distribution are *pre-data* because they are based on all possible random samples, not the specific random sample we obtained.

- An estimator of a parameter is unbiased if its expected value calculated from the sampling distribution is the true value of the parameter.

- Frequentist statistics often call the minimum variance unbiased estimator the *best* estimator.

- The mean squared error of an estimator measures its average squared distance from the true parameter value. It is the square of the bias plus the variance.

- Bayesian estimators are often better than frequentist estimators even when judged by the frequentist criteria such as mean squared error.

- Seeing how a Bayesian estimator performs using frequentist criteria for a range of possible parameter values is called a *pre-posterior* analysis, because it can be done before we obtain the data.

- A $(1 - \alpha) \times 100\%$ confidence interval for a parameter θ is an interval (l, u) such that
$$P(l \leq \theta \leq u) = 1 - \alpha,$$
where the probability is found using the sampling distribution of an estimator for θ. The correct interpretation is that $(1 - \alpha) \times 100\%$ of the random intervals calculated this way do contain the true value. When the actual data are put in and the endpoints calculated, there is nothing left to be random. The endpoints are numbers; the parameter is fixed but unknown. We say that we are $(1 - \alpha) \times 100\%$ *confident* that the calculated interval covers the true parameter. The confidence comes from our belief in the method used to calculate the interval. It does not say anything about the actual interval we got for that particular data set.

- A $(1 - \alpha) \times 100\%$ Bayesian credible interval for θ is a range of parameter values that has posterior probability $(1 - \alpha)$.

- Frequentist hypothesis testing is used to determine whether the actual parameter could be a specific value. The sample space is divided into a rejection region and an acceptance region such that the probability the test statistic lies in the rejection region if the null hypothesis is true is less than the level of significance α. If the test statistic falls into the rejection region, we reject the null hypothesis at level of significance α.

- Or we could calculate the P-value. If the P-value$< \alpha$, we reject the null hypothesis at level α.

- The P-value is not the probability the null hypothesis is true. Rather, it is the probability of observing what we observed, or even something more extreme, given that the null hypothesis is true.

- We can test a one-sided hypothesis in a Bayesian manner by computing the posterior probability of the null hypothesis. This probability is found by integrating the posterior density over the null region. If this probability is less than the level of significance α, then we reject the null hypothesis.

- We cannot test a two-sided hypothesis by integrating the posterior probability over the null region because, with a continuous prior, the prior probability of a point null hypothesis is zero, so the posterior probability will also be zero. Instead, we test the credibility of the null value by observing whether or not it lies within the Bayesian credible interval. If it does, the null value remains credible and we cannot reject it.

Exercises

9.1. Let π be the proportion of students at a university who approve the government's policy on students' allowances. The students' newspaper is going to take a random sample of $n = 30$ students at a university and ask if they approve of the governments policy on student allowances.

(a) What is the distribution of y, the number who answer "yes"?

(b) Suppose 8 out of the 30 students answered yes. What is the *frequentist* estimate of π.

(c) Find the posterior distribution $g(\pi|y)$ if we use a uniform prior.

(d) What would be the Bayesian estimate of π?

9.2. The standard method of screening for a disease fails to detect the presence of the disease in 15% of the patients who actually do have the disease. A new method of screening for the presence of the disease has been developed. A random sample of $n = 75$ patients who are known to have the disease is screened using the new method. Let π be the probability the new screening method fails to detect the disease.

(a) What is the distribution of y, the number of times the new screening method fails to detect the disease?

(b) Of these $n = 75$ patients, the new method failed to detect the disease in $y = 6$ cases. What is the frequentist estimator of π?

(c) Use a $beta(1,6)$ prior for π. Find $g(\pi|y)$, the posterior distribution of π.

(d) Find the posterior mean and variance.

(e) If $\pi \geq .15$, then the new screening method is no better than the standard method. Test

$$H_0 : \pi \geq .15 \quad \text{versus} \quad H_1 : \pi < .15$$

at the 5% level of significance in a Bayesian manner.

9.3. In the study of water quality in New Zealand streams, documented in McBride et al. (2002), a high level of *Campylobacter* was defined as a level greater than 100 per 100 ml of stream water. $n = 116$ samples were taken from streams having a high environmental impact from birds. Out of these, $y = 11$ had a high *Campylobacter* level. Let π be the true probability that a sample of water from this type of stream has a high *Campylobacter* level.

(a) Find the frequentist estimator for π.

(b) Use a *beta*$(1, 10)$ prior for π. Calculate the posterior distribution $g(\pi|y)$.

(c) Find the posterior mean and variance. What is the Bayesian estimator for π?

(d) Find a 95% credible interval for π.

(e) Test the hypothesis

$$H_0 : \pi = .10 \quad \text{versus} \quad H_1 : \pi \neq .10$$

at the 5% level of significance.

9.4. In the same study of water quality, $n = 145$ samples were taken from streams having a high environmental impact from dairying. Out of these $y = 9$ had a high *Campylobacter* level. Let π be the true probability that a sample of water from this type of stream has a high *Campylobacter* level.

(a) Find the frequentist estimator for π.

(b) Use a *beta*$(1, 10)$ prior for π. Calculate the posterior distribution $g(\pi|y)$.

(c) Find the posterior mean and variance. What is the Bayesian estimator for π?

(d) Find a 95% credible interval for π.

(e) Test the hypothesis

$$H_0 : \pi = .10 \qquad \text{versus} \qquad H_1 : \pi \neq .10$$

at the 5% level of significance.

9.5. In the same study of water quality, $n = 176$ samples were taken from streams having a high environmental impact from sheep farming. Out of these $y = 24$ had a high *Campylobacter* level. Let π be the true probability that a sample of water from this type of stream has a high *Campylobacter* level.

(a) Find the frequentist estimator for π.

(b) Use a $beta(1, 10)$ prior for π. Calculate the posterior distribution $g(\pi|y)$.

(c) Find the posterior mean and variance. What is the Bayesian estimator for π?

(d) Test the hypothesis

$$H_0 : \pi \geq .15 \quad \text{versus} \quad H_1 : \pi < .15$$

at the 5% level of significance.

9.6. In the same study of water quality, $n = 87$ samples were taken from streams in municipal catchments. Out of these $y = 8$ had a high *Campylobacter* level. Let π be the true probability that a sample of water from this type of stream has a high *Campylobacter* level.

(a) Find the frequentist estimator for π.

(b) Use a $beta(1, 10)$ prior for π. Calculate the posterior distribution $g(\pi|y)$.

(c) Find the posterior mean and variance. What is the Bayesian estimator for π?

(d) Test the hypothesis

$$H_0 : \pi \geq .10 \quad \text{versus} \quad H_1 : \pi < .10$$

at the 5% level of significance.

Monte Carlo Exercises

9.1. **Comparing Bayesian and frequentist estimators for π.** In Chapter 1 we learned that the frequentist procedure for evaluating a statistical procedure, namely looking at how it performs in the long-run, for a (range of) fixed but unknown parameter values can also be used to evaluate a Bayesian statistical procedure. This "what if the parameter has this value " type of analysis would be done before we obtained the data and is called a *pre-posterior* analysis. It evaluates the procedure by seeing how it performs over all possible random samples, given that parameter value. In Chapter 8 we found that the posterior mean used as a Bayesian estimator minimizes the posterior mean squared error. Thus it has optimal post-data properties, in other words after making use of the actual data. We will see that Bayesian estimators have excellent pre-data (frequentist) properties as well, often better than the corresponding frequentist estimators.

We will perform a Monte Carlo study approximating the sampling distribution of two estimators of π. The frequentist estimator we will use is $\hat{\pi}_f = \frac{y}{n}$, the sample proportion. The Bayesian estimator we will use is $\hat{\pi}_B = \frac{y+1}{n+2}$, which equals the posterior mean when we used a uniform prior for π. We will compare the sampling distributions (in terms of bias, variance, and mean squared error) of the two estimators over a range of π values from 0 to 1. However, unlike the exact analysis we did in Section 9.3, here we will do a Monte Carlo study. For each of the parameter values, we will approximate the sampling distribution of the estimator by an empirical distribution based on 5,000 samples drawn when that is the parameter value. The true characteristics of the sampling distribution (mean, variance, mean squared error) are approximated by the sample equivalent from the empirical distribution. You can use either Minitab or R for your analysis.

(a) For $\pi = .1, .2, \ldots, .9$

     i. Draw 5,000 random samples from $binomial(n = 10, \pi)$.

     ii. Calculate the frequentist estimator $\hat{\pi}_f = \frac{y}{n}$ for each of the 5,000 samples.

     iii. Calculate the Bayesian estimator $\hat{\pi}_B = \frac{y+1}{n+2}$ for each of the 5,000 samples.

     iv. Calculate the means of these estimators over the 5,000 samples, and subtract π to give the biases of the two estimators. Note that this is a function of π.

     v. Calculate the variances of these estimators over the 5,000 samples. Note that this is also a function of π.

     vi. Calculate the mean squared error of these estimators over the 5,000 samples. The first way is

$$\mathrm{MSE}[\hat{\pi}] = (bias(\hat{\pi}))^2 + \mathrm{Var}[\hat{\pi}].$$

     The second way is to take the sample mean of the squared distance the estimator is away from the true value over all 5,000 samples. Do it both ways, and see that they give the same result.

(b) Plot the biases of the two estimators versus π at those values and connect the adjacent points. (Put both estimators on the same graph.)

     i. Does the frequentist estimator appear to be unbiased over the range of π values?

     ii. Does the Bayesian estimator appear to be unbiased over the range of the π values?

(c) Plot the mean squared errors of the two estimators versus π over the range of π values, connecting adjacent points. (Put both estimators on the same graph.)

     i. Does your graph resemble Figure 9.2?

ii. Over what range of π values does the Bayesian estimator have smaller mean squared error than that of the frequentist estimator?

CHAPTER 10

BAYESIAN INFERENCE FOR POISSON

The Poisson distribution is used to count the number of occurrences of rare events which are occurring randomly through time (or space) at a constant rate. The events must occur one at a time. The Poisson distribution could be used to model the number of accidents on a highway over a month. However, it could not be used to model the number of fatalities occurring on the highway, since some accidents have multiple fatalities.

Bayes' Theorem for Poisson Parameter with a Continuous Prior

We have a random sample $y_1, \ldots, y_n$ from a $Poisson(\mu)$ distribution. The proportional form of Bayes' theorem is given by *posterior* $\propto$ *prior* $\times$ *likelihood*

$$g(\mu|y_1, \ldots, y_n) \propto g(\mu) \times f(y_1, \ldots, y_n|\mu).$$

The parameter μ can have any positive value, so we should use a continuous prior defined on all positive values. The proportional form of Bayes' theorem gives the shape of the posterior. We need to find the scale factor to make it a density. The actual posterior is given by

$$g(\mu|y_1,\ldots,y_n) = \frac{g(\mu) \times f(y_1,\ldots,y_n|\mu)}{\int_0^\infty g(\mu) \times f(y_1,\ldots,y_n|\mu)\,d\mu}. \qquad (10.1)$$

This equation holds for any continuous prior $g(\mu)$. However, the integration would have to be done numerically except for the few special cases which we will investigate.

Likelihood of Poisson parameter. The likelihood of a single draw from a *Poisson*(μ) distribution is given by

$$f(y|\mu) = \frac{\mu^y e^{-\mu}}{y!}$$

for $y = 0,1,\ldots$ and $\mu > 0$. The part that determines the shape of the likelihood is

$$f(y|\mu) \propto \mu^y e^{-\mu}.$$

When $y_1,\ldots,y_n$ is a random sample from a *Poisson*(μ) distribution, the likelihood of the random sample is the product of the original likelihoods. This simplifies to

$$f(y_1,\ldots,y_n|\mu) = \prod_{i=1}^n f(y_i|\mu)$$
$$\propto \mu^{\sum y_i} e^{-n\mu}.$$

We recognize this as the likelihood where $\sum y_i$ is a single draw from a *Poisson*$(n\mu)$ distribution. It has the shape of a *gamma*(r',v') density where $r' = \sum y_i + 1$ and $v' = n$

10.1   Some Prior Distributions for Poisson

In order to use Bayes' theorem, we will need the prior distribution of the Poisson parameter μ. In this section we will look at several possible prior distributions of μ for which we can work out the posterior density without having to do the numerical integration.

Positive uniform prior density. Suppose we have no idea what the value of μ is prior to looking at the data. In that case, we would consider that we should give all positive values of μ equal weight. So we let the positive uniform prior density be

$$g(\mu) = 1 \text{ for } \mu > 0.$$

Clearly this prior density is improper since its integral over all possible values is infinite. Nevertheless, the posterior will be proper in this case[1] and we can

---

[1]There are cases where an improper prior will result in an improper posterior, so no inference is possible.

use it for making inference about μ. The posterior will be proportional to prior times likelihood, so in this case the proportional posterior will be

$$g(\mu|y_1, \ldots, y_n) \propto g(\mu)\, f(y_1, \ldots, y_n|\mu)$$

$$\propto 1 \times \mu^{\sum y_i} e^{-n\mu}.$$

The posterior is the same shape as the likelihood function so we know that it is a *gamma*(r', v') density where $r' = \sum y + 1$ and $v' = n$. Clearly the posterior is proper despite starting from an improper prior.

Jeffreys' prior for Poisson. The parameter indexes all possible observation distributions. Any one-to-one continuous function of the parameter would give an equally valid index.[2] Jeffreys' method gives us priors which are objective in the sense that they are invariant under any continuous transformation of the parameter. The Jeffreys' prior for the Poisson is

$$g(\mu) \propto \frac{1}{\sqrt{\mu}} \text{ for } \mu > 0.$$

This also will be an improper prior, since its integral over the whole range of possible values is infinite. However, it is not non-informative since it gives more weight to small values. The proportional posterior will be the prior times likelihood. Using the Jeffreys' prior the proportional posterior will be

$$g(\mu|y_1, \ldots, y_n) \propto g(\mu)\, f(y_1, \ldots, y_n|\mu)$$

$$\propto \frac{1}{\sqrt{\mu}} \times \mu^{\sum y_i} e^{-n\mu}$$

$$\propto \mu^{\sum y - \frac{1}{2}} e^{-n\mu},$$

which we recognize as the shape of a *gamma*(r', v') density where $r' = \sum y + \frac{1}{2}$ and $v' = n$. Again, we have a proper posterior despite starting with an improper prior.

Conjugate family for Poisson observations is the gamma family. The conjugate prior for the observations from the *Poisson* distribution with parameter (μ) will have the same form as the likelihood. Hence it has shape given by

$$g(\mu) \propto e^{-k\mu} e^{\log \mu \times l}$$

$$\propto \mu^l e^{-k\mu}.$$

---

[2]If $\psi = h(\theta)$ is a continuous function of the parameter θ, then $g_\psi(\psi)$, the prior for ψ that corresponds to $g_\theta(\theta)$ is found by the change of variable formula $g_\psi(\psi) = g_\theta(\theta(\psi)) \times \frac{d\theta}{d\psi}$.

The distribution having this shape is known as the *gamma(r, v)* distribution and has density given by

$$g(\mu; r, v) = \frac{v^r \mu^{r-1} e^{-v\mu}}{\Gamma(r)} ,$$

where $r - 1 = l$ and $v = k$ and $\frac{v^r}{\Gamma(r)}$ is the scale factor needed to make this a density. When we have a single *Poisson(μ)* observation, and use *gamma(r, v)* prior for μ, the shape of the posterior is given by

$$g(\mu|y) \propto g(\mu) \times f(y|\mu)$$

$$\propto \frac{v^r \mu^{r-1} e^{-v\mu}}{\Gamma(r)} \times \frac{\mu^y e^{-\mu}}{y!}$$

$$\propto \mu^{r-1+y} e^{-(v+1)\mu} .$$

We recognize this to be a *gamma(r', v')* density where the constants are updated by the simple formulas $r' = r + y$ and $v' = v + 1$. We add the observation y to r, and we add 1 to v. Hence when we have a random sample $y_1, \ldots, y_n$ from a *Poisson(μ)* distribution, and use a *gamma(r, v)* prior , we repeat the updating after each observation, using the posterior from the ith observation as the prior for the $i + 1^{\text{st}}$ observation. We end up with a a *gamma(r', v')* posterior where $r' = r + \sum y$ and $v' = v + n$. The simple updating rules are "add the sum of the observations to r" , and "add the number of observations to v." Note: these same updating rules work for the positive uniform prior, and the Jeffreys' prior for the Poisson.[3] We use Equation 7.10 and Equation 7.11 to find the posterior mean and variance. They are:

$$E[\mu|y] = \frac{r'}{v'} \text{ and } \text{Var}[\mu|y] = \frac{r'}{(v')^2} ,$$

respectively.

Choosing a conjugate prior. The *gamma(r, v)* family of distributions is the conjugate family for *Poisson(μ)* observations. It is advantageous to use a prior from this family, as the posterior will also be from this family and can be found by the simple updating rules. This avoids having to do any numerical integration. We want to find the *gamma(r, v)* that matches our prior belief.

We suggest that you summarize your prior belief into your prior mean m, and your prior standard deviation s. Your prior variance will be the square of

---

[3]The positive uniform prior $g(\mu) = 1$ has the form of a *gamma$(1, 0)$* prior, and the Jeffreys' prior for the Poisson $g(\mu) = u^{-\frac{1}{2}}$ has the form of a *gamma$(\frac{1}{2}, 0)$* prior. They can be considered limiting cases of the *gamma(r, v)* family where $v \to 0$.

your prior standard deviation. Then we find the gamma conjugate prior that matches those first two prior moments. That means that r and v will be the simultaneous solutions of the two equations

$$m = \frac{r}{v} \text{ and } s^2 = \frac{r}{v^2}.$$

Hence

$$v = \frac{m}{s^2}.$$

Substitute this into the first equation and solve for r. We find

$$r = \frac{m^2}{s^2}.$$

This gives your *gamma*(r, v) prior.

Precautions before using your conjugate prior.

1. Graph your prior. If the shape looks reasonably close to your prior belief then use it. Otherwise you can adjust your prior mean m and prior standard deviation s until you find a prior with shape matching your prior belief.

2. Calculate the *equivalent sample size* of your prior. This is the size of a random sample of *Poisson*(μ) variables that matches the amount of prior information about μ that you are putting in with your prior. We note that if $y_1, \ldots, y_n$ is a random sample from *Poisson*(μ), then $\bar{y}$ will have mean μ and variance $\frac{\mu}{n}$. The equivalent sample size will be the solution of

$$\frac{\mu}{n_{eq}} = \frac{r}{v^2}.$$

Setting the mean equal to the prior mean $\mu = \frac{r}{v}$ the equivalent sample size of the *gamma*(r, v) prior for μ is $n_{eq} = v$. We check to make sure this is not too large. Ask yourself "Is my prior knowledge about μ really equal to the knowledge I would get about μ if I took a random sample of size n_{eq} from the *Poisson*(μ) distribution?" If the answer is no, then you should increase your prior standard deviation and recalculate your prior. Otherwise you are putting in too much prior information relative to the amount you will be getting from the data.

EXAMPLE 10.1

The weekly number of traffic accidents on a highway has the *Poisson*(μ) distribution. Four students are going to count the number of traffic accidents for each of the next eight weeks. They are going to analyze this in a Bayesian manner, so they each need a prior distribution. Aretha

Table 10.1   Diana's relative prior weights. The shape of her continuous prior is found by linearly interpolating between those values. The constant gets canceled when finding the posterior using Equation 10.1.

Value	Weight
0	0
2	2
4	2
8	0
10	0

says she has no prior information, so will assume all possible values are equally likely. Thus she will use the positive uniform prior $g(\mu) = 1$ for $\mu > 0$, which is improper. Byron also says he has no prior information, but he wants his prior to be invariant if the parameter is multiplied by a constant. Thus, he uses the Jeffreys' prior for the Poisson which is $g(\mu) = \mu^{-1/2}$ which will also be improper. Chase decides that he believes the prior mean should be 2.5, and the prior standard deviation is 1. He decides to use the $gamma(r, v)$ that matches his prior mean and standard deviation, and finds that $v = 2.5$ and $r = 6.25$. His equivalent sample size is $n_{eq} = 2.5$, which he decides is acceptable since he will be putting information worth 2.5 observations and there will be 8 observations from the data. Diana decides that her prior distribution has a trapezoidal shape found by interpolating the prior weights given in Table 10.1. The shapes of the four prior distributions are shown in Figure 1.1. The number of accidents on the highway over the next 8 weeks are:

$$3, 2, 0, 8, 2, 4, 6, 1.$$

Aretha will have a $gamma(27, 8)$ posterior, Byron will have a $gamma(26.5, 8)$ posterior, and Chase will have a $gamma(32.25, 10.5)$ posterior. Diana finds her posterior numerically using Equation 10.1. The four posterior distributions are shown in Figure 10.2. We see that the four posterior distributions are similarly shaped, despite the very different shape priors. ■

Summarizing the Posterior Distribution

The posterior density explains our complete belief about the parameter given the data. It shows the relative belief weights we give each possible parameter value, taking into account both our prior belief and the data, through the likelihood. However, a posterior distribution is hard to interpret, and we like to summarize it with some numbers.

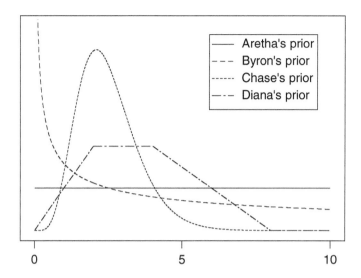

Figure 10.1   The shapes of Aretha's, Byron's, Chase's, and Diana's prior distributions.

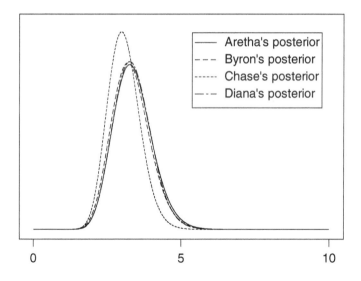

Figure 10.2   Aretha's, Byron's, Chase's, and Diana's posterior distributions.

When we are summarizing a distribution, the most important summary number would be a measure of location, which characterizes where the distribution is located along the number line. Three possible measures of location are the posterior mode, the posterior median, and the posterior mean. The posterior mode is the found by setting the derivative of the posterior density equal to zero, and solving. When the posterior distribution is $gamma(r', v')$, its derivative is given by

$$g'(\mu|y) = (r' - 1)\mu^{r'-2}e^{-v'\mu} - v'e^{-v'\mu}\mu^{r'-1}$$
$$= \mu^{r'-2}e^{-v'\mu}[r' - 1 - v'\mu].$$

When we set that equal to zero and solve, we find the posterior mode is

$$mode = \frac{r' - 1}{v'}.$$

When the posterior distribution is $gamma(r', v')$ the posterior median can be found using Minitab or R. The posterior mean will be

$$m' = \frac{r'}{v'}.$$

If the posterior distribution has been found numerically, then both the posterior median and mean will both have to be found numerically using the Minitab macro *tintegral* or the R functions `mean` and `median`.

The second most important summary number would be a measure of spread, that characterizes how spread out the distribution is. Some possible measures of spread include the interquartile range $IQR = Q_3 - Q_1$ and the standard deviation s'. When the posterior distribution is $gamma(r', v')$, the IQR can be found using Minitab or R. The posterior standard deviation will be the square root of the posterior variance. If the posterior distribution has been found numerically, then the IQR and the posterior variance can be found numerically.

◨ EXAMPLE 10.1 (continued)

The four students calculate measures of location and spread to summarize their posteriors. Aretha, Byron, and Chase have gamma posteriors, so they can calculate them easily using the formulas, and Diana has a numerical posterior so she has to calculate them numerically using the Minitab macro *tintegral* or the R `sintegral` function. The results are shown in Table 10.2.   ∎

10.2   Inference for Poisson Parameter

The posterior distribution is the complete inference in the Bayesian approach. It explains our complete belief about the parameter given the data. It shows the relative belief weights we can give every possible parameter value. However, in the frequentist approach there are several types of inference about the parameter we can make. These are point estimation, interval estimation, and hypothesis testing. In this section we see how we can do these inferences on the parameter μ of the *Poisson* distribution using the Bayesian approach, and we compare these to the corresponding frequentist inferences.

Table 10.2   Measures of location and spread of posterior distributions

Person	Posterior	Mean	Median	Mode	St.Dev.	IQR
Aretha	$gamma(27,8)$	3.375	3.333	3.25	.6495	.8703
Byron	$gamma(26\frac{1}{2},8)$	3.313	3.271	3.187	.6435	.8622
Chase	$gamma(32\frac{1}{4},10\frac{1}{2})$	3.071	3.040	2.976	.5408	.7255
Diana	Numerical	3.353	3.318		.6266	.8502

Point Estimation

We want to find the value of the parameter μ that best represents the posterior and use it as the point estimate. The posterior mean square of $\hat{\mu}$, an estimator of the Poisson mean, measures the average squared distance away from the true value with respect to the posterior.[4] It is given by

$$\text{PMSE}[\hat{\mu}] = \int_0^\infty (\hat{\mu} - \mu)^2 g(\mu|y_1, \ldots, y_n) \, d\mu$$

$$= \int_0^\infty (\hat{\mu} - m' + m' - \mu)^2 g(\mu|y_1, \ldots, y_n) \, d\mu,$$

where m' is the posterior mean. Squaring the term and separating the integral into three integrals, we see that

$$\text{PMSE}[\hat{\mu}] = \text{Var}[\mu|y] + 0 + (m' - \hat{\mu})^2.$$

We see that the last term is always nonnegative, so that the estimator that has smallest posterior mean square is the posterior mean. On the average the squared distance the true value is away from the posterior mean is smaller than for any other possible estimator.[5] That is why we recommend the posterior mean

$$\hat{\mu}_B = \frac{r'}{v'}$$

as the Bayesian point estimate of the Poisson parameter. The frequentist point estimate is $\hat{\mu}_f = \bar{y}$, the sample mean.

Comparing estimators for the Poisson parameter. Bayesian estimators can have superior properties, despite being biased. They often perform better than frequentist estimators, even when judged by frequentist criteria. The mean squared error of an estimator

$$\text{MSE}[\hat{\mu}] = \text{Bias}[\hat{\mu}]^2 + \text{Var}[\hat{\mu}] \tag{10.2}$$

---

[4]The estimator that minimizes the average absolute distance away from the true value is the posterior median.
[5]This is the squared-error loss function approach.

measures the average squared distance the estimator is from the true value. The averaging is over all possible values of the sample, so it is a frequentist criterion. It combines the bias and the variance of the estimator into a single measure. The frequentist estimator of the Poisson parameter is

$$\hat{\mu}_f = \frac{\sum y_i}{n} .$$

This is unbiased, so its mean square equals its variance

$$\mathrm{MSE}[\hat{\mu}_f] = \frac{\mu}{n} .$$

When we use a *gamma*(r, v) prior the posterior will be a *gamma*(r', v'). The bias will be

$$\mathrm{Bias}[\hat{\mu}_B, \mu_B] = \mathrm{E}[\hat{\mu}_B] - \mu$$
$$= \mathrm{E}\left[\frac{r + \sum y_i}{v + n}\right] - \mu$$
$$= \frac{r - v\mu}{v + n} .$$

The variance will be

$$\mathrm{Var}[\hat{\mu}_B] = \left(\frac{1}{v + n}\right)^2 \sum \mathrm{Var}[y_i]$$
$$= \frac{n\mu}{(v + n)^2} .$$

Often we can find a Bayesian estimator that has smaller mean squared error over the range where we believe the parameter lies.

Suppose we are going to observe the number of chocolate chips in a random sample of six chocolate chip cookies. We know that the number of chocolate chips in a single cookie is a *Poisson*(μ) random variable and we want to estimate μ. We know that μ should be close to 2. The frequentist estimate $\hat{\mu}_f = \bar{y}$ will be unbiased and its mean squared error will be

$$\mathrm{MSE}[\hat{\mu}_f] = \frac{\mu}{6} .$$

Suppose we decide to use a *gamma*(2, 1) prior, which has prior mean 2 and prior variance 2. Using Equation 9.2, we find the mean squared error of the Bayesian estimator will be

$$\mathrm{MSE}[\hat{\mu}_B] = \left(\frac{2 - \mu}{1 + 6}\right)^2 + \frac{6\mu}{(1 + 6)^2} .$$

The mean squared errors of the two estimators are shown in Figure 10.3. We see that, on average, the Bayesian estimator is closer to the true value than the frequentist estimator in the range from .7 to 5. Since that is the range in which we believe that μ lies, the Bayesian estimator would be preferable to the frequentist one.

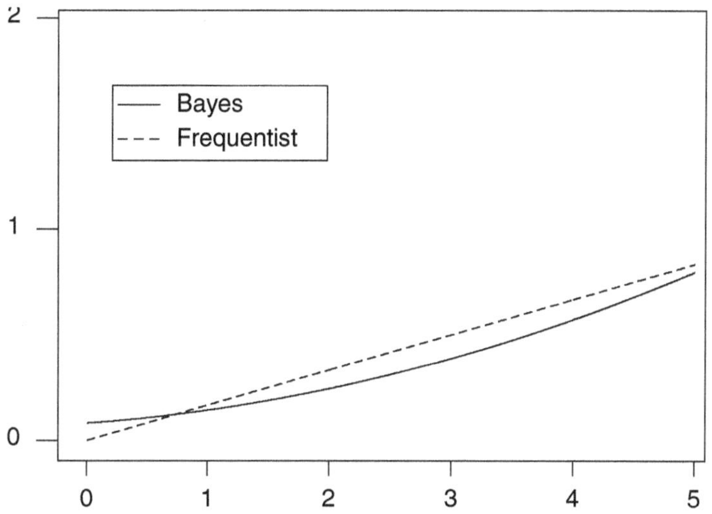

Figure 10.3 The mean squared error for the two estimators.

Bayesian Credible Interval for μ

An equal tail area 95% Bayesian credible interval for μ can be found by obtaining the difference between the 97.5^{th} and the 2.5^{th} percentiles of the posterior. When we used either the $gamma(r, v)$ prior, the positive uniform prior $g(\mu) = 1$ for $\mu > 0$, or the Jeffreys' prior $g(\mu) = \mu^{-\frac{1}{2}}$ the posterior is $gamma(r', v')$. Using Minitab, pull down the *Calc* menu to *Probability Distributions* and over to *Gamma...* and fill in the dialog box.

If we had started with a general continuous prior, the posterior would not be a gamma. The Bayesian credible interval would still be the difference between the 97.5^{th} and the 2.5^{th} percentiles of the posterior, but we would find these percentiles numerically.

EXAMPLE 10.1 (continued)

The four students calculated their 95% Bayesian credible intervals for μ. Aretha, Byron, and Chase all had $gamma(r', v')$ posteriors, with different values of r' and v' because of their different priors. Chase has a shorter credible interval because he put in more prior information than the others. Diana used a general continuous prior so she had to find the credible interval numerically. They are shown in Table 10.3. ∎

Bayesian Test of a One-Sided Hypothesis

Sometimes we have a null value of the Poisson parameter, μ_0. This is the value that the parameter has had before in the past. For instance, the random

Table 10.3   Exact 95% credible intervals

Person	Posterior	Credible Interval	
		Lower	Upper
Aretha	$gamma(27,8)$	2.224	4.762
Byron	$gamma(26\frac{1}{2},8)$	2.174	4.688
Chase	$gamma(32\frac{1}{4},10\frac{1}{2})$	2.104	4.219
Diana	$numerical$	2.224	4.666

variable Y may be the number of defects occurring in a bolt of cloth, and μ is the mean number of defects per bolt. The null value μ_0 is the mean number of defects when the machine manufacturing the cloth is under control. We are interested in determining if the Poisson parameter value has got larger than the null value. This means the rate of defects has increased. We set this up as a one-sided hypothesis test

$$H_0 : \mu \leq \mu_0 \quad \text{versus} \quad H_1 : \mu > \mu_0 .$$

Note: The alternative is in the direction we wish to detect. We test this hypothesis in the Bayesian manner by computing the posterior probability of the null hypothesis. This is found by integrating the posterior density over the correct region

$$P(\mu \leq \mu_0) = \int_0^{\mu_0} g(\mu|y_1,\ldots,y_n)\, d\mu . \tag{10.3}$$

If the posterior distribution is $gamma(r,s)$ we can find this probability using Minitab. Pull down the *Calc* menu to the *Probability Distributions* and over to *Gamma...* and fill in the dialog box. Otherwise, we can evaluate this probability numerically. We compare this probability with the level of significance α. If the posterior probability of the null hypothesis is less than α, then we reject the null hypothesis at the α level of significance.

■ **EXAMPLE 10.1   (continued)**

The four students decide to test the null hypothesis

$$H_0 : \mu \leq 3 \quad \text{versus} \quad H_1 : \mu > 3$$

at the 5% level of significance. Aretha, Byron, and Chase all have $gamma(r',v')$ posteriors each with their own values of the constants. They each calculate the posterior probability of the null hypothesis using Minitab. Diana has a numerical prior, so she must evaluate the integral numerically. The results are shown in Table 10.4.   ■

Table 10.4   Posterior probability of null hypothesis

Person	Posterior	$P(\mu \leq 3.0\|y_1, \ldots, y_n)$ $= \int_0^3 g(\mu\|y_1, \ldots, y_n)\, d\mu$
Aretha	$gamma(27, 8)$	.2962
Byron	$gamma(26\frac{1}{2}, 8)$	.3312
Chase	$gamma(32\frac{1}{4}, 10\frac{1}{2})$	.4704
Diana	$numerical$	.3012

Bayesian Test of a Two-Sided Hypothesis

Sometimes we want to test whether or not the Poisson parameter value has changed from its null value in either direction. We would set that up as a two-sided hypothesis

$$H_0 : \mu = \mu_0 \quad \text{versus} \quad H_1 : \mu \neq \mu_0$$

Since we started with a continuous prior, we will have a continuous posterior. The probability that the continuous parameter taking on the null value will be 0, so we cannot test the hypothesis by calculating its posterior probability. Instead, we test the *credibility* of the null hypothesis by observing whether or not the null value μ_0 lies inside the $(1 - \alpha) \times 100\%$ credible interval for μ. If it lies outside, we can reject the null hypothesis and conclude $\mu \neq \mu_0$. If it lies inside the credible interval, we cannot reject the null hypothesis. We conclude μ_0 remains a credible value.

Main Points

- The Poisson distribution counts the number of occurrence of a rare events which occur randomly through time (or space) at a constant rate. The events must occur one at a time.

- The *posterior* $\propto$ *prior* $\times$ *likelihood* is the key relationship. We cannot use this for inference because it only has the shape of the posterior, and is not an exact density.

- The constant $k = \int prior \times likelihood$ is needed to find the exact posterior density

$$posterior = \frac{prior \times likelihood}{\int prior \times likelihood}$$

so that inference is possible.

- The *gamma* family of priors is the *conjugate* family for *Poisson* observations.

- If the prior is $gamma(r, v)$, then the posterior is $gamma(r', v')$ where the constants are updated by the simple rules $r' = r + \sum y$ (add sum of observations to r, and $v' = v + n$ (add number of observations to v.

- It makes sense to use prior from conjugate family if possible. Determine your prior mean and prior standard deviation. Choose the $gamma(r, v)$ prior that has this prior mean and standard deviation. Graph it to make sure it looks similar to your prior belief.

- If you have no prior knowledge, you can use a positive uniform prior density $g(\mu) = 1$ for $\mu > 0$, which has the form of a $gamma(1, 0)$. Or, you can use the Jeffreys' prior for the Poisson $g(\mu) \propto \mu^{-\frac{1}{2}}$ for $\mu > 0$, which has the form of a $gamma(\frac{1}{2}, 0)$. Both of these are improper priors (their integral over the whole range is infinite). Nevertheless, the posteriors will work out to be proper, and can be found from the same simple rules.

- If you cannot find a member of the conjugate family that matches your prior belief, construct a discrete prior using your belief weights at several values over the range. Interpolate between them to make your general continuous prior. You can ignore the constant needed to make this an exact density since it will get canceled out when you divide by $\int prior \times likelihood$.

- With a good choice of prior the Bayesian posterior mean performs better than the frequentist estimator when judged by the frequentist criterion of mean squared error.

- The $(1 - \alpha) \times 100\%$ Bayesian credible interval gives a range of values for the parameter μ that has posterior probability of $1 - \alpha$.

- We test a one-sided hypothesis in a Bayesian manner by calculating the posterior probability of the null hypothesis. If this is less than the level of significance $alpha$, then we reject the null hypothesis.

- We cannot test a two-sided hypothesis by calculating the posterior probability of the null hypothesis, since it must equal 0 whenever we use a continuous prior. Instead, we test the *credibility* of the null hypothesis value by observing whether or not the null value lies inside the $(1 - \alpha) \times 100\%$ credible interval. If it lies outside the credible interval, we reject the null hypothesis at the level of significance α. Otherwise, we accept that the null value remains credible.

Exercises

10.1. The number of particles emitted by a radioactive source during a ten second interval has the *Poisson*(μ) distribution. The radioactive source is observed over five non-overlapping intervals of ten seconds each. The number of particles emitted during each interval are: 4, 1, 3, 1, 3.

(a) Suppose a prior uniform distribution is used for μ.

   i. Find the posterior distribution for μ.
   ii. What are the posterior mean, median, and variance in this case?

(b) Suppose Jeffreys' prior is used for μ.

   i. Find the posterior distribution for μ.
   ii. What are the posterior mean, median, and variance in this case?

10.2. The number of claims received by an insurance company during a week follows a *Poisson*(μ) distribution. The weekly number of claims observed over a ten week period are: 5, 8, 4, 6, 11, 6, 6, 5, 6, 4.

(a) Suppose a prior uniform distribution is used for μ.

   i. Find the posterior distribution for μ.
   ii. What are the posterior mean, median, and variance in this case?

(b) Suppose Jeffreys' prior is used for μ.

   i. Find the posterior distribution for μ.
   ii. What are the posterior mean, median, and variance in this case?

10.3. The Russian mathematician Ladislaus Bortkiewicz noted that the Poisson distribution would apply to low-frequency events in a large population, even when the probabilities for individuals in the population varied. In a famous example he showed that the number of deaths by horse kick per year in the cavalry corps of the Prussian army follows the Poisson distribution. The following data is reproduced from Hoel (1984).

y	(deaths)	0	1	2	3	4
$n(y)$	(frequency)	109	65	22	3	1

(a) Suppose a prior uniform distribution is used for μ.

   i. Find the posterior distribution for μ.

ii. What are the posterior mean, median, and variance in this case?

(b) Suppose Jeffreys' prior is used for μ.

i. Find the posterior distribution for μ.

ii. What are the posterior mean, median, and variance in this case?

10.4. The number of defects per 10 meters of cloth produced by a weaving machine has the *Poisson* distribution with mean μ. You examine 100 meters of cloth produced by the machine and observe 71 defects.

(a) Your prior belief about μ is that it has mean 6 and standard deviation 2. Find a $gamma(r, v)$ prior that matches your prior belief.

(b) Find the posterior distribution of μ given that you observed 71 defects in 100 meters of cloth.

(c) Calculate a 95% Bayesian credible interval for μ.

Computer Exercises

10.1. We will use the Minitab macro *PoisGamP*, or `poisgamp` function in R, to find the posterior distribution of the Poisson probability μ when we have a random sample of observations from a *Poisson*(μ) distribution and we have a $gamma(r, v)$ prior for μ. The *gamma* family of priors is the conjugate family for *Poisson* observations. That means that if we start with one member of the family as the prior distribution, we will get another member of the family as the posterior distribution. The simple updating rules are "add sum of observations to r" and "add sample size to v. When we start with a $gamma(r, v)$ prior, we get a $gamma(r', v')$ posterior where $r' = r + \sum(y)$ and $v' = v + n$.

Suppose we have a random sample of five observations from a *Poisson*(μ) distribution. They are:

3	4	3	0	1

(a) Suppose we start with a positive uniform prior for μ. What $gamma(r, v)$ prior will give this form?

(b) [**Minitab:**] Find the posterior distribution using the Minitab macro *PoisGamP* or the R function `poisgamp`.

[**R:**] Find the posterior distribution using the R function `poisgamp`.

(c) Find the posterior mean and median.

(d) Find a 95% Bayesian credible interval for μ.

10.2. Suppose we start with a Jeffreys' prior for the Poisson parameter μ.

$$g(\mu) = \mu^{-\frac{1}{2}}$$

(a) What $gamma(r, v)$ prior will give this form?

(b) Find the posterior distribution using the macro *PoisGamP* in Minitab or or the function `poisgamp` in R.

(c) Find the posterior mean and median.

(d) Find a 95% Bayesian credible interval for μ.

10.3. Suppose we start with a $gamma(6, 2)$ prior for μ. Find the posterior distribution using the macro *PoisGamP* in Minitab or or the function `poisgamp` in R.

(a) Find the posterior mean and median.

(b) Find a 95% Bayesian credible interval for μ.

10.4. Suppose we take an additional five observations from the *Poisson*(μ). They are:

1	2	3	3	6

(a) Use the posterior from Computer Exercise 10.3 as the prior for the new observations and find the posterior distribution using the macro *PoisGamP* in Minitab or or the function `poisgamp` in R.

(b) Find the posterior mean and median.

(c) Find a 95% Bayesian credible interval for μ.

10.5. Suppose we use the entire sample of ten *Poisson*(μ) observations as a single sample. We will start with the original prior from Computer Exercise 10.3.

(a) Find the posterior given all ten observations using the Minitab macro *PoisGamP* or the R function `poisgamp`.

(b) What do you notice from Computer Exercises 10.3–10.5?

(c) Test the null hypothesis $H_0 : \mu \leq 2$ vs $H_1 : \mu > 2$ at the 5% level of significance.

10.6. We will use the Minitab macro *PoisGCP*, or the R function `poisgcp`, to find the posterior when we have a random sample from a *Poisson*(μ) distribution and general continuous prior. Suppose we use the data from Computer Exercise 10.4, and the prior distribution is given by

$$g(\mu) = \begin{cases} \mu & \text{for } 0 < \mu \leq 2 \\ 2 & \text{for } 2 < \mu \leq 4 \\ 4 - \frac{\mu}{2} & \text{for } 4 < \mu \leq 8 \\ 0 & \text{for } 8 < \mu \end{cases}$$

[**Minitab:**] Store the values of μ and prior $g(\mu)$ in columns c1 and c2.
[**R:**]

```
g = createPrior(c(0 ,2, 4, 8), c(0, 2, 2, 0))
mu = seq(0, 8, length = 100)
y = c(1, 2, 3,3, 6)
results = poisgcp(y, "user", mu = mu, mu.prior = g(mu))
```

(a) Use *PoisGCP* in Minitab, or the function **poisgcp** in R, to determine the posterior distribution $g(\mu|y_1 : \ldots, y_n)$.

(b) Use Minitab macro *tintegral* to find the posterior mean, median, and standard deviation, or the R functions **mean, median** and **sd.**

(c) Find a 95% Bayesian credible interval for μ by using *tintegral* in Minitab or the function **quantile** applied to the results of **poisgcp** in R.

CHAPTER 11

BAYESIAN INFERENCE FOR NORMAL MEAN

Many random variables seem to follow the normal distribution, at least approximately. The reasoning behind the central limit theorem suggests why this is so. Any random variable that is the sum of a large number of similar-sized random variables from independent causes will be approximately normal. The shapes of the individual random variables "average out" to the normal shape. Sample data from the sum distribution will be well approximated by a normal. The most widely used statistical methods are those that have been developed for random samples from a normal distribution. In this chapter we show how Bayesian inference on a random sample from a normal distribution is done.

11.1 Bayes' Theorem for Normal Mean with a Discrete Prior

For a Single Normal Observation

We are going to take a single observation from the conditional density $f(y|\mu)$ that is known to be normal with known variance σ^2. The standard deviation, σ, is the square root of the variance. There are only m possible values

$\mu_1, \ldots, \mu_m$ for the mean. We choose a discrete prior probability distribution over these values, which summarizes our prior belief about the parameter, before we take the observation. If we really do not have any prior information, we would give all values equal prior probability. We only need to choose the prior probabilities up to a multiplicative constant, since it is only the relative weights we give to the possible values that are important.

The likelihood gives relative weights to all the possible parameter values according to how likely the observed value was given each parameter value. It looks like the conditional observation distribution given the parameter, μ, but instead of the parameter being fixed and the observation varying, we fix the observation at the one that actually occurred, and vary the parameter over all possible values. We only need to know it up to a multiplicative constant since the *relative weights* are all that is needed to apply Bayes' theorem. The posterior is proportional to prior times likelihood, so it equals

$$g(\mu|y) = \frac{prior \times likelihood}{\sum prior \times likelihood}.$$

Any multiplicative constant in either the prior or likelihood would cancel out.

Likelihood of Single Observation

The conditional observation distribution of $y|\mu$ is normal with mean μ and variance σ^2, which is known. Its density is

$$f(y|\mu) = \frac{1}{\sqrt{2\pi}\,\sigma} e^{-\frac{1}{2\sigma^2}(y-\mu)^2}.$$

The likelihood of each parameter value is the value of the observation distribution at the observed value. The part that does not depend on the parameter μ is the same for all parameter values, so it can be absorbed into the proportionality constant. The part that gives the shape as a function of the parameter μ is the important part. Thus the likelihood shape is given by

$$f(y|\mu) \propto e^{-\frac{1}{2\sigma^2}(y-\mu)^2}, \tag{11.1}$$

where y is held constant at the observed value and μ is allowed to vary over all possible values.

Table for Performing Bayes' Theorem

We set up a table to help us find the posterior distribution using Bayes' theorem. The first and second columns contain the possible values of the parameter μ and their prior probabilities. The third column contains the likelihood, which is the observation distribution evaluated for each of the possible values μ_i where y is held at the observed value. This puts a weight on each possible value μ_i proportional to the probability of getting the value actually observed if μ_i is the parameter value. There are two methods we can use to evaluate the likelihood.

Table 11.1   Method 1: Finding posterior using likelihood from Table B.3 "ordinates of normal distribution"

μ	Prior	z	Likelihood	Prior × Likelihood	Posterior
2.0	.2	-1.2	.1942	.03884	.1238
2.5	.2	-.7	.3123	.06246	.1991
3.0	.2	-.2	.3910	.07820	.2493
3.5	.2	.3	.3814	.07628	.2431
4.0	.2	.8	.2897	.05794	.1847
				.31372	1.0000

Finding likelihood from the "ordinates of normal distribution" table.   The first method is to find the likelihood from the "ordinates of the normal distribution" table. Let

$$z = \frac{y - \mu}{\sigma}$$

for each possible value of μ. Z has a standardized $normal(0,1)$ distribution. The likelihood can be found by looking up $f(z)$ in the "ordinates of the standard normal distribution" given in Table B.3 in Appendix B. Note that $f(-z) = f(z)$ because of standard normal distribution is symmetric about 0 .

Finding the likelihood from the normal density function.   The second method is to use the normal density formula given in Equation 11.1, holding y fixed at the observed value and varying μ over all possible values.

■ **EXAMPLE 11.1**

Suppose $y|\mu$ is normal with mean μ and known variance $\sigma^2 = 1$. We know there are only five possible values for μ. They are 2.0, 2.5, 3.0, 3.5, and 4. We let them be equally likely for our prior. We take a single observation of y and obtain the value $y = 3.2$. Let

$$z = \frac{y - \mu}{\sigma} .$$

The values for the likelihood $f(z)$ are found in Table B.3, "ordinates of normal distribution," in Appendix B. Note that $f(-z) = f(z)$ because of standard normal density is symmetric about 0. The posterior probability is the prior × likelihood divided by sum of prior × likelihood. The results are shown in Table 11.1.

If we evaluate the likelihood using the normal density formula, the likelihood is proportional to

$$e^{-\frac{1}{2\sigma^2}(y-\mu)^2} ,$$

Table 11.2   Method 2: Finding posterior using likelihood from normal density formula

μ	Prior	Likelihood (ignoring constant)		Prior × Likelihood	Posterior
2.0	.2	$e^{-\frac{1}{2}(3.2-2.0)^2}$	$=.4868$	.0974	.1239
2.5	.2	$e^{-\frac{1}{2}(3.2-2.5)^2}$	$=.7827$	.1565	.1990
3.0	.2	$e^{-\frac{1}{2}(3.2-3.0)^2}$	$=.9802$	.1960	.2493
3.5	.2	$e^{-\frac{1}{2}(3.2-3.5)^2}$	$=.9560$	.1912	.2432
4.0	.2	$e^{-\frac{1}{2}(3.2-4.0)^2}$	$=.7261$	.1452	.1846
				.7863	1.0000

where y is held at 3.2 and μ varies over all possible values. Note, we are absorbing everything that does not depend on μ into the proportionality constant. The posterior probability is the prior × likelihood divided by sum of prior × likelihood. The results are shown in Table 11.2. We note that the results agree with what we found before except for small round-off errors.

∎

For a Random Sample of Normal Observations

Usually we have a random sample $y_1, \ldots, y_n$ of observations instead of a single observation. The posterior is always proportional to the prior × likelihood. The observations in a random sample are all independent of each other, so the joint likelihood of the sample is the product of the individual observation likelihoods. This gives

$$f(y_1, \ldots, y_n|\mu) = f(y_1|\mu) \times f(y_2|\mu) \times \cdots \times f(y_n|\mu).$$

Thus given a random sample,[1] Bayes' theorem with a discrete prior is given by

$$g(\mu|y_1, \ldots, y_n) \propto g(\mu) \times f(y_1|\mu) \times \ldots \times f(y_n|\mu).$$

We are considering the case where the distribution of each observation $y_j|\mu$ is normal with mean μ and variance σ^2, which is known.

---

[1]de Finetti introduced a condition weaker than independence called exchangeability. Observations are exchangeable if the conditional density of the sample $f(y_1, \ldots, y_n)$ is the unchanged for any permutation of the subscripts. In other words, the order the observations were taken has no useful information. de Finetti (1991) shows that when the observations are exchangeable, $f(y_1, \ldots, y_n) = \int v(\theta) \times w(y_1|\theta) \times w(y_n|\theta) \, d\theta$, for some parameter θ where $v(\theta)$ is some prior distribution and $w(y|\theta)$ is some conditional distribution. The observations are conditionally independent, given θ. The posterior $g(\theta) \propto v(\theta) \times w(y_1|\theta) \times w(y_n|\theta)$. This allows us to treat the exchangeable observations as if they come from a random sample.

Finding the posterior probabilities analyzing observations sequentially one at a time. We could analyze the observations one at a time, in sequence $y_1, \ldots, y_n$, letting the posterior from the previous observation become the prior for the next observation. The likelihood of a single observation y_j is the column of values of the observation distribution at each possible parameter value at that observed value. The posterior is proportional to prior times likelihood.

■ EXAMPLE 11.2

Suppose we take a random sample of four observations from a normal distribution having mean μ and known variance $\sigma^2 = 1$. The observations are 3.2, 2.2, 3.6, and 4.1.

The possible values of μ are 2.0, 2.5, 3.0, 3.5, and 4.0. Again, we will use the prior that gives them all equal weight. We want to use Bayes' theorem to find our posterior belief about μ given the whole random sample. The posterior equals

$$g(\mu|y) = \frac{prior \times likelihood}{\sum prior \times likelihood}.$$

The results of analyzing the observations one at a time are shown in Table 11.3. This is clearly a lot of work for a large sample. We will see that it is much easier to use the whole sample together.

■

Finding the posterior probabilities analyzing the sample all together in a single step. The posterior is proportional to the prior $\times$ likelihood, and the joint likelihood of the sample is the product of the individual observation likelihoods. Each observation is normal, so it has a normal likelihood. This gives the joint likelihood

$$f(y_1, \ldots, y_n|\mu) \propto e^{-\frac{1}{2\sigma^2}(y_1-\mu)^2} \times e^{-\frac{1}{2\sigma^2}(y_2-\mu)^2} \times \cdots \times e^{-\frac{1}{2\sigma^2}(y_n-\mu)^2}.$$

Adding the exponents gives

$$f(y_1, \ldots, y_n|\mu) \propto e^{-\frac{1}{2\sigma^2}\left[(y_1-\mu)^2+(y_2-\mu)^2+\cdots+(y_n-\mu)^2\right]}.$$

We look at the term in brackets

$$[(y_1 - \mu)^2 + \cdots + (y_n - \mu)^2] = y_1^2 - 2y_1\mu + \mu^2 + \cdots + y_n^2 - 2y_n\mu + \mu^2$$

and combine similar terms to get

$$= (y_1^2 + \cdots + y_n^2) - 2\mu(y_1 + \cdots + y_n) + n\mu^2.$$

Table 11.3 Analyzing observations one at a time[a]

μ	Prior$_1$	Likelihood$_1$ (ignoring constant)		Prior$_1$ × Likelihood$_1$	Posterior$_1$
2.0	.2	$e^{-\frac{1}{2}(3.2-2.0)^2}$	$=.4868$	.0974	.1239
2.5	.2	$e^{-\frac{1}{2}(3.2-2.5)^2}$	$=.7827$	.1565	.1990
3.0	.2	$e^{-\frac{1}{2}(3.2-3.0)^2}$	$=.9802$	.1960	.2493
3.5	.2	$e^{-\frac{1}{2}(3.2-3.5)^2}$	$=.9560$	.1912	.2432
4.0	.2	$e^{-\frac{1}{2}(3.2-4.0)^2}$	$=.7261$	.1452	.1846
				.7863	1.0000

μ	Prior$_2$	Likelihood$_2$ (ignoring constant)		Prior$_2$ × Likelihood$_2$	Posterior$_2$
2.0	.1239	$e^{-\frac{1}{2}(2.2-2.0)^2}$	$=.9802$	.1214	.1916
2.5	.1990	$e^{-\frac{1}{2}(2.2-2.5)^2}$	$=.9560$	.1902	.3002
3.0	.2493	$e^{-\frac{1}{2}(2.2-3.0)^2}$	$=.7261$	.1810	.2857
3.5	.2432	$e^{-\frac{1}{2}(2.2-3.5)^2}$	$=.4296$	.1045	.1649
4.0	.1846	$e^{-\frac{1}{2}(2.2-4.0)^2}$	$=.1979$	.0365	.0576
				.6336	1.0000

μ	Prior$_3$	Likelihood$_3$ (ignoring constant)		Prior$_3$ × Likelihood$_3$	Posterior$_3$
2.0	.1916	$e^{-\frac{1}{2}(3.6-2.0)^2}$	$=.2780$	.0533	.0792
2.5	.3002	$e^{-\frac{1}{2}(3.6-2.5)^2}$	$=.5461$	.1639	.2573
3.0	.2857	$e^{-\frac{1}{2}(3.6-3.0)^2}$	$=.8353$	.2386	.3745
3.5	.1649	$e^{-\frac{1}{2}(3.6-3.5)^2}$	$=.9950$	.1641	.2576
4.0	.0576	$e^{-\frac{1}{2}(3.6-4.0)^2}$	$=.9231$	.0532	.0835
				.6731	1.0000

μ	Prior$_4$	Likelihood$_4$ (ignoring constant)		Prior$_4$ × Likelihood$_4$	Posterior$_4$
2.0	.0792	$e^{-\frac{1}{2}(4.1-2.0)^2}$	$=.1103$	.0087	.0149
2.5	.2573	$e^{-\frac{1}{2}(4.1-2.5)^2}$	$=.2780$	.0715	.1226
3.0	.3745	$e^{-\frac{1}{2}(4.1-3.0)^2}$	$=.5461$	.2045	.3508
3.5	.2576	$e^{-\frac{1}{2}(4.1-3.5)^2}$	$=.8352$	.2152	.3691
4.0	.0835	$e^{-\frac{1}{2}(4.1-4.0)^2}$	$=.9950$	.0838	.1425
				.5830	1.0000

[a]Note: The prior for observation i is the posterior after previous observation $i-1$.

When we substitute this back in, factor out n, and complete the square we get

$$f(y_1,\ldots,y_n|\mu) \propto e^{-\frac{n}{2\sigma^2}\left[\mu^2-2\mu\bar{y}+\bar{y}^2-\bar{y}^2+\frac{y_1^2+\cdots+y_n^2}{n}\right]}$$

$$\propto e^{-\frac{n}{2\sigma^2}\left[\mu^2-2\mu\bar{y}+\bar{y}^2\right]} \times e^{-\frac{n}{2\sigma^2}\left[\frac{y_1^2+\cdots+y_n^2}{n}-\bar{y}^2\right]}.$$

The likelihood of the normal random sample $y_1,\ldots,y_n$ is proportional to the likelihood of the sample mean $\bar{y}$. When we absorb the part that does not involve μ into the proportionality constant we get

$$f(y_1,\ldots,y_n|\mu) \propto e^{-\frac{1}{2\sigma^2/n}(\bar{y}-\mu)^2}.$$

We recognize that this likelihood has the shape of a normal distribution with mean μ and variance $\frac{\sigma^2}{n}$. We know $\bar{y}$, the sample mean, is normally distributed with mean μ and variance $\frac{\sigma^2}{n}$. So the joint likelihood of the random sample is proportional to the likelihood of the sample mean, which is

$$f(\bar{y}|\mu) \propto e^{-\frac{1}{2\sigma^2/n}(\bar{y}-\mu)^2}. \tag{11.2}$$

We can think of this as drawing a single value, $\bar{y}$, the sample mean, from the normal distribution with mean μ and variance $\frac{\sigma^2}{n}$. This will make analyzing the random sample much easier.

We substitute in the observed value of $\bar{y}$, the sample mean, and calculate its likelihood. Then we just find the posterior probabilities using Bayes' theorem in only one table. This is much less work!

■ **EXAMPLE 11.2** (continued)

In the preceding example the sample mean was $\bar{y} = 3.275$. We use the likelihood of $\bar{y}$ which is proportional to the likelihood of the whole sample. The results are shown in Table 11.4. We see that they agree with the previous results to three figures. The slight discrepancy in the fourth decimal place is due to the accumulation of round off errors when we analyze the observations one at a time. It is clearly easier to use $\bar{y}$ to summarize the sample, and perform the calculations for Bayes' theorem only once.[2]   ■

---

[2] $\bar{y}$ is said to be a sufficient statistic for the parameter μ. The likelihood of a random sample $y_1,\ldots,y_n$ can be replaced by the likelihood of a single statistic only if the statistic is sufficient for the parameter. One-dimensional sufficient statistics only exist for some distributions, notably those that come from the one-dimensional exponential family.

Table 11.4   Analyze the observations all together using likelihood of sample mean

μ	Prior$_1$	Likelihood$_{\bar{y}}$	Prior$_1$ × Likelihood$_{\bar{y}}$	Posterior$_{\bar{y}}$
2.0	.2	$e^{-\frac{1}{2\times 1/4}(3.275-2.0)^2}$ =.0387	.0077	.0157
2.5	.2	$e^{-\frac{1}{2\times 1/4}(3.275-2.5)^2}$ =.3008	.0602	.1228
3.0	.2	$e^{-\frac{1}{2\times 1/4}(3.275-3.0)^2}$ =.8596	.1719	.3505
3.5	.2	$e^{-\frac{1}{2\times 1/4}(3.275-3.5)^2}$ =.9037	.1807	.3685
4.0	.2	$e^{-\frac{1}{2\times 1/4}(3.275-4.0)^2}$ =.3495	.0699	.1425
			.4904	1.0000

11.2 Bayes' Theorem for Normal Mean with a Continuous Prior

We have a random sample $y_1, \ldots, y_n$ from a normal distribution with mean μ and known variance σ^2. It is more realistic to believe that all values of μ are possible, at least all those in an interval. This means we should use a continuous prior. We know that Bayes' theorem can be summarized as *posterior proportional to prior times likelihood*

$$g(\mu|y_1, \ldots, y_n) \propto g(\mu) \times f(y_1, \ldots, y_n|\mu).$$

Here we allow $g(\mu)$ to be a continuous prior density. When the prior was discrete, we evaluated the posterior by dividing the *prior × likelihood* by the *sum* of the *prior × likelihood* over all possible parameter values. Integration for continuous variables is analogous to summing for discrete variables. Hence we can evaluate the posterior by dividing the *prior × likelihood* by the *integral* of the *prior × likelihood* over the whole range of possible parameter values. Thus

$$g(\mu|y_1, \ldots, y_n) = \frac{g(\mu) \times f(y_1, \ldots, y_n|\mu)}{\int g(\mu) \times f(y_1, \ldots, y_n|\mu)\, d\mu}. \qquad (11.3)$$

For a normal distribution, the likelihood of the random sample is proportional to the likelihood of the sample mean, $\bar{y}$. So

$$g(\mu|y_1, \ldots, y_n) = \frac{g(\mu) \times e^{-\frac{1}{2\sigma^2/n}(\bar{y}-\mu)^2}}{\int g(\mu) \times e^{-\frac{1}{2\sigma^2/n}(\bar{y}-\mu)^2}\, d\mu}.$$

This works for any continuous prior density $g(\mu)$. However, it requires an integration, which may have to be done numerically. We will look at some special cases where we can find the posterior without having to do the integration. For these cases, we have to be able to recognize when a density must be normal from the shape given in Equation 11.1.

Flat Prior Density for μ (Jeffrey's Prior for Normal Mean)

We know that the actual values the prior gives to each possible value is not important. Multiplying all the values of the prior by the same constant would multiply the integral of the prior times likelihood by the same constant, so it would cancel out, and we would obtain the same posterior. What is important is that the prior gives the *relative* weights to all possible values that we believe before looking at the data.

The flat prior gives each possible value of μ equal weight. It does not favor any value over any other value, $g(\mu) = 1$. The flat prior is not really a proper prior distribution since $-\infty < \mu < \infty$, so it cannot integrate to 1. Nevertheless, this *improper* prior works out all right. Even though the prior is improper, the posterior will integrate to 1, so it is proper. The Jeffreys' prior for the mean of a *normal* distribution turns out to be the flat prior.

A single normal observation y. Let y be a normally distributed observation with mean μ and known variance σ^2. The likelihood is given by

$$f(y|\mu) \propto e^{-\frac{1}{2\sigma^2}(y-\mu)^2},$$

if we ignore the constant of proportionality. Since the prior always equals 1, the posterior is proportional to this. We rewrite it as

$$g(\mu|y) \propto e^{-\frac{1}{2\sigma^2}(\mu-y)^2}.$$

We recognize from this shape that the posterior is a normal distribution with mean y and variance σ^2.

A normal random sample $y_1, \ldots y_n$. In the previous section we showed that the likelihood of a random sample from a normal distribution is proportional to likelihood of the sample mean $\bar{y}$. We know that $\bar{y}$ is normally distributed with mean μ and variance $\frac{\sigma^2}{n}$. Hence the likelihood has shape given by

$$f(\bar{y}|\mu) \propto e^{-\frac{1}{2\sigma^2/n}(\bar{y}-\mu)^2},$$

where we are ignoring the constant of proportionality. Since the prior always equals 1, the posterior is proportional to this. We can rewrite it as

$$g(\mu|\bar{y}) \propto e^{-\frac{1}{2\sigma^2/n}(\mu-\bar{y})^2}.$$

We recognize from this shape that the posterior distribution is normal with mean $\bar{y}$ and variance $\frac{\sigma^2}{n}$.

Normal Prior Density for μ

Single observation. The observation y is a random variable taken from a normal distribution with mean μ and variance σ^2 which is assumed known. We have a prior distribution that is normal with mean m and variance s^2. The shape of the prior density is given by

$$g(\mu) \propto e^{-\frac{1}{2s^2}(\mu-m)^2} \,,$$

where we are ignoring the part that does not involve μ because multiplying the prior by any constant of proportionality will cancel out in the posterior. The shape of the likelihood is

$$f(y|\mu) \propto e^{-\frac{1}{2\sigma^2}(y-\mu)^2} \,,$$

where we ignore the part that does not depend on μ because multiplying the likelihood by any constant will cancel out in the posterior. The prior times likelihood is

$$g(\mu) \times f(y|\mu) \propto e^{-\frac{1}{2}\left[\frac{(\mu-m)^2}{s^2}+\frac{(y-\mu)^2}{\sigma^2}\right]} \,.$$

Putting the terms in exponent over the common denominator and expanding them out gives

$$\propto e^{-\frac{1}{2}\left[\frac{\sigma^2(\mu^2-2\mu m+m^2)+s^2(y^2-2y\mu+\mu^2)}{\sigma^2 s^2}\right]} \,.$$

We combine the like terms

$$\propto e^{-\frac{1}{2}\left[\frac{(\sigma^2+s^2)\mu^2-2(\sigma^2 m+s^2 y)\mu+m^2\sigma^2+y^2 s^2}{\sigma^2 s^2}\right]}$$

and factor out $(\sigma^2 + s^2)/(\sigma^2 s^2)$. Completing the square and absorbing the part that does not depend on μ into the proportionality constant, we have

$$\propto e^{-\frac{1}{2\sigma^2 s^2/(\sigma^2+s^2)}\left[\mu^2-2\frac{(\sigma^2 m+s^2 y)}{\sigma^2+s^2}\mu+(\frac{(\sigma^2 m+s^2 y)}{\sigma^2+s^2})^2\right]}$$

$$\propto e^{-\frac{1}{2\sigma^2 s^2/(\sigma^2+s^2)}\left[\mu-\frac{(\sigma^2 m+s^2 y)}{\sigma^2+s^2}\right]^2} \,.$$

We recognize from this shape that the posterior is a normal distribution having mean and variance given by

$$m' = \frac{(\sigma^2 m + s^2 y)}{\sigma^2 + s^2} \quad \text{and} \quad (s')^2 = \frac{\sigma^2 s^2}{(\sigma^2 + s^2)} \,, \tag{11.4}$$

respectively. We started with a *normal*(m, s^2) prior, and ended up with a *normal*$[m', (s')^2]$ posterior. This shows that the *normal*(m, s^2) distribution is the conjugate family for the normal observation distribution with known variance. Bayes' theorem moves from one member of the conjugate family to another member. Because of this we do not need to perform the integration in order to evaluate the posterior. All that is necessary is to determine the rule for updating the parameters.

Simple updating rule for normal family. The updating rules given in Equation 11.4 can be simplified. First we introduce the *precision* of a distribution that is the reciprocal of the variance. Precisions are additive. The posterior precision

$$\frac{1}{(s')^2} = \left(\frac{\sigma^2 s^2}{\sigma^2 + s^2}\right)^{-1} = \frac{\sigma^2 + s^2}{\sigma^2 s^2} = \frac{1}{s^2} + \frac{1}{\sigma^2}.$$

Thus the posterior precision equals prior precision plus the observation precision. The posterior mean is given by

$$m' = \frac{(\sigma^2 m + s^2 y)}{\sigma^2 + s^2} = \frac{\sigma^2}{\sigma^2 + s^2} \times m + \frac{s^2}{\sigma^2 + s^2} \times y.$$

This can be simplified to

$$m' = \frac{1/s^2}{1/\sigma^2 + 1/s^2} \times m + \frac{1/\sigma^2}{1/\sigma^2 + 1/s^2} \times y.$$

Thus the posterior mean is the weighted average of the prior mean and the observation, where the weights are the proportions of the precisions to the posterior precision.

This updating rule also holds for the flat prior. The flat prior has infinite variance, so it has zero precision. The posterior precision will equal the observation precision

$$1/\sigma^2 = 0 + 1/\sigma^2,$$

and the posterior variance equals the observation variance σ^2. The flat prior does not have a well-defined prior mean. It could be anything. We note that

$$\frac{0}{1/\sigma^2} \times \text{anything} + \frac{1/\sigma^2}{1/\sigma^2} \times y = y,$$

so the posterior mean using flat prior equals the observation y

A random sample $y_1, \ldots, y_n$. A random sample $y_1, \ldots, y_n$ is taken from a normal distribution with mean μ and variance σ^2, which is assumed known. We have a prior distribution that is normal with mean m and variance s^2 given by

$$g(\mu) \propto e^{-\frac{1}{2s^2}(\mu - m)^2},$$

where we are ignoring the part that does not involve μ because multiplying the prior by any constant will cancel out in the posterior.

We use the likelihood of the sample mean, $\bar{y}$ which is normally distributed with mean μ and variance $\frac{\sigma^2}{n}$. The precision of $\bar{y}$ is $\left(\frac{n}{\sigma^2}\right)$. We see that this is the sum of all the observation precisions for the random sample.

We have reduced the problem to updating given a single normal observation of $\bar{y}$, which we have already solved. Posterior precision equals the prior precision plus the precision of $\bar{y}$.

$$\frac{1}{(s')^2} = \frac{1}{s^2} + \frac{n}{\sigma^2} = \frac{\sigma^2 + ns^2}{\sigma^2 s^2}. \tag{11.5}$$

The posterior variance equals the reciprocal of posterior precision. The posterior mean equals the weighted average of the prior mean and $\bar{y}$ where the weights are the proportions of the posterior precision:

$$m' = \frac{1/s^2}{n/\sigma^2 + 1/s^2} \times m + \frac{n/\sigma^2}{n/\sigma^2 + 1/s^2} \times \bar{y}. \qquad (11.6)$$

11.3   Choosing Your Normal Prior

The prior distribution you choose should match your prior belief. When the observation is from a normal distribution with known variance, the conjugate family of priors for μ is the $normal(m, s^2)$. If you can find a member of this family that matches your prior belief, it will make finding the posterior using Bayes' theorem very easy. The posterior will also be a member of the same family where the parameters have been updated by the simple rules given in Equations 11.5 and 11.6. You will not need to do any numerical integration.

First, decide on your prior mean m. This is the value your prior belief is centered on. Then decide on your prior standard deviation s. Think of the points above and below that you consider to be the upper and lower bounds of possible values of μ. Divide the distance between these two points by 6 to get your prior standard deviation s. This way you will get reasonable probability over all the region you believe possible.

A useful check on your prior is to consider the "equivalent sample size". Set your prior variance $s^2 = \sigma^2/n_{eq}$ and solve for n_{eq}. This relates your prior precision to the precision from a sample. Your belief is of equal importance to a sample of size n_{eq}. If n_{eq} is large, it shows you have very strong prior belief about μ. It will take a lot of sample data to move your posterior belief far from your prior belief. If it is small, your prior belief is not strong, and your posterior belief will be strongly influenced by a more modest amount of sample data.

If you cannot find a prior distribution from the conjugate family that corresponds to your prior belief, then you should determine your prior belief for a selection of points over the range you believe possible, and linearly interpolate between them. Then you can determine your posterior distribution by

$$g(\mu|y_1, \ldots, y_n) = \frac{f(y_1, \ldots, y_n|\mu) \times g(\mu)}{\int f(y_1, \ldots, y_n|\mu) \times g(\mu) d\mu}.$$

■ EXAMPLE 11.3

Arnie, Barb, and Chuck are going to estimate the mean length of one-year-old rainbow trout in a stream. Previous studies in other streams have shown the length of yearling rainbow trout to be normally distributed with known standard deviation of 2 cm. Arnie decides his prior mean is 30 cm.

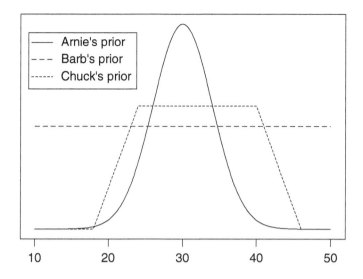

Figure 11.1   The shapes of Arnie's, Barb's, and Chuck's priors.

He decides that he does not believe it is possible for a yearling rainbow to be less than 18 cm or greater than 42 cm. Thus his prior standard deviation is 4 cm. Thus he will use a $normal(30, 4^2)$ prior. Barb does not know anything about trout, so she decides to use the "flat" prior. Chuck decides his prior belief is not normal. His prior has a trapezoidal shape. His prior gives zero weight at 18 cm. It gives weight one at 24 cm, and is level up to 40 cm, and then goes down to zero at 46 cm. He linearly interpolates between those values. The shapes of the three priors are shown in Figure 11.1.

They take a random sample of 12 yearling trout from the stream and find the sample mean $\bar{y} = 32$cm. Arnie and Barb find their posterior distributions using the simple updating rules for the normal conjugate family given by Equations 11.5 and 11.6. For Arnie

$$\frac{1}{(s')^2} = \frac{1}{4^2} + \frac{12}{2^2}.$$

Solving for this gives his posterior variance $(s')^2 = .3265$. His posterior standard deviation is $s' = .5714$. His posterior mean is found by

$$m' = \frac{\frac{1}{4^2}}{\frac{1}{.5714^2}} \times 30 + \frac{\frac{12}{2^2}}{\frac{1}{.5714^2}} \times 32 = 31.96\,.$$

Barb is using the "flat" prior, so her posterior variance is

$$(s')^2 = \frac{2^2}{12} = .3333$$

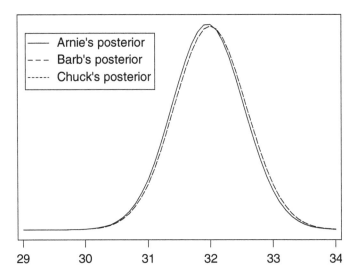

Figure 11.2   Arnie's, Barb's, and Chuck's posteriors. (Barb and Chuck have nearly identical posteriors.)

and her posterior standard deviation is $s' = .5774$. Her posterior mean $m' = 32$, the sample mean. Both Arnie and Barb have normal posterior distributions.

Chuck finds his posterior using Equation 11.3 which requires numerical integration. The three posteriors are shown in Figure 11.2. Since Chuck used a prior that was flat over the whole region where the likelihood was appreciable, his posterior is virtually indistinguishable from Barb's who used the flat improper prior. Arnie who used an informative prior has a posterior that is also close to Barb's. This shows that given the data, the posteriors are similar despite starting from quite different priors.   ∎

11.4   Bayesian Credible Interval for Normal Mean

The posterior distribution $g(\mu|y_1, \ldots, y_n)$ is the inference we make for μ given the observations. It summarizes our entire belief about the parameter given the data. Sometimes we want to summarize our posterior belief into a range of values that we believe cannot be ruled out at some probability level, given the sample data. An interval like this is called a Bayesian credible interval. It summarizes the range of possible values that are credible at that level. There are many possible credible intervals for a given probability level. Generally, the shortest one is preferred. However, in some cases it is easier to find the credible interval with equal tail probabilities.

Known Variance

When $y_1, \ldots, y_n$ is a random sample from a *normal* (μ, σ^2) distribution, the sampling distribution of $\bar{y}$, the sample mean, is *normal*$(\mu, \sigma^2/n)$. Its mean equals that for a single observation from the distribution, and its variance equals the variance of single observation divided by sample size. Using either a "flat" prior, or a *normal* (m, s^2) prior, the posterior distribution of μ given $\bar{y}$ is *normal*$[m', (s')^2]$, where we update according to the rules:

1. Precision is the reciprocal of the variance.

2. Posterior precision equals prior precision plus the precision of sample mean.

3. Posterior mean is weighted sum of prior mean and sample mean, where the weights are the proportions of the precisions to the posterior precision.

Our $(1 - \alpha) \times 100\%$ Bayesian credible interval for μ is

$$m' \pm z_{\frac{\alpha}{2}} \times s', \tag{11.7}$$

which is the *posterior mean* plus or minus the *z-value* times the *posterior standard deviation*, where the *z-value* is found in the standard normal table. Our posterior *probability* that the true mean μ lies outside the credible interval is α. Since the posterior distribution is *normal* and thus symmetric, the credible interval found using Equation 11.7 is the shortest, as well as having equal tail probabilities.

Unknown Variance

If we do not know the variance, we do not know the precision, so we cannot use the updating rules directly. The obvious thing to do is to calculate the sample variance

$$\hat{\sigma}^2 = \frac{1}{n-1} \sum_{i=1}^{n} (y_i - \bar{y})^2$$

from the data. Then we use Equations 11.5 and 11.6 to find $(s')^2$ and m' where we use the sample variance $\hat{\sigma}^2$ in place of the unknown variance σ^2.

There is extra uncertainty here, the uncertainty in estimating σ^2. We should widen the credible interval to account for this added uncertainty. We do this by taking the values from the table for the *Student's t* distribution instead of the *standard normal* table. The correct Bayesian credible interval is

$$m' \pm t_{\frac{\alpha}{2}} \times s'. \tag{11.8}$$

The t value is taken from the row labeled $df = n - 1$ (degrees of freedom equals number of observations minus 1).[3]

Nonnormal Prior

When we start with a nonnormal prior, we find the posterior distribution for μ using Bayes' theorem where we have to integrate numerically. The posterior distribution will be nonnormal. We can find a $(1 - \alpha) \times 100\%$ credible interval by finding a lower value μ_l and an upper value μ_u such that

$$\int_{\mu_l}^{\mu_u} g(\mu|y_1, \ldots, y_n) \, d\mu = 1 - \alpha \, .$$

There are many such values. The best choice μ_l and μ_u would give us the shortest possible credible interval. These values also satisfy

$$g(\mu_l|y_1, \ldots, y_n) = g(\mu_u|y_1, \ldots, y_n) \, .$$

Sometimes it is easier to find the credible interval with lower and upper tail areas that are equal.

■ EXAMPLE 11.3 (continued)

Arnie, and Barb each calculated their 95% credible interval from their respective posterior distributions using Equation 11.7.

[Minitab:] Chuck had to calculate his credible interval numerically from his numerical posterior using the Minitab macro *normgcp*.

[R:] Chuck had to calculate his credible numerically from his numerical posterior using the **quantile** function on the results of the **normgcp** function in R.

The credible intervals are shown in Table 11.5. Arnie, Barb, and Chuck end up with slightly different credible intervals because they started with different prior beliefs. But the effect of the data was much greater than the effect of their priors and their credible intervals are quite similar. ■

---

[3]The resulting Bayesian credible interval is exactly the same one that we would find if we did the full Bayesian analysis with σ^2 as a nuisance parameter, using the joint prior distribution for μ and σ^2 made up of the same prior for $\mu|\sigma^2$ that we used before ["flat" or *normal*(m, s^2)]times the prior for σ^2 given by $g(\sigma^2) \propto (\sigma^2)^{-1}$. We would find the joint posterior by Bayes' theorem. We would find the marginal posterior distribution of μ by marginalizing out σ^2. We would get the same Bayesian credible interval using *Student's t* critical values.

Table 11.5   95% credible intervals

Person	Posterior	Credible Interval	
	Distribution	Lower	Upper
Arnie	$normal(31.96, .3265)$	30.84	33.08
Barb	$normal(32.00, .3333)$	30.87	33.13
Chuck	numerical	30.82	33.07

11.5   Predictive Density for Next Observation

Bayesian statistics has a general method for developing the conditional distribution of the next random observation, given the previous random sample. This is called the predictive distribution. This is a clear advantage over frequentist statistics, which can only determine the predictive distribution for some situations. The problem is how to combine the uncertainty from the previous sample with the uncertainty in the observation distribution. The Bayesian approach is called *marginalization*. It entails finding the joint posterior for the next observation and the parameter, given the random sample. The parameter is treated as a *nuisance parameter*, and the marginal distribution of the next observation given the random sample is found by integrating the parameter out of the joint posterior distribution.

Let y_{n+1} be the next random variable drawn after the random sample $y_1, \ldots, y_n$. The predictive density of $y_{n+1} | y_1, \ldots, y_n$ is the conditional density

$$f(y_{n+1} | y_1, \ldots, y_n).$$

This can be found by Bayes' theorem. $y_1, \ldots, y_n, y_{n+1}$ is a random sample from $f(y|\mu)$, which is a normal distribution with mean μ and known variance σ^2. The conditional distribution of the random sample $y_1, \ldots, y_n$ and the next random observation y_{n+1} given the parameter μ is

$$f(y_1, \ldots, y_n, y_{n+1} | \mu) = f(y_1|\mu) \times \cdots \times f(y_n|\mu) \times f(y_{n+1}|\mu).$$

Let the prior distribution be $g(\mu)$ (either flat prior or $normal(m, s^2)$ prior). The joint distribution of the observations and the parameter μ is

$$g(\mu) \times f(y_1|\mu) \times \ldots \times f(y_n|\mu) \times f(y_{n+1}|\mu).$$

The conditional density of y_{n+1} and μ given $y_1, \ldots, y_n$ is

$$f(y_{n+1}, \mu | y_1, \ldots, y_n) = f(y_{n+1} | \mu, y_1, \ldots, y_n) \times g(\mu | y_1, \ldots, y_n).$$

We have already found that the posterior $g(\mu|y_1, \ldots, y_n,)$ is normal with posterior precision equal to prior precision plus the precision of $\bar{y}$ and mean equal to the weighted average of the prior mean and $\bar{y}$ where the weights are proportions of the precisions to the posterior precision. Say it is normal with mean m_n and variance s_n^2. The distribution of y_{n+1} given μ and $y_1, \ldots, y_n$ only depends on μ, because y_{n+1} is another random draw from the distribution $f(y|\mu)$. Thus the joint posterior (to first n observations) distribution is

$$f(y_{n+1}, \mu|y_1, \ldots, y_n) = f(y_{n+1}|\mu) \times g(\mu|y_1, \ldots, y_n).$$

The conditional distribution we want is found by integrating μ out of the joint posterior distribution. This is the marginal posterior distribution

$$f(y_{n+1}|y_1, \ldots, y_n) = \int f(y_{n+1}, \mu|y_1, \ldots, y_n)\, d\mu$$

$$= \int f(y_{n+1}|\mu) \times g(\mu|y_1, \ldots, y_n)\, d\mu.$$

These are both normal under our assumed model, so

$$f(y_{n+1}|y_1, \ldots, y_n) \propto \int e^{-\frac{1}{2\sigma^2}(y_{n+1}-\mu)^2} e^{-\frac{1}{2s_n^2}(\mu - m_n)^2}\, d\mu.$$

Adding the exponents and combining like terms.

$$f(y_{n+1}|y_1, \ldots, y_n) \propto \int e^{-\frac{1}{2}\left[\frac{(\mu^2 - 2\mu y_{n+1} + y_{n+1}^2)}{\sigma^2} + \frac{(\mu^2 - 2\mu m_n + m_n^2)}{s_n^2}\right]}\, d\mu$$

$$\propto \int e^{-\frac{1}{2}\left[(\frac{1}{\sigma^2} + \frac{1}{s_n^2})\mu^2 - 2(\frac{y_{n+1}}{\sigma^2} + \frac{m_n}{s_n^2})\mu + \frac{y_{n+1}^2}{\sigma^2} + \frac{m_n^2}{s_n^2}\right]}\, d\mu.$$

Factoring out $(\frac{1}{\sigma^2} + \frac{1}{s_n^2})$ of the exponent and completing the square

$$\propto \int e^{-\frac{1}{2(\sigma^2 s_n^2)/(\sigma^2 + s_n^2)}\left[\mu - \frac{(s_n^2 y_{n+1} + \sigma^2 m_n)}{\sigma^2 + s_n^2}\right]^2}$$

$$\times e^{-\frac{1}{2(\sigma^2 s_n^2)/(\sigma^2 + s_n^2)}\left[-\left(\frac{s_n^2 y_{n+1} + \sigma^2 m_n}{\sigma^2 + s_n^2}\right)^2 + \frac{s_n^2 y_{n+1}^2 + \sigma^2 m_n^2}{s_n^2 + \sigma^2}\right]}\, d\mu.$$

The first line is the only part that depends on μ, and we recognize that it is proportional to a normal density, so integrating it over its whole range gives a constant. Reorganizing the second part gives

$$\propto e^{-\frac{1}{2(\sigma^2 s_n^2)/(\sigma^2 + s_n^2)}\left[\frac{(s_n^2 y_{n+1}^2 + \sigma^2 m_n^2)(\sigma^2 + s_n^2) - (s_n^4 y_{n+1}^2 + 2s_n^2 \sigma^2 y_{n+1} m_n + \sigma^4 m_n^2)}{(\sigma^2 + s_n^2)^2}\right]},$$

which simplifies to

$$\propto e^{-\frac{1}{2(\sigma^2 + s_n^2)}(y_{n+1} - m_n)^2}. \tag{11.9}$$

We recognize this as a normal density with mean $m' = m_n$ and variance $(s')^2 = \sigma^2 + s_n^2$. The predictive mean for the observation y_{n+1} is the posterior mean of μ given the observations $y_1, \ldots, y_n$. The predictive variance is the observation variance σ^2 plus the posterior variance of μ given the observations $y_1, \ldots, y_n$. (Part of the uncertainty in the prediction is due to the uncertainty in estimating the posterior mean.)

This is one of the advantages of the Bayesian approach. It has a single clear approach (marginalization) that is always used to construct the predictive distribution. There is no single clear-cut way this can be done in frequentist statistics, although in many problems such as the normal case we just did, they can come up with similar results.

Main Points

- Analyzing the observations sequentially one at a time, using the posterior from the previous observation as the next prior, gives the same results as analyzing all the observations at once using the initial prior.

- The likelihood of a random sample of normal observations is proportional to the likelihood of the sample mean.

- The conjugate family of priors for *normal* observations with known variance is the *normal(m, s^2)* family.

- If we have a random sample of normal observations and use a *normal(m, s^2)* prior the posterior is *normal(m', $(s')^2$)*, where m' and $(s')^2$ are found by the simple updating rules:

  o The precision is the reciprocal of the variance.

  o Posterior precision is the sum of the prior precision and the precision of the sample.

  o The posterior mean is the weighted average of the prior mean and the sample mean, where the weights are the proportions of their precisions to the posterior precision.

- The same updating rules work for the flat prior, remembering the flat prior has precision equal to zero.

- A Bayesian credible interval for μ can be found using the posterior distribution.

- If the variance σ^2 is not known, we use the estimate of the variance calculated from the sample, $\hat{\sigma}^2$, and use the critical values from the *Student's t* table where the degrees of freedom is $n-1$, the sample size minus 1. Using the *Student's t* critical values compensates for the extra uncertainty due to not knowing σ^2. (This actually gives the correct credible interval if we used a prior $g(\sigma^2) \propto \frac{1}{\sigma^2}$ and marginalized σ^2 out of the joint posterior.)

- The predictive distribution of the next observation is $normal(m', (s')^2)$ where the mean $m' = m_n$, the posterior mean, and $(s')^2 = \sigma^2 + s_n^2$, the observation variance plus the posterior variance. (The posterior variance s_n^2 allows for the uncertainty in estimating μ.) The predictive distribution is found by marginalizing μ out of the joint distribution $f(y_{n+1}, \mu | y_1, \ldots, y_n)$.

Exercises

11.1. You are the statistician responsible for quality standards at a cheese factory. You want the probability that a randomly chosen block of cheese labelled "1 kg" is actually less than 1 kilogram (1,000 grams) to be 1% or less. The weight (in grams) of blocks of cheese produced by the machine is $normal\,(\mu, \sigma^2)$ where $\sigma^2 = 3^2$. The weights (in grams) of 20 blocks of cheese are:

994	997	999	1003	994	998	1001	998	996	1002
1004	995	994	995	998	1001	995	1006	997	998

You decide to use a discrete prior distribution for μ with the following probabilities:

$$g(\mu) = \begin{cases} .05 & \text{for } \mu \in \{991, 992, \ldots, 1010\}, \\ 0 & \text{otherwise.} \end{cases}$$

(a) Calculate your posterior probability distribution.

(b) Calculate your posterior probability that $\mu < 1,000$.

(c) Should you adjust the machine?

11.2. The city health inspector wishes to determine the mean bacteria count per liter of water at a popular city beach. Assume the number of bacteria per liter of water is $normal$ with mean μ and standard deviation known to be $\sigma = 15$. She collects 10 water samples and found the bacteria counts to be:

175	190	215	198	184
207	210	193	196	180

She decides that she will use a discrete prior distribution for μ with the following probabilities:

$$g(\mu) = \begin{cases} .125 & \text{for } \mu \in \{160, 170, \ldots, 230\} \\ 0 & \text{otherwise} \end{cases}$$

Calculate her posterior distribution.

11.3. The standard process for making a polymer has mean yield 35%. A chemical engineer has developed a modified process. He runs the process on 10 batches and measures the yield (in percent) for each batch. They are:

38.7	40.4	37.2	36.6	35.9
34.7	37.6	35.1	37.5	35.6

Assume that yield is $normal(\mu, \sigma^2)$ where the standard deviation $\sigma = 3$ is known.

(a) Use a $normal(30, 10^2)$ prior for μ. Find the posterior distribution.

(b) The engineer wants to know if the modified process increases the mean yield. Set this up as a hypothesis test stating clearly the null and alternative hypotheses.

(c) Perform the test at the 5% level of significance.

11.4. An engineer takes a sample of 5 steel I beams from a batch, and measures the amount they sag under a standard load. The amounts in mm are:

5.19	4.72	4.81	4.87	4.88

It is known that the sag is $normal(\mu, \sigma^2)$ where the standard deviation $\sigma = .25$ is known.

(a) Use a $normal(5, .5^2)$ prior for μ. Find the posterior distribution.

(b) For a batch of I beams to be acceptable, the mean sag under the standard load must be less than 5.20. ($\mu < 5.20$). Set this up as a hypothesis test stating clearly the null and alternative hypotheses.

(c) Perform the test at the 5% level of significance.

11.5. New Zealand was the last major land mass to be settled by human beings. The Shag River Mouth in Otago (lower South Island), New Zealand, is one of the sites of early human inhabitation that New Zealand archeologists have investigated, in trying to determine when the Polynesian migration to New Zealand occurred and documenting local adaptations to New Zealand conditions. Petchey and Higham (2000) describe the radiocarbon dating of well-preserved barracouta *thyrsites atun* bones found at the Shag River Mouth site. They obtained four acceptable samples, which were analyzed by the Waikato University Carbon Dating Unit. Assume that the conventional radiocarbon age (CRA) of a sample follows the $normal(\mu, \sigma^2)$ distribution, where the standard deviation $\sigma = 40$ is known. The observations are:

Observation	1	2	3	4
CRA	940	1040	910	990

(a) Use a *normal*$(1000, 200^2)$ prior for μ. Find the posterior distribution $g(\mu|y_1, \ldots, y_4)$.

(b) Find a 95% credible interval for μ.

(c) To find the θ, the calibrated date, the Stuiver, Reimer, and Braziunas marine curve (Stuiver et al., 1998) was used. We will approximate this curve with the linear function

$$\theta = 2203 - .835 \times \mu.$$

Find the posterior distribution of θ given $y_1, \ldots, y_4$.

(d) Find a 95% credible interval for θ, the calibrated date.

11.6. The Houhora site in Northland (top of North Island) New Zealand is one of the sites of early human inhabitation that New Zealand archeologists have investigated, in trying to determine when the Polynesian migration to New Zealand occurred and documenting local adaptations to New Zealand conditions. Petchey (2000) describe the Radiocarbon dating of well-preserved snapper *Pagrus auratus* bones found at the Houhora site. They obtained four acceptable samples which were analyzed by the Waikato University Carbon Dating Unit. Assume that the conventional radiocarbon age (CRA) of a sample follows the *normal*(μ, σ^2) distribution where the standard deviation $\sigma = 40$ is known. The observations are:

Observation	1	2	3	4
CRA	1010	1000	950	1050

(a) Use a *normal*$(1000, 200^2)$ prior for μ. Find the posterior distribution $g(\mu|y_1, \ldots, y_4)$.

(b) Find a 95% credible interval for μ.

(c) To find the θ, the calibrated date, the Stuiver, Reimer, Braziunas marine curve (Stuiver et al., 1998) was used. We will approximate this curve with the linear function

$$\theta = 2203 - .835 \times \mu.$$

Find the posterior distribution of θ given $y_1, \ldots, y_4$.

(d) Find a 95% credible interval for θ, the calibrated date.

Computer Exercises

11.1. [**Minitab:**] Use the Minitab macro *NormDP* to find the posterior distribution of the mean μ when we have a random sample of observations from a $normal(\mu, \sigma^2)$, where σ^2 is known, and we have a *discrete* prior for μ.

[**R:**]   Use the R function **normdp** to find the posterior distribution of the mean μ when we have a random sample of observations from a *normal*(μ, σ^2), where σ^2 is known, and we have a *discrete* prior for μ.

Suppose we have a random sample of $n = 10$ observations from a *normal*(μ, σ^2) distribution where it is known $\sigma^2 = 4$. The random sample of observations are:

3.07	7.51	5.95	6.83	8.80	4.19	7.44	7.06	9.67	6.89

We only allow that there are 12 possible values for μ, 4.0, 4.5, 5.0, 5.5, 6.0, 6.5, 7.0, 7.5, 8.0, 8.5, 9.0, and 9.5. If we do not favor any possible value over another, so we give all possible values of μ probability equal to $\frac{1}{12}$. The prior distribution is:

$$g(\mu) = \begin{cases} .083333 & \text{for } \mu \in \{4.0, 4.5, \ldots, 9.0, 9.5\}, \\ 0 & \text{otherwise.} \end{cases}$$

[**Minitab:**] Use *NormDP* to find the posterior distribution $g(\mu|y_1, \ldots, y_{10})$. Details for invoking *NormDP* are in Appendix C.

[**R:**] Use **normdp** function to find the posterior distribution $g(\mu|y_1, \ldots, y_{10})$. Details for using **normdp** are in Appendix D.

11.2. Suppose another 6 random observations come later. They are:

6.22	3.99	3.67	6.35	7.89	6.13

Use *NormDP* in Minitab, or **normdp** in R, to find the posterior distribution, where we will use the posterior after the first ten observations $y_1, \ldots, y_{10}$, as the prior for the next six observations $y_{11}, \ldots, y_{16}$.

11.3. Instead, combine all the observations together to give a random sample of size $n = 16$, and use *NormDP* in Minitab, or **normdp** in R, to find the posterior distribution where we go back the original prior that had all the possible values equally likely. What do the results of the last two problems show us?

11.4. Instead of thinking of a random sample of size $n = 16$, let's think of the sample mean as a single observation from its distribution.

(a) What is the distribution of $\bar{y}$? Calculate the observed value of $\bar{y}$?

(b) Use *NormDP* in Minitab, or `normdp` in R, to find the posterior distribution $g(\mu|\bar{y})$.

(c) What does this show us?

11.5. We will use the Minitab macro *NormNP*, or the R function `normnp`, to find the posterior distribution of the normal mean μ when we have a random sample of size n from a $normal(\mu, \sigma^2)$ distribution with known σ^2, and we use a $normal(m, s^2)$ prior for μ. The *normal* family of priors is the conjugate family for *normal* observations. That means that if we start with one member of the family as the prior distribution, we will get another member of the family as the posterior distribution. It is especially easy; if we start with a $normal(m, s^2)$ prior, we get a $normal(m', (s')^2)$ posterior where $(s')^2$ and m' are given by

$$\frac{1}{(s')^2} = \frac{1}{s^2} + \frac{n}{\sigma^2}$$

and

$$m' = \frac{1/s^2}{1/(s')^2} \times m + \frac{n/\sigma^2}{1/(s')^2} \times \bar{y},$$

respectively. Suppose the $n = 15$ observations from a $normal(\mu, \sigma^2 = 4^2)$ are:

26.8	26.3	28.3	28.5	26.3
31.9	28.5	27.2	20.9	27.5
28.0	18.6	22.3	25.0	31.5

[**Minitab:**] Use *NormNP* to find the posterior distribution $g(\mu|y_1, \ldots, y_{15})$, where we choose a $normal(m = 20, s^2 = 5^2)$ prior for μ. The details for invoking *NormNP* are in Appendix C. Store the likelihood and posterior in c3 and c4, respectively.

[**R:**] Use `normnp` to find the posterior distribution $g(\mu|y_1, \ldots, y_{15})$, where we choose a $normal(m = 20, s^2 = 5^2)$ prior for μ. The details for calling `normnp` are in Appendix D. Store the results in a variable of your choice for later use.

(a) What are the posterior mean and standard deviation?

(b) Find a 95% credible interval for μ.

11.6. Repeat part (a) with a $normal(30, 4^2)$ prior, storing the likelihood and posterior in c5 and c6.

11.7. Graph both posteriors on the same graph. What do you notice? What do you notice about the two posterior means and standard deviations? What do you notice about the two credible intervals for π?

11.8. [**Minitab:**] We will use the Minitab macro *NormGCP* to find the posterior distribution of the normal mean μ when we have a random samples of size n of $normal(\mu, \sigma^2)$ observations with known $\sigma^2 = 2^2$, and we have a general continuous prior for μ.

[**R:**] We will use the R function `normgcp` to find the posterior distribution of the normal mean μ when we have a random samples of size n of $normal(\mu, \sigma^2)$ observations with known $\sigma^2 = 2^2$, and we have a general continuous prior for μ.

Suppose the prior has the shape given by

$$
g(\mu) = \begin{cases} \mu & \text{for } 0 < \mu \leq 3 \,, \\ 3 & \text{for } 3 < \mu < 5 \,, \\ 8 - \mu & \text{for } 5 < \mu \leq 8 \,, \\ 0 & \text{for } 8 < \mu \,. \end{cases}
$$

[**Minitab:**] Store the values of μ and prior $g(\mu)$ in column c1 and c2, respectively.

Suppose the random sample of size $n = 16$ is:

4.09	4.68	1.87	2.62	5.58	8.68	4.07	4.78
4.79	4.49	5.85	5.90	2.40	6.27	6.30	4.47

[**Minitab:**] Use *NormGCP* to determine the posterior distribution $g(\mu|y_1, \ldots, y_{16})$, the posterior mean and standard deviation, and a 95% credible interval. Details for invoking *NormGCP* are in Appendix C.

[**R:**] Use `normgcp` to determine the posterior distribution $g(\mu|y_1, \ldots, y_{16})$. Use `mean` to determine the posterior mean and `sd` to determine the standard deviation. Use `quantile` to compute a 95% credible interval. Details for calling `normgcp`, `mean`, `sd` and `quantile` are in Appendix D.

CHAPTER 12

COMPARING BAYESIAN AND FREQUENTIST INFERENCES FOR MEAN

Making inferences about the population mean when we have a random sample from a normally distributed population is one of the most widely encountered situations in statistics. From the Bayesian point of view, the posterior distribution sums up our entire belief about the parameter, given the sample data. It really is the complete inference. However, from the frequentist perspective, there are several distinct types of inference that can be done: point estimation, interval estimation, and hypothesis testing. Each of these types of inference can be performed in a Bayesian manner, where they would be considered summaries of the complete inference, the posterior. In Chapter 9 we compared the Bayesian and frequentist inferences about the population proportion π. In this chapter we look at the frequentist methods for point estimation, interval estimation, and hypothesis testing about μ, the mean of a normal distribution, and compare them with their Bayesian counterparts using frequentist criteria.

Introduction to Bayesian Statistics, 3rd ed.   **237**
By Bolstad, W. M. and Curran, J. M. Copyright © 2016 John Wiley & Sons, Inc.

12.1   Comparing Frequentist and Bayesian Point Estimators

A frequentist point estimator for a parameter is a statistic that we use to estimate the parameter. The simple rule we use to determine a frequentist estimator for μ is to use the statistic that is the sample analog of the parameter to be estimated. So we use the sample mean $\bar{y}$ to estimate the population mean μ.[1]

In Chapter 9 we learned that frequentist estimators for unknown parameters are evaluated by considering their sampling distribution. In other words, we look at the distribution of the estimator over all possible samples. A commonly used criterion is that the estimator be *unbiased*. That is, the mean of its sampling distribution is the true unknown parameter value. The second criterion is that the estimator have small variance in the class of all possible unbiased estimators. The estimator that has the smallest variance in the class of unbiased estimators is called the *minimum variance unbiased estimator* and is generally preferred over other estimators from the frequentist point of view.

When we have a random sample from a normal distribution, we know that the sampling distribution of $\bar{y}$ is normal with mean μ and variance $\frac{\sigma^2}{n}$. The sample mean, $\bar{y}$, turns out to be the *minimum variance unbiased estimator* of μ.

We take the mean of the posterior distribution to be the Bayesian estimator for μ:

$$\hat{\mu}_B = \mathrm{E}[\mu|y_1, \ldots, y_n] = \frac{1/s^2}{n/\sigma^2 + 1/s^2} \times m + \frac{n/\sigma^2}{n/\sigma^2 + 1/s^2} \times \bar{y}.$$

We know that the posterior mean minimizes the posterior mean square. This means that $\hat{\mu}_B$ is the optimum estimator in the *post-data* setting. In other words, it is the optimum estimator for μ given our sample data and using our prior.

We will compare its performance to that of $\hat{\mu}_f = \bar{y}$ under the frequentist assumption that the true mean μ is a fixed but unknown constant. The probabilities will be calculated from the sampling distribution of $\bar{y}$. In other words, we are comparing the two estimators for μ in the *pre-data* setting.

The posterior mean is a linear function of the random variable $\bar{y}$, so its expected value is

$$\mathrm{E}[\hat{\mu}_B] = \frac{1/s^2}{n/\sigma^2 + 1/s^2} \times m + \frac{n/\sigma^2}{n/\sigma^2 + 1/s^2} \times \mu.$$

---

[1]The maximum likelihood estimator is the value of the parameter that maximizes the likelihood function. It turns out that $\bar{y}$ is the maximum likelihood estimator of μ for a normal random sample.

The bias of the posterior mean is its expected value minus the true parameter value, which simplifies to

$$\frac{\sigma^2}{ns^2 + \sigma^2}(m - \mu).$$

The posterior mean is a biased estimator of μ. The bias could only be 0 if our prior mean coincides with the unknown true value. The probability of that happening is 0. The bias increases linearly with the distance the prior mean m is from the true unknown mean μ. The variance of the posterior mean is

$$\left[\frac{n/\sigma^2}{n/\sigma^2 + 1/s^2}\right]^2 \times \frac{\sigma^2}{n} = \left(\frac{ns^2}{ns^2 + \sigma^2}\right)^2 \times \frac{\sigma^2}{n}$$

and is seen to be clearly smaller than $\frac{\sigma^2}{n}$, which is the variance of the frequentist estimator $\hat{\mu}_f = \bar{y}$. The mean squared error of an estimator combines both the bias and the variance into a single measure:

$$\mathrm{MSE}[\hat{\mu}_B] = \mathrm{Bias}^2 + \mathrm{Var}[\hat{\mu}].$$

The frequentist estimator $\hat{\mu}_f = \bar{y}$ is an unbiased estimator of μ, so its mean squared error equals its variance:

$$\mathrm{MSE}(\hat{\mu}_f) = \frac{\sigma^2}{n}.$$

When there is prior information, we will see that the Bayesian estimator has smaller mean squared error over the range of μ values that are realistic.

■ **EXAMPLE 12.1**

Arnold, Beth, and Carol want to estimate the mean weight of "1 kg" packages of milk powder produced at a dairy company. The weight in individual packages is subject to random variation. They know that when the machine is adjusted properly, the weights are normally distributed with mean 1015 grams, and standard deviation 5 g. They are going to base their estimate on a sample of size 10. Arnold decides to use a normal prior with mean 1,000 g and standard deviation 10 g. Beth decides she will use a normal prior with mean 1,015 g and standard deviation 7.5 g. Carol decides she will use a "flat" prior. They calculate the bias, variance, and mean squared error of their estimators for various values of μ to see how well they perform.

Figure 12.1 shows that only Carol's prior will give an unbiased Bayesian estimator. Her posterior Bayesian estimator corresponds exactly to the frequentist estimator $\hat{\mu}_f = \bar{y}$, since she used the "flat" prior. In Figure 12.2 we see the ranges over which the Bayesian estimators have smaller MS than the frequentist estimator. In that range they will be closer to

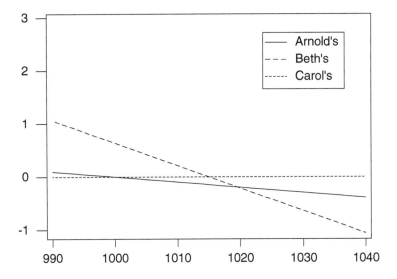

Figure 12.1   Biases of Arnold's, Beth's, and Carol's estimators.

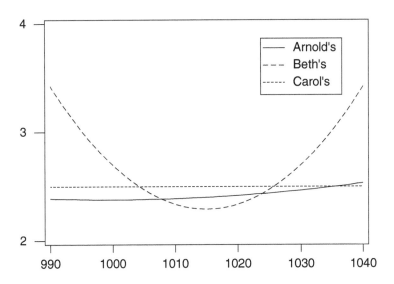

Figure 12.2   Mean-squared errors of Arnold's, Beth's, and Carol's estimators.

the true value, on average, than the frequentist estimator. The realistic range is the target mean (1,015) plus or minus 3 standard deviations (5) which is from 1,000 to 1,030.

Although both Arnold and Beth's estimators are biased since they are using the Bayesian approach, they have smaller mean squared error over most of the feasible range than Carol's estimator (which equals the ordinary frequentist estimator). Since they have smaller mean squared error, on average, they will be closer to the true value in most of the feasible range. In particular, Beth's estimator seems to offer substantially better performance over most of the feasible range, while Arnold's estimator offers somewhat better performance over the entire feasible range.   ∎

12.2   Comparing Confidence and Credible Intervals for Mean

Frequentist statisticians compute confidence intervals for the parameter μ to determine an interval that "has a high probability of containing the true value." Since they are done from the frequentist perspective, the parameter μ is considered a fixed but unknown constant. The coverage probability is found from the sampling distribution of an estimator, in this case $\bar{y}$, the sample mean. The sampling distribution of $\bar{y}$ is *normal* with mean μ and variance $\frac{\sigma^2}{n}$. We know before we take the sample that $\bar{y}$ is a random variable, so we can make the probability statement about $\bar{y}$:

$$P\left(\mu - z_{\frac{\alpha}{2}} \times \frac{\sigma}{\sqrt{n}} < \bar{y} < \mu + z_{\frac{\alpha}{2}} \times \frac{\sigma}{\sqrt{n}}\right) = 1 - \alpha,$$

where $z_{\frac{\alpha}{2}}$ is the value from the standard normal table having tail area $\frac{\alpha}{2}$. We rearrange this probability statement to have μ in the middle. The upper inequality in the first statement becomes the lower inequality in the second statement, and vice versa:

$$P\left(\bar{y} - z_{\frac{\alpha}{2}} \times \frac{\sigma}{\sqrt{n}} < \mu < \bar{y} + z_{\frac{\alpha}{2}} \times \frac{\sigma}{\sqrt{n}}\right) = 1 - \alpha.$$

The endpoints of the interval are random because they depend on $\bar{y}$, which is the random variable in this interpretation. The parameter μ is considered a fixed but unknown constant. So the correct interpretation is that $(1 - \alpha) \times 100\%$ of the intervals calculated this way will contain the true value. When we take our random sample and calculate $\bar{y}$, there is nothing random left to attach a probability to. The actual interval we calculate either contains the true value or it does not. Only we do not know which is true. So we say that we are $(1 - \alpha) \times 100\%$ *confident* that the interval we calculated using the observed value of $\bar{y}$,

$$\bar{y} \pm z_{\frac{\alpha}{2}} \times \frac{\sigma}{\sqrt{n}}, \tag{12.1}$$

does contain the true value. Our confidence comes from the sampling distribution of the statistic. It does not come from the actual sample values we used to calculate the endpoints of the confidence interval. Sometimes we write the confidence interval as

$$\left(\bar{y} - z_{\frac{\alpha}{2}} \times \frac{\sigma}{\sqrt{n}} \,,\ \bar{y} + z_{\frac{\alpha}{2}} \times \frac{\sigma}{\sqrt{n}}\right).$$

This contrasts with the Bayesian credible interval for μ that we calculated in the previous chapter. The probability statement we make is from the posterior distribution of the parameter μ given the sample data $y_1, \ldots, y_n$. It is conditional on the actual sample data we obtained. The probability given in the statement is our probability given the actual sample. It is a legitimate probability statement, since μ is considered random. But it is *subjective* because we constructed it using our *subjective* prior. Someone else who started with a different prior would end up with a (slightly) different credible interval.

Relationship between Frequentist Confidence Interval and Bayesian Credible Interval from "Flat" Prior

With a flat prior for μ, the posterior mean equals $m' = \bar{y}$, and the posterior variance equals $(s')^2 = \sigma^2/n$. So for this case the Bayesian credible interval and the frequentist confidence interval will both have the form

$$\left(\bar{y} - z_{\frac{\alpha}{2}} \times \frac{\sigma}{\sqrt{n}} < \mu < \bar{y} + z_{\frac{\alpha}{2}} \times \frac{\sigma}{\sqrt{n}} \right).$$

However, they have different interpretations.

The frequentist interpretation is that μ is fixed. The endpoints of the random interval are calculated using a probability statement on the sampling distribution of the statistic $\bar{y}$. There is no randomness left after the actual sample data have been used to calculate the endpoints. No probability statements can be made about the actual calculated interval. The confidence level $(1 - \alpha) \times 100\%$ associated with the interval means that $(1 - \alpha) \times 100\%$ of the random intervals calculated this way will contain the true unknown parameter, so we are $(1 - \alpha) \times 100\%$ *confident* that the one we calculate does.

The Bayesian interpretation lets μ be a random variable, so probability statements are allowed. The credible interval is calculated from the posterior distribution given the actual sample data that occurred. The credible interval has the stated conditional probability of containing μ, given the data.

Scientists are not interested in what would happen with hypothetical repetitions of the experiment giving all possible data sets. The only data set that matters is the one that occurred. They find direct probability statements about the parameter, conditional on their actual data set to be the most useful. Scientists often take the confidence interval given by the frequentist statistician and misinterpret it as a probability interval for the parameter given the data. The statistician knows that this interpretation is not the correct one but lets the scientist make the misinterpretation. The correct interpretation is scientifically useless.

Fortunately for frequentist statisticians, when they allow their clients to make the probability interpretation from the confidence interval for the mean of a normal distribution, μ, they can get away with it. Their interval is equivalent to the Bayesian credible interval from a "flat" prior, which allows the probability interpretation in this case

■ **EXAMPLE 12.2   (continued from Example 11.3, p. 222)**

Previous studies have determined that the length of yearling trout has a normal $(\mu, \sigma^2 = 2^2)$ distribution. Arnie, Barb, and Chuck obtained a random sample of 12 yearling trout. The sample mean $\bar{y} = 32$ cm. The 95% confidence interval for μ is given by

$$\bar{y} \pm z_{.025} \times \frac{\sigma}{\sqrt{n}} = 32 \pm 1.96 \times \frac{2}{\sqrt{12}} = (30.87, 33.13) \, .$$

Compare this with the 95% credible intervals they found in Table 11.5. We see that it is the same as the credible interval Barb found because she used the "flat" prior.   ■

12.3   Testing a One-Sided Hypothesis about a Normal Mean

Often we get data from a new population similar to a population we already know about. For instance, the new population may be the set of all possible outcomes of an experiment, where we have changed one of the experimental factors from its standard value to a new value. We know that the mean value of the standard population is μ_0. We assume that each observation from the new population is $normal(\mu, \sigma^2)$, where σ^2 is known, and that the observations are independent of each other. The question we want to answer is, Is the mean μ for the new population greater than the mean of the standard population? A one-sided hypothesis test attempts to answer that question. We consider that there are two possible explanations to any discrepancy between the observed data and μ_0.

1. The mean of the new population is less than or equal to the mean of the standard population, and any discrepancy is due to chance alone.

2. The mean of the new population is greater than the mean of the standard population and at least part of the discrepancy is due to this fact.

Hypothesis testing is a way to protect our credibility by making sure that we do not reject the first explanation unless it has probability less than our chosen level of significance α. Note that we set up the positive answer to the question we are asking as the alternative hypothesis. The null hypothesis will be the negative answer to the question. We will compare the frequentist and Bayesian approaches.

Frequentist One-Sided Hypothesis Test about μ

As we saw in Chapter 9, frequentist tests are based on the sampling distribution of a statistic. This makes the probabilities *pre-data* in that they arise from all possible random samples that could have occurred. The steps are:

1. Set up the null and alternative hypothesis

$$H_0 : \mu \leq \mu_0 \quad \text{versus} \quad H_1 : \mu > \mu_0 \,.$$

   Note the alternative hypothesis is the change in the direction we are interested in detecting. Any change in the other direction gets lumped into the null hypothesis. (We are trying to detect $\mu > \mu_0$. If $\mu < \mu_0$, it is not of any interest to us, so those values get included in the null hypothesis.)

2. The null distribution of $\bar{y}$ is $normal(\mu_0, \frac{\sigma^2}{n})$. This is the sampling distribution of $\bar{y}$ when the null hypothesis is true. Hence the null distribution of the standardized variable

$$z = \frac{\bar{y} - \mu_0}{\sigma/\sqrt{n}}$$

   will be $normal(0, 1)$.

3. Choose a level of significance α. Commonly this is .10, .05, or .01.

4. Determine the rejection region. This is a region that has probability α when the null hypothesis is true ($\mu = \mu_0$). When $\alpha = .05$, the rejection region is $z > 1.645$. This is shown in Figure 12.3.

5. Take the sample data and calculate $\bar{y}$. If the value falls in the rejection region, we reject the hypothesis at level of significance $\alpha = .05$; otherwise we cannot reject the null hypothesis.

6. Another way to perform the test is to calculate the P-value which is the probability of observing what we observed, or something even more extreme, given the null hypothesis $H_0 : \mu = \mu_0$ is true:

$$P\text{-value} = P\left(Z \geq \frac{\bar{y} - \mu_0}{\sigma/\sqrt{n}} \right). \tag{12.2}$$

   If P-value $\leq \alpha$, then we reject the null hypothesis; otherwise we cannot reject it.

Bayesian One-Sided Hypothesis Test about μ

The posterior distribution $g(\mu|y_1, \cdots, y_n)$ summarizes our entire belief about the parameter, after viewing the data. Sometimes we want to answer a specific

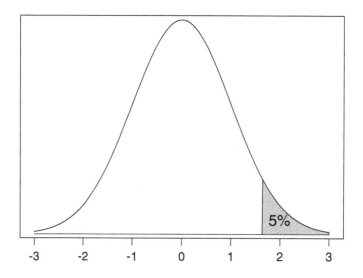

Figure 12.3   Null distribution of $z = \frac{\bar{y}-\mu_0}{\sigma/\sqrt{n}}$ with rejection region for one-sided frequentist hypothesis test at 5% level of significance.

question about the parameter. This could be, Given the data, can we conclude the parameter μ is greater than μ_0? The value μ_0 ordinarily comes from previous experience. If the parameter is still equal to that value, then the experiment has not demonstrated anything new that requires explaining. We would lose our scientific credibility if we go around concocting explanations for effects that may not exist. The answer to the question can be resolved by testing

$$H_0 : \mu \leq \mu_0 \quad \text{versus} \quad H_1 : \mu > \mu_0 \,.$$

This is an example of a one-sided hypothesis test. We decide on a level of significance α that we wish to use. It is the probability below which we will reject the null hypothesis. Usually α is small, for instance, .10, .05, .01, .005, or .001. Testing a one-sided hypothesis in Bayesian statistics is done by calculating the posterior probability of the null hypothesis:

$$P(H_0 : \mu \leq \mu_0 | y_1, \ldots, y_n) = \int_{-\infty}^{\mu_0} g(\mu | y_1, \ldots, y_n) \, d\mu. \qquad (12.3)$$

When the posterior distribution $g(\mu | y_1, \ldots, y_n)$ is $normal(m', (s')^2)$, this can easily be found from standard normal tables.

$$P(H_0 : \mu \leq \mu_0 | y_1, \ldots, y_n) = P\left(\frac{\mu - m'}{s'} \leq \frac{\mu_0 - m'}{s'}\right)$$

$$= P\left(Z \leq \frac{\mu_0 - m'}{s'}\right), \qquad (12.4)$$

where Z is a standard normal random variable. If the probability is less than our chosen α, we reject the null hypothesis and can conclude that $\mu > \mu_0$. Only then can we search for an explanation of why μ is now larger than μ_0.

■ **EXAMPLE 12.3**   **(continued from Example 11.3, p. 222)**

Arne, Barb, and Chuck read in a journal that the mean length of yearling rainbow trout in a typical stream habitat is 31 cm. They each decide to determine if the mean length of trout in the stream they are researching is greater than that by testing

$$H_0 : \mu \leq 31 \quad \text{versus} \quad H_1 : \mu > 31$$

at the $\alpha = 5\%$ level. For one-sided Bayesian hypothesis tests, they calculate the posterior probability of the null hypothesis. Arnie and Barb have normal posteriors, so they use Equation 12.4. Chuck has a nonnormal posterior that he calculated numerically.

[**Minitab:**] He calculates the posterior probability of the null hypothesis using Equation 12.3, and he evaluates it numerically using the Minitab macro *tintegral*.

[**R:**] He calculates the posterior probability of the null hypothesis using Equation 12.3, and he evaluates it numerically using the R function `cdf`.

The results of the Bayesian hypothesis tests are shown in Table 12.1.
They also decide that they will perform the corresponding frequentist hypothesis test of

$$H_0 : \mu \leq 31 \quad \text{versus} \quad H_1 : \mu > 31$$

and compare the results. The null distribution of $z = \frac{\bar{y}-31}{\sigma/\sqrt{n}}$ and the correct rejection region are given in Figure 12.3. For this data, $z = \frac{32-31}{2/\sqrt{12}} = 1.732$. This lies in the rejection region; hence the null hypothesis is rejected at the 5% level. The other way we could perform this frequentist hypothesis test is to calculate the P-value. For these data,

$$P\text{-value} = P\left(Z > \frac{32-31}{2/\sqrt{12}} \right)$$
$$= P(Z > 1.732),$$

which equals .0416 from the standard normal table in Appendix B (Table B.2). This is less than the level of significance α, so the null hypothesis is rejected, same as before.[2]

■

---

[2]We note that in this case the P-value equals Barb's probability of the null hypothesis because she used the "flat" prior. For the *normal* case, the P-value can be interpreted

Table 12.1   Results of Bayesian one-sided hypothesis tests

Person	Posterior	$P(\mu \leq 31 \vert y_1, \ldots, y_n)$	
Arnie	$normal(31.96, .5714^2)$	$P(Z \leq \frac{31-31.96}{.5714})$ = .0465	reject
Barb	$normal(32.00, .5774^2)$	$P(Z \leq \frac{31-32}{.5774})$ = .0416	reject
Chuck	$numerical$	$\int_{-\infty}^{31} g(\mu \vert y_1, \ldots, y_n) d\mu$ = .0489	reject

12.4   Testing a Two-Sided Hypothesis about a Normal Mean

Sometimes the question we want to have answered is, Is the mean for the new population μ the same as the mean for the standard population which we know equals μ_0? A two-sided hypothesis test attempts to answer this question. We are interested in detecting a change in the mean, in either direction. We set this up as

$$H_0 : \mu = \mu_0 \quad \text{versus} \quad H_1 : \mu \neq \mu_0 \,. \tag{12.5}$$

The null hypothesis is known as a *point hypothesis*. This means that it is true only for the exact value μ_0. This is only a single point along the number line. At all the other values in the parameter space the null hypothesis is false. When we think of the infinite number of possible parameter values in an interval of the real line, we see that the it is impossible for the null hypothesis to be literally true. There are an infinite number of values that are extremely close to μ_0 but eventually differ from μ_0 when we look at enough decimal places. So rather than testing whether we believe the null hypothesis to actually be true, we are testing whether the null hypothesis is in the range that could be true.

Frequentist Two-Sided Hypothesis Test About μ

1. The null and alternative hypothesis are set up as in Equation 12.5. Note that we are trying to detect a change in either direction.

as the posterior probability of the null hypothesis when the noninformative "flat" prior was used. However, it is not generally true that P-value has any meaning in the Bayesian perspective.

2. The null distribution of the standardized variable

$$z = \frac{\bar{y} - \mu_0}{\sigma/\sqrt{n}}$$

will be *normal*$(0, 1)$.

3. Choose α, the level of significance. This is usually a low value such as .10, .05, .01, or .001.

4. Determine the rejection region. This is a region that has probability = α when the null hypothesis is true. For a two-sided hypothesis test, we have a two-sided rejection region. When $\alpha = .05$, the rejection region is $|z| > 1.96$. This is shown in Figure 12.4.

5. Take the sample and calculate $z = \frac{\bar{y} - \mu_0}{\sigma/\sqrt{n}}$. If it falls in the rejection region, reject the null hypothesis at level of significance α; otherwise we cannot reject the null hypothesis.

6. Another way to do the test is to calculate the P-value which is the probability of observing what we observed, or something even more extreme than what we observed, given the null hypothesis is true. Note that the P-value includes probability of two tails:

$$P\text{-value} = P\left(Z < -\left|\frac{\bar{y} - \mu_0}{\sigma/\sqrt{n}}\right|\right) + P\left(Z > \left|\frac{\bar{y} - \mu_0}{\sigma/\sqrt{n}}\right|\right).$$

If P-value $\leq \alpha$, then we can reject the null hypothesis; otherwise we cannot reject it.

Relationship between two-sided hypothesis test and confidence interval. We note that the rejection region for the two-sided test at level α is

$$z = \left|\frac{\bar{y} - \mu_0}{\sigma/\sqrt{n}}\right| > z_{\frac{\alpha}{2}},$$

and this can be manipulated to give either

$$\mu_0 < \bar{y} - z_{\frac{\alpha}{2}} \times \frac{\sigma}{\sqrt{n}} \quad \text{or} \quad \mu_0 > \bar{y} + z_{\frac{\alpha}{2}} \times \frac{\sigma}{\sqrt{n}}.$$

We see that if we reject $H_0 : \mu = \mu_0$ at the level α, then μ_0 lies outside the $(1 - \alpha) \times 100\%$ confidence interval for μ. Similarly, we can show that if we accept $H_0 : \mu = \mu_0$ at level α, then μ_0 lies inside $(1 - \alpha) \times 100\%$ confidence interval for μ. So the confidence interval contains all those values of μ_0 that would be accepted if tested for.

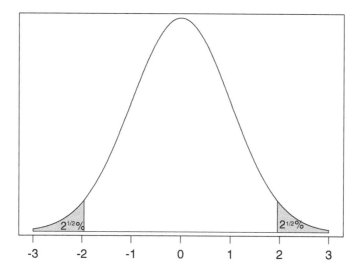

Figure 12.4   Null distribution of $z = \frac{\bar{y} - \mu_0}{\sigma/\sqrt{n}}$ with rejection region for two-sided frequentist hypothesis test at 5% level of significance.

Bayesian Two-Sided Hypothesis Test about μ

If we wish to test the two-sided hypothesis

$$H_0 : \mu = \mu_0 \quad \text{versus} \quad H_1 : \mu \neq \mu_0$$

in a Bayesian manner, and we have a continuous prior, we cannot calculate the posterior probability of the null hypothesis as we did for the one-sided hypothesis. Since we have a continuous prior, we have a continuous posterior. We know that the probability of any specific value of a continuous random variable always equals 0. The posterior probability of the null hypothesis $H_0 : \mu = \mu_0$ will equal zero. This means we cannot test this hypothesis by calculating the posterior probability of the null hypothesis and comparing it to α.

Instead, we calculate a $(1 - \alpha) \times 100\%$ credible interval for μ using our posterior distribution. If μ_0 lies inside the credible interval, we conclude that μ_0 still has credibility as a possible value. In that case we will not reject the null hypothesis $H_0 : \mu = \mu_0$, so we consider that it is credible that there is no effect. (However, we realize it has zero probability of being exactly true if we look at enough decimal places.) There is no need to search for an explanation of a nonexistent effect. However, if μ_0 lies outside the credible interval, we conclude that μ_0 does not have credibility as a possible value, and we will reject the null hypothesis. Then it is reasonable to attempt to explain why the mean has shifted from μ_0 for this experiment.

Main Points

- When we have prior information on the values of the parameter that are realistic, we can find a prior distribution so that the mean of the posterior distribution of μ (the Bayesian estimator) has a smaller mean squared error than the sample mean (the frequentist estimator) over the range of realistic values. This means that on the average, it will be closer to the true value of the parameter.

- A confidence interval for μ is found by inverting a probability statement for $\bar{y}$, and then plugging in the sample value to compute the endpoints. It is called a confidence interval because there is nothing left to be random, so no probability statement can be made after the sample value is plugged in.

- The interpretation of a $(1 - \alpha) \times 100\%$ frequentist confidence interval for μ is that $(1-\alpha) \times 100\%$ of the random intervals calculated this way would cover the true parameter, so we are $(1 - \alpha) \times 100\%$ *confident* that the interval we calculated does.

- A $(1 - \alpha) \times 100\%$ Bayesian credible interval is an interval such that the posterior probability it contains the random parameter is $(1-\alpha) \times 100\%$.

- This is more useful to the scientist because he/she is only interested in his/her particular interval.

- The $(1 - \alpha) \times 100\%$ frequentist confidence interval for μ corresponds to the $(1 - \alpha) \times 100\%$ Bayesian credible interval for μ when we used the "flat prior." So, in this case, frequentist statisticians can get away with misinterpreting their confidence interval for μ as a probability interval.

- In the general, misinterpreting a frequentist confidence interval as a probability interval for the parameter will be wrong.

- Hypothesis testing is how we protect our credibility, by not attributing an effect to a cause if that effect could be due to chance alone.

- If we are trying to detect an effect in one direction, say $\mu > \mu_0$, we set this up as the one-sided hypothesis test

$$H_0 : \mu \le \mu_0 \quad \text{versus} \quad H_1 : \mu > \mu_0 \, .$$

Note that the alternative hypothesis contains the effect we wish to detect. The null hypothesis is that the mean is still at the old value (or is changed in the direction we are not interested in detecting).

- If we are trying to detect an effect in either direction, we set this up as the two-sided hypothesis test

$$H_0 : \mu = \mu_0 \quad \text{versus} \quad H_1 : \mu \ne \mu_0 \, .$$

The null hypothesis contains only a single value μ_0 and is called a point hypothesis.

- Frequentist hypothesis tests are based on the sample space.

- The level of significance α is the low probability we allow for rejecting the null hypothesis when it is true. We choose α.

- A frequentist hypothesis test divides the sample space into a rejection region, and an acceptance region such that the probability the test statistic lies in the rejection region if the null hypothesis is true is less than the level of significance α. If the test statistic falls into the rejection region we reject the null hypothesis at level of significance α.

- Or we could calculate the P-value. If the P-value$< \alpha$, we reject the null hypothesis at level α.

- The P-value is not the probability the null hypothesis is true. Rather, it is the probability of observing what we observed, or even something more extreme, given that the null hypothesis is true.

- We can test a one-sided hypothesis in a Bayesian manner by computing the posterior probability of the null hypothesis by integrating the posterior density over the null region. If this probability is less than the level of significance α, then we reject the null hypothesis.

- We cannot test a two-sided hypothesis by integrating the posterior probability over the null region because with a continuous prior, the prior probability of a point null hypothesis is zero, so the posterior probability will also be zero. Instead, we test the credibility of the null value by observing whether or not it lies within the Bayesian credible interval. If it does, the null value remains credible and we cannot reject it.

Exercises

12.1. A statistician buys a pack of 10 new golf balls, drops each golf ball from a height of one meter, and measures the height in centimeters it returns on the first bounce. The ten values are:

| 79.9 | 80.0 | 78.9 | 78.5 | 75.6 | 80.5 | 82.5 | 80.1 | 81.6 | 76.7 |

Assume that y, the height (in cm) a golf ball bounces when dropped from a one-meter height, is $normal(\mu, \sigma^2)$, where the standard deviation $\sigma = 2$.

(a) Assume a $normal(75, 10^2)$ prior for μ. Find the posterior distribution of μ.

(b) Calculate a 95% Bayesian credible interval for μ.

(c) Perform a Bayesian test of the hypothesis

$$H_0 : \mu \geq 80 \quad \text{versus} \quad H_1 : \mu < 80$$

at the 5% level of significance.

12.2. The statistician buys ten used balls that have been recovered from a water hazard. He drops each from a height of one meter and measures the height in centimeters it returns on the first bounce. The values are:

73.1	71.2	69.8	76.7	75.3	68.0	69.2	73.4	74.0	78.2

Assume that y, the height (in cm) a golf ball bounces when dropped from a one-meter height, is $normal(\mu, \sigma^2)$, where the standard deviation $\sigma = 2$.

(a) Assume a emphnormal$(75, 10^2)$ prior for μ. Find the posterior distribution of μ.

(b) Calculate a 95% Bayesian credible interval for μ.

(c) Perform a Bayesian test of the hypothesis

$$H_0 : \mu \geq 80 \quad \text{versus} \quad H_1 : \mu < 80$$

at the 5% level of significance.

12.3. The local consumer watchdog group was concerned about the cost of electricity to residential customers over the New Zealand winter months (Southern Hemisphere). They took a random sample of 25 residential electricity accounts and looked at the total cost of electricity used over the three months of June, July, and August. The costs were:

514	536	345	440	427
443	386	418	364	483
506	385	410	561	275
306	294	402	350	343
480	334	324	414	296

Assume that the amount of electricity used over the three months by a residential account is $normal(\mu, \sigma^2)$, where the known standard deviation $\sigma = 80$.

(a) Use a $normal(325, 80^2)$ prior for μ. Find the posterior distribution for μ.

(b) Find a 95% Bayesian credible interval for μ.

(c) Perform a Bayesian test of the hypothesis

$$H_0 : \mu = 350 \quad \text{versus} \quad H_1 : \mu \neq 350$$

at the 5% level.

(d) Perform a Bayesian test of the hypothesis

$$H_0 : \mu \leq 350 \quad \text{versus} \quad H_1 : \mu > 350$$

at the 5% level.

12.4. A medical researcher collected the systolic blood pressure reading for a random sample of $n = 30$ female students under the age of 21 who visited the Student's Health Service. The blood pressures are:

120	122	121	108	133	119	136	108	106	105
122	139	133	115	104	94	118	93	102	114
123	125	124	108	111	134	107	112	109	125

Assume that systolic blood pressure comes from a *normal* (μ, σ^2) distribution where the standard deviation $\sigma = 12$ is known.

(a) Use a *normal*$(120, 15^2)$ prior for μ. Calculate the posterior distribution of μ.

(b) Find a 95% Bayesian credible interval for μ.

(c) Suppose we had not actually known the standard deviation σ. Instead, the value $\hat{\sigma} = 12$ was calculated from the sample and used in place of the unknown true value. Recalculate the 95% Bayesian credible interval.

CHAPTER 13

BAYESIAN INFERENCE FOR DIFFERENCE BETWEEN MEANS

Comparisons are the main tool of experimental science. When there is uncertainty present due to observation errors or experimental unit variation, comparing observed values cannot establish the existence of a difference because of the uncertainty within each of the observations. Instead, we must compare the means of the two distributions the observations came from. In many cases the distributions are normal, so we are comparing the means of two normal distributions. There are two experimental situations that the data could arise from.

The most common experimental situation is where there are independent random samples from each distribution. The treatments have been applied to different random samples of experimental units. The second experimental situation is where the random samples are paired. It could be that the two treatments have been applied to the same set of experimental units (at separate times). The two measurements on the same experimental unit cannot be considered independent. Or it could be that the experimental units were formed into pairs of similar units, with one of each pair randomly assigned to each treatment group. Again, the two measurements in the same pair cannot

Introduction to Bayesian Statistics, 3rd ed.
By Bolstad, W. M. and Curran, J. M. Copyright © 2016 John Wiley & Sons, Inc.

be considered independent. We say the observations are paired. The random samples from the two populations are dependent.

In Section 13.1 we look at how to analyze data from independent random samples. If the treatment effect is an additive constant, we get equal variances for the two distributions. If the treatment effect is random, not constant, we get unequal variances for the two distributions. In Section 13.2 we investigate the case where we have independent random samples from two normal distributions with equal variances. In Section 13.3, we investigate the case where we have independent random samples from two normal distributions with unequal variances. In Section 13.4 we investigate how to find the difference between proportions using the normal approximation, when we have independent random samples. In Section 13.5 we investigate the case where we have paired samples.

13.1   Independent Random Samples from Two Normal Distributions

We may want to determine whether or not a treatment is effective in increasing growth rate in lambs. We know that lambs vary in their growth rate. Each lamb in a flock is randomly assigned to either the treatment group or the control group that will not receive the treatment. The assignments are done independently. This is called a completely randomized design, and we discussed it in Chapter 2. The reason the assignments are done this way is that any differences among lambs enters the treatment group and control group randomly. There will be no bias in the experiment. On average, both groups have been assigned similar groups of lambs over the whole range of the flock. The distribution of underlying growth rates for lambs in each group is assumed to be normal with the same means and variances σ^2. The means and variances for the two groups are equal because the assignment is done randomly.

The mean growth rate for a lamb in the treatment group, μ_1, equals the mean underlying growth rate plus the treatment effect for that lamb. The mean growth rate for a lamb in the control group, μ_2, equals the mean underlying growth rate plus zero, since the control group does not receive the treatment. Adding a constant to a random variable does not change the variance, so if the treatment effect is constant for all lambs, the variances of the two groups will be equal. We call that an *additive* model. If the treatment effect is different for different lambs, the variances of the two groups will be unequal. This is called a *nonadditive* model.

If the treatment is effective, μ_1 will be greater than μ_2. In this chapter we will develop Bayesian methods for inference about the difference between means $\mu_1 - \mu_2$ for both additive and nonadditive models.

13.2   Case 1: Equal Variances

We often assume the treatment effect is the same for all units. The observed value for a unit given the treatment is the mean for that unit plus the constant treatment effect. Adding a constant does not change the variance, so the variance of the treatment group is equal to the variance of the control group. That sets up an *additive* model.

When the Variance Is Known

Suppose we know the variance σ^2. Since we know the two samples are independent of each other we will use independent priors for both means. They can either be *normal* (m_1, s_1^2) and *normal*(m_2, s_2^2) priors, or we can use *flat* priors for one or both of the means.

Because the priors are independent, and the samples are independent, the posteriors are also independent. The posterior distributions are

$$\mu_1 | y_{11}, \ldots, y_{n_{11}} \sim Normal(m_1', (s_1')^2)$$

and

$$\mu_2 | y_{12}, \ldots, y_{n_{22}} \sim Normal(m_2', (s_2')^2),$$

where the $m_1', (s_1')^2, m_2',$ and $(s_2')^2$ are found using the simple updating formulas given by Equations 11.5 and 11.6.

Since $\mu_1 | y_{11}, \ldots, y_{n_{11}}$ and $\mu_2 | y_{12}, \ldots, y_{n_{22}}$ are independent of each other, we can use the rules for mean and variance of a difference between independent random variables. This gives the posterior distribution of $\mu_d = \mu_1 - \mu_2$. It is

$$\mu_d | y_{11}, \ldots, y_{n_{11}}, y_{12}, \ldots, y_{n_{22}} \sim Normal(m_d', (s_d')^2),$$

where $m_d' = m_1' - m_2'$, and $(s_d')^2 = (s_1')^2 + (s_2')^2$. We can use this posterior distribution to make further inferences about the difference between means $\mu_1 - \mu_2$.

Credible interval for difference between means, known equal variance case. The general rule for finding a $(1 - \alpha) \times 100\%$ Bayesian credible interval when the posterior distribution is emphnormal$(m', (s')^2)$ is to take the *posterior mean ± critical value × posterior standard deviation*. When the observation variance (or standard deviation) is assumed known, the critical value comes from the standard normal table. In that case the $(1 - \alpha) \times 100\%$ Bayesian credible interval for $\mu_d = \mu_1 - \mu_2$ is

$$m_d' \pm z_{\frac{\alpha}{2}} \times s_d'. \tag{13.1}$$

This can be written as

$$m_1' - m_2' \pm z_{\frac{\alpha}{2}} \times \sqrt{(s_1')^2 + (s_2')^2}. \tag{13.2}$$

Thus, given the data, the probability that $\mu_1 - \mu_2$ lies between the endpoints of the credible interval equals $(1 - \alpha) \times 100\%$.

Confidence interval for difference between means, known equal variance case.
The frequentist confidence interval for $\mu_d = \mu_1 - \mu_2$ when the two distributions
have equal known variance is given by

$$\bar{y}_1 - \bar{y}_2 \pm z_{\frac{\alpha}{2}} \times \sigma \sqrt{\frac{1}{n_1} + \frac{1}{n_2}}. \tag{13.3}$$

This is the same formula as the Bayesian credible interval would be if we
had used independent "flat" priors for μ_1 and μ_2, but the interpretations are
different. The endpoints of the confidence interval are what is random under
the frequentist viewpoint. $(1 - \alpha) \times 100\%$ of the intervals calculated using
this formula would contain the fixed, but unknown, value $\mu_1 - \mu_2$. We would
have that confidence that the particular interval we calculated using our data
contains the true value.

◪ EXAMPLE 13.1

In Example 3.2 (Chapter 3, p. 40), we looked at two series of measurements Michelson made on the speed of light in 1879 and 1882, respectively. The data are shown in Table 3.3. (The measurements are figures
given plus 299,000.) Suppose we assume each speed of light measurement
is normally distributed with known standard deviation 100. Let us use
independent $normal(m, s^2)$ priors for the 1879 and 1882 measurements,
where $m = 300,000$ and $s^2 = 500^2$.

The posterior distributions of μ_{1879} and μ_{1882} can be found using the
updating rules. For μ_{1879} they give

$$\frac{1}{(s'_{1879})^2} = \frac{1}{500^2} + \frac{20}{100^2} = .002004\,,$$

so $(s'_{1879})^2 = 499$, and

$$m'_{1879} = \frac{\frac{1}{500^2}}{.002004} \times 300,000 + \frac{\frac{20}{100^2}}{.002004} \times (299,000 + 909) = 299,909\,.$$

Similarly, for μ_{1882} they give

$$\frac{1}{(s'_{1882})^2} = \frac{1}{500^2} + \frac{23}{100^2} = .002304\,,$$

so $(s'_{1882})^2 = 434$, and

$$m'_{1882} = \frac{\frac{1}{500^2}}{.002304} \times 300,000 + \frac{\frac{23}{100^2}}{.002304} \times (299,000 + 756) = 299,757\,.$$

The posterior distribution of $\mu_d = \mu_{1879} - \mu_{1882}$ will be $normal(m'_d, (s'_d)^2)$
where

$$m'_d = 299,909 - 299,757 = 152$$

and

$$(s'_d)^2 = 499 + 434 = 30.5^2 \,.$$

The 95% Bayesian credible interval for $\mu_d = \mu_{1879} - \mu_{1882}$ is

$$152 \pm 1.96 \times 30.5 = (92.1, 211.9) \,.$$

∎

One-sided Bayesian hypothesis test.   If we wish to determine whether or not the treatment mean μ_1 is greater than the control mean μ_2, we will use hypothesis testing. We test the null hypothesis

$$H_0 : \mu_d \leq 0 \quad \text{versus} \quad H_1 : \mu_d > 0 \,,$$

where $\mu_d = \mu_1 - \mu_2$ is the difference between the two means. To do this test in a Bayesian manner, we calculate the posterior probability of the null hypothesis $P(\mu_d \leq 0 | data)$ where $data$ includes the observations from both samples $y_{11}, \ldots, y_{n_{11}}$ and $y_{12}, \ldots, y_{n_{22}}$. Standardizing by subtracting the mean and dividing by the standard deviation gives

$$P(\mu_d \leq 0 | data) = P\left(\frac{\mu_d - m'_d}{s'_d} \leq \frac{0 - m'_d}{s'_d} \right)$$

$$= P\left(Z \leq \frac{0 - m'_d}{s'_d} \right), \tag{13.4}$$

where Z has the standard normal distribution. We find this probability in Table B.2 in Appendix B. If it is less than α, we can reject the null hypothesis at that level. Then we can conclude that μ_1 is indeed greater than μ_2 at that level of significance.

Two-sided Bayesian hypothesis test.   We cannot test the two-sided hypothesis

$$H_0 : \mu_1 - \mu_2 = 0 \quad \text{versus} \quad H_1 : \mu_1 - \mu_2 \neq 0$$

in a Bayesian manner by calculating the posterior probability of the null hypothesis. It is a point null hypothesis since it is only true for a single value $\mu_d = \mu_1 - \mu_2 = 0$. When we used the continuous prior, we got a continuous posterior, and the probability that any continuous random variable takes on any particular value always equals 0.

Instead, we use the credible interval for μ_d. If 0 lies in the interval, we cannot reject the null hypothesis and 0 remains a credible value for the difference between the means. However, if 0 lies outside the interval, then 0 is no longer a credible value at the significance level α.

■ **EXAMPLE 13.1**   (continued)

The 95% Bayesian credible interval for $\mu_d = \mu_{1879} - \mu_{1882}$ is $(92.1, 211.9)$. 0 lies outside the interval; hence we reject the null hypothesis that the means for the two measurement groups were equal and conclude that they are different. This shows that there was a bias in Michelson's first group of measurements, which was very much reduced in the second group of measurements.   ■

When the Variance Is Unknown and Flat Priors Are Used

Suppose we use independent "flat" priors for μ_1 and μ_2. Then $(s_1')^2 = \frac{\sigma^2}{n_1}$, $(s_2')^2 = \frac{\sigma^2}{n_2}$, $m_1' = \bar{y}_1$ and $m_2' = \bar{y}_2$.

Credible interval for difference between means, unknown equal variance case.   If we knew the variance σ^2, the credible interval could be written as

$$\bar{y}_1 - \bar{y}_2 \pm z_{\frac{\alpha}{2}} \times \sigma \sqrt{\frac{1}{n_1} + \frac{1}{n_2}}.$$

However, we do not know σ^2. We will have to estimate it from the data. We can get an estimate from each of the samples. The best thing to do is to combine these estimates to get the pooled variance estimate

$$\hat{\sigma}_p^2 = \frac{\sum_{i=1}^{n_1}(y_{i1} - \bar{y}_1)^2 + \sum_{j=1}^{n_2}(y_{j2} - \bar{y}_2)^2}{n_1 + n_2 - 2}. \tag{13.5}$$

Since we used the estimated $\hat{\sigma}_p^2$ instead of the unknown true variance σ^2, the credible interval should be widened to allow for the additional uncertainty. We will get the critical value from the *Student's t* table with $n_1 + n_2 - 2$ degrees of freedom. The approximate $(1 - \alpha) \times 100\%$ Bayesian credible interval for $\mu_1 - \mu_2$ is

$$\bar{y}_1 - \bar{y}_2 \pm t_{\frac{\alpha}{2}} \times \hat{\sigma}_p \sqrt{\frac{1}{n_1} + \frac{1}{n_2}}, \tag{13.6}$$

where the critical value comes from the *Student's t* table with $n_1 + n_2 - 2$ degrees of freedom.[1]

Confidence interval for difference between means, unknown equal variance case. The frequentist confidence interval for $\mu_d = \mu_1 - \mu_2$ when the two distributions

---

[1] Actually, we are treating the unknown σ^2 as a nuisance parameter and are using an independent prior $g(\sigma^2) \propto \frac{1}{\sigma^2}$ for it. We find the marginal posterior distribution of $\mu_1 - \mu_2$ from the joint posterior of $\mu_1 - \mu_2$ and σ^2 by integrating out the nuisance parameter. The marginal posterior will be *Student's t* with $n_1 + n_2 - 2$ degrees of freedom instead of normal. This gives us the credible interval with the z critical value replaced by the t critical value. We see that our approximation gives us the correct credible interval for these assumptions.

have equal unknown variance is

$$\bar{y}_1 - \bar{y}_2 \, \pm t_{\frac{\alpha}{2}} \times \hat{\sigma}_p \sqrt{\frac{1}{n_1} + \frac{1}{n_2}}, \tag{13.7}$$

where the critical value again comes from the *Student's* t table with $n_1 + n_2 - 2$ degrees of freedom. The confidence interval has exactly the same form as the Bayesian credible interval when we use independent "flat" priors for μ_1 and μ_2. Of course, the interpretations are different.

The frequentist has $(1 - \alpha) \times 100\%$ confidence that the interval contains the true value of the difference because $(1-\alpha) \times 100\%$ of the random intervals calculated this way do contain the true value. The Bayesian interpretation is that given the data from the two samples, the posterior probability the random parameter $\mu_1 - \mu_2$ lies in the interval is $(1 - \alpha)$.

In this case the scientist who misinterprets the confidence interval for a probability statement about the parameter gets away with it, because it actually is a probability statement using independent *flat* priors. It is fortunate for frequentist statisticians that their most commonly used techniques (confidence intervals for means and proportions) are equivalent to Bayesian credible intervals for some specific prior.[2] Thus a scientist who misinterpret his/her confidence interval as a probability statement, can do so in this case, but he/she is implicitly assuming independent flat priors. The only loss that the scientist will have incurred is he/she did not get to use any prior information he/she may have had.[3]

One-sided Bayesian hypothesis test.   If we want to test

$$H_0 : \mu_d \leq 0 \quad \text{versus} \quad H_1 : \mu_d > 0$$

when we assume that the two random samples come from *normal* distributions having the same unknown variance σ^2, and we use the pooled estimate of the variance $\hat{\sigma}_p^2$ in place of the unknown σ^2 and assume independent "flat" priors for the means μ_1 and μ_2, we calculate the posterior probability of the null hypothesis using Equation 13.5; but instead of finding the probability in the standard normal table, we find it from the *Student's* t distribution with $n_1 + n_2 - 2$ degrees of freedom. We could calculate it using Minitab or R. Alternatively, we could find values that bound this probability in the *Student's* t table.

---

[2] In the case of a single random sample from a *normal* distribution, frequentist confidence intervals are equivalent to Bayesian credible intervals with *flat* prior for μ. In the case of independent random samples from *normal* distributions having equal unknown variance σ^2, confidence intervals for the difference between means are equivalent to Bayesian credible intervals using independent flat priors for μ_1 and μ_2, along with the improper prior $g(\sigma) \propto \sigma^{-1}$ for the nuisance parameter.

[3] Frequentist techniques such as the confidence intervals used in many other situations do not have Bayesian interpretations. Interpreting the confidence interval as the basis for a probability statement about the parameter would be completely wrong in those situations.

Two-sided Bayesian hypothesis test.   When we assume that both samples come from *normal* distributions with equal unknown variance σ^2 and we use the pooled estimate of the variance $\hat{\sigma}_p^2$ in place of the unknown variance σ^2 and assume independent "flat" priors, we can test the two-sided hypothesis

$$H_0 : \mu_1 - \mu_2 = 0 \quad \text{versus} \quad H_1 : \mu_1 - \mu_2 \neq 0$$

using the credible interval for $\mu_1 - \mu_2$ given in Equation 13.7. There are $n_1 + n_2 - 2$ degrees of freedom. If 0 lies in the credible interval, we cannot reject the null hypothesis, and 0 remains a credible value for the difference between the means. However, if 0 lies outside the interval, then 0 is no longer a credible value at the significance level α.

13.3   Case 2: Unequal Variances

When the Variances Are Known

In this section we will look at a nonadditive model, but with known variances. Let $y_{11}, \ldots, y_{n_{11}}$ be a random sample from normal distribution having mean μ_1 and known variance σ_1^2. Let $y_{12}, \ldots, y_{n_{22}}$ be a random sample from normal distribution having mean μ_2 and known variance σ_2^2. The two random samples are independent of each other.

We use independent priors for μ_1 and μ_2. They can be either normal priors or "flat" priors. Since the samples are independent and the priors are independent, we can find each posterior independently of the other. We find these using the simple updating formulas given in Equations 11.5 and 11.6. The posterior of $\mu_1 | y_{11}, \ldots, y_{n_{11}}$ is $normal[m_1', (s_1')^2]$. The posterior of $\mu_2 | y_{12}, \ldots, y_{n_{22}}$ is $normal[m_2', (s_2')^2]$. The posteriors are independent since the priors are independent and the samples are independent. The posterior distribution of $\mu_d = \mu_1 - \mu_2$ is normal with mean equal to the *difference* of the posterior means, and variance equal to the *sum* of the posterior variances.

$$(\mu_d | y_{11}, \ldots, y_{n_{11}}, y_{12}, \ldots, y_{n_{22}}) \sim normal(m_d', (s_d')^2),$$

where $m_d' = m_1' - m_2'$ and $(s_d')^2 = (s_1')^2 + (s_2')^2$

Credible interval for difference between means, known unequal variance case.   A $(1 - \alpha) \times 100\%$ Bayesian credible interval for $\mu_d = \mu_1 - \mu_2$, the difference between means is

$$m_d' \pm z_{\frac{\alpha}{2}} \times (s_d'), \tag{13.8}$$

which can be written as

$$m_1' - m_2' \pm z_{\frac{\alpha}{2}} \times \sqrt{(s_1')^2 + (s_2')^2}. \tag{13.9}$$

Note these are identical to Equations 13.1 and 13.2.

Confidence interval for difference between means, known unequal variance case.
The frequentist confidence interval for $\mu_d = \mu_1 - \mu_2$ in this case would be

$$\bar{y}_1 - \bar{y}_2 \pm z_{\frac{\alpha}{2}} \times \sqrt{\frac{\sigma_1^2}{n_1} + \frac{\sigma_2^2}{n_2}} . \tag{13.10}$$

Note that this has the same formula as the Bayesian credible interval we would
get if we had used flat priors for both μ_1 and μ_2. However, the intervals have
very different interpretations.

When the Variances Are Unknown

When the variances are unequal and unknown, each of them will have to be
estimated from the sample data

$$\hat{\sigma}_1^2 = \frac{1}{n_1 - 1} \sum_{i=1}^{n_1} (y_{i1} - \bar{y}_1)^2 \quad \text{and} \quad \hat{\sigma}_2^2 = \frac{1}{n_2 - 1} \sum_{i=1}^{n_2} (y_{i2} - \bar{y}_2)^2 .$$

These estimates will be used in place of the unknown true values in the simple updating formulas. This adds extra uncertainty. To allow for this, we
should use the *Student's* t table to find the critical values. However, it is no
longer straightforward what degrees of freedom should be used. Satterthwaite
suggested that the adjusted degrees of freedom be

$$\frac{\left(\frac{\hat{\sigma}_1^2}{n_1} + \frac{\hat{\sigma}_2^2}{n_2}\right)^2}{\frac{(\hat{\sigma}_1^2/n_1)^2}{n_1-1} + \frac{(\hat{\sigma}_2^2/n_2)^2}{n_2-1}}$$

rounded down to the nearest integer.

Credible interval for difference between means, unequal unknown variances. When
we use the sample estimates of the variances in place of the true unknown
variances in Equations 11.5 and 11.6, an approximate $(1-\alpha) \times 100\%$ credible
interval for $\mu_d = \mu_1 - \mu_2$ is given by

$$m_1' - m_2' \pm t_{\frac{\alpha}{2}} \times \sqrt{(s_1')^2 + (s_2')^2} ,$$

where we find the degrees of freedom using Satterthwaite's adjustment. In
the case where we use independent "flat" priors for μ_1 and μ_2, this can be
written as

$$m_1' - m_2' \pm t_{\frac{\alpha}{2}} \times \sqrt{\frac{\hat{\sigma}_1^2}{n_1} + \frac{\hat{\sigma}_2^2}{n_2}} . \tag{13.11}$$

Confidence interval for difference between means, unequal unknown variances.
An approximate $(1 - \alpha) \times 100\%$ confidence interval for $\mu_d = \mu_1 - \mu_2$ is given

by

$$m_1' - m_2' \pm t_{\frac{\alpha}{2}} \times \sqrt{\frac{\hat{\sigma}_1^2}{n_1} + \frac{\hat{\sigma}_2^2}{n_2}} \,. \tag{13.12}$$

We see this is the same form as the $(1-\alpha) \times 100\%$ credible interval found when we used independent flat priors.[4] However, the interpretations are different.

Bayesian hypothesis test of $H_0 : \mu_1 - \mu_2 \leq 0$ versus $H_1 : \mu_1 - \mu_2 > 0$. To test

$$H_0 : \mu_1 - \mu_2 \leq 0 \quad \text{versus} \quad H_1 : \mu_1 - \mu_2 > 0$$

at the level α in a Bayesian manner, we calculate the posterior probability of the null hypothesis. We would use Equation 13.5. If the variances σ_1^2 and σ_2^2 are known, we get the critical value from the standard normal table. However, when we use estimated variances instead of the true unknown variances, we will find the probabilities using the *Student's t* distribution with degrees of freedom given by Satterthwaite's approximation. If this probability is less than α, then we reject the null hypothesis and conclude that $\mu_1 > \mu_2$. In other words, that the treatment is effective. Otherwise, we cannot reject the null hypothesis.

---

[4]Finding the posterior distribution of $\mu_1 - \mu_2 - (\bar{y}_1 - \bar{y}_2)|y_{11}, \ldots, y_{n_{11}}, y_{12}, \ldots, y_{n_{22}}$ in the Bayesian paradigm, or equivalently finding the sampling distribution of $\bar{y}_1 - \bar{y}_2 - (\mu_1 - \mu_2)$ in the frequentist paradigm when the variances are both unknown and not assumed equal has a long and controversial history. In the one-sample case, the sampling distribution of $\bar{y} - \mu$ is the same as the posterior distribution of $\mu - \bar{y}|y_1, \ldots, y_n$ when we use the flat prior for $g(\mu) = 1$ and the noninformative prior $g(\sigma^2) \propto \frac{1}{\sigma^2}$ and marginalize σ^2 out of the joint posterior. This leads to the equivalence between the confidence interval and the credible interval for that case. Similarly, in the two-sample case with equal variances, the sampling distribution of $\bar{y}_1 - \bar{y}_2$ equals the posterior distribution of $\mu_1 - \mu_2|y_{11}, \ldots, y_{n_{11}}, y_{12}, \ldots, y_{n_{22}}$ where we use flat priors for μ_1 and μ_2 and the noninformative prior $g(\sigma^2) \propto \frac{1}{\sigma^2}$, and marginalized σ^2 out of the joint posterior. Again, that led to the equivalence between the confidence interval and the credible interval for that case. One might be led to believe this pattern would hold in general. However, it does not hold in the two sample case with unknown unequal variances. The Bayesian posterior distribution in this case is known as the *Behrens–Fisher* distribution. The frequentist distribution depends on the ratio of the unknown variances. Both of the distributions can be approximated by *Student's t* with an adjustment made to the degrees of freedom. Satterthwaite suggested that the adjusted degrees of freedom be

$$\frac{\left(\frac{\hat{\sigma}_1^2}{n_1} + \frac{\hat{\sigma}_2^2}{n_2}\right)^2}{\frac{(\hat{\sigma}_1^2/n_1)^2}{n_1-1} + \frac{(\hat{\sigma}_2^2/n_2)^2}{n_2-1}}$$

rounded down to the nearest integer.

13.4 Bayesian Inference for Difference Between Two Proportions Using Normal Approximation

Often we want to compare the proportions of a certain attribute in two populations. The true proportions in population 1 and population 2 are π_1 and π_2, respectively. We take a random sample from each of the populations and observe the number of each sample having the attribute. The distribution of $y_1|\pi_1$ is $binomial(n_1, \pi_1)$ and the distribution of $y_2|\pi_2$ is $binomial(n_2, \pi_2)$, and they are independent of each other

We know that if we use independent prior distributions for π_1 and π_2, we will get independent posterior distributions. Let the prior for π_1 be $beta(a_1, b_1)$ and for π_2 be $beta(a_2, b_2)$. The posteriors are independent beta distributions. The posterior for π_1 is $beta(a_1', b_1')$, where $a_1' = a_1 + y_1$ and $b_1' = b_1 + n_1 - y_1$. Similarly the posterior for π_2 is $beta(a_2', b_2')$, where $a_2' = a_2 + y_2$ and $b_2' = b_2 + n_2 - y_2$

Approximate each posterior distribution with the normal distribution having same mean and variance as the beta. The posterior distribution of $\pi_d = \pi_1 - \pi_2$ is approximately $normal(m_d', (s_d')^2)$ where the posterior mean is given by

$$m_d' = \frac{a_1'}{a_1' + b_1'} - \frac{a_2'}{a_2' + b_2'}$$

and the posterior variance is given by

$$(s_d')^2 = \frac{a_1' b_1'}{(a_1' + b_1')^2 (a_1' + b_1' + 1)} + \frac{a_2' b_2'}{(a_2' + b_2')^2 (a_2' + b_2' + 1)}.$$

Credible interval for difference between proportions. We find the $(1-\alpha) \times 100\%$ Bayesian credible interval for $\pi_d = \pi_1 - \pi_2$ using the general rule for the (approximately) normal posterior distribution. It is

$$m_d' \pm z_{\frac{\alpha}{2}} \times s_d'. \tag{13.13}$$

One-sided Bayesian hypothesis test for difference between proportions. Suppose we are trying to detect whether $\pi_d = \pi_1 - \pi_2 > 0$. We set this up as a test of

$$H_0 : \pi_d \leq 0 \quad \text{versus} \quad H_1 : \pi_d > 0.$$

Note, the alternative hypothesis is what we are trying to detect. We calculate the approximate posterior probability of the null distribution by

$$P(\pi_d \leq 0) = P\left(\frac{\pi_d - m_d'}{s_d'} \leq \frac{0 - m_d'}{s_d'}\right)$$

$$= P\left(Z \leq \frac{0 - m_d'}{s_d'}\right). \tag{13.14}$$

If this probability is less than the level of significance α that we chose, we would reject the null hypothesis at that level and conclude $\pi_1 > \pi_2$. Otherwise, we cannot reject the null hypothesis.

Two-sided Bayesian hypothesis test for difference between proportions.   To test the hypothesis

$$H_0 : \pi_1 - \pi_2 = 0 \quad \text{versus} \quad H_1 : \pi_1 - \pi_2 \neq 0$$

in a Bayesian manner, check whether the null hypothesis value (0) lies inside the credible interval for π_d given in Equation 13.13. If it lies inside the interval, we cannot reject the null hypothesis $H_0 : \pi_1 - \pi_2 = 0$ at the level α. If it lies outside the interval, we can reject the null hypothesis at the level α and accept the alternative $H_1 : \pi_1 - \pi_2 \neq 0$.

■ **EXAMPLE 13.2**

The student newspaper wanted to write an article on the smoking habits of students. A random sample of 200 students (100 males and 100 females) between ages of 16 and 21 were asked about whether they smoked cigarettes. Out of the 100 males, 22 said they were regular smokers, and out of the 100 females, 31 said they were regular smokers. The editor of the paper asked Donna, a statistics student, to analyze the data.

Donna considered the male and female samples would be independent. Her prior knowledge was that a minority of students smoked cigarettes, so she decided to use independent beta(1,2) priors for π_m and π_f, the male and female proportions respectively. Her posterior distribution of π_m will be beta(23,80), and her posterior distribution of π_f will be beta(32,71). Hence, her posterior distribution of the difference between proportions, $\pi_d = \pi_m - \pi_f$, will be approximately $normal(m'_d, (s'_d)^2)$ where

$$m'_d = \frac{23}{23 + 80} - \frac{32}{32 + 71}$$
$$= -.087$$

and

$$(s'_d)^2 = \frac{23 * 80}{(23 + 80)^2 * (23 + 80 + 1)} + \frac{32 * 71}{(32 + 71)^2 * (32 + 71 + 1)}$$
$$= .061^2 .$$

Her 95% credible interval for π_d will be (-.207, .032) which contains 0. She cannot reject the null hypothesis $H_0 : \pi_m - \pi_f = 0$ at the 5% level, so she tells the editor that the data does not conclusively show that there is any difference between the proportions of male and female students who smoke.   ■

13.5   Normal Random Samples from Paired Experiments

Variation between experimental units often is a major contributor to the variation in the data. When the two treatments are administered to two independent random samples of the experimental units, this variation makes it harder to detect any difference between the treatment effects, if one exists.

Often designing a paired experiment makes it much easier to detect the difference between treatment effects. For a paired experiment, the experimental units are matched into pairs of similar units. Then one of the units from each pair is assigned to the first treatment, and the other in that pair is assigned the second treatment. This is a *randomized block* experimental design, where the pairs are blocks. We discussed this design in Chapter 2. For example, in the dairy industry, identical twin calves are often used for experiments. They are exact genetic copies. One of each pair is randomly assigned to the first treatment, and the other is assigned to the second treatment.

Paired data can arise other ways. For instance, if the two treatments are applied to the same experimental units (at different times) giving the first treatment effect time to dissipate before the second treatment is applied. Or, we can be looking at "before treatment" and "after treatment" measurements on the same experimental units.

Because of the variation between experimental units, the two observations from units in the same pair will be more similar than two observations from units in different pairs. In the same pair, the only difference between the observation given treatment A and the observation given treatment B is the treatment effect plus the measurement error. In different pairs, the difference between the observation given treatment A and the observation given treatment B is the treatment effect plus the experimental unit effect plus the measurement error. Because of this we cannot treat the paired random samples as independent of each other. The two random samples come from *normal* populations with means μ_A and μ_B, respectively. The populations will have equal variances σ^2 when we have an additive model. We consider that the variance comes from two sources: measurement error plus random variation between experimental units.

Take Differences within Each Pair

Let y_{i1} be the observation from pair i given treatment A, and let y_{i2} be the observation from pair i given treatment B. If we take the difference between the observations within each pair, $d_i = y_{i1} - y_{i2}$, then these d_i will be a random sample from a *normal* population with mean $\mu_d = \mu_A - \mu_B$, and variance σ_d^2. We can treat this (differenced) data as a sample from a single *normal* distribution and do inference using techniques found in Chapters 11 and 12.

■ EXAMPLE 13.3

An experiment was designed to determine whether a mineral supplement was effective in increasing annual yield in milk. Fifteen pairs of identical twin dairy cows were used as the experimental units. One cow from

Table 13.1 Milk annual yield

Twin Set	Milk Yield: Control (liters)	Milk Yield: Treatment (liters)
1	3525	3340
2	4321	4279
3	4763	4910
4	4899	4866
5	3234	3125
6	3469	3680
7	3439	3965
8	3658	3849
9	3385	3297
10	3226	3124
11	3671	3218
12	3501	3246
13	3842	4245
14	3998	4186
15	4004	3711

each pair was randomly assigned to the treatment group that received the supplement. The other cow from the pair was assigned to the control group that did not receive the supplement. The annual yields are given in Table 13.1. Assume that the annual yields from cows receiving the treatment are $normal(\mu_t, \sigma_t^2)$, and that the annual yields from the cows in the control group are $normal(\mu_c, \sigma_c^2)$. Aleece, Brad, and Curtis decided that since the two cows in the same pair share identical genetic background, their responses will be more similar than two cows that were from different pairs. There is natural pairing. As the samples drawn from the two populations cannot be considered independent of each other, they decided to take differences $d_i = y_{i1} - y_{i2}$. The differences will be $normal(\mu_d, \sigma_d^2)$, where $\mu_d = \mu_t - \mu_c$ and we will assume that $\sigma_d^2 = 270^2$ is known.

Aleece decided she would use a "flat" prior for μ_d. Brad decided he would use a $normal(m, s^2)$ prior for μ_d where he let $m = 0$ and $s = 200$. Curtis decided that his prior for μ_d matched a triangular shape. He set up a numerical prior that interpolated between the heights given in Table 13.2 The shapes of the priors are shown in Figure 13.1. Aleece used a "flat" prior, so her posterior will be normal $[m', (s')^2]$ where $m' = \bar{y} = 7.067$ and $(s')^2 = 270^2/15 = 4860$. Her posterior standard deviation $s' = \sqrt{4860} = 69.71$. Brad used a $normal(0, 200^2)$ prior, so his posterior will be normal $[m', (s')^2]$ where m' and s' are found by using Equations

Table 13.2   Curtis' prior weights. The shape of his continuous prior is found by linearly interpolating between them.

Value	Weight
-300	0
0	3
300	0

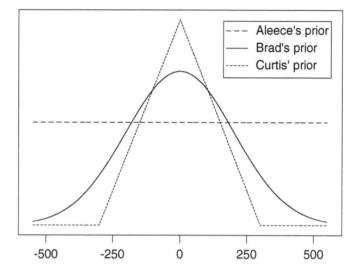

Figure 13.1   The shapes of Aleece's, Brad's, and Curtis' prior distributions.

11.5 and 11.6.

$$\frac{1}{(s')^2} = \frac{1}{200^2} + \frac{15}{270^2} = 0.000230761,$$

so his $s' = 65.83$, and

$$m' = \frac{\frac{1}{200^2}}{.000230761} \times 0 + \frac{\frac{15}{270^2}}{.000230761} \times 7.067 = 6.30.$$

Curtis has to find his posterior numerically using Equation 11.3.

[Minitab:]   He uses the Minitab macro *NormGCP* to do the numerical integration.

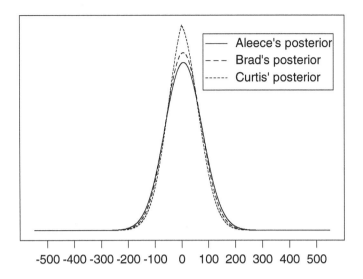

Figure 13.2   Aleece's, Brad's, and Curtis's posterior distributions.

[**R:**] He uses the R function **normgcp** calculate the posterior, and **cdf** to do the numerical integration.

The three posteriors are shown in Figure 13.2. They decided that to determine whether or not the treatment was effective in increasing the yield of milk protein, they would perform the one-sided hypothesis test

$$H_0 : \mu_d \leq 0 \quad \text{vs} \quad H_1 : \mu_d > 0$$

at the 95% level of significance. Aleece and Brad had normal posteriors, so they used Equation 13.5 to calculate the posterior probability of the null hypothesis.

[**Minitab:**]   Curtis had a numerical posterior, so he used Equation 12.3 and performed the integration using the Minitab macro *tintegral*.

[**R:**] [**Minitab:**]   Curtis had a numerical posterior, so he used Equation 12.3 and performed the integration using the **cdf** function in R.

The results are shown in Table 13.3.   ■

Main Points

- The difference between normal means are used to make inferences about the size of a treatment effect.

Table 13.3   Results of Bayesian one-sided hypothesis tests

Person	Posterior	$P(\mu_d \leq 0|d_1,\ldots,d_n)$	
Aleece	$normal(7.07, 69.71^2)$	$P(Z \leq \frac{0-7.07}{69.71})$ $\qquad =.4596$	do not reject
Brad	$normal(6.30, 65.83^2)$	$P(Z \leq \frac{0-6.30}{65.83})$ $\qquad =.4619$	do not reject
Curtis	$numerical$	$\int_{-\infty}^{0} g(\mu_d|d_1,\ldots,d_n)d\mu$ $\quad =.4684$	do not reject

- Each experimental unit is randomly assigned to the treatment group or control group. The unbiased random assignment method ensures that both groups have similar experimental units assigned to them. On average, the means are equal.

- The treatment group mean is the mean of the experimental units assigned to the treatment group, plus the treatment effect.

- If the treatment effect is constant, we call it an additive model, and both sets of observations have the same underlying variance, assumed to be known.

- If the data in the two samples are independent of each other, we use independent priors for the two means. The posterior distributions $\mu_1|y_{11},\ldots,y_{n_{11}}$ and $\mu_2|y_{12},\ldots,y_{n_{22}}$ are also independent of each other and can be found using methods from Chapter 11.

- Let $\mu_d = \mu_1-\mu_2$. The posterior distribution of $\mu_d|y_{11},\ldots,y_{n_{11}},y_{12},\ldots,y_{n_{22}}$ is $normal$ with mean $m_d' = m_1' - m_2'$ and variance $(s_d')^2 = (s_1')^2 + (s_2')^2$

- The $(1-\alpha) \times 100\%$ credible interval for $\mu_d = \mu_1 - \mu_2$ is given by

$$m_d' \pm z_{\alpha/2} \times s_d'.$$

- If the variance is unknown, use the pooled estimate from the two samples. The credible interval will have to be widened to account for the extra uncertainty. This is accomplished by taking the critical values from the *Student's t* table (with $n_1 + n_2 - 2$ degrees of freedom) instead of the *standard normal* table.

- The confidence interval for $\mu_d|y_{11},\ldots,y_{n_{11}},y_{12},\ldots,y_{n_{22}}$ is the same as the Bayesian credible interval where flat priors are used.

- If the variances are unknown, and not equal, use the sample estimates as if they were the correct values. Use the *Student's t* for critical values, with the degrees given by Satterthwaite's approximation. This is true for both credible intervals and confidence intervals.

- The posterior distribution for a difference between proportions can be found using the normal approximation. The posterior variances are known, so the critical values for credible interval come from *standard normal* table.

▪ When the observations are paired, the samples are dependent. Calculate the differences $d_i = y_{i1} - y_{i2}$ and treat them as a single sample from a $normal(\mu_d, \sigma_d^2)$, where $\mu_d = \mu_1 - \mu_2$. Inferences about μ_d are made using the single sample methods found in Chapters 11 and 12.

Exercises

13.1. The Human Resources Department of a large company wishes to compare two methods of training industrial workers to perform a skilled task. Twenty workers are selected: 10 of them are randomly assigned to be trained using method A, and the other 10 are assigned to be trained using method B. After the training is complete, all the workers are tested on the speed of performance at the task. The times taken to complete the task are:

Method A	Method B
115	123
120	131
111	113
123	119
116	123
121	113
118	128
116	126
127	125
129	128

(a) We will assume that the observations come from $normal(\mu_A, \sigma^2)$ and $normal(\mu_B, \sigma^2)$, where $\sigma^2 = 6^2$. Use independent $normal\,(m, s^2)$ prior distributions for μ_A and μ_B, respectively, where $m = 100$ and $s^2 = 20^2$. Find the posterior distributions of μ_A and μ_B, respectively.

(b) Find the posterior distribution of $\mu_A - \mu_B$.

(c) Find a 95% Bayesian credible interval for $\mu_A - \mu_B$.

(d) Perform a Bayesian test of the hypothesis

$$H_0 : \mu_A - \mu_B = 0 \quad \text{versus} \quad H_1 : \mu_A - \mu_B \neq 0$$

at the 5% level of significance. What conclusion can we draw?

13.2. A consumer testing organization obtained samples of size 12 from two brands of emergency flares and measured the burning times. They are:

Brand A	Brand B
17.5	13.4
21.2	9.9
20.3	13.5
14.4	11.3
15.2	22.5
19.3	14.3
21.2	13.6
19.1	15.2
18.1	13.7
14.6	8.0
17.2	13.6
18.8	11.8

(a) We will assume that the observations come from $normal(\mu_A, \sigma^2)$ and $normal(\mu_B, \sigma^2)$, where $\sigma^2 = 3^2$. Use independent $normal$ (m, s^2) prior distributions for μ_A and μ_B, respectively, where $m = 20$ and $s^2 = 8^2$. Find the posterior distributions of μ_A and μ_B, respectively.

(b) Find the posterior distribution of $\mu_A - \mu_B$.

(c) Find a 95% Bayesian credible interval for $\mu_A - \mu_B$.

(d) Perform a Bayesian test of the hypothesis

$$H_0 : \mu_A - \mu_B = 0 \quad \text{versus} \quad H_1 : \mu_A - \mu_B \neq 0$$

at the 5% level of significance. What conclusion can we draw?

13.3. The quality manager of a dairy company is concerned whether the levels of butterfat in a product are equal at two dairy factories which produce the product. He obtains random samples of size 10 from each of the factories' output and measures the butterfat. The results are:

Factory 1	Factory 2
16.2	16.1
12.7	16.3
14.8	14.0
15.6	16.2
14.7	15.2
13.8	16.5
16.7	14.4
13.7	16.3
16.8	16.9
14.7	13.7

(a) We will assume that the observations come from $normal(\mu_1, \sigma^2)$ and $normal(\mu_2, \sigma^2)$, where $\sigma^2 = 1.2^2$. Use independent $normal(m, s^2)$ prior distributions for μ_1 and μ_2, respectively, where $m = 15$ and $s^2 = 4^2$. Find the posterior distributions of μ_1 and μ_2, respectively.

(b) Find the posterior distribution of $\mu_1 - \mu_2$.

(c) Find a 95% Bayesian credible interval for $\mu_1 - \mu_2$.

(d) Perform a Bayesian test of the hypothesis

$$H_0 : \mu_1 - \mu_2 = 0 \quad \text{versus} \quad H_1 : \mu_1 - \mu_2 \neq 0$$

at the 5% level of significance. What conclusion can we draw?

13.4. Independent random samples of ceramic produced by two different processes were tested for hardness. The results were:

Process 1	Process 2
8.8	9.2
9.6	9.5
8.9	10.2
9.2	9.5
9.9	9.8
9.4	9.5
9.2	9.3
10.1	9.2

(a) We will assume that the observations come from $normal(\mu_1, \sigma^2)$ and $normal(\mu_2, \sigma^2)$, where $\sigma^2 = .4^2$. Use independent $normal(m, s^2)$ prior distributions for μ_1 and μ_2, respectively, where $m = 10$ and $s^2 = 1^2$. Find the posterior distributions of μ_1 and μ_2, respectively.

(b) Find the posterior distribution of $\mu_1 - \mu_2$.

(c) Find a 95% Bayesian credible interval for $\mu_1 - \mu_2$.

(d) Perform a Bayesian test of the hypothesis

$$H_0 : \mu_1 - \mu_2 \geq 0 \quad \text{versus} \quad H_1 : \mu_1 - \mu_2 < 0$$

at the 5% level of significance. What conclusion can we draw?

13.5. A thermal power station discharges its cooling water into a river. An environmental scientist wants to determine if this has adversely affected the dissolved oxygen level. She takes samples of water one kilometer upstream from the power station, and one kilometer downstream from the power station, and measures the dissolved oxygen level. The data are:

Upstream	Downstream
10.1	9.7
10.2	10.3
13.4	6.4
8.2	7.3
9.8	11.7
	8.9

(a) We will assume that the observations come from $normal(\mu_1, \sigma^2)$ and $normal(\mu_2, \sigma^2)$, where $\sigma^2 = 2^2$. Use independent $normal\,(m, s^2)$ prior distributions for μ_1 and μ_2, respectively, where $m = 10$ and $s^2 = 2^2$. Find the posterior distributions of μ_1 and μ_2, respectively.

(b) Find the posterior distribution of $\mu_1 - \mu_2$.

(c) Find a 95% Bayesian credible interval for $\mu_1 - \mu_2$.

(d) Perform a Bayesian test of the hypothesis

$$H_0 : \mu_1 - \mu_2 \leq 0 \quad \text{versus} \quad H_1 : \mu_1 - \mu_2 > 0$$

at the 5% level of significance. What conclusion can we draw?

13.6. Cattle, being ruminants, have multiple chambers in their stomachs. Stimulating specific receptors causes reflex contraction of the reticular groove and swallowed fluid then bypasses the reticulo-rumen and moves directly to the abomasum. Scientists wanted to develop a simple nonradioactive, noninvasive test to determine when this occurs. In a study to determine the fate of swallowed fluids in cattle, McLeay et al. (1997) investigate a carbon-13 (^{13}C) octanoic acid breath test as a means of detecting a reticular groove contraction in cattle. Twelve adult cows were randomly assigned to two groups of 6 cows. The first group had 200 mg of ^{13}C octanoic acid administered into the reticulum, and the second group had the same dose of ^{13}C octanoic acid administered into the reticulo-osmasal orifice. Change in the enrichment of ^{13}C in breath was measured for each cow 10 minutes later. The results are:

^{13}C Administered into Reticulum		^{13}C Administered into Reticulo-omasal Orifice	
Cow ID	x	Cow ID	y
8	1.5	14	3.5
9	1.9	15	4.7
10	0.4	16	4.8
11	-1.2	17	4.1
12	1.7	18	4.1
13	0.7	19	5.3

(a) Explain why the observations of variables x and y can be considered independent in this experiment.

(b) Suppose the change in the enrichment of ^{13}C for cows administered in the *reticulum* is *normal* (μ_1, σ_1^2), where $\sigma_1^2 = 1.00^2$. Use a emphnormal(2, 2² prior for μ_1. Calculate the posterior distribution of $\mu_1 | x_8, \ldots, x_{13}$.

(c) Suppose the change in the enrichment of ^{13}C for cows administered in the *reticulo-omasal orifice* is *normal*(μ_2, σ_2^2), where $\sigma_2^2 = 1.40^2$. Use a *normal*$(2, 2^2)$ prior for μ_2. Calculate the posterior distribution of $\mu_1 | y_{14}, \ldots, y_{19}$.

(d) Calculate the posterior distribution of $\mu_d = \mu_1 - \mu_2$, the difference between the means.

(e) Calculate a 95% Bayesian credible interval for μ_d.

(f) Test the hypothesis

$$H_0 : \mu_1 - \mu_2 = 0 \quad \text{versus} \quad H_1 : \mu_1 - \mu_2 \neq 0$$

at the 5% level of significance. What conclusion can be drawn.

13.7. Glass fragments found on a suspect's shoes or clothes are often used to connect the suspect to a crime scene. The index of refraction of the fragments are compared to the refractive index of the glass from the crime scene. To make this comparison rigorous, we need to know the variability the index of refraction is over a pane of glass. Bennett et al. (2003)

analyzed the refractive index in a pane of float glass, searching for any spatial pattern. Here are samples of the refractive index from the edge and from the middle of the pane.

Edge of Pane		Middle of Pane	
1.51996	1.51997	1.52001	1.51999
1.51998	1.52000	1.52004	1.51997
1.51998	1.52004	1.52005	1.52000
1.52000	1.52001	1.52004	1.52002
1.52000	1.51997	1.52004	1.51996

For these data, $\bar{y}_1 = 1.51999$, $\bar{y}_2 = 1.52001$, $sd_1 = .00002257$,   and $sd_2 = .00003075$.

(a) Suppose glass at the edge of the pane is $normal(\mu_1, \sigma_1^2)$, where $\sigma_1 = .00003$. Calculate the posterior distribution of μ_1 when you use a $normal(1.52000, .0001^2)$ prior for μ_1.

(b) Suppose glass in the middle of the pane is $normal(\mu_2, \sigma_2^2)$, where $\sigma_2 = .00003$. Calculate the posterior distribution of μ_2 when you use a $normal(1.52000, .0001^2)$ prior for μ_2.

(c) Find the posterior distribution of $\mu_d = \mu_1 - \mu_2$.

(d) Find a 95% credible interval for μ_d.

(e) Perform a Bayesian test of the hypothesis

$$H_0 : \mu_d = 0 \quad \text{versus} \quad H_1 : \mu_d \neq 0$$

at the 5% level of significance.

13.8. The last half of the twentieth century saw great change in the role of women in New Zealand society. These changes included education, employment, family formation, and fertility, where women took control of these aspects of their lives. During those years, phrases such as "women's liberation movement" and "the sexual revolution" were used to describe the changing role of women in society. In 1995 the Population Studies Centre at the University of Waikato sponsored the New Zealand Women Family, Employment, and Education Survey (NZFEE) to investigate these changes. A random sample of New Zealand women of all ages between 20 and 59 was taken, and the women were interviewed about their educational, employment, and personal history. The details of this survey are summarized in Marsault et al. (1997). Detailed analysis of the data from this survey is in Johnstone et al. (2001).

Have the educational qualifications of younger New Zealand women changed from those of previous generations of New Zealand women? To shed light on this question, we will compare the educational qualifications of two generations of New Zealand women 25 years apart. The women in the age group 25–29 at the time of the survey were born between 1966 and 1970. The women in the age group 50–54 at the time of the survey were born between 1941 and 1945.

(a) Out of 314 women in the age group 25–29, 234 had completed a secondary school qualification. Find the posterior distribution of π_1, the proportion of New Zealand women in that age group who have a completed a secondary school qualification. (Use a *uniform* prior for π_1.)

(b) Out of 219 women in the age group 50–54, 120 had completed a secondary school qualification. Find the posterior distribution of π_2, the proportion of New Zealand women in that age group who have a completed a secondary school qualification. (Use a *uniform* prior for π_2.)

(c) Find the approximate posterior distribution of $\pi_1 - \pi_2$.

(d) Find a 99% Bayesian credible interval for $\pi_1 - \pi_2$.

(e) What would be the conclusion if you tested the hypothesis

$$H_0 : \pi_1 - \pi_2 = 0 \quad \text{versus} \quad H_1 : \pi_1 - \pi_2 \neq 0$$

at the 1% level of significance?

13.9. Are younger New Zealand women more likely to be in paid employment than previous generations of New Zealand women? To shed light on this question, we will look at the current employment status of two generations of New Zealand women 25 years apart.

(a) Out of 314 women in the age group 25–29, 171 were currently in paid employment. Find the posterior distribution of π_1, the proportion of New Zealand women in that age group who are currently in paid employment. (Use a *uniform* prior for π_1.)

(b) Out of 219 women in the age group 50–54, 137 were currently in paid employment. Find the posterior distribution of π_2, the proportion of New Zealand women in that age group who are currently in paid employment. (Use a *uniform* prior for π_2.)

(c) Find the approximate posterior distribution of $\pi_1 - \pi_2$.

(d) Find a 99% Bayesian credible interval for $\pi_1 - \pi_2$.

(e) What would be the conclusion if you tested the hypothesis

$$H_0 : \pi_1 - \pi_2 = 0 \quad \text{versus} \quad H_1 : \pi_1 - \pi_2 \neq 0$$

at the 1% level of significance?

13.10. Are younger New Zealand women becoming sexually active at an earlier age than previous generations of New Zealand women? To shed light on this question, we look at the proportions of New Zealand women who report having experienced sexual intercourse before age 18 for the two generations of New Zealand women.

(a) Out of the 298 women in the age group 25–29 who responded to this question, 180 report having experienced sexual intercourse before reaching the age of 18. Find the posterior distribution of π_1, the proportion of New Zealand women in that age group who had experienced sexual intercourse before age 18. (Use a *uniform* prior for π_1.)

(b) Out of the 218 women in the age group 50–54 who responded to this question, 52 report having experienced sexual intercourse before reaching the age of 18. Find the posterior distribution of π_2, the proportion of New Zealand women in that age group who had experienced sexual intercourse before age 18. (Use a *uniform* prior for π_2.)

(c) Find the approximate posterior distribution of $\pi_1 - \pi_2$.

(d) Test the hypothesis

$$H_0 : \pi_1 - \pi_2 \le 0 \quad \text{versus} \quad H_1 : \pi_1 - \pi_2 > 0$$

in a Bayesian manner at the 1% level of significance. Can we conclude that New Zealand women in the generation aged 25–29 have experienced sexual intercourse at an earlier age than New Zealand women in the generation aged 50–54?

13.11. Are younger New Zealand women marrying at a later age than previous generations of New Zealand women? To shed light on this question, we look at the proportions of New Zealand women who report having been married before age 22 for the two generations of New Zealand women.

(a) Out of the 314 women in the age group 25–29, 69 report having been married before the age 22. Find the posterior distribution of π_1, the proportion of New Zealand women in that age group who have married before age 22. (Use a *uniform* prior for π_1.)

(b) Out of the 219 women in the age group 50–54, 114 report having been married before age 22. Find the posterior distribution of π_2, the proportion of New Zealand women in that age group who have been married before age 22. (Use a *uniform* prior for π_2.)

(c) Find the approximate posterior distribution of $\pi_1 - \pi_2$.

(d) Test the hypothesis

$$H_0 : \pi_1 - \pi_2 \ge 0 \quad \text{versus} \quad H_1 : \pi_1 - \pi_2 < 0$$

in a Bayesian manner at the 1% level of significance. Can we conclude that New Zealand women in the generation aged 25–29 have married at an later age than New Zealand women in the generation aged 50–54?

13.12. Family formation patterns in New Zealand have changed over the time frame covered by the survey. New Zealand society has become more accepting of couples co-habiting (living together before or instead of legally marrying). When we take this into account, are younger New Zealand women forming family-like units at a similar age to previous generations?

(a) Out of the 314 women in the age group 25–29, 199 report having formed a domestic partnership (either co-habiting or legal marriage) before age 22. Find the posterior distribution of π_1, the proportion of New Zealand women in that age group who have formed a domestic partnership before age 22. (Use a *uniform* prior for π_1.)

(b) Out of the 219 women in the age group 50–54, 116 report having formed a domestic partnership before age 22. Find the posterior distribution of π_2, the proportion of New Zealand women in that age group who have formed a domestic partnership before age 22. (Use a *uniform* prior for π_2.)

(c) Find the approximate posterior distribution of $\pi_1 - \pi_2$.

(d) Find a 99% Bayesian credible interval for $\pi_1 - \pi_2$.

(e) What would be the conclusion if you tested the hypothesis

$$H_0 : \pi_1 - \pi_2 = 0 \quad \text{versus} \quad H_1 : \pi_1 - \pi_2 \neq 0$$

at the 1% level of significance.

13.13. Are young New Zealand women having their children at a later age than previous generations?

(a) Out of the 314 women in the age group 25–29, 136 report having given birth to their first child before the age of 25. Find the posterior distribution of π_1, the proportion of New Zealand women in that age group who have given birth before age 25. (Use a *uniform* prior for π_1.)

(b) Out of the 219 women in the age group 50–54, 135 report having given birth to their first child before age 25. Find the posterior distribution of π_2, the proportion of New Zealand women in that age group who have given birth before age 25. (Use a *uniform* prior for π_2.)

(c) Find the approximate posterior distribution of $\pi_1 - \pi_2$.

(d) Test the hypothesis

$$H_0 : \pi_1 - \pi_2 \geq 0 \quad \text{versus} \quad H_1 : \pi_1 - \pi_2 < 0$$

in a Bayesian manner at the 1% level of significance. Can we conclude that New Zealand women in the generation aged 25–29 have had their first child at a later age than New Zealand women in the generation aged 50–54?

13.14. Previous research has suggested that the childhood circumcision of males may be a protective factor against the acquisition of sexually transmitted infections (STI). Fergusson et al. (2006) relate the circumcision status and self reported STI history using data from 25-year longitudinal study of a cohort of New Zealand children, known as the Christchurch Health and Development Study.

(a) Out of 356 non-circumcised males, 37 reported having had at least one STI by age 25. Find the posterior distribution of π_1, the probability a non-circumcised male reports at least one STI by age 25. (Use a $beta(1, 10)$ prior for π_1.)

(b) Out of the 154 circumcised males, 7 reported having at least one STI by age 25. Find the posterior distribution of π_2, the probability a circumcised male reports at least one STI by age 25. (Use a $beta(1, 10)$ prior for π_2.)

(c) Find the approximate posterior distribution of $\pi_1 - \pi_2$.

(d) Test the hypothesis

$$H_0 : \pi_1 - \pi_2 \leq 0 \quad \text{versus} \quad H_1 : \pi_1 - \pi_2 > 0$$

in a Bayesian manner at the 5% level of significance. What does the result say about the research hypothesis?

13.15. The experiment described in Exercise 13.6 was repeated on another set of 7 cows (McLeay et al., 1997). However, in this case, the second treatment was given to the same set of 7 cows that were given the first treatment, at a later time when the first dose of ^{13}C had been eliminated from the cow. The data are given below:

Cow ID	^{13}C Administered into Reticulum x	^{13}C Administered into Reticulo-omasal Orifice y
1	1.1	3.5
2	0.8	3.6
3	1.7	5.1
4	1.1	5.6
5	2.0	6.2
6	1.6	6.5
7	3.1	8.3

(a) Explain why the variables x and y cannot be considered independent in this experiment.

(b) Calculate the differences $d_i = x_i - y_i$ for $i = 1, \ldots, 7$.

(c) Assume that the differences come from a $normal(\mu_d, \sigma_d^2)$ distribution, where $\sigma_d^2 = 1$. Use a $normal(0, 3^2)$ prior for μ_d. Calculate the posterior for $\mu_d | d_1, \ldots, d_7$.

(d) Calculate a 95% Bayesian credible interval for μ_d.

(e) Test the hypothesis

$$H_0 : \mu_d = 0 \quad \text{versus} \quad H_1 : \mu_d \neq 0$$

at the 5% level of significance. What conclusion can be drawn?

13.16. One of the advantages of Bayesian statistics is that evidence from different sources can be combined. In Exercise 13.6 and Exercise 13.15, we found posterior distributions of μ_d using data sets from two different experiments. In the first experiment, the two treatments were given to two sets of cows, and the measurements were independent. In the second experiment, the two treatments were given to a third set of cows at different times and the measurements were paired. When we want to find the posterior distribution given data sets from two independent experiments, we should use the posterior distribution after the first experiment as the prior distribution for the second.

(a) Explain why the two data sets can be considered independent.

(b) Find the posterior distribution of $\mu_d | data$ where the $data$ include all of the measurements $x_8, \ldots, x_{13}, y_{14}, \ldots, y_{19}, d_1, \ldots, d_7$.

(c) Find a 95% credible interval for μ_d based on all the data.

(d) Test the hypothesis

$$H_0 : \mu_d = 0 \quad \text{versus} \quad H_1 : \mu_d \neq 0$$

at the 5% level of significance. Can we conclude that ^{13}C octanoic acid breath test is effective in detecting reticular groove contraction in cattle?

CHAPTER 14

BAYESIAN INFERENCE FOR SIMPLE LINEAR REGRESSION

Sometimes we want to model a relationship between two variables, x and y. We might want to find an equation that describes the relationship. Often we plan to use the value of x to help predict y using that relationship.

The data consist of n ordered pairs of points (x_i, y_i) for $i = 1, \ldots, n$. We think of x as the predictor variable (independent variable) and consider that we know it without error. We think y is a response variable that depends on x in some unknown way, but that each observed y contains an error term as well. We plot the points on a two-dimensional *scatterplot*; the predictor variable is measured along the horizontal axis, and the response variable is measured along the vertical axis.

We examine the scatterplot for clues about the nature of the relationship. To construct a regression model, we first decide on the type of equation that appears to fit the data. A *linear* relationship is the simplest equation relating two variables. This would give a straight line relationship between the predictor x and the response y. We leave the parameters of the line, the slope β, and the y-intercept α_0 unknown, so all lines are possible.

Then we determine the best estimates of the unknown parameters by some criterion. The criterion that is most frequently used is *least squares*. This

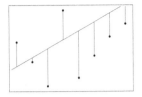 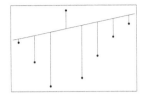 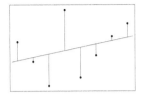

Figure 14.1   Scatterplot with three possible lines, and the residuals from each of the lines. The third line is the least squares line. It minimizes the sum of squares of the residuals.

is where we find the parameter values that minimize the sum of squares of the *residuals*, which are the vertical distances of the observed points to the fitted equation. We do this for the simple linear regression in Section 14.1. In Section 14.2 we look at how an exponential growth model can be fitted using least squares regression on the logarithm of the response variable.

At this stage no inferences are possible because there is no probability model for the data. In Section 14.3 we construct a regression model that makes assumptions on how the response variable depends on the predictor variable and how randomness enters the data. Inferences can be done on the parameters of this model. In Section 14.4 we fit a *linear* relationship between the two variables using Bayesian methods, and perform Bayesian inferences on the parameters of the model. In Section 14.5 we determine the predictive distribution of y_{n+1}, the next observation, given the data and x_{n+1}, the value of the predictor variable for the next observation.

14.1   Least Squares Regression

We could draw any number of lines on the scatterplot. Some of them would fit the data points fairly well, others would be extremely far from the points. A *residual* is the *vertical* distance from an observed point on the scatterplot to the line. We can put in any line that we like and then calculate the residuals from that line. Least squares is a method for finding the line that *best* fits the points in terms of minimizing *sum of squares of the residuals*. Figure 14.1 shows a scatterplot, three possible lines, and the residuals from each line.

The equation of a line is determined by two things: its slope β and its y-intercept α_0. Actually its slope and any other point on the line will do, for instance, $\alpha_{\bar{x}}$, the intercept of the vertical line at $\bar{x}$. Finding the least squares line is equivalent to finding its slope and the y-intercept (or another intercept).

The Normal Equations and the Least Squares Line

The sum of squares of the residuals from line $y = \alpha_0 + \beta x$ is

$$SS_{res} = \sum_{i=1}^{n} [y_i - (\alpha_0 + \beta x_i)]^2 .$$

To find values of α_0 and β that minimize SS_{res} using calculus, take derivatives with respect to each α_0 and β and set equal to 0, and solve the resulting set of simultaneous equations. First, take the derivative with respect to intercept α_0. This gives the equation,

$$\frac{\partial SS}{\partial \alpha_0} = \sum_{i=1}^{n} 2 \times [y_i - (\alpha_0 + \beta x_i)]^1 \times (-1) = 0$$

which simplifies to

$$\sum_{i=1}^{n} y_i - \sum_{i=1}^{n} \alpha_0 - \sum_{i=1}^{n} \beta x_i = 0$$

and further to

$$\bar{y} - \alpha_0 - \beta \bar{x} = 0 . \tag{14.1}$$

Second, taking the derivative with respect to the slope β gives the equation

$$\frac{\partial SS}{\partial \beta} = \sum_{i=1}^{n} 2 \times [y_i - (\alpha_0 + \beta x_i)]^1 \times (-x_i) = 0 ,$$

which simplifies to

$$\sum_{i=1}^{n} x_i y_i - \sum_{i=1}^{n} \alpha_0 x_i - \sum_{i=1}^{n} \beta x_i^2 = 0$$

and further to

$$\overline{xy} - \alpha_0 \bar{x} - \beta \overline{x^2} = 0 . \tag{14.2}$$

Equation 14.1 and Equation 14.2 are known as the *normal* equations. Here *normal* refers to right angles[1] and has nothing to do with the normal distribution. Solve Equation 14.1 for α_0 in terms of β and substitute into Equation 14.2 and solve for β

$$\overline{xy} - (\bar{y} - \beta \bar{x})\bar{x} - \beta \overline{x^2} = 0 .$$

The solution is the least squares slope[2]

$$B = \frac{\overline{xy} - \bar{x}\bar{y}}{\overline{x^2} - \bar{x}^2} . \tag{14.3}$$

---

[1]Least squares finds the projection of the (n-dimensional) observation vector onto the plane containing all possible values of (α_0, β).

[2]There are many different formulas for the least squares slope. This can be a source of confusion because many books give formulas that look quite dissimilar. However, all can be shown to be equivalent. I use this one because it is easy to remember: the average of $x \times y$ minus the average of $x \times$ the average of y all divided by the average of x^2 minus the square of the average of x.

Note that it is very important that you do not round off when calculating the least squares slope using Equation 14.3. Both the numerator and denominator are differences, and rounding off will lead to substantial error in the slope estimate! Substitute B back into Equation 14.1 and solve for the least squares y-intercept,

$$A_0 = \bar{y} - B\bar{x}. \tag{14.4}$$

Again, it is important that you do not round off when calculating the least squares intercept using Equation 14.4. The equation of the least squares line is

$$y = A_0 + Bx. \tag{14.5}$$

Alternative form for the least squares line. The slope and any other point besides y-intercept also determines the line. Say the point is $A_{\bar{x}}$, where the least squares line intercepts the vertical line at $\bar{x}$:

$$A_{\bar{x}} = A_0 + B\bar{x} = \bar{y}.$$

Thus the least squares line goes through the point $(\bar{x}, \bar{y})$. An alternative equation for the least squares line is

$$y = A_{\bar{x}} + B(x - \bar{x}) = \bar{y} + B(x - \bar{x}), \tag{14.6}$$

which is particularly useful.

Estimating the Variance around the Least Squares Line

The estimate of the variance around the least squares line is

$$\hat{\sigma}^2 = \frac{\sum_{i=1}^{n}[y_i - (A_{\bar{x}} + B(x_i - \bar{x}))]^2}{n - 2},$$

which is the sum of squares of the residuals divided by $n - 2$. The reason we use $n - 2$ is that we have used two estimates, $A_{\bar{x}}$ and B in calculating the sum of squares.[3]

◼ EXAMPLE 14.1

A company is manufacturing a food product, and must control the moisture level in the final product. It is cheaper (and hence preferable) to measure the level at an in-process stage rather than in the final product. Michael, the company statistician, recommends to the engineers running the process that a measurement of the moisture level at an in-process stage

---

[3]The general rule for finding an unbiased estimate of the variance is that the sum of squares is divided by the degrees of freedom, and we lose a degree of freedom for every estimated parameter in the sum of squares formula.

Table 14.1   In-process and final moisture levels

Batch	In-Process Level x	Final Level y	LS Fits $\hat{y} = A_0 + Bx$	Residual $y - \hat{y}$	Residual^2 $(y - \hat{y})^2$
1	14.36	13.84	14.1833	-0.343256	0.117825
2	14.48	14.41	14.3392	0.070792	0.005012
3	14.53	14.22	14.4042	-0.184188	0.033925
4	14.52	14.63	14.3912	0.238808	0.057029
5	14.35	13.95	14.1703	-0.220260	0.048514
6	14.31	14.37	14.1183	0.251724	0.063365
7	14.44	14.41	14.2872	0.122776	0.015074
8	14.23	13.99	14.0143	-0.024308	0.000591
9	14.32	13.89	14.1313	-0.241272	0.058212
10	14.57	14.59	14.4562	0.133828	0.017910
11	14.28	14.32	14.0793	0.240712	0.057942
12	14.36	14.31	14.1833	0.126744	0.016064
13	14.50	14.43	14.3652	0.064800	0.004199
14	14.52	14.44	14.3912	0.048808	0.002382
15	14.28	14.14	14.0793	0.060712	0.003686
16	14.13	13.90	13.8843	0.015652	0.000245
17	14.54	14.37	14.4172	-0.047184	0.002226
18	14.60	14.34	14.4952	-0.155160	0.024075
19	14.86	14.78	14.8331	-0.053056	0.002815
20	14.28	13.76	14.0793	-0.319288	0.101945
21	14.09	13.85	13.8324	0.017636	0.000311
22	14.20	13.89	13.9753	-0.085320	0.007280
23	14.50	14.22	14.3652	-0.145200	0.021083
24	14.02	13.80	13.7414	0.058608	0.003435
25	14.45	14.67	14.3002	0.369780	0.136737
Mean	14.3888	14.2208			

may give a good prediction of what the final moisture level will be. He organizes the collection of data from 25 batches, giving the moisture level at the in-process stage and the final moisture level for each batch. These are shown in the first three columns of Table 14.1. Summary statistics for these data are: $\bar{x} = 14.3888$, $\bar{y} = 14.2208$, $\overline{x^2} = 207.0703$, $\overline{y^2} = 202.3186$, and $\overline{xy} = 204.6628$. Note that he needs to keep all the significant figures in the squared terms. The formula for B uses subtraction, and if he rounds off too early, the differences will have too few significant figures and accuracy will be lost.

He then calculates the least squares line relating the final moisture level to the in-process moisture level. The slope is given by

$$B = \frac{\overline{xy} - \bar{x}\bar{y}}{\overline{x^2} - (\bar{x})^2} = \frac{204.6628 - 14.3888 \times 14.2208}{207.0703 - (14.3888)^2} = \frac{.0425690}{.0327546} = 1.29963 \,.$$

The equation of the least squares line is

$$y = 14.2208 + 1.29963 \times (x - 14.3888).$$

The scatterplot of final moisture level and in-process moisture level together with the least squares line is given in Figure 14.2.

He calculates the least squares fitted values $\hat{y}_i = \bar{y} + B(x_i - \bar{x})$, the residuals, and the squared residuals. They are in the last three columns of Table 14.1. The estimated variance about the least squares line is

$$\hat{\sigma}^2 = \frac{\sum_{i=1}^{n}(y_i - \hat{y}_i)^2}{n - 2} = \frac{.801882}{23} = .0348644 \,.$$

To find the estimated standard deviation about the least squares line, he takes the square root:

$$\hat{\sigma} = \sqrt{(.0348644)} = 0.18672 \,.$$

■

14.2   Exponential Growth Model

When we look at economic time series, the predictor variable is time t, and we want to see how some response variable u depends on t. Often, when we graph the response variable versus time on a scatterplot, we notice two things. First, the plotted points seem to go up not at a linear rate but at a rate that increases with time. Second, the variability of the plotted points seems to be increasing at about the same rate as the response variable. This will be shown more clearly if we graph the residuals versus time. In this case the exponential growth model will usually give a better fit:

$$u = e^{\alpha_0 + \beta \times t} \,.$$

We note that if we let $y = \log_e(u)$, then

$$y = \alpha_0 + \beta \times t$$

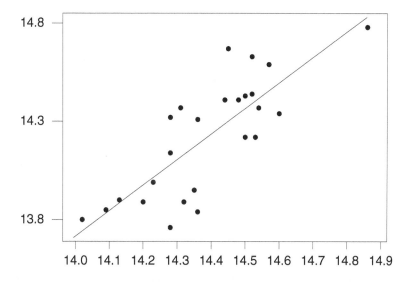

Figure 14.2   Scatterplot and least squares line for the moisture data.

is a linear relationship. We can estimate the parameters of the relationship using least squares using response variable y. The fitted exponential growth model is

$$u = e^{A_0 + B \times t},$$

where B and A_0 are the least squares slope and intercept for the logged data.

EXAMPLE 14.2

The annual New Zealand poultry production (in tonnes) for the years 1987–2001 is given in Table 14.2.

The scatterplot showing the residuals and least squares line is shown in Figure 14.3.

We see that the residuals are mostly positive at the ends of the data, and mostly negative in the center. This indicates that an exponential growth model would give a better fit. The scatterplot, along with the exponential growth model found by exponentiating the least squares line to the logged data, is shown in Figure 14.4. ■

Table 14.2   Annual poultry production in New Zealand

Year t	Poultry Production u	Linear Fitted Value	$\log_e(u)$	Fitted $\log_e u$	Exponential Fitted Value
1987	44,085	47,757	10.7739	10.7776	47,934
1988	51,646	48,725	10.8522	10.8393	50,986
1989	57,241	53,364	10.9550	10.9010	54,232
1990	56,261	58,004	10.9378	10.9628	57,686
1991	58,257	62,643	10.9726	11.0245	61,359
1992	60,944	67,283	11.0177	11.0862	65,266
1993	68,214	71,922	11.1304	11.1479	69,421
1994	74,037	76,562	11.2123	11.2097	73,842
1995	88,646	81,201	11.3924	11.2714	78,543
1996	86,869	85,841	11.3722	11.3331	83,545
1997	86,534	90,480	11.3683	11.3949	88,864
1998	95,682	95,120	11.4688	11.4566	94,522
1999	97,400	99,759	11.4866	11.5183	100,541
2000	104,927	104,398	11.5610	11.5801	106,943
2001	114,010	109,038	11.6440	11.6418	113,752

14.3   Simple Linear Regression Assumptions

The method of least squares is *nonparametric* or *distribution free*, since it makes no use of the probability distribution of the data. It is really a data analysis tool and can be applied to any bivariate data. We cannot make any inferences about the slope and intercept nor about any predictions from the least squares model, unless we make some assumptions about the probability model underlying the data. The simple linear regression assumptions are:

1. *Mean assumption.* The conditional mean of y given x is an unknown linear function of x.

$$\mu_{y|x} = \alpha_0 + \beta x \,,$$

where β is the unknown slope and α_0 is the unknown y *intercept*, the intercept of the vertical line $x = 0$. In the alternate parameterization we have

$$\mu_{y|x} = \alpha_{\bar{x}} + \beta(x - \bar{x}) \,,$$

where $\alpha_{\bar{x}}$ is the unknown intercept of the vertical line $x = \bar{x}$. In this parameterization the least squares estimates $A_{\bar{x}} = \bar{y}$ and B will be independent

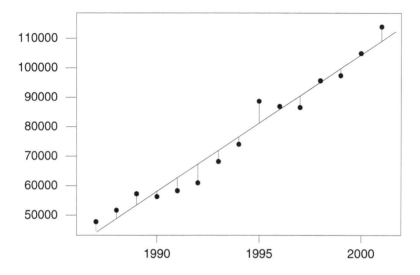

Figure 14.3   Scatterplot and least squares line for the poultry production data.

under our assumptions, so the likelihood will factor into a part depending on $\alpha_{\bar{x}}$ and a part depending on β. This greatly simplifies things, so we will use this parameterization. The mean assumption is shown in the first graph of Figure 14.5.

2. *Error assumption.* Observation equals mean plus error, which is *normally* distributed with mean 0 and *known* variance σ^2. All errors have equal variance. The equal variance assumption is shown in the second graph of Figure 14.5.

3. *Independence assumption.* The errors for all of the observations are independent of each other. The independent draw assumption is shown in the third graph of Figure 14.5.

Using the alternate parameterization we obtain

$$y_i = \alpha_{\bar{x}} + \beta \times (x_i - \bar{x}) + e_i,$$

where $\alpha_{\bar{x}}$ is the mean value for y given $x = \bar{x}$, and β is the slope. Each e_i is normally distributed with mean 0 and known variance σ^2. The e_i are all independent of each other. Therefore $y_i | x_i$ is normally distributed with mean $a_{\bar{x}} + \beta(x_i - \bar{x})$ and variance σ^2 and all the $y_i | x_i$ are all independent of each other.

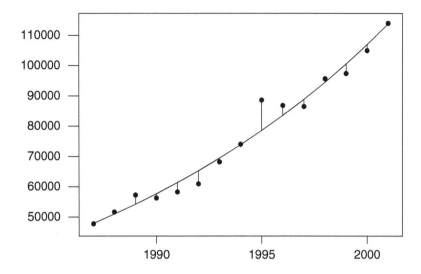

Figure 14.4   Scatterplot and fitted exponential growth model for the poultry production data.

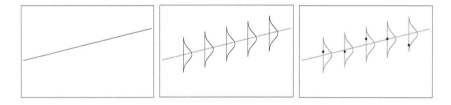

Figure 14.5   Assumptions of linear regression model. The mean of Y given X is a linear function. The observation errors are normally distributed with mean 0 and equal variances. The observations are independent of each other.

14.4   Bayes' Theorem for the Regression Model

Bayes' theorem is always summarized by

$$posterior \propto prior \times likelihood\,,$$

so we need to determine the likelihood and decide on our prior for this model.

The Joint Likelihood for β and $\alpha_{\bar{x}}$

The joint likelihood of the ith observation is its probability density function as a function of the two parameters $\alpha_{\bar{x}}$ and β, where (x_i, y_i) are fixed at the observed values. It gives relative weights to all possible values of both parameters $a_{\bar{x}}$ and β from the observation. The likelihood of observation i is

$$likelihood_i(\alpha_{\bar{x}}, \beta) \propto e^{-\frac{1}{2\sigma^2}[y_i - (\alpha_{\bar{x}} + \beta(x_i - \bar{x}))]^2},$$

since we can ignore the part not containing the parameters. The observations are all independent, so the likelihood of the whole sample of all the observations is the product of the individual likelihoods:

$$likelihood_{sample}(\alpha_{\bar{x}}, \beta) \propto \prod_{i=1}^{n} e^{-\frac{1}{2\sigma^2}[y_i - (\alpha_{\bar{x}} + \beta(x_i - \bar{x}))]^2}.$$

The product of exponentials is found by summing the exponents, so

$$likelihood_{sample}(\alpha_{\bar{x}}, \beta) \propto e^{-\frac{1}{2\sigma^2}[\sum_{i=1}^{n}[y_i - (a_{\bar{x}} + \beta(x_i - \bar{x}))]^2]}.$$

The term in brackets in the exponent equals

$$\left[\sum_{i=1}^{n} [y_i - \bar{y} + \bar{y} - (\alpha_{\bar{x}} + \beta(x_i - \bar{x}))]^2 \right].$$

Breaking this into three sums and multiplying it out gives us

$$\sum_{i=1}^{n}(y_i - \bar{y})^2 + 2\sum_{i=1}^{n}(y_i - \bar{y})(\bar{y} - (\alpha_{\bar{x}} + \beta(x_i - \bar{x})))$$

$$+ \sum_{i=1}^{n}(\bar{y} - (\alpha_{\bar{x}} + \beta(x_i - \bar{x})))^2.$$

This simplifies into

$$SS_y - 2\beta SS_{xy} + \beta^2 SS_x + n(\alpha_{\bar{x}} - \bar{y})^2,$$

where $SS_y = \sum_{i=1}^{n}(y_i - \bar{y})^2$, $SS_{xy} = \sum_{i=1}^{n}(y_i - \bar{y})(x_i - \bar{x})$, and $SS_x = \sum_{i=1}^{n}(x_i - \bar{x})^2$. Thus the joint likelihood can be written as

$$likelihood_{sample}(\alpha_{\bar{x}}, \beta) \propto e^{-\frac{1}{2\sigma^2}[SS_y - 2\beta SS_{xy} + \beta^2 SS_x + n(\alpha_{\bar{x}} - \bar{y})^2]}.$$

Writing this as a product of two exponentials gives

$$\propto e^{-\frac{1}{2\sigma^2}[SS_y - 2\beta SS_{xy} + \beta^2 SS_x]} \times e^{-\frac{1}{2\sigma^2}[n(\alpha_{\bar{x}} - \bar{y})^2]}.$$

We factor out SS_x in the first exponential, complete the square, and absorb the part that does not depend on any parameter into the proportionality constant. This gives us

$$likelihood_{sample}(\alpha_{\bar{x}}, \beta) \propto e^{-\frac{1}{2\sigma^2/SS_x}[\beta - \frac{SS_{xy}}{SS_x}]^2} \times e^{-\frac{1}{2\sigma^2/n}[(\alpha_{\bar{x}} - \bar{y})^2]}.$$

Note that $\frac{SS_{xy}}{SS_x} = B$, the least squares slope, and $\bar{y} = A_{\bar{x}}$, the least squares estimate of the intercept of the vertical line $x = \bar{x}$. We have factored the joint likelihood into the product of two individual likelihoods

$$likelihood_{sample}(\alpha_{\bar{x}}, \beta) \propto likelihood_{sample}(\alpha_{\bar{x}}) \times likelihood_{sample}(\beta),$$

where

$$likelihood_{sample}(\beta) \propto e^{-\frac{1}{2\sigma^2/SS_x}(\beta-B)^2}$$

and

$$likelihood_{sample}(\alpha_{\bar{x}}) \propto e^{-\frac{1}{\sigma^2/n}(\alpha_{\bar{x}}-A_{\bar{x}})^2}.$$

Since the joint likelihood has been factored into the product of the individual likelihoods we know the individual likelihoods are independent. We recognize that the likelihood of the slope β has the *normal* shape with mean B, the least squares slope, and variance $\frac{\sigma^2}{SS_x}$. Similarly the likelihood of $\alpha_{\bar{x}}$ has the *normal* shape with mean $A_{\bar{x}}$ and variance $\frac{\sigma^2}{n}$.

The Joint Prior for β and $\alpha_{\bar{x}}$

If we multiply the joint likelihood by a joint prior, it is proportional to the joint posterior. We will use independent priors for each parameter. The joint prior of the two parameters is the product of the two individual priors:

$$g(\alpha_{\bar{x}}, \beta) = g(\alpha_{\bar{x}}) \times g(\beta).$$

We can either use *normal* priors, or *flat* priors.

Choosing normal priors for β and $\alpha_{\bar{x}}$.   Another advantage of using this parameterization is that a person has a more intuitive prior knowledge about the $\alpha_{\bar{x}}$, the intercept of $x = \bar{x}$, than about α_0, the intercept of the y axis. Decide on what you believe the mean value of the y values to be. That will be $m_{\alpha_{\bar{x}}}$, your prior mean for $\alpha_{\bar{x}}$. Then think of the points above and below that you consider to be upper and lower bounds of the possible values of y. Divide the difference by 6 to get $s_{\alpha_{\bar{x}}}$, your prior standard deviation of $\alpha_{\bar{x}}$. This will give you reasonable probability over the whole range you believe possible.

Usually we are more interested in the slope β. Sometimes we want to determine if it could be 0. Therefore we may choose $m_\beta = 0$ as the prior mean for β. Then we think of the upper and lower bounds of the effect of an increase in x of one unit on y. Divide the difference by 6 to get s_β, your prior standard deviation of β. In other cases, we have prior belief about the slope from previous data. We would use the $normal(m_\beta, (s_\beta)^2)$ that matches that prior belief.

The Joint Posterior for β and $\alpha_{\bar{x}}$

The joint posterior then is proportional to the joint prior times the joint likelihood.

$$g(\alpha_{\bar{x}}, \beta | data) \propto g(\alpha_{\bar{x}}, \beta) \times likelihood_{sample}(\alpha_{\bar{x}}, \beta),$$

where the *data* is the set of ordered pair $(x_1, y_1), \ldots, (x_n, y_n)$. The joint prior and the joint likelihood both factor into a part depending on $\alpha_{\bar{x}}$ and a part depending on β. Rearranging them gives the joint posterior factored into the marginal posteriors

$$g(\alpha_{\bar{x}}, \beta | data) \propto g(\alpha_{\bar{x}} | data) \times g(\beta | data).$$

Since the joint posterior is the product of the marginal posteriors, they are independent. Each of these marginal posteriors can be found by using the simple updating rules for normal distributions, which works for *normal* and *flat* priors. For instance, if we use a $normal(m_\beta, s_\beta^2)$ prior for β, we get a $normal(m_\beta', (s_\beta')^2)$, where

$$\frac{1}{(s_\beta')^2} = \frac{1}{s_\beta^2} + \frac{SS_x}{\sigma^2} \qquad (14.7)$$

and

$$m_\beta' = \frac{\frac{1}{s_\beta^2}}{\frac{1}{(s_\beta')^2}} \times m_\beta + \frac{\frac{SS_x}{\sigma^2}}{\frac{1}{(s_\beta')^2}} \times B. \qquad (14.8)$$

The posterior precision equals the prior precision plus the precision of the likelihood. The posterior mean equals the weighted average of the prior mean and the likelihood mean where the weights are the proportions of the precisions to the posterior precision. And the posterior distribution is normal.

Similarly, if we use a $normal(m_{\alpha_{\bar{x}}}, s_{\alpha_{\bar{x}}}^2)$ prior for $\alpha_{\bar{x}}$, we get a $normal(m_{\alpha_{\bar{x}}}', (s_{\alpha_{\bar{x}}}')^2)$ where

$$\frac{1}{(s_{\alpha_{\bar{x}}}')^2} = \frac{1}{s_{\alpha_{\bar{x}}}^2} + \frac{n}{\sigma^2}$$

and

$$m_{\alpha_{\bar{x}}}' = \frac{\frac{1}{s_{\alpha_{\bar{x}}}^2}}{\frac{1}{(s_{\alpha_{\bar{x}}}')^2}} \times m_{\alpha_{\bar{x}}} + \frac{\frac{n}{\sigma^2}}{\frac{1}{(s_{\alpha_{\bar{x}}}')^2}} \times A_{\bar{x}}.$$

■ **EXAMPLE 14.2** (continued)

Michael, the company statistician, decides that he will use a $normal(1, (.3)^2)$ prior for β and a $normal(15, 1^2)$ prior for $\alpha_{\bar{x}}$. Since he does not know the true variance, he will use the estimated variance about the least

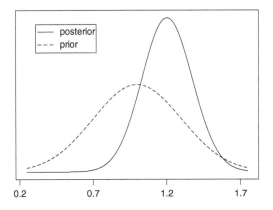

Figure 14.6   The prior and posterior distribution of the slope.

squares regression line $\hat{\sigma}^2 = .0348644$. Note that $SS_x = \sum_{i=1}^{n}(x_i - \bar{x})^2 = n(\overline{x^2} - \bar{x}^2) = 25 \times (207.0703 - 14.3888^2) = .81886$.

The posterior precision of β is

$$\frac{1}{(s'_\beta)^2} = \frac{1}{.3^2} + \frac{.81886}{.0348644} = 34.5981\,,$$

so the posterior standard deviation of β is

$$s'_\beta = 34.5981^{-\frac{1}{2}} = .17001\,.$$

The posterior mean of β is

$$m'_\beta = \frac{\frac{1}{.3^2}}{34.5981} \times 1 + \frac{\frac{.81886}{.0348644}}{34.5981} \times 1.29963 = 1.2034\,.$$

Similarly, the posterior precision of $\alpha_{\bar{x}}$ is

$$\frac{1}{(s'_{\alpha_{\bar{x}}})^2} = \frac{1}{1^2} + \frac{25}{.0348644} = 718.064\,,$$

so the posterior standard deviation is

$$s'_{\alpha_{\bar{x}}} = 718.064^{-\frac{1}{2}} = .037318\,.$$

The posterior mean of $\alpha_{\bar{x}}$ is

$$m'_{\alpha_{\bar{x}}} = \frac{\frac{1}{1^2}}{718.064} \times 15 + \frac{\frac{25}{.0348644}}{718.064} \times 14.2208 = 14.2219\,.$$

The prior and posterior distribution of the slope are shown in Figure 14.6.

■

Bayesian Credible Interval for Slope

The posterior distribution of β summarizes our entire belief about it after examining the data. We may want to summarize it by a $(1 - \alpha) \times 100\%$ Bayesian credible interval for slope β. This will be

$$m'_\beta \pm z_{\frac{\alpha}{2}} \times \sqrt{(s'_\beta)^2}. \qquad (14.9)$$

More realistically, we do not know σ^2. A sensible approach in that instance is to use the estimate calculated from the residuals

$$\hat{\sigma}^2 = \frac{\sum_{i=1}^{n}(y_i - (A_{\bar{x}} + B(x_i - \bar{x})))^2}{n - 2}.$$

We have to widen the confidence interval to account for the increased uncertainty due to not knowing σ^2. We do this by using a *Student's t* critical value with $n - 2$ degrees of freedom[4] instead of standard normal critical value. The credible interval becomes

$$m'_\beta \pm t_{\frac{\alpha}{2}} \times \sqrt{(s'_\beta)^2}. \qquad (14.10)$$

Frequentist Confidence Interval for Slope

When the variance σ^2 is unknown, the $(1 - \alpha) \times 100\%$ confidence interval for the slope β is

$$B \pm t_{\frac{\alpha}{2}} \times \frac{\hat{\sigma}}{\sqrt{SS_x}},$$

where $\hat{\sigma}^2$ is the estimate of the variance calculated from the residuals from the least squares line. The confidence interval is the same form as the Bayesian credible interval when we used *flat* priors for β and $\alpha_{\bar{x}}$. Of course the interpretation is different. Under the frequentist assumptions we are $(1 - \alpha) \times 100\%$ confident that the interval contains the true, unknown parameter value. Once again, the frequentist confidence interval is equivalent to a Bayesian credible interval, so if the scientist misinterprets it as a probability interval, he/she will get away with it. The only loss experienced will be that the scientist did not get to put in any of his/her prior knowledge.

Testing One-Sided Hypothesis about Slope

Often we want to determine whether or not the amount of increase in y associated with one unit increase in x is greater than some value, β_0. We can do this by testing

$$H_0 : \beta \leq \beta_0 \quad \text{versus} \quad H_1 : \beta > \beta_0$$

---

[4]Actually we are treating the unknown parameter σ^2 as a nuisance parameter and using the prior $g(\sigma^2) \propto (\sigma^2)^{-1}$. The marginal posterior of β is found by integrating σ^2 out of the joint posterior.

at the α level of significance in a Bayesian manner. To do the test in a Bayesian manner, we calculate the posterior probability of the null hypothesis. This is

$$P(\beta \leq \beta_0 | data) = \int_{-\infty}^{\beta_0} g(\beta | data) \, d\beta$$

$$= P\left(Z \leq \frac{\beta_0 - m'_\beta}{s'_\beta}\right). \tag{14.11}$$

If this probability is less than α, then we reject H_0 and conclude that indeed the slope β is greater than β_0. (If we used the estimate of the variance, then we would use a *Student's t* with $n-2$ degrees of freedom instead of the standard normal Z.)

Testing Two-Sided Hypothesis about Slope

If $\beta = 0$, then the mean of y does not depend on x at all. We really would like to test $H_0 : \beta = 0$ versus $H_1 : \beta \neq 0$ at the α level of significance in a Bayesian manner, before we use the regression model to make predictions. To do the test in a Bayesian manner, look where 0 lies in relation to the credible interval. If it lies outside the interval, we reject H_0. Otherwise, we cannot reject the null hypothesis, and we should not use the regression model to help with predictions.

■ **EXAMPLE 14.2   (continued)**

Since Michael used the estimated variance in place of the unknown true variance, he used Equation 14.10 to find a 95% Bayesian credible interval where there are 23 degrees of freedom. The interval is (.852,1.555). This credible interval does not contain 0, so clearly he can reject the hypothesis that the slope equals 0 and conclude that the final moisture level can be estimated using the measured in-process moisture level.   ■

14.5   Predictive Distribution for Future Observation

Making predictions of future observations for specified x values is one of the main purposes of linear regression modelling. Often, after we have established from the data that there is a linear relationship between the explanatory variable x and the response variable y, we want to use that relationship to make predictions of the next value y_{n+1}, given the next value of the explanatory variable x_{n+1}. We can make better predictions using the value of the explanatory variable than without it. The best prediction for y_{n+1} given x_{n+1} will be

$$\tilde{y}_{n+1} = \hat{\alpha}_{\bar{x}} + \hat{\beta} \times (x_{n+1} - \bar{x}),$$

where $\hat{\beta}$ is the slope estimate and $\hat{\alpha}_{\bar{x}}$ is the estimate of the intercept of the line $x = \bar{x}$.

How good is the prediction? There are two sources of uncertainty. First, we are using the estimated values of the parameters in the prediction, not the true values, which are unknown. We are considering the parameters to be random variables and have found their posterior distribution in the previous section. Second, the new observation y_{n+1} contains its own observation error e_{n+1}, which will be independent of all previous observation errors. The *predictive distribution* of the next observation y_{n+1} given the value x_{n+1} and the *data* accounts for both sources of uncertainty. It is denoted $f(y_{n+1}|x_{n+1}, data)$ and is found by Bayes' theorem.

Finding the Predictive Distribution

The predictive distribution is found by integrating the parameters $\alpha_{\bar{x}}$ and β out of the joint posterior distribution of the next observation y_{n+1} and the parameters given the next value x_{n+1} and the previous observations from the model, $(x_1, y_1), \ldots, (x_n, y_n)$, the *data*. It is

$$f(y_{n+1}|x_{n+1}, data) = \int \int f(y_{n+1}, \alpha_{\bar{x}}, \beta|x_{n+1}, data) \, d\alpha_{\bar{x}} \, d\beta \,.$$

Integrating out nuisance parameters from the joint posterior like this is known as *marginalization*. This is one of the clear advantages of Bayesian statistics. It has a single method of dealing with nuisance parameters that always works. When we find the predictive distribution, we consider all the parameters to be nuisance parameters.

First, we need to determine the joint posterior distribution of the parameters and next observation, given the value x_{n+1} and the *data*:

$$f(y_{n+1}, \alpha_{\bar{x}}, \beta|x_{n+1}, data) = f(y_{n+1}|\alpha_{\bar{x}}, \beta, x_{n+1}, data)$$
$$\times g(\alpha_{\bar{x}}, \beta|x_{n+1}, data) \,.$$

The next observation y_{n+1}, given the parameters $\alpha_{\bar{x}}$ and β and the known value x_{n+1}, is just another random observation from the regression model. Given the parameters $\alpha_{\bar{x}}$ and β, the observations are all independent of each other. This means that given the parameters, the new observation y_{n+1} does not depend on the *data*, which are the previous observations from the regression. The posterior for $\alpha_{\bar{x}}, \beta$, was calculated from the *data* alone and does not depend on the next value of the predictor x_{n+1}. So the joint distribution of new observation and parameters simplifies to

$$f(y_{n+1}, \alpha_{\bar{x}}, \beta|x_{n+1}, data) = f(y_{n+1}|\alpha_{\bar{x}}, \beta, x_{n+1}) \times g(\alpha_{\bar{x}}, \beta|data)$$

which is the distribution of the next observation given the parameters, times the posterior distribution of the parameters given the previous *data*. The next observation, given the parameters $y_{n+1}|\alpha_{\bar{x}}, \beta, x_{n+1}$, is a random observation from the regression model given the value x_{n+1}. By our assumptions it is normally distributed with mean given by the linear function of the parameters $\mu_{n+1} = \alpha_{\bar{x}} + \beta(x_{n+1} - \bar{x})$ and known variance σ^2.

The posterior distributions of the parameters given the previous data which we found using the updating rules in the previous section are independently $normal(m'_{\alpha_{\bar{x}}}, (s'_{\alpha_{\bar{x}}})^2)$ and $normal(m'_{\beta}, (s'_{\beta})^2)$, respectively. Since the next observation only depends on the parameters through the linear function

$$\mu_{n+1} = \alpha_{\bar{x}} + \beta(x_{n+1} - \bar{x}),$$

we will simplify the problem by letting μ_{n+1} be the single parameter. The two components $\alpha_{\bar{x}}$ and β are independent, so the posterior distribution of μ_{n+1} will be *normal* with mean $m'_{\mu} = m'_{\alpha_{\bar{x}}} + (x_{n+1} - \bar{x}) \times m'_{\beta}$ and variance $(s'_{\mu})^2 = (s'_{\alpha_{\bar{x}}})^2 + (x_{n+1} - \bar{x})^2 \times (s'_{\beta})^2)$ given by Equation 5.11 and Equation 5.12 respectively.

We will find the predictive distribution by marginalizing the μ_{n+1} out of the joint posterior of y_{n+1} and μ_{n+1}.

$$f(y_{n+1}|x_{n+1}, data) = \int f(y_{n+1}, \mu_{n+1}|x_{n+1}, data)\, d\mu_{n+1}$$

$$= \int f(y_{n+1}|\mu_{n+1}, x_{n+1}, data)$$
$$\times g(\mu_{n+1}|x_{n+1}, data)\, d\mu_{n+1}$$

$$= \int f(y_{n+1}|\mu_{n+1}) \times g(\mu_{n+1}|x_{n+1}, data)\, d\mu_{n+1}$$

$$\propto \int e^{-\frac{1}{2\sigma^2}(y_{n+1}-\mu_{n+1})^2} \times e^{-\frac{1}{2(s'_{\mu})^2}(\mu_{n+1}-m'_{\mu})^2}\, d\mu_{n+1}$$

$$\propto \int e^{-\frac{1}{2\sigma^2(s'_{\mu})^2/(\sigma^2+(s'_{\mu})^2)}\left(\mu_{n+1}-\frac{y_{n+1}(s'_{\mu})^2+m'_{\mu}\sigma^2}{(s'_{\mu})^2+\sigma^2}\right)^2}$$
$$\times e^{-\frac{1}{2((s'_{\mu})^2+\sigma^2)}(y_{n+1}-m'_{\mu})^2}\, d\mu_{n+1}.$$

The second factor does not depend on μ_{n+1}, so it can be brought in front of the integral. We recognize that the first term integrates out, so we are left with

$$f(y_{n+1}|x_{n+1}, data) \propto e^{-\frac{1}{2((s'_{\mu})^2+\sigma^2)}(y_{n+1}-m'_{\mu})^2}. \qquad (14.12)$$

We recognize that this is a $normal(m'_y, (s'_y)^2)$, where $m'_y = m'_{\mu}$, and $(s'_y)^2 = (s'_{\mu})^2 + \sigma^2$. Thus the predictive mean of the next observation y_{n+1} taken at x_{n+1} is the posterior mean of $\mu_{n+1} = \alpha_{\bar{x}} + \beta(x_{n+1} - \bar{x})$, and the predictive variance of y_{n+1} is the posterior variance of $\mu_{n+1} = \alpha_{\bar{x}} + \beta(x_{n+1} - \bar{x})$ plus the observation variance σ^2. Thus both sources of uncertainty have been allowed for in the predictive distribution.

Credible interval for the prediction. Often we wish to find an interval that has posterior probability equal to $1 - \alpha$ of containing the next value y_{n+1} which will be observed at the value x_{n+1}. This will be a $(1 - \alpha) \times 100\%$ credible interval for the prediction. We know that the mean and the variance of the prediction distribution are m'_y and $(s'_y)^2$, respectively. The credible interval for the prediction is given by

$$m'_y \pm z_{\frac{\alpha}{2}} \times s'_y = m'_\mu \pm z_{\frac{\alpha}{2}} \times \sqrt{(s'_\mu)^2 + \sigma^2}$$
$$= m'_{\alpha_{\bar{x}}} + m'_\beta(x_{n+1} - \bar{x})$$
$$\pm z_{\frac{\alpha}{2}} \times \sqrt{(s'_{\alpha_{\bar{x}}})^2 + (s'_\beta)^2(x_{n+1} - \bar{x})^2 + \sigma^2}, \qquad (14.13)$$

when we know the observation variance σ^2. When we do not know the observation variance and instead use the variance estimate calculated from the residuals, the credible interval is given by

$$m'_y \pm t_{\frac{\alpha}{2}} \times s'_y = m'_\mu \pm t_{\frac{\alpha}{2}} \times \sqrt{(s'_\mu)^2 + \hat{\sigma}^2}$$
$$= m'_{\alpha_{\bar{x}}} + m'_\beta(x_{n+1} - \bar{x})$$
$$\pm t_{\frac{\alpha}{2}} \times \sqrt{(s'_{\alpha_{\bar{x}}})^2 + (s'_\beta)^2(x_{n+1} - \bar{x})^2 + \hat{\sigma}^2}, \qquad (14.14)$$

where we get the critical value from the *Student's t* distribution with $n - 2$ degrees of freedom. These credible intervals for the prediction are the Bayesian analogs of the frequentist prediction intervals, since they allow for both the estimation error and the observation error. The Bayesian credible intervals for the prediction generally will be shorter than the corresponding frequentist prediction intervals since the Bayesian intervals use information from the prior as well as information from the data. They give exactly the same results as the frequentist prediction interval when flat priors are used for both the slope and intercept.

■ EXAMPLE 14.2 (continued)

Michael calculated the predictive distribution for the final moisture level (y) as a function of the in-process moisture level (x), and he put 95% bounds on the prediction. The mean of the predictive distribution is given by

$$m'_y = 14.2219 + 1.2034 \times (x - 14.3888)$$

and the variance of the predictive distribution is given by

$$(s'_y)^2 = .0348644 + .037318^2 + .17001^2(x - 14.3888)^2.$$

He calculated 95% prediction intervals as

$$(m'_y - t_{.025} \times s'_y, m'_y + t_{.025} \times s'_y).$$

A graph of the predictive mean is shown in Figure 14.7, together with the 95% prediction bounds. ■

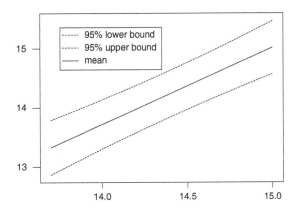

Figure 14.7 The predictive mean with 95% prediction bounds.

Main Points

- Our goal is to use one variable x, called the predictor variable, to help us predict another variable y, called the response variable.

- We think the two variables are related by a linear relationship, $y = a_0 + b \times x$. b is the slope and a_0 is the y-intercept (where the line intersects the y-axis.)

- The scatterplot of the points (x, y) would indicate a perfect linear relationship if the points lie along a straight line.

- However, the points usually do not lie perfectly along a line but are scattered around, yet still show a linear pattern.

- We could draw any line on the scatterplot. The residuals from that line would be the vertical distance from the plotted points to the line.

- Least squares is a method for finding a line that best fits a plotted points by minimizing the sum of squares of residuals from a fitted line.

- The slope and intercept of the least squares line are found by solving the *normal equations*.

- The linear regression model has three assumptions:

  1. The mean of y is an unknown linear function of x. Each observation y_i is made at a known value x_i.

  2. Each observation y_i is subject to a random error that is normally distributed with mean 0 and variance σ^2. We will assume that σ^2 is known.

  3. The observation errors are independent of each other.

- Bayesian regression is much easier if we reparameterize the model to be $y = \alpha_{\bar{x}} + \beta \times (x - \bar{x})$.

- The joint likelihood of the sample factors into a part dependent on the slope β and a part dependent on $\alpha_{\bar{x}}$.

- We use independent priors for the slope β and intercept $\alpha_{\bar{x}}$. They can be either normal priors or "flat" priors. The joint prior is the product of the two priors.

- The joint posterior is proportional to the joint prior times the joint likelihood. Since both the joint prior and joint likelihood factor into a part dependent on the slope β and a part dependent on $\alpha_{\bar{x}}$, the joint posterior is the product of the two individual posteriors. Each of them is normal where the constants can be found from the simple updating rules.

- Ordinarily we are more interested in the posterior distribution of the slope β, which is $normal(m', (s')^2)$. In particular, we are interested in knowing whether the belief $\beta = 0$ is credible, given the data. If so, we should not be using x to help predict y.

- The Bayesian credible interval for β is the *posterior mean* $\pm$ the *critical value* $\times$ the *posterior standard deviation*.

- The critical value is taken from the normal table if we assume the variance σ^2 is known. If we do not know it and use the sample estimate calculated from the residuals then we take the critical value from the *Student's t* table.

- The credible interval can be used to test the two-sided hypothesis $H_0 : \beta = 0$ versus $H_1 : \beta \neq 0$.

- We can test a one-sided hypothesis $H_0 : \beta \leq 0$ versus $H_1 : \beta > 0$ by calculating the probability of the null hypothesis and comparing it to the level of significance.

- We can compute the predictive probability distribution for the next observation y_{n+1} taken when x_{n+1}. It is the *normal* distribution with mean equal to the mean of the linear function $\mu_{n+1} = \alpha_{\bar{x}} + (x_{n+1} - \bar{x})$, and its variance is equal to the variance of the linear function plus the observation variance.

Exercises

14.1. A researcher measured heart rate (x) and oxygen uptake (y) for one person under varying exercise conditions. He wishes to determine if heart rate, which is easier to measure, can be used to predict oxygen uptake. If so, then the estimated oxygen uptake based on the measured heart rate can be used in place of the measured oxygen uptake for later experiments on the individual:

Heart Rate	Oxygen Uptake
x	y
94	.47
96	.75
94	.83
95	.98
104	1.18
106	1.29
108	1.40
113	1.60
115	1.75
121	1.90
131	2.23

(a) Plot a scatterplot of oxygen uptake y versus heart rate x.

(b) Calculate the parameters of the least squares line.

(c) Graph the least squares line on your scatterplot.

(d) Calculate the estimated variance about the least squares line.

(e) Suppose that we know that oxygen uptake given the heart rate is *emphnormal*$(\alpha_0 + \beta \times x, \sigma^2)$, where $\sigma^2 = .13^2$ is known. Use a *normal*$(0, 1^2)$ prior for β. What is the posterior distribution of β?

(f) Find a 95% credible interval for β.

(g) Perform a Bayesian test of

$$H_0 : \beta = 0 \quad \text{versus} \quad H_1 : \beta \neq 0$$

at the 5% level of significance.

14.2. A researcher is investigating the relationship between yield of potatoes
(y) and level of fertilizer (x.) She divides a field into eight plots of equal
size and applied fertilizer at a different level to each plot. The level of
fertilizer and yield for each plot is recorded below:

Fertilizer Level	Yield
x	y
1	25
1.5	31
2	27
2.5	28
3	36
3.5	35
4	32
4.5	34

(a) Plot a scatterplot of yield versus fertilizer level.

(b) Calculate the parameters of the least squares line.

(c) Graph the least squares line on your scatterplot.

(d) Calculate the estimated variance about the least squares line.

(e) Suppose that we know that yield given the fertilizer level is *normal*($\alpha_0 +$
$\beta \times x, \sigma^2$), where $\sigma^2 = 3.0^2$ is known. Use a *normal*(2, 2^2) prior for
β. What is the posterior distribution of β?

(f) Find a 95% credible interval for β.

(g) Perform a Bayesian test of

$$H_0 : \beta \leq 0 \quad \text{versus} \quad H_1 : \beta > 0$$

at the 5% level of significance.

14.3. A researcher is investigating the relationship between fuel economy and driving speed. He makes six runs on a test track, each at a different speed, and measures the kilometers traveled on one liter of fuel. The speeds (in kilometers per hour) and distances (in kilometers) are recorded below:

Speed	Distance
x	y
80	55.7
90	55.4
100	52.5
110	52.1
120	50.5
130	49.2

(a) Plot a scatterplot of distance travelled versus speed.

(b) Calculate the parameters of the least squares line.

(c) Graph the least squares line on your scatterplot.

(d) Calculate the estimated variance about the least squares line.

(e) Suppose that we know distance travelled, given that the speed is emphnormal($\alpha_0 + \beta \times x, \sigma^2$) where $\sigma^2 = .57^2$ is known. Use a *normal*$(0, 1^2)$ prior for β. What is the posterior distribution of β?

(f) Perform a Bayesian test of

$$H_0 : \beta \geq 0 \quad \text{versus} \quad H_1 : \beta < 0$$

at the 5% level of significance.

14.4. The Police Department is interested in determining the effect of alcohol consumption on driving performance. Twelve male drivers of similar weight, age, and driving experience were randomly assigned to three groups of four. The first group consumed two cans of beer within 30 minutes, the second group consumed four cans of beer within 30 minutes, and the third group was the control and did not consume any beer. Twenty minutes later, each of the twelve took a driving test under the same conditions, and their individual scores were recorded. (The higher the score, the better the driving performance.) The results were:

Cans	Score
x	y
0	78
0	82
0	75
0	58
2	75
2	42
2	50
2	55
4	27
4	48
4	49
4	39

(a) Plot a scatterplot of score versus cans.

(b) Calculate the parameters of the least squares line.

(c) Graph the least squares line on your scatterplot.

(d) Calculate the estimated variance about the least squares line.

(e) Suppose we know that the driving score given the number of cans of beer drunk is $normal(\alpha_0 + \beta \times x, \sigma^2)$, where $\sigma^2 = 12^2$ is known. Use a $normal(0, 10^2)$ prior for β. What is the posterior distribution of β?

(f) Find a 95% credible interval for β.

(g) Perform a Bayesian test of

$$H_0 : \beta \geq 0 \quad \text{versus} \quad H_1 : \beta < 0$$

at the 5% level of significance.

(h) Find the predictive distribution for the y_{13} the driving score of the next male who will be tested after drinking $x_{13} = 3$ cans of beer.

(i) Find a 95% credible interval for the prediction.

14.5. A textile manufacturer is concerned about the strength of cotton yarn. In order to find out whether fiber length is an important factor in determining the strength of yarn, the quality control manager checked the fiber length (x) and strength (y) for a sample of 10 segments of yarn. The results are:

Fiber Length	Strength
x	y
85	99
82	93
75	103
73	97
76	91
73	94
96	135
92	120
70	88
74	92

(a) Plot a scatterplot of strength versus fiber length.

(b) Calculate the parameters of the least squares line.

(c) Graph the least squares line on your scatterplot.

(d) Calculate the estimated variance about the least squares line.

(e) Suppose we know that the strength given the fiber length is *emphnormal*(α_0 $\beta \times x, \sigma^2$), where $\sigma^2 = 7.7^2$ is known. Use a *normal*$(0, 10^2)$ prior for β. What is the posterior distribution of β.

(f) Find a 95% credible interval for β.

(g) Perform a Bayesian test of

$$H_0 : \beta \leq 0 \quad \text{versus} \quad H_1 : \beta > 0$$

at the 5% level of significance.

(h) Find the predictive distribution for y_{11}, the strength of the next piece of yarn which has fiber length $x_{11} = 90$.

(i) Find a 95% credible interval for the prediction.

14.6. In Chapter 3, Exercise 3.7, we were looking at the relationship between log(*mass*) and log(*length*) for a sample of 100 New Zealand slugs of the species *Limax maximus* from a study conducted by Barker and McGhie (1984). These data are in the Minitab worksheet *slug.mtw*. We identified observation 90, which did not appear to fit the pattern. It is likely that this observation is an outlier that was recorded incorrectly, so remove it from the data set. The summary statistics for the 99 remaining observations are. Note: x is log(*length*), and y is log(*weight*)

$$\sum x = 352.399 \qquad \sum y = -33.6547 \qquad \sum x^2 = 1292.94$$

$$\sum xy = -18.0147 \qquad \sum y^2 = 289.598.$$

(a) Calculate the least squares line for the regression of y on x from the formulas.

(b) Using Minitab, calculate the least squares line. Plot a scatterplot of log weight on log length. Include the least squares line on your scatterplot.

(c) Using Minitab, calculate the residuals from the least squares line, and plot the residuals versus x. From this plot, does it appear the linear regression assumptions are satisfied?

(d) Using Minitab, calculate the estimate of the standard deviation of the residuals.

(e) Suppose we use a $normal(3, .5^2)$ prior for β, the regression slope coefficient. Calculate the posterior distribution of $\beta|data$. (Use the standard deviation you calculated from the residuals as if it is the true observation standard deviation.)

(f) Find a 95% credible interval for the true regression slope β.

(g) If the slugs stay the same shape as they grow (allotropic growth), the height and width would both be proportional to the length, so the weight would be proportional to the cube of the length. In that case the coefficient of log(weight) on log(length) would equal 3. Test the hypothesis

$$H_0 : \beta = 3 \quad \text{versus} \quad H_1 : \beta \neq 3$$

at the 5% level of significance. Can you conclude this slug species shows allotropic growth?

14.7. Endophyte is a fungus *Neotyphodium lolli*, which lives inside ryegrass plants. It does not spread between plants, but plants grown from endophyte-infected seed will be infected. One of its effects is that it produces a range of compounds that are toxic to Argentine stem weevil *Listronotus bonariensis*, which feeds on ryegrass. AgResearch New Zealand did a study on the persistence of perennial ryegrass at four rates of Argentine stem weevil infestation. For ryegrass that was infected with endophyte the following data were observed:

Infestation Rate x	Number of Ryegrass Plants (n)	$\log_e(n+1)$ y
0	19	2.99573
0	23	3.17805
0	2	1.09861
0	0	0.00000
0	24	3.21888
5	20	3.04452
5	18	2.94444
5	10	2.39790
5	6	1.94591
5	6	1.94591
10	12	2.56495
10	2	1.09861
10	11	2.48491
10	7	2.07944
10	6	1.94591
20	3	1.38629
20	16	2.83321
20	14	2.70805
20	9	2.30259
20	12	2.56495

(a) Plot a scatterplot of number of ryegrass plants versus the infestation rate.

(b) The relationship between infestation rate and number of ryegrass plants is clearly nonlinear. Look at the transformed variable $y = \log_e(n+1)$. Plot y versus x on a scatterplot. Does this appear to be more linear?

(c) Find the least squares line relating y to x. Include the least squares line on your scatterplot.

(d) Find the estimated variance about the least squares line.

(e) Assume that the observed y_i are normally distributed with mean $\alpha_{\bar{x}} + \beta \times (x_i - \bar{x})$ and known variance σ^2 equal to that calculated in part (d.) Find the posterior distribution of $\beta|(x_1, y_1), \ldots, (x_{20}, y_{20})$. Use a *normal*$(0, 1^2)$ prior for β.

14.8. For ryegrass that was not infected with endophyte the following data were observed:

Infestation Rate	Number of Ryegrass Plants (n)	$\log_e(n+1)$
x		y
0	16	2.83321
0	23	3.17805
0	2	1.09861
0	16	2.83321
0	6	1.94591
5	8	2.19722
5	6	1.94591
5	1	0.69315
5	2	1.09861
5	5	1.79176
10	5	1.79176
10	0	0.00000
10	6	1.94591
10	2	1.09861
10	2	1.09861
20	1	0.69315
20	0	0.00000
20	0	0.00000
20	1	0.69315
20	0	0.00000

(a) Plot a scatterplot of number of ryegrass plants versus the infestation rate.

(b) The relationship between infestation rate and number of ryegrass plants is clearly nonlinear. Look at the transformed variable $y = \log_e(n+1)$. Plot y versus x on a scatterplot. Does this appear to be more linear?

(c) Find the least squares line relating y to x.

(d) Find the estimated variance about the least squares line.

(e) Assume that the observed y_i are normally distributed with mean $\alpha_{\bar{x}} + (x_i - \bar{x}) \times \beta$ and variance equal to that calculated in part (b.) Find the posterior distribution of $\beta | (x_1, y_1), \ldots, (x_{20}, y_{20})$. Use a $normal(0, 1^2)$ prior for β.

14.9. In the previous two problems we found the posterior distribution of the slope of y on x, the rate of weevil infestation for endophyte infected and noninfected ryegrass. Let β_1 be the slope for noninfected ryegrass, and let β_2 be the slope for infected ryegrass

(a) Find the posterior distribution of $\beta_1 - \beta_2$.

(b) Calculate a 95% credible interval for $\beta_1 - \beta_2$.

(c) Test the hypothesis

$$H_0 : \beta_1 - \beta_2 \leq 0 \quad \text{versus} \quad H_1 : \beta_1 - \beta_2 > 0$$

at the 10% level of significance.

Computer Exercises

14.1. We will use the Minitab macro *BayesLinReg*, or the R function `bayes.lin.reg`, to find the posterior distribution of the slope β given a random sample $(x_1, y_1), \ldots, (x_n, y_n)$ from the simple linear regression model

$$y_i = \alpha_0 + \beta \times x_i + e_i \,,$$

where the observation errors e_i are independent $normal(0, \sigma^2)$ random variables and σ^2 is known. We will use independent $normal(m_\beta, s_\beta^2)$ and $normal(m_{\alpha_{\bar{x}}}, s_{\alpha_{\bar{x}}}^2)$ priors for the slope β and the intercept of the line $y = \bar{x}$ respectively. This parameterization will give independent normal posteriors where the simple updating rules are "posterior precision equals the prior precision plus the precision of the least squares estimate" and "posterior mean equals the weighted sum of prior mean plus the least squares estimate where the weights are the proportions of the precisions to the posterior precision." The following eight observations come from a simple linear regression model where the variance $\sigma^2 = 1^2$ is known.

x	11	9	9	9	9	12	11	9
y	-21.6	-16.2	-19.5	-16.3	-18.3	-24.6	-22.6	-17.7

(a) **[Minitab:]** Use *BayesLinReg* to find the posterior distribution of the slope β when we use a $normal(0, 3^2)$ prior for the slope. Details for invoking *BayesLinReg* are given in Appendix C.

   [R:] Use `bayes.lin.reg` to find the posterior distribution of the slope β when we use a $normal(0, 3^2)$ prior for the slope. Details for calling `bayes.lin.reg` are given in Appendix D. **Note:** There is a shorthand alias for `bayes.lin.reg` called `blr` which removes the burden of typing `bayes.lin.reg` correctly every time you wish to use it.

(b) Find a 95% Bayesian credible interval for the slope β.

(c) Test the hypothesis $H_0 : \beta \leq -3$ vs. $H_1 : \beta > -3$ at the 5% level of significance.

(d) Find the predictive distribution of y_9 which will be observed at $x_9 = 10$.

(e) Find a 95% credible interval for the prediction.

14.2. The following 10 observations come from a simple linear regression model where the variance $\sigma^2 = 3^2$ is known.

x	30	30	29	21	37	28	26	38	32	21
y	22.4	16.3	16.2	30.6	12.1	17.9	25.5	9.8	20.5	29.8

(a) [**Minitab:**] Use *BayesLinReg* to find the posterior distribution of the slope β when we use a *normal*$(0, 3^2)$ prior for the slope.

[**R:**] Use `bayes.lin.reg` to find the posterior distribution of the slope β when we use a *normal*$(0, 3^2)$ prior for the slope.

(b) Find a 95% Bayesian credible interval for the slope β.

(c) Test the hypothesis $H_0 : \beta \geq 1$ vs. $H_1 : \beta < 1$ at the 5% level of significance.

(d) Find the predictive distribution of y_{11} which will be observed at $x_{11} = 36$.

(e) Find a 95% credible interval for the prediction.

14.3. The following 10 observations come from a simple linear regression model where the variance $\sigma^2 = 3^2$ is known.

x	22	31	21	23	19	26	27	16	28	21
y	24.2	25.4	23.9	22.8	22.6	29.7	24.8	22.3	28.2	30.7

(a) Use *BayesLinReg* in Minitab, `bayes.lin.reg` in R, to find the posterior distribution of the slope β when we use a *normal*$(0, 3^2)$ prior for the slope and a *normal*$(25, 3^2)$ prior for the intercept $\alpha_{\bar{x}}$.

(b) Find a 95% Bayesian credible interval for the slope β.

(c) Test the hypothesis $H_0 : \beta \geq 1$ vs. $H_1 : \beta < 1$ at the 5% level of significance.

(d) Find the predictive distribution of y_{11} which will be observed at $x_{11} = 25$.

(e) Find a 95% credible interval for the prediction.

14.4. The following 8 observations come from a simple linear regression model where the variance $\sigma^2 = 2^2$ is known.

x	54	47	44	47	55	50	52	48
y	1.7	4.5	4.6	8.9	0.9	1.4	5.2	6.4

(a) Use *BayesLinReg* in Minitab, or `bayes.lin.reg` in R, to find the posterior distribution of the slope β when we use a *normal*$(0, 3^2)$ prior for the slope β and a *normal*$(4, 2^2)$ prior for the intercept $\alpha_{\bar{x}}$.

(b) Find a 95% Bayesian credible interval for the slope β.

(c) Test the hypothesis $H_0 : \beta \geq 1$ vs. $H_1 : \beta < 1$ at the 5% level of significance.

(d) Find the predictive distribution of y_9 which will be observed at $x_9 = 51$.

(e) Find a 95% credible interval for the prediction.

CHAPTER 15

BAYESIAN INFERENCE FOR STANDARD DEVIATION

When dealing with any distribution, the parameter giving its location is the most important, with the parameter giving the spread of secondary importance. For the normal distribution, these are the mean and the standard deviation (or its square, the variance), respectively. Usually we will be doing inference on the unknown mean, with the standard deviation and hence the variance either assumed known, or treated as a *nuisance parameter*. In Chapter 11, we looked at making Bayesian inferences on the mean, where the observations came from a normal distribution with known variance. We also saw that when the variance was unknown, inferences about the mean could be adjusted by using the sample estimate of the variance in its place and taking critical values from the *Student's t* distribution. The resulting inferences would be equivalent to the results we would have obtained if the unknown variance was a nuisance parameter and was integrated out of the joint posterior.

However, sometimes we want to do inferences on the standard deviation of the normal distribution. In this case, we reverse the roles of the parameters. We will assume that the mean is known, or else we treat it as the nuisance parameter and make the necessary adjustments to the inference. We will use

Bayes' theorem on the variance. However, the variance is in squared units, and it is hard to visualize our belief about it. So for graphical presentation, we will make the transformation to the corresponding prior and posterior density for the standard deviation.

15.1   Bayes' Theorem for Normal Variance with a Continuous Prior

We have a random sample $y_1, \ldots, y_n$ from a $normal(\mu, \sigma^2)$ distribution where the mean μ is assumed known, but the variance σ^2 is unknown. Bayes' theorem can be summarized by *posterior proportional to prior times likelihood*

$$g(\sigma^2|y_1, \ldots, y_n) \propto g(\sigma^2) \times f(y_1, \ldots, y_n|\sigma^2).$$

It is realistic to consider that the variance can have any positive value, so the continuous prior we use should be defined on all positive values. Since the prior is continuous, the actual posterior is evaluated by

$$g(\sigma^2|y_1, \ldots, y_n) = \frac{g(\sigma^2) \times f(y_1, \ldots, y_n|\sigma^2)}{\int g(\sigma^2) \times f(y_1, \ldots, y_n|\sigma^2)\, d\sigma^2}, \tag{15.1}$$

where the denominator is the *integral* of the *prior* × *likelihood* over its whole range. This will hold true for any continuous prior density. However, the integration would have to be done numerically, except for a few special prior densities which we will investigate later.

The inverse chi-squared distribution.   The distribution with shape given by

$$g(x) \propto \frac{1}{x^{\frac{\kappa}{2}+1}} e^{-\frac{1}{2x}}$$

for $0 < x < \infty$ is called the *inverse chi-squared* distribution with κ degrees of freedom. To make this a probability density function we multiply by the constant $c = \frac{1}{2^{\frac{\kappa}{2}}\Gamma(\kappa/2)}$. The exact density function of the *inverse chi-squared* distribution with κ degrees of freedom is

$$g(x) = \frac{1}{2^{\frac{\kappa}{2}}\Gamma(\kappa/2)x^{\frac{\kappa}{2}+1}} e^{-\frac{1}{2x}} \tag{15.2}$$

for $0 < x < \infty$. When the shape of the density is given by

$$g(x) \propto \frac{1}{x^{\frac{\kappa}{2}+1}} e^{-\frac{S}{2x}}.$$

for $0 < x < \infty$ then we say x has S times an *inverse chi-squared* distribution with κ degrees of freedom. The constant $c = \frac{S^{\frac{\kappa}{2}}}{2^{\frac{\kappa}{2}}\Gamma(\kappa/2)}$ is the scale factor that makes this a density. The exact probability density function of S times an *inverse chi-squared* distribution with κ degrees of freedom[1] is

$$g(x) = \frac{S^{\frac{\kappa}{2}}}{2^{\frac{\kappa}{2}}\Gamma(\kappa/2)} \times \frac{1}{x^{\frac{\kappa}{2}+1}} e^{-\frac{S}{2x}} \tag{15.3}$$

for $0 < x < \infty$. When U has S times an *inverse chi-squared* distribution with κ degrees of freedom, then $W = S/U$ has the *chi-squared* distribution with κ degrees of freedom. This transformation allows us to find probabilities for the *inverse chi-squared* random variables using Table B.6, the upper tail area of the *chi-squared* distribution.

A random variable X having S times an *inverse chi-squared* distribution with κ degrees of freedom has mean

$$E[X] = \frac{S}{\kappa - 2}$$

provided $\kappa > 2$ and variance given by

$$\text{Var}[X] = \frac{2S^2}{(\kappa - 2)^2 \times (\kappa - 4)}$$

provided $\kappa > 4$.

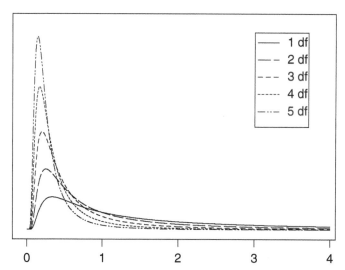

Figure 15.1 Inverse chi-squared distribution with for $1, \ldots, 5$ degrees of freedom. As the degrees of freedom increase, the probability gets more concentrated at smaller values. Note: $S = 1$ for all these graphs.

Some *inverse chi-squared* distributions with $S = 1$ are shown in Figure 15.1.

---

[1] This is also known as the *inverse Gamma*(r, S) distribution where $r = \frac{\kappa}{2}$.

Likelihood of variance for normal random sample   The likelihood of the variance for a single random draw from a *normal*(μ, σ^2) where μ is known is the density of the observation at the observed value taken as a function of the variance σ^2.

$$f(y|\sigma^2) = \frac{1}{\sqrt{(2\pi)\sigma^2}} e^{-\frac{1}{2\sigma^2}(y-\mu)^2}$$

We can absorb any part not depending on the parameter σ^2 into the constant. This leaves

$$f(y|\sigma^2) \propto (\sigma^2)^{-\frac{1}{2}} e^{-\frac{1}{2\sigma^2}(y-\mu)^2}.$$

as the part that determines the shape. The likelihood of the variance for the random sample $y_1, \ldots, y_n$ from a *normal*(μ, σ^2) where μ is known is product of the likelihoods of the variance for each of the observations. The part that gives the shape is

$$f(y_1, \ldots, y_n|\sigma^2) \propto \prod_{i=1}^{n} (\sigma^2)^{-\frac{1}{2}} e^{-\frac{1}{2\sigma^2}(y_i-\mu)^2}$$

$$\propto (\sigma^2)^{-\frac{n}{2}} e^{-\frac{1}{2\sigma^2}\sum_{i=1}^{n}(y_i-\mu)^2}$$

$$\propto \frac{1}{(\sigma^2)^{\frac{n}{2}}} e^{-\frac{SS_T}{2\sigma^2}}, \tag{15.4}$$

where $SS_T = \sum_{i=1}^{n}(y_i - \mu)^2$ is the total sum of squares about the mean. We see that the likelihood of the variance has the same shape as SS_T times an *inverse chi-squared* distribution with $\kappa = n - 2$ degrees of freedom.[2]

15.2   Some Specific Prior Distributions and the Resulting Posteriors

Since we are using Bayes' theorem on the normal variance σ^2, we will need its prior distribution. However, the variance is in squared units, not the same units as the mean. This means that they are not directly comparable so that it is much harder understand a prior density for σ^2. The standard deviation σ is in the same units as the mean, so it is much easier to understand. Generally, we will do the calculations to find the posterior for the variance σ^2, but we will graph the corresponding posterior for the standard deviation σ since it is more easily understood. For the rest of the chapter, we will use the subscript on the prior and posterior densities to denote which parameter σ or σ^2 we are using. The variance is a function of the standard deviation, so we can use the chain rule from Appendix 1 to get the prior density for σ^2 that corresponds

---

[2]When the mean is not known but considered a nuisance parameter, use the marginal likelihood for σ^2 which has same shape as SS_y times an *inverse chi-squared* distribution with $\kappa = n - 3$ degrees of freedom where $SS_y = \sum(y - \bar{y})^2$.

to the prior density for σ. This gives the change of variable formula[3] which in this case is given by

$$g_{\sigma^2}(\sigma^2) = g_\sigma(\sigma) \times \frac{1}{2\sigma}. \tag{15.5}$$

Similarly, if we have the prior density for the variance, we can use the change of variable formula to find the corresponding prior density for the standard deviation

$$g_\sigma(\sigma) = g_{\sigma^2}(\sigma^2) \times 2\sigma. \tag{15.6}$$

Positive Uniform Prior Density for Variance

Suppose we decide that we consider all positive values of the variance σ^2 to be equally likely and do not wish to favor any particular value over another. We give all positive values of σ^2 equal prior weight. This gives the positive uniform prior density for the variance

$$g_{\sigma^2}(\sigma^2) = 1 \quad \text{for} \quad \sigma^2 > 0.$$

This is an improper prior since its integral over the whole range would be ∞; however that will not cause a problem here. The corresponding prior density for the standard deviation would be $g_\sigma(\sigma) = 2\sigma$ is also clearly improper. (Giving equal prior weight to all values of the variance gives more weight to larger values of the standard deviation.) The shape of the posterior will be given by

$$g_{\sigma^2}(\sigma^2 | y_1, \dots, y_n) \propto 1 \times \frac{1}{(\sigma^2)^{\frac{n}{2}}} e^{-\frac{SS_T}{2\sigma^2}}$$

$$\propto \frac{1}{(\sigma^2)^{\frac{n}{2}}} e^{-\frac{SS_T}{2\sigma^2}}.$$

which we recognize to be $SS_T \times$ an *inverse chi-squared* distribution with $n-2$ degrees of freedom.

Positive Uniform Prior Density for Standard Deviation

Suppose we decide that we consider all positive values of the standard deviation σ to be equally likely and do not wish to favor any particular value over another. We give all positive values all equal prior weight. This gives the positive uniform prior density for the standard deviation

$$g_\sigma(\sigma) = 1 \quad \text{for} \quad \sigma > 0.$$

---

[3]In general, when $g_\theta(\theta)$ is the prior density for parameter θ and if $\psi(\theta)$ is a *one-to-one* function of θ, then ψ is another possible parameter. The prior density of ψ is given by $g_\psi(\psi) = g_\theta(\theta(\psi)) \times \frac{d\theta}{d\psi}$.

This prior is clearly an improper prior since when we integrate it over the whole range we get ∞, however that will not cause trouble in this case. Using Equation 15.6 we find the corresponding prior for the variance is

$$g_{\sigma^2}(\sigma^2) = 1 \times \frac{1}{2\sigma}.$$

(We see that giving equal prior weight to all values of the standard deviation gives more weight to smaller values of the variance.) The posterior will be proportional to the prior times likelihood. We can absorb the part not containing the parameter into the constant. The shape of the posterior will be given by

$$g_{\sigma^2}(\sigma^2 | y_1, \dots, y_n) \propto \frac{1}{\sigma} \times \frac{1}{(\sigma^2)^{\frac{n}{2}}} e^{-\frac{SS_T}{2\sigma^2}}$$

$$\propto \frac{1}{(\sigma^2)^{\frac{n+1}{2}}} e^{-\frac{SS_T}{2\sigma^2}}.$$

We recognize this to be $SS_T \times$ an *inverse chi-squared* distribution with $n-1$ degrees of freedom.

Jeffreys' Prior Density

If we think of a parameter as an index of all the possible densities we are considering, any continuous function of the parameter will give an equally valid index. Jeffreys' wanted to find a prior that would be invariant for a continuous transformation of the parameter.[4] In the case of the *normal*(μ, σ^2) distribution where μ is known, Jeffreys' rule gives

$$g_{\sigma^2}(\sigma^2) \propto \frac{1}{\sigma^2} \quad \text{for} \quad \sigma^2 > 0.$$

This prior is also improper, but again in the single sample case this will not cause any problem. (Note that the corresponding prior for the standard deviation is $g_\sigma(\sigma) \propto \sigma^{-1}$.) The shape of the posterior will be given by

$$g_{\sigma^2}(\sigma^2 | y_1, \dots, y_n) \propto \frac{1}{\sigma^2} \times \frac{1}{(\sigma^2)^{\frac{n}{2}}} e^{-\frac{SS_T}{2\sigma^2}}$$

$$\propto \frac{1}{(\sigma^2)^{\frac{n}{2}+1}} e^{-\frac{SS_T}{2\sigma^2}},$$

which we recognize to be $SS_T \times$ an *inverse chi-squared* with n degrees of freedom.

---

[4] Jeffreys' invariant prior for parameter θ is given by $g(\theta) \propto \sqrt{I(\theta|y)}$ where $I(\theta|y)$ is known as Fisher's information and is given by $I(\theta|y) = -E\left[\frac{\partial^2 \log f(y|\theta)}{\partial \theta^2}\right]$.

Inverse Chi-squared Prior

Suppose we decide to use S times an *inverse chi-squared* with κ degrees of freedom as the prior for σ^2. In this case the shape of the prior is given by

$$g_{\sigma^2}(\sigma^2) \propto \frac{1}{(\sigma^2)^{\frac{\kappa}{2}+1}} e^{-\frac{S}{2\sigma^2}}$$

for $0 < \sigma^2 < \infty$. Note the shape of the corresponding prior density for σ found using the change of variable formula would be

$$g_{\sigma}(\sigma) \propto \frac{1}{(\sigma^2)^{\frac{\kappa-1}{2}+1}} e^{-\frac{S}{2\sigma^2}}$$

for $0 < \sigma^2 < \infty$. The prior densities for σ corresponding to inverse chi-squared prior with $S = 1$ for variance σ^2 for $\kappa = 1, 2, 3, 4$, and 5 degrees of freedom are shown in Figure 15.2. We see that as the degrees of freedom increase, the probability gets more concentrated at smaller values of σ. This suggests that to allow for the possibility of a large standard deviation, we should use low degrees of freedom when using an *inverse chi-squared* prior for the variance.

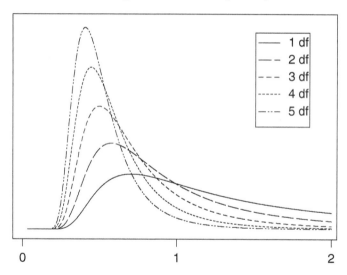

Figure 15.2   Prior for standard deviation σ corresponding to inverse chi square prior for variance σ^2 where $S = 1$.

The posterior density for σ^2 will have shape given by

$$g_{\sigma^2}(\sigma^2|y_1,\ldots,y_n) \propto \frac{1}{(\sigma^2)^{\frac{\kappa}{2}+1}} e^{-\frac{S}{2\sigma^2}} \times \frac{1}{(\sigma^2)^{\frac{n}{2}}} e^{-\frac{SS_T}{2\sigma^2}}$$

$$\propto \frac{1}{(\sigma^2)^{\frac{n+\kappa}{2}+1}} e^{-\frac{S+SS_T}{2\sigma^2}},$$

which we recognize as S' times an *inverse chi-squared* distribution with κ' degrees of freedom, where $S' = S + SS_T$ and $\kappa' = \kappa + n$. So when observations come from $normal(\mu, \sigma^2)$ with known mean μ, the conjugate family is the S times an *inverse chi-squared* distribution and the simple updating rule is "add total sum of squares about known mean to constant S" and "add sample size to degrees of freedom."

The corresponding priors for the standard deviation and the variance are shown in Table 15.1. All these priors will yield S' times an *inverse chi-squared* with κ' posteriors.[5]

Table 15.1   Corresponding priors for standard deviation and variance, and S' and κ' for the resulting *inverse chi-squared* posterior

Prior	$g_\sigma(\sigma) \propto$	$g_{\sigma^2}(\sigma^2) \propto$	S'	κ'
Pos. unif. for var.	σ	1	SS_T	$n - 2$
Pos. unif. for st. dev.	1	$\frac{1}{\sigma}$	SS_T	$n - 1$
Jeffreys'	$\frac{1}{\sigma}$	$\frac{1}{\sigma^2}$	SS_T	n
$S \times$ *inv. chi-sq κ df*	$\frac{1}{(\sigma^2)^{\frac{\kappa-1}{2}+1}} e^{-\frac{S}{2\sigma^2}}$	$\frac{1}{(\sigma^2)^{\frac{\kappa}{2}+1}} e^{-\frac{S}{2\sigma^2}}$	$S + SS_T$	$\kappa + n$

Choosing an inverse chi-squared prior.   Frequently, our prior belief about σ is fairly vague. Before we look at the data, we believe that we decide on a value c such that we believe $\sigma < c$ and $\sigma > c$ are equally likely. This means that c is our prior median.

We want to choose $S \times$ an *inverse chi-squared* distribution with κ degrees of freedom that fits our prior median. Since we have only vague prior knowledge about σ, we would like the prior to have as much spread as possible, given that it has the prior median. $W = \frac{S}{\sigma^2}$ has a *chi-squared* distribution with κ degrees of freedom.

$$.50 = P(\sigma > c)$$

$$= P\left(\frac{\sigma^2}{S} > \frac{c^2}{S}\right)$$

$$= P\left(W < \frac{S}{c^2}\right),$$

where W has the *chi-squared* distribution with κ degree of freedom. We look in Table B.6 to find the 50% point for the *chi-squared* distribution with κ degree of freedom and solve the resulting equation for S. Figure 15.3 shows

---

[5]The positive uniform prior for st. dev., the positive uniform prior for the variance, and the Jeffreys' prior have the form of an *inverse chi-squared* with $S = 0$ and $\kappa = -1, -2$, and 0, respectively. They can be considered limiting cases of the S times an *inverse chi-squared* family as $S \to 0$.

the prior densities having the same median for $\kappa = 1, \ldots, 5$ degrees of freedom. We see that the prior with $\kappa = 1$ degree of freedom has more weight on the

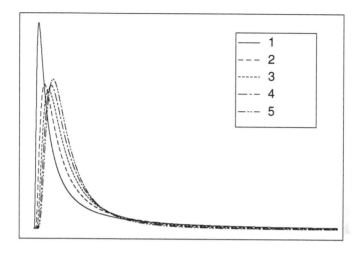

Figure 15.3   *Inverse chi-squared* prior densities having same prior medians, for $\kappa = 1, \ldots, 5$ degrees of freedom.

lower tail. The upper tail is hard to tell as all are squeezed towards 0. We take logarithms of the densities to spread out the upper tail. These are shown in Figure 15.4, and clearly the prior with $\kappa = 1$ degrees of freedom shows the most weight in both tails. Thus, the *inverse chi-squared* prior with 1 degree

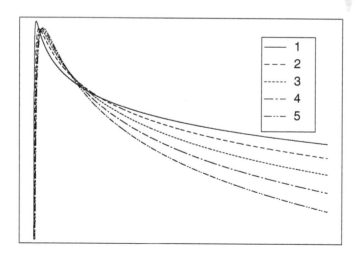

Figure 15.4   Logarithms of *Inverse chi-squared* prior densities having same prior medians, for $\kappa = 1, \ldots, 5$ degrees of freedom.

of freedom matching the prior median has maximum spread out of all other *inverse chi-squared* priors that also match the prior median.

Inverse gamma priors

The inverse chi-squared distribution is in fact a special case of the inverse gamma distribution. The inverse gamma distribution has pdf

$$g(\sigma^2; \alpha, \beta) = \frac{\beta^\alpha}{\Gamma(\alpha)} (\sigma^2)^{-\alpha-1} e^{-\frac{\beta}{\sigma^2}}.$$

We can see that if $\alpha = \frac{\kappa}{2}, \beta = \frac{1}{2}$, then this is the same density as a inverse chi-squared distribution with κ degrees of freedom. If $\alpha = \frac{\kappa}{2}, \beta = \frac{S}{2}$, then this is the same as S times an inverse chi-squared distribution with κ degrees of freedom. If we use this prior, then the posterior is

$$g(\sigma^2|y_1, y_2, \ldots, y_n) \propto \frac{1}{(\sigma^2)^{\alpha+1}} e^{-\frac{\beta}{\sigma^2}} \times \frac{1}{(\sigma^2)^{\frac{n}{2}}} e^{-\frac{SS_T}{2\sigma^2}}$$

$$\propto \frac{1}{(\sigma^2)^{\left[\frac{n}{2}+\alpha+1\right]}} e^{\frac{-(SS_T+2\beta)}{2\sigma^2}}.$$

This is proportional to an inverse gamma density with parameters $\alpha' = \frac{n}{2} + \alpha$ and $\beta' = \frac{SS_T + 2\beta}{2}$. Gelman et al. (2003) showed that we can reparameterize the *inverse gamma*(α, β) distribution as a scaled inverse chi-squared distribution with scale $S = \frac{\beta}{\alpha}$ and $\kappa = 2\alpha$ degrees of freedom. This parameterization is helpful in understanding the ensuing arguments about choices of parameters.

It is common, especially amongst users of the BUGS modelling language, to choose an *inverse gamma*(ϵ, ϵ) prior distribution for σ^2 where ϵ is a small value such as 1, or .1, or .001 (Gelman, 2006). The difficulty with this choice of prior is that it can lead to an improper posterior distribution. This is unlikely to occur in the examples discussed in this book, but it can occur in hierarchical models where it may be reasonable to believe that very small values of σ^2 are possible. The advice we offer here is to choose $\alpha \geq .5$ so that $2\alpha \geq 1$. That is, in the scaled inverse chi-squared parameterization, the prior distribution will have at least $\kappa = 1$ degree of freedom, and hence will be a proper prior.

◼ EXAMPLE 15.1

Aroha, Bernardo, and Carlos are three statisticians employed at a dairy factory who want to do inference on the standard deviation of the content weights of "1 kg" packages of dried milk powder coming off the production line. The three employees consider that the weights of the packages will be *normal*(μ, σ^2) where μ is known to be at the target which is 1015 grams. Aroha decides that she will use the positive uniform prior for the standard deviation, $g(\sigma) = 1$ for $\sigma > 0$. Bernardo decides he will use

Jeffreys' prior $g(\sigma) \propto \frac{1}{\sigma}$. Carlos decides that his prior belief about the standard deviation distribution is that its median equals 5. He looks in Table B.6 and finds that the 50% point for the chi-squared distribution with 1 degree of freedom equals .4549, and he calculates $S = .4549 \times 5^2 = 11.37$. Therefore his prior for σ^2 will be 11.37 times an inverse chi-squared distribution with 1 degree of freedom. He converts this to the equivalent prior density for σ using the change of variable formula. The shapes of the three prior densities for σ are shown in Figure 15.5. We see that

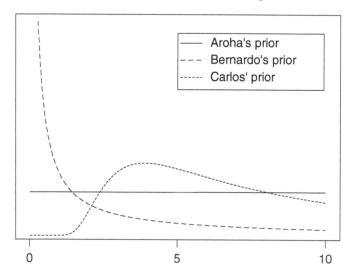

Figure 15.5 The shapes of Aroha's, Bernardo's, and Carlos' prior distributions for the standard deviation σ.

Aroha's prior does not go down as σ increases, and that both Bernardo's and Carlos' prior only goes down very slowly as σ increase. This means that all three priors will be satisfactory if the data shows much more variation than was expected. We also see that Bernardo's prior increases towards infinity as σ goes to zero. This means his prior gives a great deal of weight to very small values.[6] Carlos' prior does not give much weight to small values, but this does not cause a problem, since overestimating the variance is more conservative than underestimating it. They take a random sample of size 10 and measure the content weights in grams. They are:

---

[6]The prior $g(\theta) \propto \theta^{-1}$ is improper two ways. Its limit of its integral from a to 1 as a approaches 0 is infinite. This can cause problems in more complicated models where posterior may also be improper because the data cannot force the corresponding integral for the posterior to be finite. However, it will not cause any problem in this particular case. The limit of the integral from 1 to b of Bernardo's prior as b increases without bounds is also infinite. However, this will not cause any problems, as the data can always force the corresponding integral for the posterior to be finite.

1011	1009	1019	1012	1011	1016	1018	1021	1016	1012

The calculations for SS_T are

Value	Subtract mean	Squared
1011	-4	16
1009	-6	36
1019	4	16
1012	-3	9
1011	-4	16
1016	1	1
1018	3	9
1021	6	36
1016	1	1
1012	-3	9
SS_T		149

Each employee has $S' \times$ an inverse chi-squared with κ' degrees of freedom for the posterior distribution for the variance. Aroha's posterior will be $149 \times$ an inverse chi-squared with 9 degrees of freedom, Bernardo's posterior will be $149 \times$ an inverse chi-squared with 10 degrees of freedom, and Carlos' posterior will be $11.37 + 149 = 160.37 \times$ an inverse chi-squared with $10 + 1 = 11$ degrees of freedom. The corresponding posterior densities for the variance σ^2 and the standard deviation σ are shown in Figure 15.6 and Figure 15.7, respectively. We see that Aroha's posterior has a somewhat longer upper tail than the others since her prior gave more weight for large values of σ. ∎

15.3 Bayesian Inference for Normal Standard Deviation

The posterior distribution summarizes our belief about the parameter taking into account our prior belief and the observed data. We have seen in the previous section, that the posterior distribution of the variance $g(\sigma^2|y_1, \ldots, y_n)$ is $S' \times$ an *inverse chi-squared* with κ' degrees of freedom.

Bayesian Estimators for σ

Sometimes we want an estimator about the parameter which summarizes the posterior distribution into a single number. We will base our Bayesian estimators for σ on measures of location from the posterior distribution of the variance σ^2. Calculate the measure of location from the posterior distribution

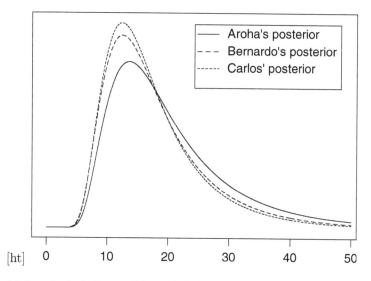

Figure 15.6   Aroha's, Bernardo's, and Carlos' posterior distributions for variance σ^2.

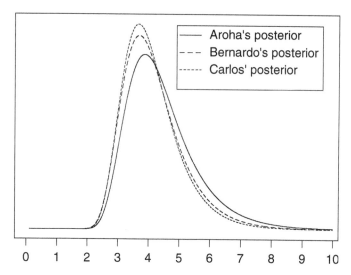

Figure 15.7   Aroha's, Bernardo's, and Carlos' posterior distributions for standard deviation σ.

of the variance $g(\sigma^2|y_1, \ldots, y_n)$ and then take the square root for our estimator of the standard deviation σ. Three possible measures of location are the posterior mean, posterior mode, and posterior median.

Posterior mean of variance σ^2.   The posterior mean is found by taking the expectation $\mathrm{E}[\sigma^2 g(\sigma^2|y_1, \ldots, y_n)]$. Lee (1989) showed that when $\kappa' > 2$ the posterior mean is given by

$$m' = \frac{S'}{\kappa' - 2}.$$

The first possible Bayesian estimator for the standard deviation would be its square root,

$$\hat{\sigma} = \sqrt{\frac{S'}{\kappa' - 2}}.$$

Posterior mode of variance σ^2. The posterior distribution of the variance σ^2 is given by $S' \times$ an *inverse chi-squared* distribution with κ' degrees of freedom. The posterior mode is found by setting the derivative of $g(\sigma^2|y_1, \ldots, y_n)$ equal to 0, and solving the resulting equation. It is given by

$$mode = \frac{S'}{\kappa' + 2}.$$

The second possible Bayesian estimator for the standard deviation would be its square root,

$$\hat{\sigma} = \sqrt{\frac{S'}{\kappa' + 2}}.$$

Posterior median of variance σ^2. The posterior median is the value that has 50% of the posterior distribution below it, and 50% above it. It is the solution of

$$\int_0^{median} g(\sigma^2|y_1, \ldots, y_n) d\sigma^2 = .5$$

which can be found numerically. The third possible Bayesian estimator for the standard deviation would be its square root

$$\hat{\sigma} = \sqrt{median}.$$

◼ **EXAMPLE 15.1** **(continued)**

The three employees decide to find their estimates of the standard deviation σ. They are shown in Table 15.2. Since the posterior density of the standard deviation can seen to be positively skewed with a somewhat heavy right tail, the estimates found using the posterior mean would be the best, followed by the estimates found using the posterior median. The estimates found using the posterior mode would tend to underestimate the standard deviation. ◼

Bayesian Credible Interval for σ

The posterior distribution of the variance σ^2 given the sample data is $S' \times$ an *inverse chi-squared* with κ' degrees of freedom. Thus $W = S'/\sigma^2$ has the

Table 15.2 Posterior estimates of the standard deviation σ

Person	Posterior Parameters		Estimator Found using Posterior		
	S'	κ'	Mode	Mean	Median
Aroha	149	9	3.680	4.614	4.226
Bernardo	149	10	3.524	4.316	3.994
Carlos	160.37	11	3.512	4.221	3.938

chi-squared distribution with κ' degrees of freedom. We set up a probability statement about W, and invert it to find the credible interval for σ^2. Let u be the chi-squared value with κ' degrees of freedom having upper tail area $1 - \frac{\alpha}{2}$ and let l be the chi-squared value having upper tail area $\frac{\alpha}{2}$. These values are found in Table B.6.

$$P\left(u < \frac{S'}{\sigma^2} < l\right) = 1 - \alpha,$$

$$P\left(\frac{S'}{l} < \sigma^2 < \frac{S'}{u}\right) = 1 - \alpha.$$

We take the square roots of the terms inside the brackets to convert this to a credible interval for the standard deviation σ

$$P\left(\sqrt{\frac{S'}{l}} < \sigma < \sqrt{\frac{S'}{u}}\right) = 1 - \alpha. \tag{15.7}$$

■ **EXAMPLE 15.1 (continued)**

Each of the three employees has $S' \times$ a inverse chi-squared distribution with κ' degrees of freedom. They calculate their 95% credible intervals for σ and put them in Table 15.3. We see that Aroha's credible interval is shifted slightly upwards and has a somewhat larger upper value than the others, which makes sense since her posterior distribution has a longer upper tail as seen in Figure 15.7. ■

Testing a One-Sided Hypothesis about σ

Usually we want to determine whether or not the standard deviation is less than or equal to some value. We can set this up as a one-sided hypothesis test about σ,

$$H_0 : \sigma \leq \sigma_0 \text{ versus } H_1 : \sigma > \sigma_0.$$

Table 15.3   Credible intervals for the standard deviation σ

Person	Posterior Parameters		95% Credible Interval	
	S'	κ'	Lower Limit	Upper Limit
Aroha	149	9	2.80	7.43
Bernardo	149	10	2.70	6.77
Carlos	160.37	11	2.70	6.48

We will test this by calculating the posterior probability of the null hypothesis and comparing this to the level of significance α that we chose. Let $W = \frac{S'}{\sigma^2}$.

$$P(H_0 \text{ is true } |y_1, \ldots, y_n) = P(\sigma \leq \sigma_0 | y_1, \ldots, y_n)$$
$$= P(\sigma^2 \leq \sigma_0^2 | y_1, \ldots, y_n)$$
$$= P(W \geq W_0),$$

where $W_0 = \frac{S'}{\sigma_0^2}$. When the null hypothesis is true, W has the *chi-squared* distribution with κ' degrees of freedom. This probability can be bounded by values from Table B.6, or alternatively it can be calculated using Minitab or R.

■ EXAMPLE 15.1   (continued)

The three employees want to determine if the standard deviation is greater than 5.00. They set this up as the one-sided hypothesis test

$$H_0 : \sigma \leq 5.00 \text{ versus } H_1 : \sigma > 5.00$$

and choose a level of significance $\alpha = .10$. They each calculate their posterior probability of the null hypothesis. The results are in the Table 15.4. None of their posterior probabilities of the null hypothesis are below $\alpha = .10$, so each employee accepts the null hypothesis at that level.

Table 15.4   Results of Bayesian one-sided hypothesis tests

Person	Posterior	$P(\sigma \leq 5 | y_1, \ldots, y_n)$	
Aroha	$149 \times$ inv. chi-sq. 9 df	$P(W \geq \frac{149}{5^2}) = .7439$	Accept
Bernardo	$149 \times$ inv. chi-sq. 10 df	$P(W \geq \frac{149}{5^2}) = .8186$	Accept
Carlos	$160.37 \times$ inv. chi-sq. 11 df	$P(W \geq \frac{160.37}{5^2}) = .8443$	Accept

Main Points

- The shape of the S times an *inverse chi-squared* distribution with κ degrees of freedom is given by

$$g(x) \propto \frac{1}{x^{\frac{\kappa}{2}+1}} e^{\frac{S}{2x}} .$$

- If U has S times an *inverse chi-squared* distribution with κ degrees of freedom, then $W = \frac{S}{U}$ has the *chi-squared* distribution with κ degrees of freedom. Hence inverse chi-squared probabilities can be calculated using the chi-squared distribution table.

- When X is random variable having S times an *inverse chi-squared* distribution with κ degrees of freedom then its mean and variance are given by

$$E[X] = \frac{S}{\kappa - 2} \quad \text{and} \quad \text{Var}[X] = \frac{2S^2}{(\kappa - 2)^2 \times (\kappa - 4)} ,$$

provided that $\kappa > 2$ and $\kappa > 4$, respectively.

- The likelihood of the variance for a random sample from a *normal*(μ, σ^2) when μ is known has the shape of SS_T times an *inverse chi-squared* distribution with $n - 2$ degrees of freedom.

- We use Bayes' theorem on the variance, so we need the prior distribution of the variance σ^2.

- It is much easier to understand and visualize the prior distribution of the standard deviation σ.

- The prior for the standard deviation can be found from the prior for the variance using the change of variable formula, and vice versa.

- Possible priors include

  1. Positive uniform prior for variance
  2. Positive uniform prior for standard deviation
  3. Jeffreys' prior (same for standard deviation and variance)
  4. S times an *inverse chi-squared* distribution with κ degrees of freedom. (This is the conjugate family of priors for the variance.) Generally it is better to choose a conjugate prior with low degrees of freedom.

- Find Bayesian estimators for standard deviation σ by calculating a measure of location such as the mean, median, or mode from the posterior distribution of the variance σ^2, and taking the square root. Generally, using the posterior mean as the measure of location works best because the posterior distribution has a heavy tail, and it is more conservative to overestimate the variance.

- Bayesian credible intervals for σ can be found by converting the posterior distribution of σ^2 (which is S' times an *inverse chi-squared* with κ' degrees of freedom) to the posterior distribution of $W = \frac{S'}{\sigma^2}$ which is *chi-squared* with κ' degrees of freedom. We can find the upper and lower values for W, and convert them back to find the lower and upper values for the credible interval for σ.

- One-sided hypothesis tests about he standard deviation σ can be performed by calculating the posterior probability of the null hypothesis and comparing it to the chosen level of significance α

Exercises

15.1. The strength of an item is known to be normally distributed with mean 200 and unknown variance σ^2. A random sample of ten items is taken and their strength measured. The strengths are:

215	186	216	203	221
188	202	192	208	195

(a) What is the equation for the shape of the likelihood function of the variance σ^2?

(b) Use a positive uniform prior distribution for the variance σ^2. Change the variable from the variance to the standard deviation to find the prior distribution for the standard deviation σ.

(c) Find the posterior distribution of the variance σ^2.

(d) Change the variable from the variance to the standard deviation to find the posterior distribution of the standard deviation.

(e) Find a 95% Bayesian credible interval for the standard deviation σ.

(f) Test $H_0 : \sigma \leq 8$ vs. $H_1 : \sigma > 8$ at the 5% level of significance.

15.2. The thickness of items produced by a machine is normally distributed with mean $\mu = .001$ cm and unknown variance σ^2. A random sample of ten items are taken and measured. They are:

.00110	.00146	.00102	.00066	.00139
.00121	.00053	.00144	.00146	.00075

(a) What is the equation for the shape of the likelihood function of the variance σ^2?

(b) Use a positive uniform prior distribution for the variance σ^2. Change the variable from the variance to the standard deviation to find the prior distribution for the standard deviation σ.

(c) Find the posterior distribution of the variance σ^2.

(d) Change the variable from the variance to the standard deviation to find the posterior distribution of the standard deviation.

(e) Find a 95% Bayesian credible interval for the standard deviation σ.

(f) Test $H_0 : \sigma \leq .0003$ vs. $H_1 : \sigma > .0003$ at the 5% level of significance.

15.3. The moisture level of a dairy product is normally distributed with mean 15% and unknown variance σ^2. A random sample of size 10 is taken and the moisture level measured. They are:

15.01	14.95	14.99	14.09	16.63
13.98	15.78	15.07	15.64	16.98

(a) What is the equation for the shape of the likelihood function of the variance σ^2?

(b) Use Jeffreys' prior distribution for the variance σ^2. Change the variable from the variance to the standard deviation to find the prior distribution for the standard deviation σ.

(c) Find the posterior distribution of the variance σ^2.

(d) Change the variable from the variance to the standard deviation to find the posterior distribution of the standard deviation.

(e) Find a 95% Bayesian credible interval for the standard deviation σ.

(f) Test $H_0 : \sigma \leq 1.0$ vs. $H_1 : \sigma > 1.0$ at the 5% level of significance.

15.4. The level of saturated fats in a brand of cooking oil is normally distributed with mean $\mu = 15\%$ and unknown variance σ^2. The percentages of saturated fat in a random sample of ten bottles of the cooking oil are:

13.65	14.31	14.73	13.88	14.66
15.53	15.36	15.16	15.76	18.55

(a) What is the equation for the shape of the likelihood function of the variance σ^2?

(b) Use Jeffreys' prior distribution for the variance σ^2. Change the variable from the variance to the standard deviation to find the prior distribution for the standard deviation σ.

(c) Find the posterior distribution of the variance σ^2.

(d) Change the variable from the variance to the standard deviation to find the posterior distribution of the standard deviation.

(e) Find a 95% Bayesian credible interval for the standard deviation σ.

(f) Test $H_0 : \sigma \le .05$ vs. $H_1 : \sigma > .05$ at the 5% level of significance.

15.5. Let a random sample of 5 observations from a *normal*(μ, σ^2) distribution (where it is known that the mean $\mu = 25$) be

26.05   29.39   23.58   23.95   23.38

(a) What is the equation for the shape of the likelihood function of the variance σ^2?

(b) We believe (before looking at the data) that the standard deviation is as likely to be above 4 as it is to be below 4. (Our prior belief is that the distribution of the standard deviation has median 4.) Find the inverse chi-squared prior with 1 degree of freedom that fits our prior belief about the median.

(c) Change the variable from the variance to the standard deviation to find the prior distribution for the standard deviation σ.

(d) Find the posterior distribution of the variance σ^2.

(e) Change the variable from the variance to the standard deviation to find the posterior distribution of the standard deviation.

(f) Find a 95% Bayesian credible interval for the standard deviation σ.

(g) Test $H_0 : \sigma \le 5$ vs. $H_1 : \sigma > 5$ at the 5% level of significance.

15.6. The weight of milk powder in a "1 kg" package is *normal*(μ, σ^2) distribution (where it is known that the mean $\mu = 1015$ g). Let a random sample of 10 packages be taken and weighed. The weights are

1019   1023   1014   1027   1017   1031   1004   1018   1004   1025

(a) What is the equation for the shape of the likelihood function of the variance σ^2?

(b) We believe (before looking at the data) that the standard deviation is as likely to be above 5 as it is to be below 5. (Our prior belief is that the distribution of the standard deviation has median 5.) Find the inverse chi-squared prior with 1 degree of freedom that fits our prior belief about the median.

(c) Change the variable from the variance to the standard deviation to find the prior distribution for the standard deviation σ.

(d) Find the posterior distribution of the variance σ^2.

(e) Change the variable from the variance to the standard deviation to find the posterior distribution of the standard deviation.

(f) Find a 95% Bayesian credible interval for the standard deviation σ.

(g) If there is evidence that the standard deviation is greater than 8, then the machine will be stopped and adjusted. Test $H_0 : \sigma \le 8$ vs. $H_1 : \sigma > 8$ at the 5% level of significance. Is there evidence that the packaging machine needs to be adjusted?

Computer Exercises

15.1. We will use the Minitab macro *NVarICP*, or the function `nvaricp` in R, to find the posterior distribution of the standard deviation σ when we have a random sample of size n from a $normal(\mu, \sigma^2)$ distribution and the mean μ is known. We have $S\times$ an *inverse chi-squared*(κ) prior for the variance σ^2. This is the conjugate family for *normal* observations with known mean. Starting with one member of the family as the prior distribution, we will get another member of the family as the posterior distribution. The simple updating rules are

$$S' = S + SS_T \text{ and } \kappa' = \kappa + n,$$

where $SS_T = \sum (y_i - \mu)^2$. Suppose we have five observations from a $normal(\mu, \sigma^2)$ distribution where $\mu = 200$ is known. They are:

206.4	197.4	212.7	208.5	203.4

(a) Suppose we start with a positive uniform prior for the standard deviation σ. What value of $S\times$ an *inverse chi-squared*(κ) will we use?

(b) Find the posterior using the macro *NVarICP* in Minitab or the function `nvaricp` in R.

(c) Find the posterior mean and median.

(d) Find a 95% Bayesian credible interval for σ.

15.2. Suppose we start with a Jeffreys' prior for the standard deviation σ. What value of $S\times$ an *inverse chi-squared*(κ) will we use?

(a) Find the posterior using the macro *NVarICP* in Minitab or the function `nvaricp` in R.

(b) Find the posterior mean and median.

(c) Find a 95% Bayesian credible interval for σ.

15.3. Suppose our prior belief is σ is just as likely to be below 8 as it is to be above 8. (Our prior distribution $g(\sigma)$ has median 8.) Determine an $S\times$ an *inverse chi-squared*(κ) that matches our prior median where we use $\kappa = 1$ degree of freedom.

(a) Find the posterior using the macro *NVarICP* in Minitab or the function nvaricp in R.

(b) Find the posterior mean and median.

(c) Find a 95% Bayesian credible interval for σ.

15.4. Suppose we take five additional observations from the *normal*(μ, σ^2) distribution where $\mu = 200$ is known. They are:

211.7	205.4	206.0	206.5	201.7

(a) Use the posterior from Exercise 15.3 as the prior for the new observations and find the posterior using the Minitab macro *NVarICP*, or the nvaricp function in R.

(b) Find the posterior mean and median.

(c) Find a 95% Bayesian credible interval for σ.

15.5. Suppose we take the entire sample of ten *normal*(μ, σ^2) observations as a single sample. We will start with the original prior we found in Exercise 15.3.

(a) Find the posterior using the macro *NVarICP* in Minitab or the function nvaricp in R.

(b) What do you notice from Exercises 15.3–15.5?

(c) Test the hypothesis $H_0 : \sigma \leq 5$ vs. $H_1 : \sigma > 5$ at the 5% level of significance.

CHAPTER 16

ROBUST BAYESIAN METHODS

Many statisticians hesitate to use Bayesian methods because they are reluctant to let their prior belief into their inferences. In almost all cases they have some prior knowledge, but they may not wish to formalize it into a prior distribution. They know that some values are more likely than others, and some are not realistically possible. Scientists are studying and measuring something they have observed. They know the scale of possible measurements. We saw in previous chapters that all priors that have reasonable probability over the range of possible values will give similar, although not identical, posteriors. And we saw that Bayes' theorem using the prior information will give better inferences than frequentist ones that ignore prior information, even when judged by frequentist criteria. The scientist would be better off if he formed a prior from his prior knowledge and used Bayesian methods.

However, it is possible that a scientist could have a strong prior belief, yet that belief could be incorrect. When the data are taken, the likelihood is found to be very different from that expected from the prior. The posterior would be strongly influenced by the prior. Most scientists would be very reluctant to use that posterior. If there is a strong disagreement between the

Introduction to Bayesian Statistics, 3rd *ed.*
By Bolstad, W. M. and Curran, J. M. Copyright © 2016 John Wiley & Sons, Inc.

prior and the likelihood, the scientist would want to go with the likelihood, since it came from the data.

In this chapter we look at how we can make Bayesian inference more robust against a poorly specified prior. We find that using a *mixture* of conjugate priors enables us to do this. We allow a small prior probability that our prior is misspecified. If the likelihood is very different from what would be expected under the prior, the posterior probability of misspecification is large, and our posterior distribution will depend mostly on the likelihood.

16.1   Effect of Misspecified Prior

One of the main advantages of Bayesian methods is that it uses your prior knowledge, along with the information from the sample. Bayes' theorem combines both prior and sample information into the posterior. Frequentist methods only use sample information. Thus Bayesian methods usually perform better than frequentist ones because they are using more information. The prior should have relatively high values over the whole range where the likelihood is substantial.

However, sometimes this does not happen. A scientist could have a strong prior belief, yet it could be wrong. Perhaps he (wrongly) bases his prior on some past data that arose from different conditions than the present data set. If a strongly specified prior is incorrect, it has a substantial effect on the posterior. This is shown in the following two examples.

◼ EXAMPLE 16.1

Archie is going to conduct a survey about how many Hamilton voters say they will attend a casino if it is built in town. He decides to base his prior on the opinions of his friends. Out of the 25 friends he asks, 15 say they will attend the casino. So he decides on a $beta(a, b)$ prior that matches those opinions. The prior mean is .6, and the equivalent samples size is 25. Thus $a + b + 1 = 25$ and $\frac{a}{a+b} = .6$. Thus $a = 14.4$ and $b = 9.6$. Then he takes a random sample of 100 Hamilton voters and finds that 25 say they will attend the casino. His posterior distribution is $beta(39.4, 84.60)$. Archie's prior, the likelihood, and his posterior are shown in Figure 16.1. We see that the prior and the likelihood do not overlap very much. The posterior is in between. It gives high posterior probability to values that are not supported strongly by the data (likelihood) and are not strongly supported by prior either. This is not satisfactory.   ◼

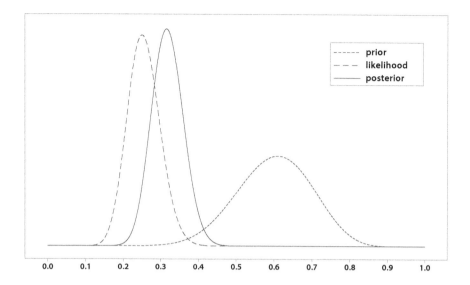

Figure 16.1   Archie's prior, likelihood, and posterior.

■ EXAMPLE 16.2

Andrea is going to take a sample of measurements of dissolved oxygen
level from a lake during the summer. Assume that the dissolved oxygen
level is approximately normal with mean μ and known variance $\sigma^2 = 1$.
She had previously done a similar experiment from the river that flowed
into the lake. She considered that she had a pretty good idea of what
to expect. She decided to use a $normal(8.5, .7^2)$ prior for μ, which was
similar to her river survey results. She takes a random sample of size 5
and the sample mean is 5.45. The parameters of the posterior distribution
are found using the simple updating rules for normal. The posterior is
$normal(6.334, .3769^2)$. Andrea's prior, likelihood, and posterior are shown
in Figure 16.2. The posterior density is between the prior and likelihood,
and gives high probability to values that are not supported strongly either
by the data or by the prior, which is a very unsatisfactory result.   ■

These two examples show how an incorrect prior can arise. Both Archie and
Andrea based their priors on past data, each judged to arise from a situation
similar the one to be analyzed. They were both wrong. In Archie's case he
considered his friends to be representative of the population. However, they
were all similar in age and outlook to him. They do not constitute a good
data set to base a prior on. Andrea considered that her previous data from
the river survey would be similar to data from the lake. She neglected the
effect of water movement on dissolved oxygen. She is basing her prior on data

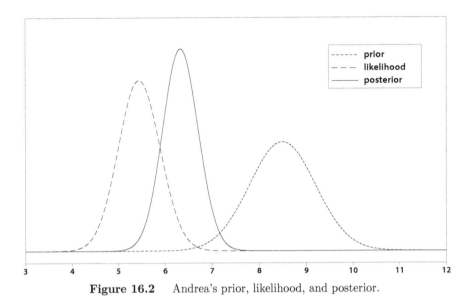

Figure 16.2   Andrea's prior, likelihood, and posterior.

obtained from an experiment under different conditions than the one she is now undertaking.

16.2   Bayes' Theorem with Mixture Priors

Suppose our prior density is $g_0(\theta)$ and it is quite precise, because we have substantial prior knowledge. However, we want to protect ourselves from the possibility that we misspecified the prior by using prior knowledge that is incorrect. We do not consider it likely, but concede that it is possible that we failed to see the reason why our prior knowledge will not applicable to the new data. If our prior is misspecified, we do not really have much of an idea what values θ should take. In that case the prior for θ is $g_1(\theta)$, which is either a very vague conjugate prior or a flat prior. Let $g_0(\theta|y_1, \ldots, y_n)$ be the posterior distribution of θ given the observations when we start with $g_0(\theta)$ as the prior. Similarly, we let $g_1(\theta|y_1, \ldots, y_n)$ be the posterior distribution of θ, given the observations when we start with $g_1(\theta)$ as the prior:

$$g_i(\theta|y_1, \ldots, y_n) \propto g_i(\theta)f(y_1, \ldots, y_n|\theta).$$

These are found using the simple updating rules, since we are using priors that are either from the conjugate family or are flat.

The Mixture Prior

We introduce a new parameter, I, that takes two possible values. If $i = 0$, then θ comes from $g_0(\theta)$. However, if $i = 1$, then θ comes from $g_1(\theta)$. The conditional prior probability of θ given i is

$$g(\theta|i) = \begin{cases} g_0(\theta) & \text{if } i = 0, \\ g_1(\theta) & \text{if } i = 1. \end{cases}$$

We let the prior probability distribution of I be $P(I = 0) = p_0$, where p_0 is some high value like .9, .95, or .99, because we think our prior $g_0(\theta)$ is correct. The prior probability that our prior is misspecified is $p_1 = 1 - p_0$. The joint prior distribution of θ and I is

$$g(\theta, i) = p_i \times (1 - i) \times g_0(\theta) + (1 - p_i) \times (i) \times g_1(\theta)$$

We note that this joint distribution is continuous in the parameter θ and discrete in the parameter I. The marginal prior density of the random variable θ is found by marginalizing (summing I over all possible values) the joint density. It has a *mixture* prior distribution since its density

$$g(\theta) = .95 \times (i - 1) \times g_0(\mu) + .05 \times (i) \times g_1(\mu) \tag{16.1}$$

is a mixture of the two prior densities.

The Joint Posterior

The joint posterior distribution of θ, I given the observations $y_1, \ldots, y_n$ is proportional to the joint prior times the joint likelihood. This gives

$$g(\theta, i|y_1, \ldots, y_n) = c \times g(\theta, i) \times f(y_1, \ldots, y_n|\theta, i) \quad \text{for} \quad i = 0, 1$$

for some constant c. But the sample only depends on θ, not on i, so the joint posterior

$$\begin{aligned} g(\theta, i|y_1, \ldots, y_n) &= c \times p_i g_i(\theta) f(y_1, \ldots, y_n|\theta) && \text{for} \quad i = 0, 1 \\ &= c \times p_i h_i(\theta, y_1, \ldots, y_n) && \text{for} \quad i = 0, 1, \end{aligned}$$

where $h_i(\theta, y_1, \ldots, y_n) = g_i(\theta) f(y_1, \ldots, y_n|\theta)$ is the joint distribution of the parameter and the data, when $g_i(\theta)$ is the correct prior. The marginal posterior probability $P(I = i|y_1, \ldots, y_n)$ is found by integrating θ out of the joint posterior:

$$\begin{aligned} P(I = i|y_1, \ldots, y_n) &= \int g(\theta, i|y_1, \ldots, y_n) \, d\theta \\ &= c \times p_i \int h_i(\theta, y_1, \ldots, y_n) \, d\theta \\ &= c \times p_i f_i(y_1, \ldots, y_n) \end{aligned}$$

for $i = 0, 1$, where $f_i(y_1, \ldots, y_n)$ is the marginal probability (or probability density) of the data when $g_i(\theta)$ is the correct prior. The posterior probabilities sum to 1, and the constant c cancels, so

$$P(I = i | y_1, \ldots, y_n) = \frac{p_i f_i(y_1, \ldots, y_n)}{\sum_{i=0}^{1} p_i f_i(y_1, \ldots, y_n)}.$$

These can be easily evaluated.

The Mixture Posterior

We find the marginal posterior of θ by summing all possible values of i out of the joint posterior:

$$g(\theta | y_1, \ldots, y_n) = \sum_{i=0}^{1} g(\theta, i | y_1, \ldots, y_n).$$

But there is another way the joint posterior can be rearranged from conditional probabilities:

$$g(\theta, i | y_1, \ldots, y_n) = g(\theta | i, y_1, \ldots, y_n) \times P(I = i | y_1, \ldots, y_n),$$

where $g(\theta | i, y_1, \ldots, y_n) = g_i(\theta | y_1, \ldots, y_n)$ is the posterior distribution when we started with $g_i(\theta)$ as the prior. Thus the marginal posterior of θ is

$$g(\theta | y_1, \ldots, y_n) = \sum_{i=0}^{1} g_i(\theta | y_1, \ldots, y_n) \times P(I = i | y_1, \ldots, y_n). \qquad (16.2)$$

This is the mixture of the two posteriors, where the weights are the posterior probabilities of the two values of i given the data.

◨ EXAMPLE 16.2 (continued)

One of Archie's friends, Ben, decided that he would reanalyze Archie's data with a mixture prior. He let g_0 be the same $beta(14.4, 9.6)$ prior that Archie used. He let g_1 be the (uniform) $beta(1, 1)$ prior. He let the prior probability $p_0 = .95$. Ben's mixture prior and its components are shown in Figure 16.3. His mixture prior is quite similar to Archie's. However, it has heavier weight in the tails. This gives makes his prior robust against prior misspecification. In this case, $h_i(\pi, y)$ is a product of a beta times a binomial. Of course, we are only interested in $y = 25$, the value that

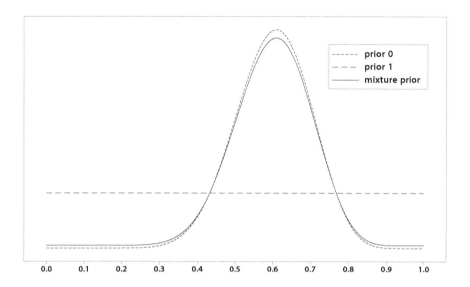

Figure 16.3   Ben's mixture prior and components.

occurred:

$$h_0(\pi, y = 25) = \frac{\Gamma(24)}{\Gamma(14.4)\Gamma(9.6)}\pi^{13.4}(1-\pi)^{8.6} \times \left(\frac{100!}{25!75!}\right)\pi^{25}(1-\pi)^{75}$$

$$= \frac{\Gamma(24)}{\Gamma(14.4)\Gamma(9.6)} \times \left(\frac{100!}{25!75!}\right) \times \pi^{38.4}(1-\pi)^{83.6}$$

and

$$h_1(\pi, y = 25) = \pi^0(1-\pi)^0 \times \left(\frac{100!}{25!75!}\right)\pi^{25}(1-\pi)^{75}$$

$$= \left(\frac{100!}{25!75!}\right)\pi^{25}(1-\pi)^{75}.$$

We recognize each of these as a constant times a beta distribution. So integrating them with respect to π gives

$$\int_0^1 h_0(\pi, y = 25)\,d\pi = \frac{\Gamma(24)}{\Gamma(14.4)\Gamma(9.6)} \times \left(\frac{100!}{25!75!}\right) \times \int_0^1 \pi^{38.4}(1-\pi)^{83.6}\,d\pi$$

$$= \frac{\Gamma(24)}{\Gamma(14.4)\Gamma(9.6)} \times \left(\frac{100!}{25!75!}\right) \times \frac{\Gamma(39.4)\Gamma(84.6)}{\Gamma(124)}$$

and

$$\int_0^1 h_1(\pi, y = 25)\,d\pi = \left(\frac{100!}{25!75!}\right) \times \int_0^1 \pi^{25}(1-\pi)^{75}\,d\pi$$

$$= \left(\frac{100!}{25!75!}\right) \times \frac{\Gamma(26)\Gamma(76)}{\Gamma(102)}.$$

Remember that $\Gamma(a) = (a-1) \times \Gamma(a-1)$ and if a is an integer, $\Gamma(a) = (a-1)!$. The second integral is easily evaluated and gives

$$f_1(y = 25) = \int_0^1 h_1(\pi, y = 25) \, d\pi = \frac{1}{101} = 9.90099 \times 10^{-3}.$$

We can evaluate the first integral numerically:

$$f_0(y = 25) = \int_0^1 h_0(\pi, y = 25) \, d\pi = 2.484 \times 10^{-4}.$$

So the posterior probabilities are $P(I = 0|25) = 0.323$ and $P(I = 1|25) = 0.677$. The posterior distribution is the mixture $g(\pi|25) = .323 \times g_0(\pi|25) + .677 \times g_1(\pi|25)$, where $g_0(\pi|y)$ and $g_1(\pi|y)$ are the conjugate posterior distributions found using g_0 and g_1 as the respective priors. Ben's mixture posterior distribution and its two components is shown in Figure 16.4. Ben's prior and posterior, together with the likelihood, is shown in Fig-

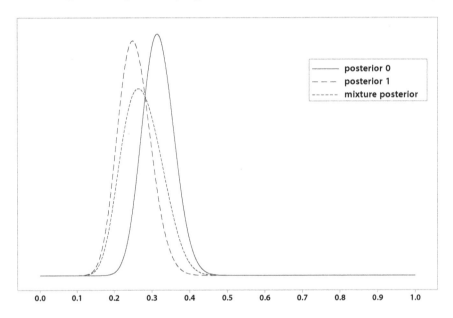

posterior 0
posterior 1
mixture posterior

Figure 16.4   Ben's mixture posterior and its two components.

ure 16.5. When the prior and likelihood disagree, we should go with the likelihood because it is from the data. Superficially, Ben's prior looks very similar to Archie's prior. However, it has a heavier tail allowed by the mixture, and this has allowed his posterior to be very close to the likelihood. We see that this is much more satisfactory than Archie's analysis shown in Figure 16.1.   ∎

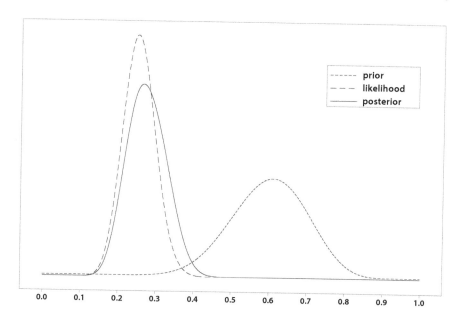

Figure 16.5   Ben's mixture prior, likelihood, and mixture posterior.

◾ EXAMPLE 16.2 (continued)

Andrea's friend Caitlin looked at Figure 16.2 and told her it was not satisfactory. The values given high posterior probability were not supported
strongly either by the data or by the prior. She considered it likely that
the prior was misspecified. She said to protect against that, she would do
the analysis using a mixture of normal priors. $g_0(\theta)$ was the same as Andrea's, $normal(8.5, .7^2)$, and $g_1(\theta)$ would be $normal(8.5, (4 \times .7)^2)$, which
has the same mean as Andrea's prior, but with the standard deviation 4
times as large. She allows prior probability .05 that Andrea's prior was
misspecified. Caitlin's mixture prior and its components are shown in Figure 16.6. We see that her mixture prior appears very similar to Andrea's
except there is more weight in the tail regions. Caitlin's posterior $g_0(\theta|\bar{y})$
is $normal(6.334, .3769^2)$, the same as for Andrea. Caitlin's posterior when
the original prior was misspecified $g_1(\theta|\bar{y})$ is $normal(5.526, .4416^2)$, where
the parameters are found by the simple updating rules for the normal. In
the normal case

$$h_i(\mu, y_1, \ldots, y_n) \propto g_i(\mu) \times f(\bar{y}|\mu)$$
$$\propto e^{-\frac{1}{2s_i^2}(\mu - m_i)^2} \times e^{-\frac{1}{2\sigma^2/n}(\bar{y} - \mu)^2},$$

where m_i and s_i^2 are the mean and variance of the prior distribution $g_i(\mu)$.
The integral $\int h_i(\mu, y_1, \ldots, y_n)\, d\mu$ gives the unconditional probability of

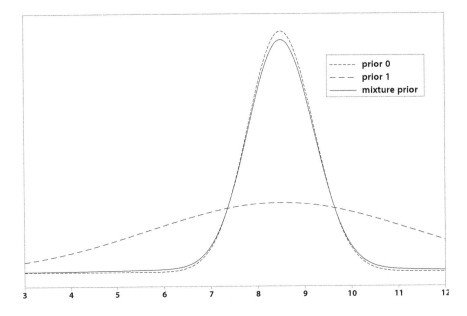

Figure 16.6   Caitlin's mixture prior and its components.

the sample, when g_i is the correct prior. We multiply out the two terms and then rearrange all the terms containing μ, which is normal and integrates. The terms that are left simplify to

$$f_i(\bar{y}) = \int h_i(\mu, \bar{y}) \, d\mu$$

$$\propto \frac{1}{\sqrt{s_i^2 + \sigma^2/n}} \times e^{-\frac{1}{2(s_i^2 + \sigma^2/n)}(\bar{y} - m_i)^2},$$

which we recognize as a normal density with mean m_i and variance $\frac{\sigma^2}{n} + s_i^2$. In this example, $m_0 = 8.5, s_0^2 = .7^2, m_1 = 8.5, s_1^2 = (4 \times .7)^2, \sigma^2 = 1$, and $n = 5$. The data are summarized by the value $\bar{y} = 5.45$ that occurred in the sample. Plugging in these values, we get $P(I = 0|\bar{y} = 5.45) = .12$ and $P(I = 1|\bar{y} = 5.45) = .88$. Thus Caitlin's posterior is the mixture $.12 \times g_0(\mu|\bar{y}) + .88 \times g_1(\mu|\bar{y})$. Caitlin's mixture posterior and its components are given in Figure 16.7. Caitlin's prior, likelihood, and posterior are shown in Figure 16.8. Comparing this with Andrea's analysis shown in Figure 16.2, we see that using mixtures has given her a posterior that is much closer to the likelihood than the one obtained with the original misspecified prior. This is a much more satisfactory result. ∎

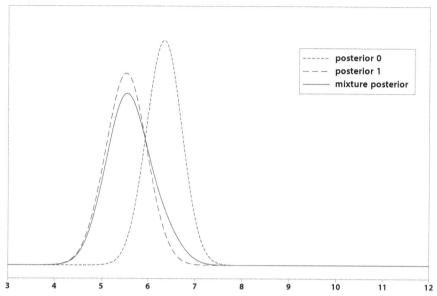

Figure 16.7   Caitlin's mixture posterior and its two components.

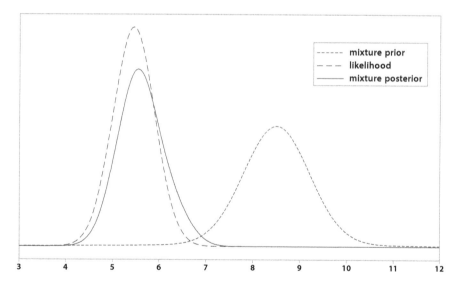

Figure 16.8   Caitlin's mixture prior, the likelihood, and her mixture posterior.

Summary

Our prior represents our prior belief about the parameter before looking at the data from this experiment. We should be getting our prior from past data from similar experiments. However, if we think an experiment is similar, but it is not, our prior can be quite misspecified. We may think we know a lot about the parameter, but what we think is wrong. That makes the prior quite precise, but wrong. It will be quite a distance from the likelihood. The posterior will be in between, and will give high probability to values neither supported by the data or the prior. That is not satisfactory. If there is a conflict between the prior and the data, we should go with the data.

We introduce a indicator random variable that we give a small prior probability of indicating our original prior is misspecified. The mixture prior we use is $P(I = 0) \times g_0(\theta) + P(I = 1) \times g_1(\theta)$, where g_0 and g_1 are the original prior and a more widely spread prior, respectively. We find the joint posterior of distribution of I and θ given the data. The marginal posterior distribution of θ, given the data, is found by marginalizing the indicator variable out. It will be the mixture distribution

$$g_{mixture}(\theta|y_1, \ldots, y_n) = P(I = 0|y_1, \ldots, y_n)g_0(\theta|y_1, \ldots, y_n)$$
$$+ P(I = 1|y_1, \ldots, y_n)g_1(\theta|y_1, \ldots, y_n).$$

This posterior is very robust against a misspecified prior. If the original prior is correct, the mixture posterior will be very similar to the original posterior. However, if the original prior is very far from the likelihood, the posterior probability $p(i = 0|y_1, \ldots, y_n)$ will be very small, and the mixture posterior will be close to the likelihood. This has resolved the conflict between the original prior and the likelihood by giving much more weight to the likelihood.

Main Points

- If the prior places high probability on values that have low likelihood, and low probability on values that have high likelihood, the posterior will place high probability on values that are not supported either by the prior or by the likelihood. This is not satisfactory.

- This could have been caused by a misspecified prior that arose when the scientist based his/her prior on past data, which had been generated by a process that differs from the process that will generate the new data in some important way that the scientist failed to take into consideration.

- Using mixture priors protects against this possible misspecification of the prior. We use mixtures of conjugate priors. We do this by introducing a mixture index random variable that takes on the values 0 or 1. The mixture prior is

$$g(\theta) = p_0 \times g_0(\theta) + p_1 \times g_1(\theta),$$

  where $g_0(\theta)$ is the original prior we believe in, and g_1 is another prior that has heavier tails and thus allows for our original prior being wrong. The respective posteriors that arise using each of the priors are $g_0(\theta|y_1, \ldots, y_n)$ and $g_1(\theta|y_1, \ldots, y_n)$.

- We give the original prior g_0 high prior probability by letting the prior probability $p_0 = P(I = 0)$ be high and the prior probability $p_1 = (1 - p_0) = P(I = 1)$ is low. We think the original prior is correct, but have allowed a small probability that we have it wrong.

- Bayes' theorem is used on the mixture prior to determine a mixture posterior. The mixture index variable is a nuisance parameter and is marginalized out.

- If the likelihood has most of its value far from the original prior, the mixture posterior will be close to the likelihood. This is a much more satisfactory result. When the prior and likelihood are conflicting, we should base our posterior belief mostly on the likelihood, because it is based on the data. Our prior was based on faulty reasoning from past data that failed to note some important change in the process we are drawing the data from.

- The mixture posterior is a mixture of the two posteriors, where the mixing proportions $P(I = i)$ for $i = 0, 1$, are proportional to the prior probability times the the marginal probability (or probability density) evaluated at the data that occurred.

$$P(I = i) \propto p_i \times f_i(y_1, \ldots y_n) \text{ for } i = 0, 1.$$

- They sum to 1, so

$$P(I = i) = \frac{p_i \times f_i(y_1, \ldots y_n)}{\sum_{i=0}^{1} p_i \times f_i(y_1, \ldots y_n)} \text{ for } i = 0, 1.$$

Exercises

16.1. You are going to conduct a survey of the voters in the city you live in. They are being asked whether or not the city should build a new convention facility. You believe that most of the voters will disapprove the proposal because it may lead to increased property taxes for residents. As a resident of the city, you have been hearing discussion about this proposal, and most people have voiced disapproval. You think that only about 35% of the voters will support this proposal, so you decide that a $beta(7, 13)$ summarizes your prior belief. However, you have a nagging doubt that the group of people you have heard voicing their opinions is representative of the city voters. Because of this, you decide to use a mixture prior:

$$g(\pi|i) = \begin{cases} g_0(\pi) & \text{if } i = 0, \\ g_1(\pi) & \text{if } i = 1. \end{cases}$$

where $g_0(\pi)$ is the $beta(7, 13)$ density, and $g_1(\pi)$ is the $beta(1, 1)$ (uniform) density. The prior probability $P(I = 0) = .95$. You take a random sample of $n = 200$ registered voters who live in the city. Of these, $y = 10$ support the proposal.

(a) Calculate the posterior distribution of π when $g_0(\pi)$ is the prior.

(b) Calculate the posterior distribution of π when $g_1(\pi)$ is the prior.

(c) Calculate the posterior probability $P(I = 0|Y)$.

(d) Calculate the marginal posterior $g(\pi|Y)$.

16.2. You are going to conduct a survey of the students in your university to find out whether they read the student newspaper regularly. Based on your friends opinions, you think that a strong majority of the students do read the paper regularly. However, you are not sure your friends are representative sample of students. Because of this, you decide to use a mixture prior.

$$g(\pi|i) = \begin{cases} g_0(\pi) & \text{if } i = 0, \\ g_1(\pi) & \text{if } i = 1. \end{cases}$$

where $g_0(\pi)$ is the $beta(20, 5)$ density, and $g_1(\pi)$ is the $beta(1, 1)$ (uniform) density. The prior probability $P(I = 0) = .95$. You take a random sample of $n = 100$ students. Of these, $y = 41$ say they read the student newspaper regularly.

(a) Calculate the posterior distribution of π when $g_0(\pi)$ is the prior.

(b) Calculate the posterior distribution of π when $g_1(\pi)$ is the prior.

(c) Calculate the posterior probability $P(I = 0|Y)$.

(d) Calculate the marginal posterior $g(\pi|Y)$.

16.3. You are going to take a sample of measurements of specific gravity of a chemical product being produced. You know the specific gravity measurements are approximately $normal(\mu, \sigma^2)$ where $\sigma^2 = .005^2$. You have precise $normal(1.10, .001^2)$ prior for μ because the manufacturing process is quite stable. However, you have a nagging doubt about whether the process is correctly adjusted, so you decide to use a mixture prior. You let $g_0(\mu)$ be your precise $normal(1.10, .001^2)$ prior, you let $g_1(\mu)$ be a $normal(1.10, .01^2)$, and you let $p_0 = .95$. You take a random sample of product and measure the specific gravity. The measurements are

1.10352   1.10247   1.10305   1.10415   1.10382   1.10187

(a) Calculate the posterior distribution of μ when $g_0(\mu)$ is the prior.

(b) Calculate the posterior distribution of μ when $g_1(\mu)$ is the prior.

(c) Calculate the posterior probability $P(I = 0|y_1, \ldots, y_6)$.

(d) Calculate the marginal posterior $g(\mu|y_1, \ldots, y_6)$.

16.4. You are going to take a sample of 500 g blocks of cheese. You know they are approximately $normal(\mu, \sigma^2)$, where $\sigma^2 = 2^2$. You have a precise $normal(502, 1^2)$ prior for μ because this is what the the process is set for. However, you have a nagging doubt that maybe the machine needs adjustment, so you decide to use a mixture prior. You let $g_0(\mu)$ be your precise $normal(502, 1^2)$ prior, you let $g_1(\mu)$ be a $normal(502, 2^2)$, and you let $p_0 = .95$. You take a random sample of ten blocks of cheese and weigh them. The measurements are

501.5	499.1	498.5	499.9	500.4
498.9	498.4	497.9	498.8	498.6

(a) Calculate the posterior distribution of μ when $g_0(\mu)$ is the prior.

(b) Calculate the posterior distribution of μ when $g_1(\mu)$ is the prior.

(c) Calculate the posterior probability $P(I = 0|y_1, \ldots, y_{10})$.

(d) Calculate the marginal posterior $g(\mu|y_1, \ldots, y_{10})$.

Computer Exercises

16.1. We will use the Minitab macro *BinoMixP*, or function `binomixp` in R, to find the posterior distribution of π given an observation y from the $binomial(n, \pi)$ distribution when we use a mixture prior for π. Suppose our prior experience leads us to believe a $beta(7, 13)$ prior would be appropriate. However we have a nagging suspicion that our experience was under different circumstances, so our prior belief may be quite incorrect and we need a fallback position. We decide to use a mixture prior where $g_0(\pi)$ is the $beta(7, 13)$ and $g_1(\pi)$ is the $beta(1, 1)$ distribution, and the prior probability $P(I = 0) = .95$. Suppose we take a random sample of $n = 100$ and observe $y = 76$ successes.

(a) [**Minitab:**] Use *BinoMixp* to find the posterior distribution $g(\pi|y)$.

[**R:**] Use the function `binomixp` to find the posterior distribution $g(\pi|y)$.

(b) Find a 95% Bayesian credible interval for π.

(c) Test the hypothesis $H_0 : \pi \leq .5$ vs. $H_1 : \pi > .5$ at the 5% level of significance.

16.2. We are going to observe the number of "successes" in $n = 100$ independent trials. We have prior experience and believe that a $beta(6, 14)$ summarizes our prior experience. However, we consider that our prior experience may have occurred under different conditions, so our prior may be bad. We decide to use a mixture prior where $g_0(\pi)$ is the $beta(6, 14)$ and $g_1(\pi)$ is the $beta(1, 1)$ distribution, and the prior probability $P(I = 0) = .95$. Suppose we take a random sample of $n = 100$ and observe $y = 36$ successes.

(a) Use *BinoMixp* in Minitab, or `binomixp` in R, to find the posterior distribution $g(\pi|y)$.

(b) Find a 95% Bayesian credible interval for π.

(c) Test the hypothesis $H_0 : \pi \leq .5$ vs. $H_1 : \pi > .5$ at the 5% level of significance.

16.3. We will use the Minitab macro *NormMixP*, or `normmixp` in R, to find the posterior distribution of μ given a random sample $y_1, \ldots, y_n$ from the $normal(\mu, \sigma^2)$ distribution where we know the standard deviation $\sigma = 5$. when we use a mixture prior for μ. Suppose that our prior experience in similar situations leads us to believe that the prior distribution should be $normal(1000, 5^2)$. However, we consider that the prior experience may have been under different circumstances, so we decide to use a mixture prior where $g_0(\mu)$ is the $normal(1000, 5^2)$ and $g_1(\mu)$ is the $normal(1000, 15^2)$ distribution, and the prior probability $P(I = 0) = .95$. We take a random sample of $n = 10$ observations. They are

1030	1023	1027	1022	1023
1023	1030	1018	1015	1011

(a) [**Minitab:**] Use *NormMixp* to find the posterior distribution $g(\mu|y)$.

[**R:**] Use `normmixp` to find the posterior distribution $g(\mu|y)$.

(b) Find a 95% Bayesian credible interval for μ.

(c) Test the hypothesis $H_0 : \mu \leq 1,000$ vs. $H_1 : \mu > 1,000$ at the 5% level of significance.

16.4. We are taking a random sample from the $normal(\mu, \sigma^2)$ distribution where we know the standard deviation $\sigma = 4$. Suppose that our prior experience in similar situations leads us to believe that the prior distribution should be $normal(255, 4^2)$. However, we consider that the prior experience may have been under different circumstances, so we decide to use a mixture prior where $g_0(\mu)$ is the $normal(255, 4^2)$ and $g_1(\mu)$ is the $normal(255, 12^2)$ distribution, and the prior probability $P(I = 0) = .95$. We take a random sample of $n = 10$ observations. They are

249	258	255	261	259
254	261	256	253	254

(a) Use *NormMixp* in Minitab, or `normmixp` in R, to find the posterior distribution $g(\mu|y)$.

(b) Find a 95% Bayesian credible interval for μ.

(c) Test the hypothesis $H_0 : \mu \leq 1,000$ vs. $H_1 : \mu > 1,000$ at the 5% level of significance.

CHAPTER 17

BAYESIAN INFERENCE FOR NORMAL WITH UNKNOWN MEAN AND VARIANCE

The $normal(\mu, \sigma^2)$ distribution has two parameters, the mean μ and the variance σ^2. Usually we are more interested in making our inferences about the mean μ, and regard the variance σ^2 as a nuisance parameter.

In Chapter 11 we looked at the case where we had a random sample of observations from a $normal(\mu, \sigma^2)$ distribution where the mean μ was the only unknown parameter. That is, we assumed that the variance σ^2 was a known constant. This observation distribution is a member of the *one-dimensional exponential family*[1] of distributions. We saw that when we used a $normal(m, s^2)$ conjugate prior for μ, we could find the $normal(m', (s')^2)$ conjugate posterior easily using the simple updating rules[2] that are appropriate for this case. We also saw that, as a rule of thumb, when the variance is

---

[1] A random variable Y with parameter θ is a member of the one-dimensional exponential family of distributions means that its probability or probability density function can be written as $f(y|\theta) = a(\theta) \times b(y) \times e^{c(\theta \times t(y))}$ for some functions $a(\theta)$, $b(y)$, $c(\theta)$, and $t(y)$. For this case, $a(\mu) = e^{-\frac{\mu^2}{2\sigma^2}}$, $b(y) = \frac{1}{\sigma} e^{-\frac{y^2}{2\sigma^2}}$, $c(\mu) = \frac{\mu}{\sigma^2}$, and $t(y) = y$

[2] Distributions in the one-dimensional exponential family always have a family of conjugate priors, and simple updating rules to find the posterior.

---

not known, we can use the estimated variance $\hat{\sigma}^2$ calculated from the sample. This introduces a certain amount of uncertainty because we do not know the actual value of σ^2. We allow for this extra uncertainty by using *Student's t* distribution instead of the *standard normal* distribution to find critical values for our credible intervals and to do probability calculations.

In Chapter 15 we looked at the case where we had a random sample of observations from the $normal(\mu, \sigma^2)$ distribution where the only unknown parameter is the variance σ^2. The mean μ was assumed to be a known constant. This observation distribution is also a member of the one-dimensional exponential family.[3] We saw that when we used S times an *inverse chi-squared* with κ degrees of freedom conjugate prior for σ^2, we could find the S' times an *inverse chi-squared* with κ' degrees of freedom conjugate posterior by using the simple updating rules appropriate for this case. We used the sample mean $\hat{\mu} = \bar{y}$ when μ was not known. This costs us one degree of freedom, which in turn means that there is more uncertainty in our calculations. We calculate the sum of squares around $\hat{\mu}$ and find the critical values from the inverse chi-squared distribution with one less degree of freedom.

In each of these earlier chapters, we suggested the following procedure as an approximation to be used when we do not know the value of the nuisance parameter. We estimate the nuisance parameter from the sample, and plug that estimated value into the model. We then perform our inference on the parameter of interest as if the nuisance parameter has that value, and we change how we find the critical values for credible intervals and probability calculations, *Student's t* instead of *standard normal* for μ, and *inverse chi-squared* with one less degree of freedom for σ^2.

The assumption that we know one of the two parameters and not the other is a little artificial. Realistically, if we do not know the mean μ, then how can we know the variance, and vice versa. In this chapter we look at the more realistic case where we have a random sample of observations from the $normal(\mu, \sigma^2)$ distribution where both parameters are unknown. We will use a joint prior for the two parameters and find the joint posterior distribution using Bayes' theorem. We find the marginal posterior distribution of the parameter of interest (usually μ) by integrating the nuisance parameter out of the joint posterior. Then we do the inference on the parameter of interest using its marginal posterior.

In Section 17.1 we look at the joint likelihood function of the $normal(\mu, \sigma^2)$ distribution where both the mean μ and the variance σ^2 are unknown parameters. It factors into the product of a conditional normal shape likelihood for μ given σ^2 times an inverse chi-squared shaped likelihood for σ^2.

In Section 17.2 we look at inference in the case where we use independent Jeffrey's priors for μ and σ^2. We find the marginal posterior distribution for μ by integrating the nuisance parameter σ^2 out of the joint posterior. We will

---

[3]In this case, $a(\sigma^2) = (\sigma^2)^{-\frac{1}{2}} e^{-\frac{\mu^2}{\sigma^2}}$, $c(\sigma^2) = \frac{\mu}{\sigma^2}$, and $t(y) = y$.

find that for this case, the Bayesian results will be the same as the results using frequentist assumptions.

In Section 17.3 we find that the joint conjugate prior for the two parameters is not the product of independent conjugate priors for the two parameters since it must have the same form as the joint likelihood: a conditional normal prior for μ given σ^2 times an inverse chi-squared prior for σ^2. We find the joint posterior when we use the joint conjugate prior for the two parameters. We integrate the nuisance parameter σ^2 out of the joint posterior to get a *Student's* t shaped marginal posterior of μ. Misspecification of the prior mean leads to inflation of the posterior variance. This may mask the real problem which is the misspecified prior mean. As an alternative, we present an approximate Bayesian approach that does not have this variance inflation. It uses the fact that the joint prior and the joint likelihood factor into a conditional normal part for μ given σ^2 times a multiple of an inverse chi-squared part for σ^2. We find an approximation to the joint posterior that has that same form. However, this method does not use all the information about the variance in the prior and likelihood, specifically the distance between the prior mean and the sample mean. We find a *Student's* t shaped posterior distribution for μ using an estimate of the variance that incorporates the prior for the variance and the sample variance estimate. This shows why the approximation we suggested in Chapter 11 holds. We should, when using either the exact or approximate approach, examine graphs similar to those shown in Chapter 16 to decide if we have made a poor choice of prior distribution for the mean.

In Section 17.4 we find the posterior distribution of the difference between means $\mu_d = \mu_1 - \mu_2$ for two independent random samples from *normal* distributions having the same unknown variance σ^2. We look at two cases. In the first case we use independent Jeffreys' priors for all three parameters μ_1, μ_2, and σ^2. We simplify to the joint posterior of μ_d and σ^2, and then we integrate σ^2 out to find the *Student's* t shaped marginal posterior distribution of μ_d. In the second case we use the joint conjugate prior for all three parameters. Again, we find the joint posterior of μ_1, μ_2, and σ^2. Then we simplify it to the joint posterior of μ_d and σ^2. We integrate the nuisance parameter σ^2 out of the joint posterior to find the marginal posterior of μ_d which has a *Student's* t-shape. We also give an approximate method based on factoring the joint posterior. We find the joint posterior distribution of μ_d and σ^2 also factors. A theorem gives us a *Student's* t-shaped posterior for μ_d. Again, this approximation does not use all information about the variance from the prior and likelihood.

In Section 17.5, we find the posterior distribution for the difference between the means $\mu_d = \mu_1 - \mu_2$, when we have independent random samples from $normal(\mu_1, \sigma_1^2)$ and $normal(\mu_2, \sigma_2^2)$, respectively, where both variances are unknown.

17.1   The Joint Likelihood Function

The joint likelihood for a single observation from the $normal(\mu, \sigma^2)$ distribution has shape given by

$$f(y|\mu, \sigma^2) \propto \frac{1}{\sqrt{\sigma^2}} e^{-\frac{1}{2\sigma^2} \sum (y-\mu)^2} .$$

Note: We include the factor $\frac{1}{\sqrt{\sigma^2}}$ in the likelihood because σ^2 is no longer a constant and is now considered to be one of the parameters in this model. The likelihood for a random sample $y_1, \dots, y_n$ from the $normal(\mu, \sigma^2)$ distribution is the product of the individual likelihoods. Its shape is given by

$$f(y_1, \dots, y_n|\mu, \sigma^2) \propto \prod_{i=1}^{n} f(y_i|\mu, \sigma^2)$$

$$\propto \prod_{i=1}^{n} \frac{1}{\sqrt{\sigma^2}} e^{-\frac{1}{2\sigma^2}(y_i - \mu)^2}$$

$$\propto \frac{1}{(\sigma^2)^{\frac{n}{2}}} e^{-\frac{1}{2\sigma^2} \sum (y_i - \mu)^2}$$

We subtract and add the sample mean $\bar{y}$ in the exponent

$$f(y_1, \dots, y_n|\mu, \sigma^2) \propto \frac{1}{(\sigma^2)^{\frac{n}{2}}} e^{-\frac{1}{2\sigma^2} \sum [(y_i - \bar{y}) + (\bar{y} - \mu)]^2}$$

We expand the exponent, break it into three sums, and simplify. The middle term sums to zero and drops out. We get

$$f(y_1, \dots, y_n|\mu, \sigma^2) \propto \frac{1}{(\sigma^2)^{\frac{n}{2}}} e^{-\frac{1}{2\sigma^2} [n(\bar{y} - \mu)^2 + SS_y]}$$

where $SS_y = \sum (y_i - \bar{y})^2$ is the sum of squares away from the sample mean. Note that the joint likelihood factors into two parts:

$$f(y_1, \dots, y_n|\mu, \sigma^2) \propto \frac{1}{(\sigma^2)^{\frac{1}{2}}} e^{-\frac{n}{2\sigma^2}(\bar{y} - \mu)^2} \times \frac{1}{(\sigma^2)^{\frac{n-1}{2}}} e^{-\frac{SS_y}{2\sigma^2}} . \qquad (17.1)$$

The first part shows $\mu|\sigma^2$ has a $normal(\bar{y}, \frac{\sigma^2}{n})$ distribution. The second part shows σ^2 has SS_y times an *inverse chi-squared* distribution. Since the joint posterior factors, the conditional random variable $\mu|\sigma^2$ is independent of the random variable σ^2.

17.2 Finding the Posterior when Independent Jeffreys' Priors for μ and σ^2 Are Used

In Chapter 11 we saw that Jeffreys' prior for the normal mean μ is the improper flat prior

$$g_\mu(\mu) = 1 \quad \text{for} \quad -\infty < \mu < \infty$$

In Chapter 15 we observed that the Jeffreys' prior for σ^2 is

$$g_{\sigma^2}(\sigma^2) = \frac{1}{\sigma^2} \quad \text{for} \quad 0 < \sigma^2 < \infty$$

which also is improper. The joint prior for the two parameters when we are using independent Jeffreys' priors is their product which is given by

$$g_{\mu,\sigma^2}(\mu,\sigma^2) = \frac{1}{\sigma^2} \quad \text{for} \quad \begin{cases} -\infty < \mu < \infty, \\ 0 < \sigma^2 < \infty. \end{cases} \tag{17.2}$$

The joint posterior will be proportional to the joint prior times the joint likelihood given by

$$g_{\mu,\sigma^2}(\mu,\sigma^2|y_1,\dots,y_n) \propto g_{\mu,\sigma^2}(\mu,\sigma^2) \times f(y_1,\dots,y_n|\mu,\sigma^2)$$

$$\propto \frac{1}{\sigma^2} \times \frac{1}{(\sigma^2)^{\frac{1}{2}}} e^{-\frac{n}{2\sigma^2}(\bar{y}-\mu)^2} \times \frac{1}{(\sigma^2)^{\frac{n-1}{2}}} e^{-\frac{SS_y}{2\sigma^2}}$$

$$\propto \frac{1}{(\sigma^2)^{\frac{n}{2}+1}} e^{-\frac{1}{2\sigma^2}[n(\mu-\bar{y})^2+SS_y]}. \tag{17.3}$$

We see when we look at the joint posterior as a function of σ^2 as the only parameter that it is in the form of a constant $(n(\mu-\bar{y})^2 + SS_y)$ times an *inverse chi-squared* distribution with n degrees of freedom.

Finding the Marginal Posterior for μ

Usually we are interested in doing inference about the parameter μ and consider the variance σ^2 to be a nuisance parameter. The general way to eliminate a nuisance parameter is to marginalize it out of the joint posterior to find the marginal posterior of the parameter of interest. In this case, the marginal posterior for μ has shape given by

$$g_\mu(\mu|y_1,\dots,y_n) \propto \int_0^\infty g_{\mu,\sigma^2}(\mu,\sigma^2|y_1,\dots,y_n)\, d\sigma^2$$

$$\propto \int_0^\infty \frac{1}{(\sigma^2)^{\frac{n}{2}+1}} e^{-\frac{1}{2\sigma^2}[n(\mu-\bar{y})^2+SS_y]}\, d\sigma^2$$

$$\propto [n(\mu-\bar{y})^2 + SS_y]^{-\frac{n}{2}} \tag{17.4}$$

since we are integrating an *inverse chi-squared* density over its whole range. The details of evaluating the integral are found in the appendix at the end of the chapter. We change variables to

$$t = \frac{\mu - \bar{y}}{\sqrt{\frac{SS_y}{n(n-1)}}}.$$

Let the updated constants be $\kappa' = n - 1$, $n' = n$, $m' = \bar{y}$, and $S' = SS_y$. Then

$$t = \frac{\mu - m'}{\frac{\hat{\sigma}_B}{\sqrt{n'}}},$$

where $\hat{\sigma}_B^2 = \frac{SS_y}{\kappa'}$ is the unbiased estimator of the variance calculated from the sample. By the change of variable formula, the the density of t has shape given by

$$g(t) \propto g_\mu(\mu(t)) \times \frac{d\mu(t)}{dt}$$

$$\propto \left[1 + \frac{t^2}{n-1}\right]^{-\frac{n}{2}},$$

where we absorb the term $\frac{d\mu(t)}{dt}$ into the constant of proportionality. Thus we say that the marginal posterior distribution of μ given in Equation 17.4 has the $ST_{\kappa'}(m', \frac{\hat{\sigma}_B^2}{n'})$ distribution. It has the shape of a *Student's t* with κ' degrees of freedom and is centered at m' with spread parameter $\frac{\hat{\sigma}^2}{n'}$.

Another Way to Find the Marginal Posterior

In this case, there is an easier way to find the marginal posterior of μ that does not require us to integrate σ^2 out of the joint posterior. We make use of the following theorem:

Theorem 17.1 *If z and w are independent random variables having the* normal$(0, 1^2)$ *distribution and the chi-squared distribution with κ degrees of freedom respectively, then*

$$u = \frac{z}{\sqrt{\left(\frac{w}{\kappa}\right)}}$$

will have the Student's t distribution with κ degrees of freedom. In words, a normal random variable with mean 0 and variance 1 divided by the square root of an independent chi-squared random variable over its degrees of freedom will have the Student's t distribution.

A proof of the theorem is given in Mood, Graybill, and Boes (1974). We note that we can factor the joint posterior

$$g_{\mu,\sigma^2}(\mu, \sigma^2 | y_1, \ldots, y_n) \propto \frac{1}{(\sigma^2)^{\frac{1}{2}}} e^{-\frac{n}{2\sigma^2}[\mu - \bar{y}]^2} \times \frac{1}{(\sigma^2)^{\frac{n-1}{2}+1}} e^{-\frac{SS_y}{2\sigma^2}}$$

The conditional random variable $\mu | \sigma^2$ has the *normal* $(\bar{y}, \frac{\sigma^2}{n})$ distribution and the random variable σ^2 has the SS_y times an *inverse chi-squared* distribution with $n-1$ degrees of freedom, and the two components are independent. We know that given σ^2,

$$\frac{\mu - \bar{y}}{\sqrt{\frac{\sigma^2}{n}}} \quad \text{is} \quad normal(0, 1^2)$$

and

$$\frac{SS_y}{\sigma^2} \quad \text{is } chi\text{-}squared \text{ with } n - 1 \text{ df,}$$

so

$$t = \frac{\frac{\mu - \bar{y}}{\sqrt{\frac{\sigma^2}{n}}}}{\sqrt{\frac{SS_y}{\sigma^2(n-1)}}}$$

$$= \frac{\mu - m'}{\frac{\hat{\sigma}_B}{\sqrt{n'}}} \tag{17.5}$$

will have the *Student's* t distribution with κ' degrees of freedom, where $\hat{\sigma}_B^2 = \frac{SS_y}{\kappa'}$ is the sample estimator of the variance. This means that the posterior density of $\mu = m' + \frac{\hat{\sigma}_B}{\sqrt{n'}} \times t$ has the $ST_{\kappa'}(m', \frac{\hat{\sigma}_B^2}{n'})$ distribution. This is the same result we found previously by integrating out the nuisance parameter, σ^2.

This means we can do the inferences for μ treating the unknown variance σ^2 as if it had the value $\hat{\sigma}_B^2$ but using the *Student's* t table instead of the standard normal table. We see that the same rule we suggested as an approximation in Chapter 11 holds exactly.

17.3 Finding the Posterior when a Joint Conjugate Prior for μ and σ^2 Is Used

In Chapter 11 we found that the conjugate prior for μ, the mean of a normal observation with known variance σ^2, is the $normal(m, s^2)$ prior distribution. In Chapter 15 we found that the conjugate prior for σ^2, the variance of a normal observation with known mean μ, is S times an *inverse chi-squared* with κ degrees of freedom. We might think that the joint conjugate prior for both parameters μ and σ^2 of a normal observation would be the product of the independent conjugate priors for each parameter. However, this is not the case. The product of independent conjugate priors is a perfectly acceptable prior, but it is not jointly conjugate. If we used that prior[4], then there is no exact formula for the posterior that can be found by simple updating rules. Instead, the posterior would have to be found numerically. Later, in Chapter 20, we will see how we can draw random samples from this posterior using the computational Bayesian approach to inference. In this section, we will see what form the actual joint conjugate prior takes, and how to do inference when we use it.

The Joint Conjugate Prior

The joint conjugate prior must have the same form as the joint likelihood function found in Equation 17.1. We saw it is the product of a part that only depends on σ^2 which we recognize has the shape of SS_y times an *inverse chi-squared* distribution and a part for μ given σ^2 which we recognize as having the shape of a $normal(\bar{y}, \frac{\sigma^2}{n})$. The joint prior will have this same form. It is the product of S times an *inverse chi-squared* distribution with κ df for σ^2 times a $normal(m, \frac{\sigma^2}{n_0})$ distribution for μ given σ^2. We can think of n_0 as the prior sample size for our prior for μ. It represents the sample size of normal observations that would have the same precision as our prior belief about μ. The joint conjugate prior is given by

$$g_{\mu,\sigma^2}(\mu,\sigma^2) \propto \frac{1}{(\sigma^2)^{\frac{1}{2}}} e^{-\frac{n_0}{2\sigma^2}(\mu-m)^2} \times \frac{1}{(\sigma^2)^{\frac{\kappa}{2}+1}} e^{-\frac{S}{2\sigma^2}} \tag{17.6}$$

for $-\infty < \mu < \infty$ and $0 < \sigma^2 < \infty$. The joint posterior will be proportional to the joint prior times the joint likelihood. Its shape is given by

$$g_{\mu,\sigma^2}(\mu,\sigma^2|y_1,\ldots,y_n) \propto g_{\mu,\sigma^2}(\mu,\sigma^2) \times f(y_1,\ldots,y_n|\mu,\sigma^2)$$

$$\propto \frac{1}{(\sigma^2)^{\frac{1}{2}}} e^{-\frac{n_0}{2\sigma^2}(\mu-m)^2} \frac{1}{(\sigma^2)^{\frac{\kappa}{2}+1}} e^{-\frac{S}{2\sigma^2}}$$

$$\times \frac{1}{(\sigma^2)^{\frac{1}{2}}} e^{-\frac{n}{2\sigma^2}(\bar{y}-\mu)^2} \frac{1}{(\sigma^2)^{\frac{n-1}{2}}} e^{-\frac{SS_y}{2\sigma^2}}$$

$$\propto \frac{1}{(\sigma^2)^{\frac{\kappa'+1}{2}+1}} e^{-\frac{1}{2\sigma^2}[n'(\mu-m')^2+S'+\left(\frac{n_0 n}{n_0+n}\right)(\bar{y}-m)^2]},$$

$$\tag{17.7}$$

---

[4]It has the shape of a mixture of *Student's t* distributions with different degrees of freedom.

where $S' = S + SS_y$ and $\kappa' = \kappa + n$ and $m' = \frac{n\bar{y}+n_0 m}{n+n_0}$ and $n' = n_0 + n$ are the updated constants.

Finding the Marginal Posterior for μ

We find the marginal posterior of μ by marginalizing σ^2 out of the joint posterior.

$$g_\mu(\mu|y_1,\ldots,y_n) \propto \int_0^\infty g_{\mu,\sigma^2}(\mu,\sigma^2|y_1,\ldots,y_n)\,d\sigma^2$$

$$\propto \int_0^\infty \frac{1}{(\sigma^2)^{\frac{\kappa'+1}{2}+1}} e^{-\frac{1}{2\sigma^2}[n'(\mu-m')^2+S'+\left(\frac{n_0 n}{n_0+n}\right)(\bar{y}-m)^2]}\,d\sigma^2 \,.$$

We are integrating a constant times an inverse chi-squared density over its whole range. Thus the conditional posterior of μ given σ^2 has shape given by

$$g_\mu(\mu|y_1,\ldots,y_n) \propto [n'(\mu-m')^2+S'+\left(\frac{n_0 n}{n_0+n}\right)(\bar{y}-m)^2]^{-\frac{\kappa'+1}{2}} \,.$$

Suppose we change the variables to

$$t = \frac{\mu-m'}{\sqrt{\frac{S'+\left(\frac{n_0 n}{n_0+n}\right)(\bar{y}-m)^2}{n'\kappa'}}} = \frac{\mu-m'}{\hat{\sigma}_B/\sqrt{n'}},$$

where

$$\hat{\sigma}_B^2 = \frac{S' + \left(\frac{n_0 n}{n_0+n}\right)(\bar{y}-m)^2}{\kappa'}$$

$$= \frac{S + SS_y + \left(\frac{n_0 n}{n_0+n}\right)(\bar{y}-m)^2}{\kappa'}$$

$$= \left(\frac{\kappa}{\kappa'}\right)\left(\frac{S}{\kappa}\right) + \left(\frac{n-1}{\kappa'}\right)\left(\frac{SS_y}{n-1}\right)$$

$$+ \left(\frac{1}{\kappa'}\right)\left(\frac{n_0 n}{n_0+n}\right)(\bar{y}-m)^2$$

is a weighted average of three estimates of the variance. The first incorporates the prior distribution of σ^2, the second is the unbiased estimator of the variance from the sample data, and the third measures the distance the sample mean, $\bar{y}$, is from its prior mean, m.

We apply the change of variable formula. The posterior density of t will be

$$g(t|y_1,\ldots,y_n) \propto \left(\frac{t^2}{\kappa'}+1\right)^{-\frac{\kappa'+1}{2}}.$$

This is the density of the *Student's* t distribution with κ' degrees of freedom. This means the marginal posterior density of μ is $ST_{\kappa'}(m',\frac{\hat{\sigma}_B^2}{n'})$. It has the shape of a *Student's* t with κ' df and is centered at m' with spread parameter $\frac{\hat{\sigma}_B}{n'}$. Again, we can do the inference treating the unknown variance σ^2 as if it had the value $\hat{\sigma}_B^2$ but using the *Student's* t table to find the critical values instead of the standard normal table. The third term in the formula for $\hat{\sigma}_B^2$ shows that a misspecified prior mean inflates the spread of the posterior distribution. This may disguise the real problem which is the misspecified prior mean.

An Approximation to the Marginal Posterior for μ

We saw in Equation 17.1 that the joint likelihood factors into a *normal* part for μ conditional on σ^2 times a scaled *inverse chi-squared* part for σ^2. In Equation 17.5 we see the joint prior factors similarly. We can combine the conditional *normal* prior and conditional *normal* likelihood to get a conditional *normal* posterior for μ given σ^2. Similarly, we combine the scaled *inverse chi-squared* prior and the scaled *inverse chi-squared* likelihood for σ^2 to give a scaled *inverse chi-squared* posterior for σ^2. Thus the joint posterior factors into a product of a conditional *normal*$(m', \frac{(\sigma^2)^2)}{n'})$ posterior for $\mu|\sigma^2$ times S' times an *inverse chi-squared* with κ' degrees of freedom posterior of σ^2 where this time the degree of freedom constant is updated by $\kappa' = \kappa + n - 1$ while the others are updated as before.

$$g_{\mu,\sigma^2}(\mu,\sigma^2|y_1,\ldots,y_n) \propto \frac{1}{(\sigma^2)^{\frac{1}{2}}} e^{-\frac{n'}{2\sigma^2}(\mu-m')^2} \frac{1}{(\sigma^2)^{\frac{\kappa'}{2}+1}} e^{-\frac{S'}{2\sigma^2}}.$$

Because the joint posterior factors the two components are independent, so from Theorem 1

$$t = \frac{\mu - m'}{\sqrt{\frac{S'}{n'\kappa'}}}$$

$$= \frac{\mu - m'}{\frac{\hat{\sigma}_B}{\sqrt{n'}}} \tag{17.8}$$

will have the *Student's t* distribution with κ' degrees of freedom where

$$\hat{\sigma}_B^2 = \frac{S'}{\kappa'}$$

$$= \frac{S + SS_y}{\kappa'}$$

$$= \left(\frac{\kappa}{\kappa'}\right)\left(\frac{S}{\kappa}\right) + \left(\frac{n-1}{\kappa'}\right)\frac{SS_y}{n-1}$$

is the weighted average of two estimates of the variance. The first incorporates the estimate from the prior, and the second is the maximum likelihood estimator of the variance. This means the posterior density of μ is $ST_{\kappa'}(m', \frac{\hat{\sigma}_B^2}{n'})$. Again, this shows that if we do the inference treating the unknown variance σ^2 as if it had the value $\hat{\sigma}_B^2$ but using the *Student's t* table to find the critical values instead of the standard normal table the results are correct. It is an exact result, but it is not the full Bayesian posterior since it doesn't use all the information in the prior. That is why we refer to it as an approximation to the posterior.

When we compare the approximation with the exact result, we see the variance estimate σ_B^2 leaves out the term

$$\left(\frac{n_0 n}{n_0 + n}\right)(\bar{y} - m)^2$$

so it will be smaller. However, it is based on one less degree of freedom, so the credible intervals usually will be quite similar. All three cases (independent Jeffreys' prior, joint conjugate prior exact posterior, and joint conjugate prior approximate posterior) that we have looked at have similar *Student's t* formula

$$t = \frac{\mu - m'}{\frac{\hat{\sigma}_B}{\sqrt{n'}}}$$

where the conjugate prior constants are updated according to Table 17.1.

O'Hagan and Forster (2004) suggest the joint conjugate prior is too restrictive. If the sample mean is far from the prior mean, then this is interpreted as evidence the variance should be larger than suggested by its prior rather than giving evidence that the mean model is wrongly specified. We should graph the prior, likelihood, and posterior for μ conditional on $\sigma^2 = \hat{\sigma}_B^2$ to help us decide if the prior mean model is satisfactory. If these graphs look similar to Figure 16.2, then it indicates the (conditional) mean model is misspecified. If so, then a mixture model similar to those we discussed in Chapter 16 would be better.

Table 17.1   Updating joint conjugate prior constants for $normal(\mu, \sigma^2)$ when both parameters are unknown

Prior	S'	κ'	n'	m'	$\hat{\sigma}_B^2$
Jeffreys'	SS_y	$n-1$	n	$\bar{y}$	$\frac{S'}{\kappa'}$
Exact	$S + SS_y$	$\kappa + n$	$n_0 + n$	$\frac{n_0 m + n\bar{y}}{n_0 + n}$	$\frac{n}{\kappa'}\frac{S}{\kappa} + \frac{n-1}{\kappa'}\frac{SS_y}{n-1}$ $+ \frac{1}{\kappa'}\frac{n_0 n}{n_0 + n}(\bar{y} - m)^2$
approx.	$S + SS_y$	$\kappa + n - 1$	$n_0 + n$	$\frac{n_0 m + n\bar{y}}{n_0 + n}$	$\frac{n}{\kappa'}\frac{S}{\kappa} + \frac{n-1}{\kappa'}\frac{SS_y}{n-1}$

◼ EXAMPLE 17.1

Amber, Brett, and Chandra want to determine a 95% credible interval for the mean moisture content in a cheese product. They take a sample of 25 and measure the moisture content. The measurements are:

45.6	41.1	44.5	44.0	40.6	44.1	39.0	39.5	39.5	41.7
42.5	42.7	42.1	42.4	44.8	41.0	39.9	43.9	41.3	45.1
42.0	38.5	42.6	43.8	43.0					

For these measurements $\bar{y} = 42.208$ and $SS_y = 95.618$. They decide that the moisture level is normally distributed where both the mean μ and the variance σ^2 are not known. Amber decides she will use independent Jeffreys' priors for μ and σ^2. Brett believes the standard deviation is equally likely to be above or below 3 so its prior median is 3. He decides to use one degree of freedom. He finds $S = .4549 \times 3^2$ so his prior for σ^2 is $S = 4.094$ times an inverse chi-squared with 1 degree of freedom, He decides his prior for μ given σ^2 is normal with prior mean $m = 40$ and prior sample size $n_0 = 1$. He will use the exact solution. Chandra decides she will use the same prior as Brett, but will find the approximate solution.

Person	Prior parameters				Posterior parameters			
	S	κ	m	n_0	S'	κ'	n'	m'
Amber	na	na	na	na	SS_y 95.618	$n-1$ $= 24$	n $= 25$	$\bar{y}$ $= 42.208$
Brett (exact)	4.094	1	40	1	$S + SS_y$ 99.712	$\kappa + n$ $= 26$	$n_0 + n$ $= 26$	$\frac{n\bar{y} + n_0 m}{n + n_0}$ $= 42.12$
Chandra (approx)	4.094	1	40	4	$S + SS_y$ 99.712	$\kappa + n - 1$ $= 25$	$n_0 + n$ $= 26$	$\frac{n\bar{y} + n_0 m}{n + n_0}$ $= 42.12$

Amber's marginal posterior distribution for μ will be $ST(42.208, .3992^2)$ with 24 degrees of freedom. Brett's marginal posterior distribution for μ will be $ST(42.12, .3920^2)$ with 26 degrees of freedom. Chandra's marginal posterior for μ will be $ST(42.12, .3994^2)$ with 25 degrees of freedom. Amber's, Brett's and Chandra's priors, likelihoods, and posteriors for μ given their respective values of $\hat{\sigma}_B^2$ are shown in Figure 17.1, Figure 17.2, and Figure 17.3 respectively. ∎

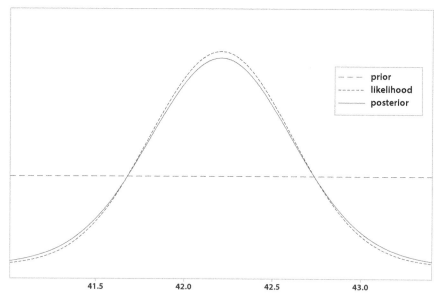

Figure 17.1   Amber's prior, likelihood and posterior distributions conditional on $\hat{\sigma}_B$.

17.4   Difference Between Normal Means with Equal Unknown Variance

Suppose we have independent random samples from two *normal* distributions having the same unknown variance σ^2, but the two distributions have different means. Let $\mathbf{y_1} = (y_{11}, \ldots, y_{1n_1})$ be the first sample which comes from a $Normal(\mu_1, \sigma^2)$ and let $\mathbf{y_2} = (y_{21} \ldots, y_{2n_2})$ be the second sample which comes from a $normal(\mu_2, \sigma^2)$. Since the two random samples are independent of each other, the joint likelihood is the product of the likelihoods of each sample. Using Equation 17.1 for the likelihood of each sample, the joint likelihood is

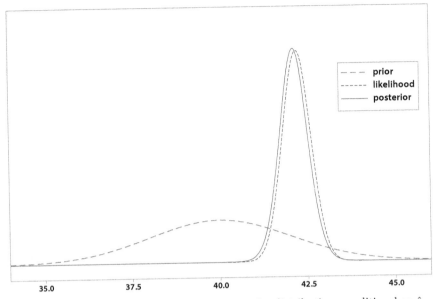

Figure 17.2   Brett's prior, likelihood and posterior distributions conditional on $\hat{\sigma}_B$.

given by

$$f(\mathbf{y_1}, \mathbf{y_2}|\mu_1, \mu_2, \sigma^2) \propto f(\mathbf{y_1}|\mu_1, \sigma^2) \times f(\mathbf{y_2}|\mu_2, \sigma^2)$$

$$\propto \frac{1}{(\sigma^2)^{\frac{1}{2}}} e^{-\frac{n_1}{2\sigma^2}(\bar{y}_1 - \mu_1)^2} \times \frac{1}{(\sigma^2)^{\frac{n_1-1}{2}}} e^{-\frac{SS_1}{2\sigma^2}}$$

$$\times \frac{1}{(\sigma^2)^{\frac{1}{2}}} e^{-\frac{n_2}{2\sigma^2}(\bar{y}_2 - \mu_2)^2} \times \frac{1}{(\sigma^2)^{\frac{n_2-1}{2}}} e^{-\frac{SS_2}{2\sigma^2}}$$

$$\propto \frac{1}{(\sigma^2)^{\frac{1}{2}}} e^{-\frac{n_1}{2\sigma^2}(\bar{y}_1 - \mu_1)^2} \times \frac{1}{(\sigma^2)^{\frac{1}{2}}} e^{-\frac{n_2}{2\sigma^2}(\bar{y}_2 - \mu_2)^2}$$

$$\times \frac{1}{(\sigma^2)^{\frac{n_1+n_2}{2}-1}} e^{-\frac{SS_p}{2\sigma^2}}, \tag{17.9}$$

where

$$SS_p = SS_1 + SS_2 = \sum_{i=1}^{n_1}(y_{1i} - \bar{y}_1)^2 + \sum_{j=1}^{n_2}(y_{2j} - \bar{y}_2)^2$$

is the pooled sum of squares.

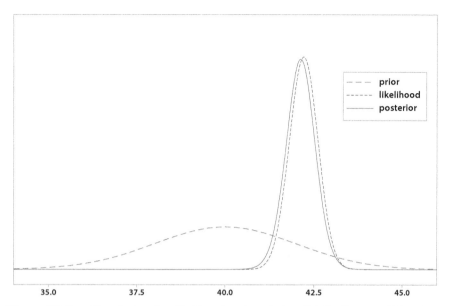

Figure 17.3   Chandra's prior, likelihood and posterior distributions conditional on $\hat{\sigma}_B$.

Finding the Posterior when Independent Jeffreys' Priors are Used for all Parameters

In this case, the joint prior is

$$g_{\mu_1,\mu_2,\sigma^2}(\mu_1, \mu_2, \sigma^2) = \frac{1}{\sigma^2} \quad \text{for} \quad \begin{cases} -\infty < \mu_1 < \infty, \\ -\infty < \mu_2 < \infty, \\ 0 < \sigma^2 < \infty. \end{cases} \tag{17.10}$$

The joint posterior is proportional to the joint prior times the joint likelihood given by

$$g_{\mu_1,\mu_2,\sigma^2}(\mu_1, \mu_2, \sigma^2 | \mathbf{y_1 y_2}) \propto g_{\mu_1,\mu_2,\sigma^2}(\mu_1, \mu_2, \sigma^2) \times f(\mathbf{y_1}, \mathbf{y_2} | \mu_1, \mu_2, \sigma^2)$$

$$\propto \frac{1}{\sigma^2} \times \frac{1}{(\sigma^2)^{\frac{1}{2}}} e^{-\frac{n_1}{2\sigma^2}(\bar{y}_1 - \mu_1)^2} \times \frac{1}{(\sigma^2)^{\frac{1}{2}}} e^{-\frac{n_2}{2\sigma^2}(\bar{y}_2 - \mu_2)^2}$$

$$\times \frac{1}{(\sigma^2)^{\frac{n_1+n_2}{2}-1}} e^{-\frac{SS_p}{2\sigma^2}}$$

$$\propto \frac{1}{(\sigma^2)^{\frac{1}{2}}} e^{-\frac{n_1(\bar{y}_1 - \mu_1)^2}{2\sigma^2}} \times \frac{1}{(\sigma^2)^{\frac{1}{2}}} e^{-\frac{n_2(\bar{y}_2 - \mu_2)^2}{2\sigma^2}}$$

$$\times \frac{1}{(\sigma^2)^{\frac{n_1+n_2}{2}}} e^{-\frac{SS_p}{2\sigma^2}}. \tag{17.11}$$

We recognize this as a product of two normal distributions for μ_1 and μ_2 respectively, given σ^2, times an SS_p times an *inverse chi-squared* with $n_1 +$

$n_2 - 2$ degrees of freedom for σ^2. Since the joint posterior factors, the parts are all independent of each other. Let $\mu_d = \mu_1 - \mu_2$ be the difference between the means and let $m'_d = \bar{y}_1 - \bar{y}_2$ be the difference between sample means. Given σ^2, the posterior distributions of the two means are independent normal random variables. Therefore given σ^2, the posterior distribution of μ_d is $normal\left[m'_d, \sigma^2\left(\frac{1}{n_1} + \frac{1}{n_2}\right)\right]$, so the shape of the joint posterior of $\mu_d|\sigma^2$ and σ^2 is given by

$$g_{\mu_d,\sigma^2}(\mu_d, \sigma^2|\mathbf{y_1}, \mathbf{y_2}) \propto \frac{1}{\left(\sigma^2\left(\frac{n_1 n_2}{n_1+n_2}\right)\right)^{\frac{1}{2}}} e^{-\frac{n_1 n_2(\mu_d - (m'_d))^2}{2\sigma^2(n_1+n_2)}}$$

$$\times \frac{1}{(\sigma^2)^{\frac{n_1+n_2}{2}}} e^{\frac{-SS_p}{2\sigma^2}}$$

$$\propto \frac{1}{(\sigma^2)^{\frac{n_1+n_2+1}{2}}} e^{-\frac{n_1 n_2(\mu_d - m'_d)^2/(n_1+n_2)+SS_p}{2\sigma^2}}, \qquad (17.12)$$

where we absorb $\left(\frac{n_1 n_2}{n_1+n_2}\right)$ into the proportionality constant. We recognize this as a function of σ^2 to be the shape of $\left(\frac{n_1 n_2}{n_1+n_2}\right)(\mu_d - m'_d)^2 + SS_p$ times an *inverse chi-squared* density with $n_1 + n_2 - 1$ degrees of freedom.

The marginal posterior density for μ_d is found by integrating the nuisance parameter σ^2 out of the joint posterior. We are integrating a constant times an *inverse chi-squared* density over its whole range, so its shape will be given by

$$g_{\mu_d}(\mu_d) \propto \int_0^\infty \frac{1}{(\sigma^2)^{\frac{n_1+n_2-1}{2}+1}} e^{-\frac{1}{2\sigma^2}\left[\left(\frac{n_1 n_2}{n_1+n_2}\right)(\mu_d - m'_d)^2 + SS_p\right]} d\sigma^2$$

$$\propto \left[\left(\frac{n_1 n_2}{n_1 + n_2}\right)(\mu_d - m'_d)^2 + SS_p\right]^{-\frac{n_1+n_2-1}{2}}$$

$$\propto \left[\left(\frac{n_1 n_2}{n_1 + n_2}\right)(\mu_d - m'_d)^2 + S'\right]^{-\frac{\kappa'+1}{2}},$$

where the updated conjugate constants are $\kappa' = n_1 + n_2 - 2$ and $S' = SS_p$. We change variables to

$$t = \frac{\mu_d - m'_d}{\sqrt{\frac{S'(n_1+n_2)}{n_1 n_2 (\kappa')}}}$$

$$= \frac{\mu_d - m'_d}{\sqrt{\hat{\sigma}_B^2 \left(\frac{1}{n_1} + \frac{1}{n_2}\right)}},$$

where

$$\hat{\sigma}_B^2 = \frac{S'}{\kappa'}$$

$$= \frac{S}{\kappa}\frac{\kappa}{\kappa'+1} + \frac{SS_p}{n_1+n_2-2}\frac{n_1+n_2-2}{\kappa+1}.$$

The shape of the density of t is

$$g(t) \propto \left(1 + \frac{t^2}{\kappa'}\right)^{-\frac{\kappa'+1}{2}},$$

which is the *Student's t* density with κ' degrees of freedom. Hence, the marginal posterior of μ_d is $ST(m'_d, \frac{\hat{\sigma}_B^2}{\kappa'})$ with κ' degrees of freedom.

On the other hand, we could show this using Theorem 1. We see $\mu_d | \sigma^2$ is $normal[m'_d, \sigma^2\left(\frac{n_1+n_2}{n_1 n_2}\right)]$ and σ^2 has SS_p times an *inverse chi-squared* distribution with $n_1 + n_2 - 2$ degrees of freedom, and the two components are independent. Hence

$$t = \frac{\mu_d - (m'_d)}{\left[\frac{SS_p}{n_1+n_2-2}\left(\frac{n_1+n_2}{n_1 n_2}\right)\right]^{\frac{1}{2}}}$$

$$= \frac{\mu_d - (m'_d)}{\left[\frac{S'}{\kappa'}\left(\frac{n_1+n_2}{n_1 n_2}\right)\right]^{\frac{1}{2}}}$$

$$= \frac{\mu_d - (m'_d)}{\left[\hat{\sigma}_B^2 \left(\frac{n_1+n_2}{n_1 n_2}\right)\right]^{\frac{1}{2}}} \tag{17.13}$$

has a *Student's t* distribution with κ' degrees of freedom. Again, this shows the difference between the means, μ_d, has the $ST_{\kappa'}\left[m'_d, \hat{\sigma}_B^2 \left(\frac{1}{n_1} + \frac{1}{n_2}\right)\right]$ distribution. This shows that the rule we suggested in Chapter 13 as an approximation (estimate the unknown variance by the pooled variance from the two samples, then get critical values from *Student's t* table instead of the normal table) holds true exactly.

Finding the Exact Posterior when the Joint Conjugate Prior Is Used for All Parameters

The product of independent conjugate priors for each parameter will not be jointly conjugate for all the parameters together just as in the single sample case. In this case, we saw that the form of the joint likelihood is a product of two normal densities for μ_1 and μ_2 respectively, each conditional on σ^2, multiplied by SS_p times an *inverse chi-squared* density with κ degrees of freedom for σ^2. The joint conjugate prior will have the same form as the joint likelihood and is given by

$$g_{\mu_1,\mu_2,\sigma^2}(\mu_1,\mu_2,\sigma^2) \propto \frac{1}{(\sigma^2)^{\frac{1}{2}}} e^{-\frac{n_{10}}{2\sigma^2}(\mu_1-m_1)^2} \times \frac{1}{(\sigma^2)^{\frac{1}{2}}} e^{-\frac{n_{20}}{2\sigma^2}(\mu_2-m_2)^2}$$

$$\times \frac{1}{(\sigma^2)^{\frac{\kappa}{2}+1}} e^{-\frac{S}{2\sigma^2}}, \tag{17.14}$$

where m_1 and m_2 are the prior means, n_{10} and n_{20} are the equivalent sample sizes for the respective *normal* μ_1 and μ_2 priors conditional on σ^2, S is the prior multiplicative constant and κ is the prior degrees of freedom for σ^2. The joint posterior is proportional to the product of the joint prior times the joint likelihood. It is given by

$$g_{\mu_1,\mu_2,\sigma^2}(\mu_1,\mu_2,\sigma^2|\mathbf{y_1},\mathbf{y_2}) \propto \frac{1}{(\sigma^2)^{\frac{1}{2}}} e^{-\frac{n_{10}}{2\sigma^2}(\mu_1-m_1)^2} \times \frac{1}{(\sigma^2)^{\frac{1}{2}}} e^{-\frac{n_{20}}{2\sigma^2}(\mu_2-m_2)^2}$$

$$\times \frac{1}{(\sigma^2)^{\frac{\kappa}{2}+1}} e^{-\frac{S}{2\sigma^2}} \times \frac{1}{(\sigma^2)^{\frac{1}{2}}} e^{-\frac{n_1}{2\sigma^2}(\bar{y}_1-\mu_1)^2}$$

$$\times \frac{1}{(\sigma^2)^{\frac{1}{2}}} e^{-\frac{n_2}{2\sigma^2}(\bar{y}_2-\mu_2)^2} \times \frac{1}{(\sigma^2)^{\frac{n_1+n_2}{2}-1}} e^{-\frac{SS_p}{2\sigma^2}}$$

$$\propto \frac{1}{(\sigma^2)^{\frac{1}{2}}} e^{-\frac{n_1'}{2\sigma^2}(\mu_1-m_1')^2} \times \frac{1}{(\sigma^2)^{\frac{1}{2}}} e^{-\frac{n_2'}{2\sigma^2}(\mu_2-m_2')^2}$$

$$\times \frac{1}{(\sigma^2)^{\frac{\kappa'}{2}+1}} e^{\frac{1}{2\sigma^2}\left(S'+\frac{n_{10}n_1(\bar{y}_1-m_1)^2}{n_{10}+n_1}+\frac{n_{20}n_2(\bar{y}_2-m_2)^2}{n_{20}+n_2}\right)}$$

$$\tag{17.15}$$

where

$$\kappa' = \kappa + n_1 + n_2, \quad n_1' = n_1 + n_{10}, \quad n_2' = n_2 + n_{20}, \quad S' = S + SS_p,$$

$$m_1' = \frac{n_1 \bar{y}_1 + n_{10} m_1}{n_1 + n_{10}}, \text{ and } m_2' = \frac{n_2 \bar{y}_2 + n_{20} m_2}{n_2 + n_{20}}.$$

Since the joint posterior factors, we see that the posterior distributions of μ_1 and μ_2, each conditional on σ^2, are normal distributions and that the posterior distribution of σ^2 is $S' + c_1 + c_2$ times an *inverse chi-squared* with κ' degrees of freedom, where

$$c_i = \frac{n_{i0} n_i}{n_{i0} + n_i} (\bar{y}_i - m_i)^2, \quad i = 1, 2.$$

Therefore given σ^2, the distribution of $\mu_d = \mu_1 - \mu_2$ is *normal* with mean $m_d' = m_1' - m_2'$ and variance equal to $\sigma^2(\frac{1}{n_1'} + \frac{1}{n_2'}))$. The joint posterior of μ_d and σ^2 is given by

$$g_{\mu_d, \sigma^2}(\mu_d, \sigma^2 | \mathbf{y_1}, \mathbf{y_2}) \propto \frac{1}{(\sigma^2)^{\frac{1}{2}}} e^{-\frac{1}{2\sigma^2} \left(\frac{n_1' n_2'}{n_1' + n_2'} \right)(\mu_d - m_d')^2}$$

$$\times \frac{1}{(\sigma^2)^{\frac{\kappa'}{2}+1}} e^{\frac{1}{2\sigma^2}(S' + c_1 + c_2)}. \tag{17.16}$$

The marginal posterior of μ_d is found by integrating σ^2 out of the joint posterior and is

$$g_{\mu_d}(\mu_d | \mathbf{y_1}, \mathbf{y_2}) \propto \int_0^\infty \frac{1}{(\sigma^2)^{\frac{\kappa'+1}{2}+1}}$$

$$\times e^{\frac{1}{2\sigma^2} \left(\frac{n_1' n_2'}{n_1' n_2'}(\mu_d - m_d')^2 + S' + c_1 + c_2 \right)} d\sigma^2$$

$$\propto \left[\frac{n_1' n_2'}{n_1' + n_2'}(\mu_d - m_d')^2 + S' + c_1 + c_2 \right]^{-\frac{\kappa'+1}{2}}.$$

Let

$$t = \frac{\mu_d - m_d'}{\sqrt{\frac{n_1' + n_2'}{n_1' n_2' \kappa'}(S' + c_1 + c_2)}}$$

$$\propto \frac{\mu_d - m_d'}{\hat{\sigma}_B \sqrt{\frac{1}{n_1'} + \frac{1}{n_2'}}},$$

where

$$\hat{\sigma}_B^2 = \frac{S' + \frac{n_{10}n_1}{n_{10}+n_1}(\bar{y}_1 - m_1)^2 + \frac{n_{20}n_2}{n_{20}+n_2}(\bar{y}_2 - m_2)^2}{\kappa'}$$

$$= \frac{\kappa}{\kappa'}\frac{S}{\kappa} + \frac{n_1 + n_2 - 2}{\kappa'}\frac{SS_p}{n_1 + n_2 - 2} + \frac{1}{\kappa'}\frac{n_{10}n_1}{n_{10}+n_1}(\bar{y}_1 - m_1)^2$$
$$+ \frac{1}{\kappa'}\frac{n_{20}n_2}{n_{20}+n_2}(\bar{y}_2 - m_2)^2$$

is the weighted average of four estimates of the variance. The terms come from the prior for σ^2, the pooled estimate from the likelihood, and the distance the prior means are from their respective sample means. The last two terms increase the posterior variance due to misspecification of the prior means. This can increase the width of credible intervals for μ_d when the real problem may be the prior misspecification.

Finding the Approximate Posterior when the Joint Conjugate Prior Is Used for All Parameters

The joint posterior is proportional to the product of the joint prior times the joint likelihood and is given by

$$g_{\mu_1,\mu_2,\sigma^2}(\mu_1, \mu_2, \sigma^2 | \mathbf{y_1}, \mathbf{y_2}) \propto \frac{1}{(\sigma^2)^{\frac{1}{2}}}e^{-\frac{n_{10}}{2\sigma^2}(\mu_1 - m_1)^2} \times \frac{1}{(\sigma^2)^{\frac{1}{2}}}e^{-\frac{n_{20}}{2\sigma^2}(\mu_2 - m_2)^2}$$

$$\times \frac{1}{(\sigma^2)^{\frac{\kappa}{2}+1}}e^{-\frac{S}{2\sigma^2}} \times \frac{1}{(\sigma^2)^{\frac{1}{2}}}e^{-\frac{n_1}{2\sigma^2}(\bar{y}_1 - \mu_1)^2}$$

$$\times \frac{1}{(\sigma^2)^{\frac{1}{2}}}e^{-\frac{n_2}{2\sigma^2}(\bar{y}_2 - \mu_2)^2} \times \frac{1}{(\sigma^2)^{\frac{n_1+n_2}{2}-1}}e^{-\frac{SS_p}{2\sigma^2}}.$$

$$(17.17)$$

We combine each pair of conditional normal prior times likelihood and the inverse chi-squared prior times likelihood separately to get the approximate posterior

$$g_{\mu_1,\mu_2,\sigma^2}(\mu_1, \mu_2, \sigma^2 | \mathbf{y_1}, \mathbf{y_2}) \propto \frac{1}{(\sigma^2)^{\frac{1}{2}}}e^{-\frac{n_1'}{2\sigma^2}(\mu_1 - m_1')^2} \times \frac{1}{(\sigma^2)^{\frac{1}{2}}}e^{-\frac{n_2'}{2\sigma^2}(\mu_2 - m_2')^2}$$

$$\times \frac{1}{(\sigma^2)^{\frac{\kappa'}{2}+1}}e^{\frac{S'}{2\sigma^2}},$$

where the updated constants are

$$\kappa' = \kappa + n_1 + n_2 - 2, \quad n_1' = n_1 + n_{10}, \quad n_2' = n_2 + n_{20}, \quad S' = S + SS_p,$$

$$m_1' = \frac{n_1\bar{y}_1 + n_{10}m_1}{n_1 + n_{10}}, \text{ and } m_2' = \frac{n_2\bar{y}_2 + n_{20}m_2}{n_2 + n_{20}}.$$

Since the joint posterior factors, we see that the conditional normal posterior distributions of μ_1 and μ_2 given σ^2 are independent of each other and are independent of the posterior distribution of σ^2, which is S' times an *inverse chi-squared* with κ' degrees of freedom. Therefore, the distribution of $\mu_d = \mu_1 - \mu_2$ given σ^2 is *normal* with mean $m_d' = m_1' - m_2'$ and variance equal to $\sigma^2\left(\frac{1}{n_1'} + \frac{1}{n_2'}\right)$. The joint posterior of μ_d and σ^2 is given by

$$g_{\mu_d,\sigma^2}(\mu_d, \sigma^2 | \mathbf{y_1}, \mathbf{y_2}) \propto \frac{1}{(\sigma^2)^{\frac{1}{2}}} e^{-\frac{1}{2\sigma^2}\left(\frac{n_1'n_2'}{n_1'+n_2'}\right)(\mu_d - m_d')^2}$$

$$\times \frac{1}{(\sigma^2)^{\frac{\kappa'}{2}+1}} e^{\frac{1}{2\sigma^2}S'}, \tag{17.18}$$

The marginal posterior of μ_d is found by integrating σ^2 out of the joint posterior and is

$$g_{\mu_d}(\mu_d | \mathbf{y_1}, \mathbf{y_2}) \propto \int_0^\infty \frac{1}{(\sigma^2)^{\frac{\kappa'+1}{2}+1}} e^{\frac{1}{2\sigma^2}\left(\frac{n_1'n_2'}{n_1'n_2'}(\mu_d-m_d')^2 + S'\right)} d\sigma^2$$

$$\propto \left[\frac{n_1'n_2'}{n_1'+n_2'}(\mu_d - m_d')^2 + S'\right]^{-\frac{\kappa'+1}{2}}$$

Let

$$t = \frac{\mu_d - m_d'}{\sqrt{\frac{n_1'+n_2'}{n_1'n_2'\kappa'}(S')}}$$

$$\propto \frac{\mu_d - m_d'}{\hat{\sigma}_B \sqrt{\frac{1}{n_1'} + \frac{1}{n_2'}}},$$

where this time

$$\hat{\sigma}_B^2 = \frac{S'}{\kappa'}$$

$$= \frac{\kappa}{\kappa'}\frac{S}{\kappa} + \frac{n_1 + n_2 - 2}{\kappa'}\frac{SS_p}{n_1 + n_2 - 2}$$

is the weighted average of only two estimates of the variance. The terms come from the prior for σ^2 and the pooled estimate of the variance from the likelihood, respectively. Thus there is no inflation of the variance estimate due to misspecification of the mean.

◪ EXAMPLE 17.2

In Example 3.2 (Chapter 3, p. 40) we looked at two series of measurements Michelson made on the speed of light in 1879 and 1882 respectively. In Example 13.1 (Chapter 13, p. 258) we found a 95% Bayesian credible interval for $\mu_d = \mu_{1879} - \mu_{1882}$ under the assumptions that the known underlying variance $\sigma^2 = 100^2$ and we used a $normal(300000, 500^2)$ prior for each of the means. David, Esther, and Fiona decided to each fine a 95% Bayesian credible interval for μ_d where we assume the variance is unknown. David decides to use independent Jeffreys' priors for all the parameters. Esther and Fiona decide to use priors that are comparable to that used in Example 13.1 so comparisons can be made with that result. In Example 13.1 the standard deviation was assumed to be 100. Taking that as the median for the inverse chi-squared prior distribution with $\kappa = 1$ degree of freedom for the variance gives $S = 4549$ (See choosing an inverse chi-squared prior on page 322). The prior distributions for μ_{1879} and μ_{1882} were $normal(300000, 500^2)$. This gives $\frac{\sigma}{n_0} = 500$, so $n_0 = .04$. Note that the equivalent sample size does not have to be an integer. Esther decides to find the exact posterior, and Fiona decides to find the approximate one. Their results are given below.

Person	Posterior constants		95% credible interval	
(Method)	m_d	σ_d	lower	upper
David (Jeffreys)	152.783	32.441	(87.27,	218.30)
Esther (Exact)	152.541	31.531	(88.99,	216.09)
Fiona (Approx.)	152.541	32.180	(87.60,	217.48)

Comparing the credible intervals to the approximate one we found in Example 13.1 we see that these credible intervals are slightly wider. This is because in Example 13.1 we used the standard normal table to get the critical value because we assumed the standard deviation was known. In this example we used the more reasonable assumption that the standard deviation is not known. ∎

17.5 Difference Between Normal Means with Unequal Unknown Variances

In the case of a single sample from a *normal* distribution with unknown variance σ^2, and the case of two independent samples from *normal* distributions with equal unknown variances σ^2, we found that frequentist confidence intervals have the same form (although different interpretations) as Bayesian credible intervals when independent Jeffreys' priors are used for the mean(s) and the variance. One might think that this is true in general. However, the case of two independent samples from *normal* distributions with unknown unequal variances σ_1^2 and σ_2^2, respectively, will show that this supposition is not true in general. Let $\mathbf{y_1} = (y_{11}, \ldots, y_{1n_1})$ be a random sample from *normal*(μ_1, σ_1^2) where both parameters are unknown, let $\mathbf{y_2} = (y_{21}, \ldots, y_{2n_2})$ be a random sample from *normal*(μ_2, σ_2^2) where both parameters are unknown, and let the two samples be independent. We want to do inferences on the difference between the means, $\mu_1 - \mu_2$, and treat the unknown variances σ_1^2 and σ_2^2 as nuisance parameters. This is known as the Behrens Fisher problem. Fisher (1935) developed an approach known as fiducial inference which derived a probability distribution for the parameter from the sampling distribution of the statistic. Its success requires using a pivotal quantity.[5] These do not always exist which limits the application of fiducial inference. When it works, the fiducial approach gives similar results to the Bayesian approach using non-informative priors, which are widely applicable. However, Fisher denied that fiducial inference was in principal a Bayesian approach. The fiducial intervals found for this problem do not have the same form as the confidence interval should have and do not have the frequency interpretation of confidence intervals.[6] We will look at the Bayesian approach to the Behrens Fisher problem first proposed by Jeffreys (1961).

Since the two random samples are independent of each other, the joint likelihood is the product of the two likelihoods given by

$$f(\mathbf{y_1}, \mathbf{y_2} | \mu_1, \mu_2, \sigma_1^2, \sigma_2^2) \propto f(\mathbf{y_1} | \mu_1, \sigma_1^2) \times f(\mathbf{y_2} | \mu_2, \sigma_2^2).$$

We are using independent Jeffreys' priors so the joint prior is

$$g_{\mu_1, \mu_2, \sigma_1^2, \sigma_2^2}(\mu_1, \mu_2, \sigma_1^2, \sigma_2^2) \propto \frac{1}{\sigma_1^2} \times \frac{1}{\sigma_2^2}.$$

Since both the likelihood and the prior factor, the joint posterior

$$g_{\mu_1, \mu_2, \sigma_1^2, \sigma_2^2}(\mu_1, \mu_2, \sigma_1^2, \sigma_2^2 | \mathbf{y_1}, \mathbf{y_2}) \propto g_{\mu_1, \sigma_1^2}(\mu_1, \sigma_1^2 | \mathbf{y_1}) \times g_{\mu_2, \sigma_2^2}(\mu_2, \sigma_2^2 | , \mathbf{y_2})$$

---

[5] A function of the parameter and the statistic that does not depend on any unknown parameters.
[6] The fiducial intervals would have the (post-data) probability interpretation for this case similar to the Bayesian interpretation instead of the (pre-data) long-run frequency interpretation of confidence intervals.

also factors. It simplifies to

$$g_{\mu_1,\mu_2,\sigma_1^2,\sigma_2^2}(\mu_1,\mu_2,\sigma_1^2,\sigma_2^2|\mathbf{y_1},\mathbf{y_2}) \propto \frac{1}{(\sigma_1^2)^{\frac{1}{2}}}e^{-\frac{n_1}{2\sigma_1^2}(\bar{y}_1-\mu_1)^2}\frac{1}{(\sigma_1^2)^{\frac{n_1-1}{2}+1}}e^{\frac{SS_{y_1}}{2\sigma_1^2}}$$

$$\times \frac{1}{(\sigma_2^2)^{\frac{1}{2}}}e^{-\frac{n_2}{2\sigma_2^2}(\bar{y}_2-\mu_2)^2}\frac{1}{(\sigma_2^2)^{\frac{n_2-1}{2}+1}}e^{\frac{SS_{y_2}}{2\sigma_2^2}}.$$

This is the product of a *normal*$(\bar{y}_1,\frac{\sigma_1^2}{n_1})$ conditional distribution for μ_1 given σ_1^2, a *normal*$(\bar{y}_2,\frac{\sigma_2^2}{n_2})$ conditional distribution for μ_2 given σ_2^2, an SS_{y_1} times an *inverse chi-squared* distribution with n_1-1 for σ_1^2, and an SS_{y_2} times an *inverse chi-squared* distribution with n_2-1 for σ_2^2, respectively, and the components are all independent. By Theorem 1,

$$t_1 = \frac{\mu_1-\bar{y}_1}{\sqrt{\frac{\hat{\sigma}_1^2}{n_1}}} \text{ and } t_2 = \frac{\mu_2-\bar{y}_2}{\sqrt{\frac{\hat{\sigma}_2^2}{n_2}}},$$

where $\hat{\sigma}_1^2 = \frac{SS_{y_1}}{n_1-1}$ and $\hat{\sigma}_2^2 = \frac{SS_{y_2}}{n_2-1}$ will be independent *Student's t* random variables having n_1-1 and n_2-1 degrees of freedom respectively. Given σ_1^2 and σ_2^2, the distribution of $\mu_1-\mu_2$ will be *normal* $(\bar{y}_1-\bar{y}_2,\frac{\sigma_1^2}{n_1}+\frac{\sigma_2^2}{n_2})$. We want to find the marginal posterior distribution of $\mu_1-\mu_2$. Note

$$\tau_1 = \frac{\mu_1-\mu_2-(\bar{y}_1-\bar{y}_2)}{\sqrt{\frac{\hat{\sigma}_1^2}{n_1}+\frac{\hat{\sigma}_2^2}{n_2}}}$$

$$= \frac{\mu_1-\bar{y}_1}{\sqrt{\frac{\hat{\sigma}_1^2}{n_1}}} \times \frac{\sqrt{\frac{\hat{\sigma}_1^2}{n_1}}}{\sqrt{\frac{\hat{\sigma}_1^2}{n_1}+\frac{\hat{\sigma}_2^2}{n_2}}} + \frac{\mu_2-\bar{y}_2}{\sqrt{\frac{\hat{\sigma}_2^2}{n_2}}} \times \frac{\sqrt{\frac{\hat{\sigma}_2^2}{n_2}}}{\sqrt{\frac{\hat{\sigma}_1^2}{n_1}+\frac{\hat{\sigma}_2^2}{n_2}}}$$

$$= t_1\cos\hat{\phi} - t_2\sin\hat{\phi}$$

where $\hat{\phi}$ is the angle such that

$$\tan\hat{\phi} = \frac{\sqrt{\frac{\hat{\sigma}_1^2}{n_1}}}{\sqrt{\frac{\hat{\sigma}_2^2}{n_2}}}.$$

Thus the marginal posterior distribution of $\mu_1-\mu_2$ is a linear function of t_1 and t_2. Let

$$\tau_2 = t_1\sin\hat{\phi} + t_2\cos\hat{\phi}$$

Then

$$\begin{pmatrix} \tau_1 \\ \tau_2 \end{pmatrix} = \begin{pmatrix} \cos\hat{\phi} & \sin\hat{\phi} \\ -\sin\hat{\phi} & \cos\hat{\phi} \end{pmatrix} \begin{pmatrix} t_1 \\ t_2 \end{pmatrix}$$

The vector $\boldsymbol{\tau} = \begin{pmatrix} \tau_1 \\ \tau_2 \end{pmatrix}$ is a linear transformation[7] of the vector $\mathbf{t} = \begin{pmatrix} t_1 \\ t_2 \end{pmatrix}$. The joint posterior of $\boldsymbol{\tau}$ is

$$g(\tau_1, \tau_2 | \mathbf{y_1}, \mathbf{y_2}) = g(t_1(\tau_1, \tau_2), t_2(\tau_1, \tau_2) | \hat{\phi}, \mathbf{y_1}, \mathbf{y_2}) \times |J|,$$

where the Jacobian is

$$|J| = \begin{vmatrix} \frac{\partial t_1}{\partial \tau_1} & \frac{\partial t_1}{\partial \tau_2} \\ \frac{\partial t_2}{\partial \tau_1} & \frac{\partial t_2}{\partial \tau_2} \end{vmatrix} = \begin{vmatrix} \cos\hat{\phi} & \sin\hat{\phi} \\ -\sin\hat{\phi} & \cos\hat{\phi} \end{vmatrix} = 1.$$

The marginal posterior density of τ_1 can be found by integrating τ_2 out of their joint posterior. It has shape given by

$$g(\tau_1 | \mathbf{y_1}, \mathbf{y_2}) \propto \int_{-\infty}^{\infty} g(\tau_1, \tau_2 | \mathbf{y_1}, \mathbf{y_2}) \, d\tau_2$$

$$\propto \int_{-\infty}^{\infty} \left(1 + \frac{(\tau_1 \cos\hat{\phi} + \tau_2 \sin\hat{\phi})^2}{n_1 - 1}\right)^{-\frac{n_1}{2}}$$

$$\times \left(1 + \frac{(-\tau_1 \sin\hat{\phi} + \tau_2 \cos\hat{\phi})^2}{n_2 - 1}\right)^{-\frac{n_2}{2}} d\tau_2.$$

This distribution, known as the *Behrens Fisher* distribution, depends on the three constants n_1, n_2, and $\hat{\phi}$. The critical values for the *Behrens Fisher* distribution can be calculated numerically, but no closed form exists. It is symmetric about 0 like a *Student's t* and it has tail weight similar to a *Student's t*, but it is not exactly a *Student's t*. Fisher (1935) used $\hat{\phi}$, the ratio calculated from the sample variances as if it was the true value ϕ. Welch (1938) used Satterthwaite's approximation to find the degrees of freedom for a *Student's t* distribution that gives the closest match to the *Behrens Fisher* distribution.

---

[7]This transformation is a rotation.

Main Points

Normal Distribution with Both Parameters Unknown

- For a $normal(\mu, \sigma^2)$ with both parameters unknown, the joint likelihood factors into a part that only depends on σ^2 times a part that depends on μ given σ^2. The likelihood has shape given by

$$f(y_1, \ldots, y_n | \mu, \sigma^2) \propto \frac{1}{(\sigma^2)^{\frac{1}{2}}} e^{-\frac{n}{2\sigma^2}(\bar{y}-\mu)^2} \times \frac{1}{(\sigma^2)^{\frac{n-1}{2}}} e^{-\frac{SS_y}{2\sigma^2}}.$$

- When we use independent Jeffreys' priors for both parameters, the joint posterior is

$$g_{\mu,\sigma^2}(\mu, \sigma^2 | y_1, \ldots, y_n) \propto \frac{1}{(\sigma^2)^{\frac{1}{2}}} e^{-\frac{n}{2\sigma^2}(\mu-\bar{y})^2} \times \frac{1}{(\sigma^2)^{\frac{n-1}{2}+1}} e^{-\frac{SS_y}{2\sigma^2}}.$$

We find the marginal posterior distribution of μ by integrating the variance σ^2 out of the joint posterior. The marginal posterior of μ is $ST_{\kappa'}(m', \frac{\hat{\sigma}_B^2}{n'})$, where $\kappa' = n - 1$, $n' = n$, $m' = \bar{y}$, and $\hat{\sigma}_B^2 = \frac{SS_y}{\kappa'}$ is the sample variance. It has the shape of a *Student's t* with κ' degrees of freedom and is entered about m' and has spread $\frac{\sigma_B^2}{n'}$. This means we can do the inferences for μ treating the unknown variance σ^2 as if it had the value $\hat{\sigma}^2$ but using the *Student's t* table instead of the standard normal table. In this case the rule we suggested as an approximation in Chapter 11 holds exactly.

- The joint conjugate prior is not a product of independent joint priors. It has the same form as the joint likelihood. It is the product of an S times an *inverse chi-squared* prior with κ degrees of freedom for σ^2 and a conditional $normal(m, \frac{\sigma^2}{n_0})$ prior for μ given σ^2.

- The joint posterior is given by

$$g_{\mu,\sigma^2}(\mu, \sigma^2 | y_1, \ldots, y_n) \propto \frac{1}{(\sigma^2)^{\frac{\kappa'+1}{2}+1}} e^{-\frac{n'}{2\sigma^2}(\mu-m')^2 + S' + \left(\frac{n_0 n}{n_0+n}\right)(\bar{y}-m)^2},$$

where $S' = S + SS_y$ and $\kappa' = \kappa + n$ and $m' = \frac{n\bar{y}+n_0 m}{n+n_0}$ and $n' = n_0 + n$. The joint posterior factors into a product of a conditional $normal(m', (s')^2)$ posterior of μ given σ^2 times S' times an *inverse chi-squared* with κ' degrees of freedom posterior of σ^2 and the two components are independent. We find the exact marginal posterior for μ by integrating the variance σ^2 out of the joint posterior. The marginal posterior of μ is $ST_{\kappa'}(m', \frac{\hat{\sigma}_B^2}{n'})$, where $\kappa' = n - 1$, $n' = n$, $m' = \bar{y}$, and

$$\hat{\sigma}_B^2 = \frac{S' + \left(\frac{n_0 n}{n_0 + n}\right)(\bar{y} - m)^2}{\kappa'}$$

$$= \left(\frac{\kappa}{\kappa'}\right)\left(\frac{S}{\kappa}\right) + \left(\frac{n-1}{\kappa'}\right)\frac{SS_y}{n-1} + \left(\frac{1}{\kappa'}\right)\left(\frac{n_0 n}{n_0 + n}\right)(\bar{y} - m)^2$$

is the weighted average of three estimates of the variance. The first incorporates the estimate from the prior, the second is the maximum likelihood estimator of the variance, and the third is the variance inflation term due to misspecification of the prior mean. Its effect is to widen the credible interval whenever the prior mean is misspecified.

- Both the joint likelihood and the joint conjugate prior factor into a conditional *normal* distribution for μ given σ^2 times a scaled *inverse chi-squared* distribution for σ^2. We can use Bayes' theorem to find a conditional *normal* posterior distribution for μ given σ^2 using the simple updating rules for the normal distribution. Similarly we can use Bayes' theorem to find a *inverse chi-squared* posterior for $sigma^2$ also using simple updating rules for the inverse chi-squared distribution. Multiplying these together gives an approximation to the joint posterior for the two parameters

$$g_{\mu,\sigma}(\mu, \sigma^2 | y_1, \ldots, y_n) \propto \frac{1}{(\sigma^2)^{\frac{1}{2}}} e^{-\frac{n'}{2\sigma^2}(\mu - m')^2} \frac{1}{(\sigma^2)^{\frac{\kappa'}{2}+1}} e^{-\frac{S'}{2\sigma^2}},$$

where this time $\kappa' = \kappa + n - 1$ and the other constants are updated as before. The marginal posterior for μ is $ST_{\kappa'}(m', \frac{\sigma_B^2}{n'})$, where

$$\sigma_B^2 = \left(\frac{\kappa}{\kappa'}\right)\left(\frac{S}{\kappa}\right) + \left(\frac{n-1}{\kappa'}\right)\left(\frac{SS_y}{n-1}\right)$$

is the weighted average of two estimates of the variance. The first is from the prior, and the second is the maximum likelihood estimator of variance. Note: This variance does not include a term due to the misspecification of the prior mean as the in the exact case. It also has one less degree of freedom. This shows that the approximation we introduced in Chapter 11 where we use the sample variance in place of the unknown true variance and get the critical values from the *Student's t* table holds exactly.

Two Normal Samples with Same Unknown Variance

Two Normal Samples with Same Unknown Variance

- We have independent random samples from $normal(\mu_1, \sigma^2)$ and $normal(\mu_2, \sigma^2)$ distributions respectively where the two distributions have equal unknown variance.

- When we use independent Jeffreys' prior for all parameters, the marginal posterior of the difference between means, $\mu_1 - \mu_2$ is
  $ST(\bar{y}_1 - \bar{y}_2, \hat{\sigma}_p^2 \left(\frac{1}{n_1} + \frac{1}{n_2} \right))$ with $n_1 + n_2 - 2$ degrees of freedom, where
  $\sigma_p^2 = \frac{SS_p}{n_1 + n_2 - 2}$ is the pooled estimate of the variance. The approximation we gave in Chapter 12 holds exactly.

- When we use the joint conjugate prior for all parameters, the exact marginal posterior of the difference between means, $\mu_1 - \mu_2$, is
  $ST(m_d', \hat{\sigma}_B^2 \left(\frac{1}{n_1'} + \frac{1}{n_2'} \right))$ with $\kappa' = \kappa + n_1 + n_2$ degrees of freedom where
  $m_d' = m_1' - m_2'$ and

$$\hat{\sigma}_B^2 = \frac{S'}{\kappa'}$$

$$= \left(\frac{\kappa}{\kappa'} \right) \left(\frac{S}{\kappa} \right) + \left(\frac{n_1 + n_2 - 2}{\kappa'} \right) \left(\frac{SS_p}{n_1 + n_2 - 2} \right)$$

$$+ \left(\frac{1}{\kappa'} \right) \left(\frac{n_{10} n_1}{n_{10} + n_1 (\bar{y}_1 - m_1)^2} \right) + \left(\frac{1}{\kappa'} \right) \left(\frac{n_{20} n_2}{n_{20} + n_2 (\bar{y}_2 - m_2)^2} \right)$$

is an estimate of the variance incorporating the prior and the data. Note the last two terms inflate the posterior variance if the prior mean is misspecified.

- The approximate marginal posterior of μ_d can be found by first using Bayes' theorem on the conditional normal parts of the respective likelihoods and priors for μ_1 and μ_2 given σ^2 and then using Bayes' theorem on the inverse chi-square prior and likelihood for σ^2. Then we find the conditional posterior for $\mu_d = \mu_1 - \mu_2$ given σ^2. Multiplying this by the posterior for σ^2 gives an approximation to the joint posterior of μ_d and σ^2. The approximate marginal posterior for μ_d is found by integrating σ^2 out of the joint posterior, and is $ST(m_d', \hat{\sigma}_B^2 \left(\frac{1}{n_1'} + \frac{1}{n_2'} \right))$ with

$\kappa' = \kappa + n_1 + n_2 - 2$ degrees of freedom where $m'_d = m'_1 - m'_2$ and

$$\hat{\sigma}_B^2 = \frac{S'}{\kappa'}$$

$$= \left(\frac{\kappa}{\kappa'}\right)\left(\frac{S}{\kappa}\right) + \left(\frac{n_1 + n_2 - 2}{\kappa'}\right)\left(\frac{SS_p}{n_1 + n_2 - 2}\right)$$

$$+ \left(\frac{1}{\kappa'}\right)\left(\frac{n_{10}n_1}{n_{10} + n_1(\bar{y}_1 - m_1)^2}\right) + \left(\frac{1}{\kappa'}\right)\left(\frac{n_{20}n_2}{n_{20} + n_2(\bar{y}_2 - m_2)^2}\right)$$

When we have two independent random samples from *normal* distributions with unknown means μ_1 and μ_2 and unknown unequal variances σ_1^2 and σ_2^2 respectively, then

- When we use independent Jeffreys' prior for all parameters the posterior distribution of
$$\tau_1 = \frac{\mu_1 - \mu_2 - (\bar{y}_1 - \bar{y}_2)}{\sqrt{\frac{\hat{\sigma}_1^2}{n_1} + \frac{\hat{\sigma}_2^2}{n_2}}}$$

depends on ϕ, which is related to the ratio of the standard deviations. It is called the *Behrens Fisher* distribution and is somewhat similar to a *Student's t* distribution. It can be approximated by a *Student's t* distribution by using Satterthwaite's approximation for the degrees of freedom.

Computer Exercises

17.1. The strength of an item is known to be normally distributed with an unknown mean and unknown variance σ^2. A random sample of ten items is taken and their strength measured. The strengths are:

215	186	216	203	221
188	202	192	208	195

Use the Minitab macro *Bayesttest.mac*, or the R function `bayes.t.test`, to answer the following questions.

(a) Test $H_0 : \mu \leq 200$ vs. $H_1 : \mu > 200$ at the 5% level of significance using independent Jeffreys' priors for μ and σ.

(b) Test the hypothesis again, this time using the joint conjugate prior, with a prior mean of $m = 200$, and a prior median value of $\sigma = 5$. Set the prior sample size to $n_0 = 1$ initially. How do your results change when you increase the value of n_0? How do they change when you decrease the value of n_0 (i.e. set $0 < n_0 < 1$)?

17.2. Wild and Seber (1999) describe a data set collected by the New Zealand Air Force. After purchasing a batch of flight helmets that did not fit the heads of many pilots, the NZ Air Force decided to measure the head sizes of all recruits. Before this was carried out, information was collected to determine the feasibility of using cheap cardboard calipers to make the measurements, instead of metal ones which were expensive and uncomfortable. The data lists the head diameters of 18 recruits measured once using cardboard calipers and again using metal calipers.

Cardboard	146	151	163	152	151	151	149	166	149
(mm)	155	155	156	162	150	156	158	149	163
Metal	145	153	161	151	145	150	150	163	147
(mm)	154	150	156	161	152	154	154	147	160

The measurements are *paired* so that 146 mm and 145 mm belong to recruit 1, 151 mm and 153 mm belong to recruit 2 and so on. This places us in a (potentially) special situation which in Frequentist statistics usually calls for the *paired t-test*. In this situation we believe that measurements made on the same individual, or object, are more likely to be similar to each other than those made on different subjects. If we ignore this relationship then our estimate of the variance, σ^2 could be inflated by the inherent differences between individuals. We did not cover this situation explicitly in the theory because it really is a special case of the single unknown mean and variance case. We believe that the two measurements made on each individual are related to each other; therefore it makes sense to look at the *differences* between each pair of measurements rather than the measurements themselves. The differences are:

Difference	1	-2	2	1	6	1	-1	3	2
(mm)	1	5	0	1	-2	2	4	2	3

Use the Minitab macro *Bayesttest.mac*, or the R function bayes.t.test, to answer the following questions.

(a) Test $H_0 : \mu_{difference} \leq 0$ vs. $H_1 : \mu_{difference} > 0$ at the 5% level of significance using independent Jeffreys' priors for μ and σ.

(b) Test the hypothesis again, this time using the joint conjugate prior, with a prior mean of $m = 0$, and a prior median value of $\sigma = 1$. Set the prior sample size to $n_0 = 1$ initially. How do your results change when you increase the value of n_0? How do they change when you decrease the value of n_0 (i.e., set $0 < n_0 < 1$)?

(c) Repeat the analysis, but this time treat the cardboard and metal caliper measurements as two independent samples. Do you come to the same conclusion?

(d) A histogram or dotplot of the differences might make us question the assumption of normality for the differences. An alternative to making this parametric assumption is simply to examine the signs of the differences rather than the magnitudes of the differences. If the differences between the pairs of measurements are truly random but centered around zero, then we would expect to see an equal number of positive and negative differences. The measurements are independent of each other, and we have a fixed sample size, so we can model this situation with a binomial distribution. If the differences are truly random then we would expect the probability of a "success" to be around about 0.5. There are 14 positive differences, 3 negative differences, and 1 case where there is no difference. If we ignore the case where there is no difference, then we say 14 in 17 trials. Use the *BinoBP.mac* in Minitab, or the `binobp` function in R, to test $H_0 : \pi \leq 0.5$ vs. $H_1 : \pi > 0$ at the 5% level of significance using a *beta*$(1, 1)$ prior for π. Does this confirm your previous conclusion? This procedure is the Bayesian equivalent of the *sign test*.

17.3. Bennett et al. (2003) measured the refractive index (RI) of a pane of glass at 49 different locations. She took a sample of 10 fragments at each location and determined the RI for each. The data from locations 1 and 3, shown below, have been rescaled by substracting 1.519 from each measurement and multiplying by 10^5 to make them easier to enter:

Location										
1	100	100	104	100	101	100	100	102	100	102
3	101	100	101	102	102	98	100	101	103	100

(a) Test $H_0 : \mu_1 - \mu_3 \leq 0$ vs. $H_1 : \mu_1 - \mu_3 > 0$ at the 5% level of significance using independent Jeffreys' priors for μ_1, μ_3 and σ.

(b) Test the hypothesis again, this time using the joint conjugate prior, with prior means of $m_1 = m_3 = 0$ and a prior median value of $\sigma = 1$. Set the prior sample size to $n_{10} = n_{20} = 0.1$.

Appendix: Proof that the Exact Marginal Posterior Distribution of μ Is *Student's t*

Using Independent Jeffreys' Priors

When we have the *normal*(μ, σ^2) distribution with both parameters unknown and use either independent Jeffreys' priors, or the joint conjugate prior, we find that the joint posterior has the *normal inverse-chi-squared* form. For the independent Jeffreys' prior case the shape of the joint posterior is given by

$$\propto \frac{1}{(\sigma^2)^{\frac{n}{2}+1}} e^{-\frac{1}{2\sigma^2}[n(\mu-\bar{y})^2+SS_y]}.$$

When we look at this as a function of σ^2 as the only parameter, we see that it is of the form of a constant $(n(\mu-\bar{y})^2+SS_y)$ times an *inverse chi-squared* distribution with n degrees of freedom.

Evaluating the Integral of an Inverse Chi-Squared Density   *Suppose x has an A times an inverse chi-squared distribution with k degrees of freedom. Then the density is*

$$g(x) = \frac{c}{x^{\frac{k}{2}+1}} e^{-\frac{A}{2x}},$$

where $c = \frac{A^{\frac{k}{2}}}{2^{\frac{k}{2}}\Gamma(\frac{k}{2})}$ is the constant needed to make this integrate to 1. Hence the rule for finding the integral is

$$\int_0^\infty \frac{1}{x^{\frac{k}{2}+1}} e^{-\frac{A}{2x}} dx \propto A^{-\frac{k}{2}},$$

where we absorb $2^{\frac{k}{2}}$ and $\Gamma(\frac{k}{2})$ into the constant of proportionality.

Thus in the case where we are using independent Jeffreys' priors for μ and σ^2, the marginal posterior for μ will have shape given by

$$g_\mu(\mu|y_1,\dots,y_n) \propto \int_0^\infty g_{\mu,\sigma^2}(\mu,\sigma^2|y_1,\dots,y_n)\, d\sigma^2$$

$$\propto \int_0^\infty \frac{1}{(\sigma^2)^{\frac{n}{2}+1}} e^{-\frac{1}{2\sigma^2}[n(\mu-\bar{y})^2+SS_y]} d\sigma^2$$

$$\propto [n(\mu-\bar{y})^2+SS_y]^{-\frac{n}{2}}$$

We divide both terms by SS_y and we get

$$g_\mu(\mu|y_1,\dots,y_n) \propto \left[1 + \frac{n(\mu-\bar{y})^2}{SS_y}\right]^{-\frac{n}{2}}.$$

We change variables to

$$t = \frac{\mu-\bar{y}}{\sqrt{\frac{SS_y}{n(n-1)}}}$$

$$= \frac{\mu-\bar{y}}{\frac{\hat{\sigma}}{\sqrt{n}}},$$

where $\hat{\sigma}^2 = \frac{SS_y}{n-1}$ is the unbiased estimator of the variance calculated from the sample. By the change of variable formula, the density of t has shape given by

$$g_t(t) \propto g_\mu(\mu(t)) \times \frac{d\mu(t)}{dt}$$

$$\propto \left[1 + \frac{t^2}{n-1}\right]^{-\frac{n}{2}},$$

where we absorb the term $\frac{d\mu(t)}{dt}$ since it is a constant. This is the shape of the *Student's t* density with $n - 1$ degrees of freedom. Thus $\mu = \bar{y} + \frac{\hat{\sigma}}{\sqrt{n}} \times t$ has the $ST(\bar{y}, \frac{\hat{\sigma}^2}{n})$ distribution with $n - 1$ degrees of freedom. It has *Student's t* with $n - 1$ degrees of freedom shape, and it is centered at $\bar{y}$ with scale factor $\sqrt{\frac{\hat{\sigma}^2}{n}}$.

Using the Joint Conjugate Prior

The other case for the $normal(\mu, \sigma^2)$ with both parameters unknown where we use the joint conjugate prior for μ and σ^2 follows the same pattern with a few changes. The marginal posterior of μ is given by

$$g_\mu(\mu|y_1, \ldots, y_n) \propto \int_0^\infty g_{\mu,\sigma^2}(\mu, \sigma^2|y_1, \ldots, y_n) \, d\sigma^2$$

$$\propto \int_0^\infty \frac{1}{(\sigma^2)^{\frac{\kappa'+1}{2}+1}} e^{-\frac{1}{2\sigma^2}[n'(\mu-m')^2+S'+\left(\frac{n_0 n}{n_0+n}\right)(\bar{y}-m)^2]} \, d\sigma^2 \,.$$

Again, when we look at this as only a function of the parameter σ^2 it is the shape of an S' times an *inverse chi-squared* with κ' degrees of freedom and we are integrating it over its whole range. So the integral can be evaluated by the same rule. Thus

$$g_\mu(\mu|y_1, \ldots, y_n) \propto [(n')(\mu - m')^2 + S' + \left(\frac{n_0 n}{n_0 + n}\right)(\bar{y} - m)^2]^{-\frac{\kappa'+1}{2}},$$

where $S' = S + SS_y$ and $\kappa' = \kappa + n$ and $m' = \frac{n\bar{y}+n_0 m}{n+n_0}$ and $n' = n_0 + n$. We change the variables to

$$t = \frac{\mu - m'}{\sqrt{\frac{S'+\left(\frac{n_0 n}{n_0+n}\right)(\bar{y}-m)^2}{n'\kappa'}}}$$

and apply the change of variable formula. We find the posterior density of t has shape given by

$$
g(t|y_1, \ldots, y_n) \propto \left(\frac{t^2}{\kappa'} + 1 \right)^{-\frac{\kappa'+1}{2}}.
$$

This is the density of the *Student's* t distribution with κ' degrees of freedom. Note that

$$
\frac{S' + \left(\frac{n_0 n}{n_0 + n} \right) (\bar{y} - m)^2}{n'} = \frac{S + SS_y + \left(\frac{n_0 n}{n_0 + n} \right) (\bar{y} - m)^2}{n'}
$$

$$
= \left(\frac{\kappa}{n'} \right) \left(\frac{S}{\kappa} \right) + \left(\frac{n-1}{n'} \right) \left(\frac{SS_y}{n-1} \right)
$$

$$
+ \left(\frac{n_0 n}{n_0 + n} \right) \left(\frac{(\bar{y} - m)^2}{n'} \right)
$$

$$
= \hat{\sigma}_B^2,
$$

which is the weighted average of three estimates of the variance. The first incorporates the prior distribution of σ^2, the second is the unbiased estimator of the variance from the sample data, and the third measures the distance the sample mean, $\bar{y}$, is from its prior mean, m. This means the posterior density of $\mu = m' + \frac{\hat{\sigma}_B}{\sqrt{\kappa'}} t$ will have the $ST(m', \frac{\hat{\sigma}_B^2}{n'})$. Again, we can do the inference treating the unknown variance σ^2 as if it had the value $\hat{\sigma}_B^2$ but using the *Student's* t table to find the critical values instead of the standard normal table.

Difference Between Means with Independent Jeffreys' Priors

We can find the exact marginal posterior for the difference between means $\mu_1 - \mu_2$ for independent random samples from $normal(\mu_1, \sigma^2)$ and $normal(\mu_2, \sigma^2)$ distributions with equal unknown variance the same way. The joint posterior of $\mu_1 - \mu_2$ and σ^2 has shape given by

$$
g_{\mu_1 - \mu_2, \sigma^2}(\mu_1 - \mu_2, \sigma^2 | \mathbf{y_1}, \mathbf{y_2}) \propto \frac{1}{(\sigma^2)^{\frac{n_1 + n_2}{2} + 1}} e^{-\frac{n_1 n_2 (\mu_1 - \mu_2 - (\bar{y}_1 - \bar{y}_2))^2 + SS_p}{2(n_1 + n_2)\sigma^2}}.
$$

The marginal posterior density for $\mu_1 - \mu_2$ is found by integrating the nuisance parameter σ^2 out of the joint posterior. Its shape will be given by

$$g_{\mu_1-\mu_2}(\mu_1 - \mu_2) \propto \int_0^\infty \frac{1}{(\sigma^2)^{\frac{n_1+n_2}{2}+1}} e^{-\frac{n_1 n_2 (\mu_1-\mu_2-(\bar{y}_1-\bar{y}_2))^2 + SS_p}{2(n_1+n_2)\sigma^2}} d\sigma^2$$

$$\propto \left[\frac{n_1 n_2}{n_1+n_2} (\mu_1 - \mu_2 - (\bar{y}_1 - \bar{y}_2))^2 + SS_p \right]^{-\frac{n_1+n_2}{2}}.$$

We change variables to

$$t = \frac{\mu_1 - \mu_2 - (\bar{y}_1 - \bar{y}_2)}{\sqrt{\frac{SS_p(n_1+n_2)}{n_1 n_2(n_1+n_2-2)}}}$$

$$= \frac{\mu_1 - \mu_2 - (\bar{y}_1 - \bar{y}_2)}{\hat{\sigma}_p \left(\frac{1}{\sqrt{n_1}} + \frac{1}{\sqrt{n_2}} \right)},$$

where $\hat{\sigma}_p^2 = \frac{SS_p}{n_1+n_2-2}$ is the unbiased estimate of the variance from the pooled sample. The shape of the density of t is

$$g(t) \propto \left(1 + \frac{t^2}{n_1+n_2-2} \right)^{-\frac{n_1+n_2}{2}},$$

which is the *Student's* t density with $n_1 + n_2 - 2$ degrees of freedom. Hence, the marginal posterior of $\mu_1 - \mu_2$ is $ST(\bar{y}_1 - \bar{y}_2, \frac{\hat{\sigma}_p^2}{n_1+n_2-2})$ with $n_1 + n_2 - 2$ degrees of freedom.

Difference Between Means with Joint Conjugate Prior

We can find the exact marginal posterior for the difference between means, $\mu_1 - \mu_2$, for independent random samples from $normal(\mu_1, \sigma^2)$ and $normal(\mu_2, \sigma^2)$ distributions with equal unknown variance the same way. When we use a joint conjugate prior for μ, σ^2, the joint posterior of $\mu_1 - \mu_2$ and σ^2 has shape given by

$$g_{\mu_1-\mu_2,\sigma^2}(\mu_1 - \mu_2, \sigma^2 | \mathbf{y_1}, \mathbf{y_2}) \propto \frac{1}{(\sigma^2)^{\frac{\kappa'+1}{2}+1}} e^{-\frac{1}{2\sigma^2} \left[\left(\frac{n_1' n_2'}{n_1'+n_2'} \right) (\mu_1-\mu_2-m_d')^2 + S' \right]}.$$

The marginal posterior density for $\mu_1 - \mu_2$ is found by integrating the nuisance parameter σ^2 out of the joint posterior. Its shape will be given by

$$g_{\mu_1-\mu_2}(\mu_1 - \mu_2) \propto \int_0^\infty \frac{1}{(\sigma^2)^{\frac{\kappa'+1}{2}+1}} e^{-\frac{1}{2\sigma^2}\left[\left(\frac{n_1'n_2'}{n_1'+n_2'}\right)(\mu_1-\mu_2-m_d')^2+S'\right]} d\sigma^2$$

$$\propto \left[\left(\frac{n_1'n_2'}{n_1'+n_2'}\right)(\mu_1 - \mu_2 - m_d')^2 + S'\right]^{-\frac{\kappa'+1}{2}}.$$

We change variables to

$$t = \frac{\mu_1 - \mu_2 - m_d'}{\sqrt{\frac{S'(n_1'+n_2')}{n_1'n_2'(\kappa'+1)}}}$$

$$= \frac{\mu_1 - \mu_2 - m_d'}{\sqrt{\hat{\sigma}_B^2\left(\frac{1}{n_1}+\frac{1}{n_2}\right)}},$$

where

$$\hat{\sigma}_B^2 = \frac{S'}{\kappa'+1}$$

$$= \frac{S}{\kappa}\frac{\kappa}{\kappa'+1} + \frac{SS_p}{n_1+n_2-2}\frac{n_1+n_2-2}{\kappa+1}.$$

The shape of the density of t is

$$g(t) \propto \left(1 + \frac{t^2}{n_1+n_2-2}\right)^{-\frac{n_1+n_2}{2}},$$

which is the *Student's t* density with $\kappa'+1$ degrees of freedom. Hence, the marginal posterior of $\mu_1 - \mu_2$ is $ST(m_d', \frac{\hat{\sigma}_B^2}{\kappa'+1})$ with $\kappa'+1$ degrees of freedom.

Table 17.2   Summary of inference on μ when both μ and σ^2 are unknown

Type	Posterior parameters				σ_B^2	t
	S'	κ'	n'	m'		
Jeffreys	SS_y	$n-1$	n	$\bar{y}$	$\dfrac{SS_y}{n-1}$	$\dfrac{\mu-m'}{\sqrt{\frac{\hat{\sigma}_B^2}{n'}}}$
Exact	$S+SS_y$	$\kappa+n$	n_0+n	$\dfrac{n_0 m+n\bar{y}}{n_0+n}$	$\dfrac{\kappa}{n'}\dfrac{S}{\kappa}+\dfrac{n-1}{n'}\dfrac{SS_y}{n-1}$	$\dfrac{\mu-m'}{\sqrt{\frac{\hat{\sigma}_B^2}{n'}}}$
					$+\dfrac{n_0 n}{n'}\dfrac{(\bar{y}-m)^2}{n'}$	
Appr.	$S+SS_y$	$\kappa+n-1$	n_0+n	$\dfrac{n_0 m+n\bar{y}}{n_0+n}$	$\dfrac{\kappa}{n'}\dfrac{S}{\kappa}+\dfrac{n-1}{n'}\dfrac{SS_y}{n-1}$	$\dfrac{\mu-m'}{\sqrt{\frac{\hat{\sigma}_B^2}{n'}}}$

CHAPTER 18

BAYESIAN INFERENCE FOR MULTIVARIATE NORMAL MEAN VECTOR

In this chapter we will introduce the *multivariate normal* distribution with known covariance matrix. Instead of each observation being a random draw of a single variable from a univariate normal distribution, it will be a single simultaneous draw of k component variables, each of which has its own univariate normal distribution. Also, the different components for a simultaneous draw are related by the covariance matrix. We call this drawing a random vector from the *multivariate normal* distribution. In Section 18.1, we will start with the *bivariate normal* density where the number of components $k = 2$ and show how we can write this in matrix notation. In Section 18.2 we show how the matrix form for the *multivariate normal* density generalizes when the number of components $k \geq 2$. In Section 18.3 we use Bayes' theorem to find the posterior for the multivariate normal mean vector when the covariance matrix Σ is known. In the general case, this will require evaluating a k-dimensional integral numerically to find the scale factor needed for the posterior. We will show that we can find the exact posterior without needing to evaluate the integral for two special types of priors: a multivariate flat prior, or a *multivariate normal* prior of the correct dimension. In Section 18.4, we cover Bayesian inference for the *multivariate normal* mean parameters. We show how to find a Bayesian credible region for the mean vector. We can use the credible region for testing any point hypothesis about the mean vector. These methods can also be applied to find a credible region for any subvector of means, and hence for testing any point hypothesis about that subvector. In Section 18.5, we find the joint posterior for the mean vector and the co-variance matrix when both are unknown. We find the marginal posterior for μ which we use for inference.

18.1   Bivariate Normal Density

Let the two-dimensional random variable Y_1, Y_2 have the joint density function given by

$$f(y_1, y_2 | \mu_1, \mu_2, \sigma_1^2, \sigma_2^2, \rho) = \frac{1}{2\pi\sigma_1\sigma_2\sqrt{1-\rho^2}}$$

$$\times e^{-\frac{1}{2(1-\rho^2)}\left[\left(\frac{y_1-\mu_1}{\sigma_1}\right)^2 - 2\rho\left(\frac{y_1-\mu_1}{\sigma_1}\right)\left(\frac{y_2-\mu_2}{\sigma_2}\right) + \left(\frac{y_2-\mu_2}{\sigma_2}\right)^2\right]} \quad (18.1)$$

for $-\infty < y_1 < \infty$ and $-\infty < y_2 < \infty$. The parameters μ_1 and μ_2 can take on any value, σ_1^2 and σ_2^2 can take on any positive value, and ρ must be between -1 and 1. To see that this is a density function, we must make sure that its multiple integral over the whole range equals 1. We make the substitutions

$$v_1 = \frac{y_1 - \mu_1}{\sigma_1} \quad \text{and} \quad v_2 = \frac{y_2 - \mu_2}{\sigma_2}.$$

Then the integral is given by

$$\int_{-\infty}^{\infty}\int_{-\infty}^{\infty} f(y_1, y_2 | \mu_1, \mu_2, \sigma_1^2, \sigma_2^2, \rho) dy_1\, dy_2$$

$$= \int_{-\infty}^{\infty}\int_{-\infty}^{\infty} f(y_1(v_1, v_2), y_2(v_1, v_2) | \mu_1, \mu_2, \sigma_1^2, \sigma_2^2, \rho) \times \left|\frac{\partial u}{\partial v}\right| dv_1\, dv_2$$

$$= \int_{-\infty}^{\infty}\int_{-\infty}^{\infty} \frac{1}{2\pi\sqrt{1-\rho^2}} e^{-\left(\frac{1}{2(1-\rho^2)}(v_1^2 - 2\rho v_1 v_2 + v_2^2)\right)} dv_1\, dv_2.$$

We complete the square in the exponent. The integral becomes

$$\int_{-\infty}^{\infty}\int_{-\infty}^{\infty} \frac{1}{2\pi\sqrt{1-\rho^2}} e^{-\left(\frac{1}{2(1-\rho^2)}\left((v_1 - \rho v_2)^2 + (1-\rho^2)v_2^2\right)\right)} dv_1\, dv_2$$

We substitute

$$w_1 = \frac{v_1 - \rho v_2}{\sqrt{1-\rho^2}} \quad \text{and} \quad w_2 = v_2$$

and the integral simplifies into the product of two integrals

$$\int_{-\infty}^{\infty} \frac{1}{\sqrt{2\pi}} e^{-\frac{1}{2}w_1^2} dw_1 \int_{-\infty}^{\infty} \frac{1}{\sqrt{2\pi}} e^{-\frac{1}{2}w_2^2} dw_2 = 1,$$

since each of these is the integral of a univariate $normal(0, 1)$ density over its whole range. Thus the bivariate normal density given above is a joint density.

By setting the bivariate normal density equal to a constant and taking logarithms of both sides of the equation we find a second-degree polynomial in y_1 and y_2. This means the level curves will be concentric ellipses.[1]

The Marginal Densities Of Y_1 And Y_2   We find the marginal density of y_1 by integrating y_2 out of the joint density.

$$f_{y_1}(y_1) = \int_{-\infty}^{\infty} f(y_1, y_2)\, dy_2$$

$$\stackrel{\cdot}{=} \int_{-\infty}^{\infty} \frac{e^{-\frac{1}{2(1-\rho^2)}\left[\left(\frac{y_1-\mu_1}{\sigma_1^2}\right)^2 - 2\rho\left(\frac{y_1-\mu_1}{\sigma_1}\right)\left(\frac{y_2-\mu_2}{\sigma_2}\right) + \left(\frac{y_2-\mu_2}{\sigma_2}\right)^2\right]}}{2\pi\sigma_1\sigma_2\sqrt{1-\rho^2}}\, dy_2.$$

If we make the substitution

$$v_2 = \frac{y_2 - \mu_2}{\sigma_2}$$

and complete the square, then we find that

$$f_{y_1}(y_1) = \int_{-\infty}^{\infty} \frac{1}{2\pi\sigma_1\sqrt{1-\rho^2}}\, e^{-\frac{1}{2}\left(\frac{y_1-\mu_1}{\sigma_1}\right)^2 - \frac{1}{2\sqrt{1-\rho^2}}\left(v_2 - \rho\frac{y_1-\mu_1}{\sigma_1}\right)^2}\, dv_2.$$

If we make the additional substitution

$$w_2 = \frac{v_2 - \rho(y_1 - \mu_1)}{\sigma_1\sqrt{1-\rho^2}},$$

then the marginal density becomes

$$f_{y_1}(y_1) = \int_{-\infty}^{\infty} \frac{1}{\sqrt{2\pi}\sigma_1}\, e^{-\frac{1}{2}\left(\frac{y_1-\mu_1}{\sigma_1}\right)^2} \times \frac{1}{2\pi}\, e^{-\frac{1}{2}w_2^2}\, dw_2$$

$$= \frac{1}{\sqrt{2\pi}\sigma_1}\, e^{-\frac{1}{2}\left(\frac{y_1-\mu_1}{\sigma_1}\right)^2}.$$

We recognize this as the univariate $normal(\mu_1, \sigma_1^2)$ density. Similarly, the marginal density of y_2 will be a univariate $normal(\mu_2, \sigma_2^2)$ density. Thus the parameters μ_1 and σ_1^2 and μ_2 and σ_2^2 of the bivariate normal distribution are the means and variances of the components y_1 and y_2, respectively, and the components are each normally distributed. Next we will show that the parameter ρ is the correlation coefficient between the two components.

---

[1] The ellipses will be entered at the point (μ_1, μ_2). The directions of the principal axes are given by the eigenvectors of the covariance matrix. The lengths of the axes will be the square roots of the eigenvalues of the covariance matrix. In the bivariate normal case, the major axis will be rotated by the angle $\phi = .5 \times \tan^{-1}\left(\frac{2\rho\sigma_1\sigma_2}{\sigma_1^2-\sigma_2^2}\right)$.

The Covariance of Y_1 and Y_2   The covariance is found by evaluating

$$\text{Cov}[y_1, y_2] = \int_{-\infty}^{\infty} \int_{-\infty}^{\infty} (y_1 - \mu_1)(y_2 - \mu_2) f(y_1, y_2) dy_1 \, dy_2$$

$$= \int_{-\infty}^{\infty} \int_{-\infty}^{\infty} \frac{(y_1 - \mu_1)(y_2 - \mu_2)}{2\pi\sigma_1\sigma_2\sqrt{1 - \rho^2}}$$

$$\times e^{-\frac{1}{2(1-\rho^2)}\left[\left(\frac{y_1-\mu_1}{\sigma_1^2}\right)^2 - 2\rho\left(\frac{y_1-\mu_1}{\sigma_1}\right)\left(\frac{y_2-\mu_2}{\sigma_2}\right) + \left(\frac{y_2-\mu_2}{\sigma_2}\right)^2\right]} dy_1 \, dy_2.$$

We make the substitutions

$$v_1 = \frac{y_1 - \mu_1}{\sigma_1} \qquad \text{and} \qquad v_2 = \frac{y_2 - \mu_2}{\sigma_2}.$$

Then the covariance is given by

$$\text{Cov}[y_1, y_2] = \int_{-\infty}^{\infty} \int_{-\infty}^{\infty} \frac{\sigma_1\sigma_2 v_1 v_2}{2\pi\sqrt{1 - \rho^2}} e^{-\frac{1}{2(1-\rho^2)}(v_1^2 - 2\rho v_1 v_2 + v_2^2)} dv_1 \, dv_2.$$

We complete the square for v_2, reverse the order of integration, and we get

$$\text{Cov}[y_1, y_2] = \sigma_1\sigma_2 \int_{-\infty}^{\infty} \frac{v_1}{\sqrt{2\pi}} e^{-\frac{v_1^2}{2}} \left[\int_{-\infty}^{\infty} \frac{v_2}{\sqrt{2\pi}\sqrt{1 - \rho^2}} e^{-\frac{(v_2-\rho v_1)^2}{2(1-\rho^2)}} dv_2\right] dv_1.$$

If we make the substitution

$$w_2 = \frac{v_2 - \rho v_1}{\sqrt{1 - \rho^2}},$$

then the covariance becomes

$$\text{Cov}[y_1, y_2] = \sigma_1\sigma_2 \int_{-\infty}^{\infty} \frac{v_1}{\sqrt{2\pi}} e^{-\frac{v_1^2}{2}} \left[\int_{-\infty}^{\infty} \frac{w_2\sqrt{1 - \rho^2} + \rho v_1}{\sqrt{2\pi}} e^{-\frac{w_2^2}{2}} dw_2\right] dv_1$$

$$= \int_{-\infty}^{\infty} \frac{v_1}{\sqrt{2\pi}} e^{-\frac{v_1^2}{2}} [0 + \rho v_1] dv_1$$

$$= \sigma_1\sigma_2\rho.$$

Thus, the parameter ρ of the bivariate normal distribution is the correlation coefficient of the two variables y_1 and y_2.

Bivariate Normal Density in Matrix Notation

Suppose we stack the two random variables and their respective mean parameters into the vectors

$$\mathbf{y} = \begin{pmatrix} y_1 \\ y_2 \end{pmatrix} \qquad \text{and} \qquad \boldsymbol{\mu} = \begin{pmatrix} \mu_1 \\ \mu_2 \end{pmatrix},$$

respectively, and put the covariances in the matrix

$$\boldsymbol{\Sigma} = \begin{bmatrix} \sigma_1^2 & \rho\sigma_1\sigma_2 \\ \rho\sigma_1\sigma_2 & \sigma_2^2 \end{bmatrix}.$$

The inverse of the covariance matrix in the bivariate normal case is

$$\boldsymbol{\Sigma}^{-1} = \frac{1}{\sigma_1^2\sigma_2^2(1-\rho^2)} \begin{bmatrix} \sigma_2^2 & -\rho\sigma_1\sigma_2 \\ -\rho\sigma_1\sigma_2 & \sigma_1^2 \end{bmatrix}.$$

The determinant of the bivariate normal covariance matrix is given by

$$|\boldsymbol{\Sigma}| = \sigma_1^2\,\sigma_2^2 \times (1-\rho^2).$$

Thus we see that the bivariate normal joint density given in equation 18.1 can be written as

$$f(y_1, y_2) = \frac{1}{2\pi|\boldsymbol{\Sigma}|^{\frac{1}{2}}} e^{-\frac{1}{2}(\mathbf{y}-\boldsymbol{\mu})'\boldsymbol{\Sigma}^{-1}(\mathbf{y}-\boldsymbol{\mu})} \tag{18.2}$$

in matrix notation.

18.2   Multivariate Normal Distribution

The dimension of the observation $\mathbf{y}$ and the mean vector $\boldsymbol{\mu}$ can be $k \geq 2$. Then we can generalize the distribution from Equation 18.2 to give

$$f(y_1, y_2) = \frac{1}{(2\pi)^{\frac{k}{2}}|\boldsymbol{\Sigma}|^{\frac{1}{2}}} e^{-\frac{1}{2}(\mathbf{u}-\boldsymbol{\mu})'\boldsymbol{\Sigma}^{-1}(\mathbf{u}-\boldsymbol{\mu})}, \tag{18.3}$$

is called the multivariate normal distribution (MVN). Its parameters are the mean vector $\boldsymbol{\mu}$ and the covariance matrix $\boldsymbol{\Sigma}$ given by

$$\boldsymbol{\mu} = \begin{pmatrix} \mu_1 \\ \vdots \\ \mu_k \end{pmatrix} \quad \text{and} \quad \boldsymbol{\Sigma} = \begin{bmatrix} \sigma_1^2 & \cdots & \rho_{1k}\sigma_1\sigma_k \\ \vdots & \ddots & \vdots \\ \rho_{k1}\sigma_k\sigma_1 & \cdots & \sigma_k^2 \end{bmatrix},$$

respectively. Generalizing the results from the bivariate normal, we find that the level surfaces of the multivariate normal distribution will be concentric ellipsoids, centered about the mean vector and orientation determined from the covariance matrix. The marginal distribution of each component is univariate normal. Also, the marginal distribution of every subset of components is the multivariate normal. For instance, if we make the corresponding partitions in the random vector, mean vector, and covariance matrix

$$\mathbf{y} = \begin{pmatrix} \mathbf{y}_1 \\ \mathbf{y}_2 \end{pmatrix}, \quad \boldsymbol{\mu} = \begin{pmatrix} \boldsymbol{\mu}_1 \\ \boldsymbol{\mu}_2 \end{pmatrix}, \quad \text{and} \quad \boldsymbol{\Sigma} = \begin{bmatrix} \boldsymbol{\Sigma}_{11} & \boldsymbol{\Sigma}_{12} \\ \boldsymbol{\Sigma}_{21} & \boldsymbol{\Sigma}_{22} \end{bmatrix},$$

then the marginal distribution of $\mathbf{y}_1$ is $MVN(\boldsymbol{\mu}_1, \boldsymbol{\Sigma}_{11})$. Similarly, the marginal distribution of $\mathbf{y}_2$ is $MVN(\boldsymbol{\mu}_2, \boldsymbol{\Sigma}_{22})$.

18.3   The Posterior Distribution of the Multivariate Normal Mean Vector when Covariance Matrix Is Known

The joint posterior distribution of $\boldsymbol{\mu}$ will be proportional to prior times likelihood. If we have a general jointly continuous prior for $\mu_1, \ldots, \mu_k$, then it is given by

$$g(\mu_1, \ldots, \mu_k|\mathbf{y}) \propto g(\mu_1, \ldots, \mu_k) \times f(\mathbf{y}|\mu_1, \ldots, \mu_k).$$

The exact posterior will be

$$g(\mu_1, \ldots, \mu_k|\mathbf{y}) = \frac{g(\mu_1, \ldots, \mu_k) \times f(\mathbf{y}|\mu_1, \ldots, \mu_k)}{\int \ldots \int g(\mu_1, \ldots, \mu_k) \times f(\mathbf{y}|\mu_1, \ldots, \mu_k)d\mu_1 \ldots d\mu_k}.$$

To find the exact posterior we have to evaluate the denominator integral numerically, which can be complicated. We will look at two special cases where we can find the posterior without having to do the integration. These cases occur when we have a multivariate normal prior and when we have a multivariate flat prior. In both of these cases we have to be able to recognize when the density is multivariate normal from the shape given in Equation 18.2

A Single Multivariate Normal Observation

Suppose we have a single random observation from the $MVN(\boldsymbol{\mu}, \boldsymbol{\Sigma})$. The likelihood function of a single draw from the multivariate normal mean vector has the same functional form as the multivariate normal joint density function,

$$f(\boldsymbol{\mu}|\mathbf{y}) = \frac{1}{\sqrt{2\pi}|\boldsymbol{\Sigma}|^{\frac{1}{2}}}e^{-\frac{1}{2}(\mathbf{y}-\boldsymbol{\mu})'\boldsymbol{\Sigma}^{-1}(\mathbf{y}-\boldsymbol{\mu})}; \qquad (18.4)$$

however, the observation vector $\mathbf{y}$ is held at the value that occurred and the parameter vector $\boldsymbol{\mu}$ is allowed to vary. We can see this by reversing the positions of $\mathbf{y}$ and $\boldsymbol{\mu}$ in the expression above and noting that

$$(\mathbf{x} - \mathbf{y})'\mathbf{A}(\mathbf{x} - \mathbf{y}) = (-1)(\mathbf{x} - \mathbf{y})'\mathbf{A}(-1)(\mathbf{x} - \mathbf{y})$$
$$= (\mathbf{y} - \mathbf{x})'\mathbf{A}(\mathbf{y} - \mathbf{x})$$

for any symmetric matrix $\mathbf{A}$.

Multivariate Normal Prior Density for $\boldsymbol{\mu}$ Suppose we use a $MVN(\mathbf{m}_0, \mathbf{V}_0)$ prior for the mean vector $\boldsymbol{\mu}$. The posterior is given by

$$g(\boldsymbol{\mu}|\mathbf{y}) \propto g(\boldsymbol{\mu})f(\mathbf{y}|\boldsymbol{\mu})$$
$$\propto e^{-\frac{1}{2}(\boldsymbol{\mu}-\mathbf{m}_0)'\mathbf{V}_0^{-1}(\boldsymbol{\mu}-\mathbf{m}_0)}\,e^{-\frac{1}{2}(\mathbf{y}-\boldsymbol{\mu})'\boldsymbol{\Sigma}^{-1}(\mathbf{y}-\boldsymbol{\mu})}$$
$$\propto e^{-\frac{1}{2}(\boldsymbol{\mu}-\mathbf{m}_0)'\mathbf{V}_0^{-1}(\boldsymbol{\mu}-\mathbf{m}_0)+(\boldsymbol{\mu}-\mathbf{y})'\boldsymbol{\Sigma}^{-1}(\boldsymbol{\mu}-\mathbf{y})}.$$

We expand both terms, combine like terms, and absorb the part not containing the parameter $\boldsymbol{\mu}$ into the constant.

$$g(\boldsymbol{\mu}|\mathbf{y}) \propto e^{-\frac{1}{2}[\boldsymbol{\mu}'(\mathbf{V}_1^{-1})\boldsymbol{\mu}-\boldsymbol{\mu}'(\boldsymbol{\Sigma}^{-1}\mathbf{y}+\mathbf{V}_0^{-1}\mathbf{m}_0)-(\mathbf{y}'\boldsymbol{\Sigma}^{-1}+\mathbf{m}_0'\mathbf{V}_0^{-1})\boldsymbol{\mu}]},$$

where $\mathbf{V}_1^{-1} = \mathbf{V}_0^{-1} + \boldsymbol{\Sigma}^{-1}$. The posterior precision matrix equals the prior precision matrix plus the precision matrix of the multivariate observation. This is similar to the rule for an ordinary univariate normal observation but adapted to the multivariate normal situation. $\mathbf{V}_1^{-1}$ is a symmetric positive definite matrix of full rank, so $\mathbf{V}_1^{-1} = \mathbf{U}'\mathbf{U}$ where $\mathbf{U}$ is an triangular matrix. Both $\mathbf{U}$ and $\mathbf{U}'$ are of full rank so their inverses exist, and both $(\mathbf{U}')^{-1}\mathbf{U}'$ and $\mathbf{U}\mathbf{U}^{-1}$ equal the k-dimensional identity matrix. Simplifying the posterior, we get

$$g(\boldsymbol{\mu}|\mathbf{y}) \propto e^{-\frac{1}{2}[\boldsymbol{\mu}'(\mathbf{U}'\mathbf{U})\boldsymbol{\mu}-\boldsymbol{\mu}'(\mathbf{U}'(\mathbf{U}')^{-1})(\boldsymbol{\Sigma}^{-1}\mathbf{y}+\mathbf{V}_0^{-1}\mathbf{m}_0)-(\mathbf{y}'\boldsymbol{\Sigma}^{-1}+\mathbf{m}_0'\mathbf{V}_0^{-1})(\mathbf{U}^{-1}\mathbf{U})\boldsymbol{\mu}]}.$$

Completing the square and absorbing the part that does not contain the parameter into the constant the posterior simplifies to

$$g(\boldsymbol{\mu}|\mathbf{y}) \propto e^{-\frac{1}{2}[\boldsymbol{\mu}'\mathbf{U}'-(\mathbf{y}'\boldsymbol{\Sigma}^{-1}+\mathbf{m}_0'\mathbf{V}_0^{-1})\mathbf{U}^{-1}][\mathbf{U}\boldsymbol{\mu}-(\mathbf{U}'^{-1})(\boldsymbol{\Sigma}^{-1}\mathbf{y}+\mathbf{V}_0^{-1}\mathbf{m}_0)]}.$$

Hence

$$g(\boldsymbol{\mu}|\mathbf{y}) \propto e^{-\frac{1}{2}[\boldsymbol{\mu}'-(\mathbf{y}'\boldsymbol{\Sigma}^{-1}+\mathbf{m}_0'\mathbf{V}_0^{-1})\mathbf{V}_1]\mathbf{V}_1^{-1}[\boldsymbol{\mu}-\mathbf{V}_1'(\boldsymbol{\Sigma}^{-1}\mathbf{y}+\mathbf{V}_0^{-1}\mathbf{m}_0)]}$$

$$\propto e^{-\frac{1}{2}(\boldsymbol{\mu}'-\mathbf{m}_1')\mathbf{V}_1^{-1}(\boldsymbol{\mu}-\mathbf{m}_1)},$$

where $\mathbf{m}_1 = \mathbf{V}_1\mathbf{V}_0^{-1}\mathbf{m}_0 + \mathbf{V}_1\boldsymbol{\Sigma}^{-1}\mathbf{y}$ is the posterior mean. It is given by the rule "posterior mean vector equals the inverse of posterior precision matrix times prior precision matrix times prior mean vector plus inverse of posterior precision matrix times precision matrix of observation vector times the observation vector." This is similar to the rule for single normal observation, but adapted to vector observations. We recognize that the posterior distribution of $\boldsymbol{\mu}|\mathbf{y}$ is $MVN(\mathbf{m}_1, \mathbf{V}_1)$.

A Random Sample from the Multivariate Normal Distribution

Suppose we have a random sample $\mathbf{y}_1, \ldots, \mathbf{y}_n$ from the $MVN(\boldsymbol{\mu}, \boldsymbol{\Sigma})$ distribution where the covariance matrix $\boldsymbol{\Sigma}$ is known. The likelihood function of the random sample will be the product of the likelihood functions from each individual observation.

$$f(\mathbf{y}_1, \ldots, \mathbf{y}_n|\boldsymbol{\mu}) = \prod_{i=1}^{n} f(\mathbf{y}_i|\boldsymbol{\mu})$$

$$\propto \prod_{i=1}^{n} e^{-\frac{1}{2}(\mathbf{y}_i-\boldsymbol{\mu})'\boldsymbol{\Sigma}^{-1}(\mathbf{y}_i-\boldsymbol{\mu})}$$

$$\propto e^{-\frac{1}{2}\sum_{i=1}^{n}[(\mathbf{y}_i-\boldsymbol{\mu})'\boldsymbol{\Sigma}^{-1}(\mathbf{y}_i-\boldsymbol{\mu})]}$$

$$\propto e^{-\frac{n}{2}[(\bar{\mathbf{y}}-\boldsymbol{\mu})'\boldsymbol{\Sigma}^{-1}(\bar{\mathbf{y}}-\boldsymbol{\mu})]}.$$

We see that the likelihood of the random sample from the *multivariate normal* distribution is proportional to the likelihood of the sample mean vector, $\bar{\mathbf{y}}$. This has the *multivariate normal* distribution with mean vector $\boldsymbol{\mu}$ and covariance matrix $\frac{\boldsymbol{\Sigma}}{n}$.

Simple Updating Formulas for Multivariate Normal We can condense the random sample $\mathbf{y}_1, \ldots, \mathbf{y}_n$ from the $MVN(\boldsymbol{\mu}, \boldsymbol{\Sigma})$ into a single observation of the sample mean vector $\bar{\mathbf{y}}$ from its $MVN(\boldsymbol{\mu}, \frac{\boldsymbol{\Sigma}}{n})$ distribution. Hence the posterior precision matrix is the sum of the prior precision matrix plus the precision matrix of the sample mean vector. It is given by

$$\mathbf{V}_1^{-1} = \mathbf{V}_0^{-1} + n\boldsymbol{\Sigma}^{-1}. \tag{18.5}$$

The posterior mean vector is the weighted average of the prior mean vector and the sample mean vector, where their weights are the proportions of their precision matrices to the posterior precision matrix. It is given by

$$\mathbf{m}_1 = \mathbf{V}_1 \mathbf{V}_0^{-1} \mathbf{m}_0 + \mathbf{V}_1 n \boldsymbol{\Sigma}^{-1} \bar{\mathbf{y}}. \tag{18.6}$$

These updating rules also work in the case of *multivariate flat prior*. It will have infinite variance in each of the dimensions, so its precision matrix is a matrix of zeros. That is $\mathbf{V}_0^{-1} = \mathbf{0}$, so

$$\mathbf{V}_1^{-1} = \mathbf{0} + n\boldsymbol{\Sigma}^{-1} = n\boldsymbol{\Sigma}^{-1}$$

and

$$\begin{aligned} \mathbf{m}_1 &= \mathbf{V}_1 \mathbf{0} \mathbf{m}_0 + \mathbf{V}_1 n\boldsymbol{\Sigma}^{-1}\bar{\mathbf{y}} \\ &= \mathbf{0} + \frac{\boldsymbol{\Sigma}}{n}\frac{n}{\boldsymbol{\Sigma}}\bar{\mathbf{y}} \\ &= \bar{\mathbf{y}}. \end{aligned}$$

Thus in the case of a *multivariate flat prior*, the posterior precision matrix will equal the precision matrix of the sample mean vector, and the posterior mean vector will be the sample mean vector.

18.4 Credible Region for Multivariate Normal Mean Vector when Covariance Matrix Is Known

In the last section we found that the posterior distribution of the multivariate normal mean vector $\boldsymbol{\mu}$ is $MVN(\mathbf{m}_1, \mathbf{V}_1)$ when we used a $MVN(\mathbf{m}_0, \mathbf{V}_0)$ prior or a *multivariate flat prior*. Component μ_i of the mean vector has a univariate $normal(m_i, \sigma_i^2)$ distribution where m_i is the ith component of the posterior mean vector $\mathbf{m}_1$ and σ_i^2 is the ith diagonal element of the posterior covariance matrix $\mathbf{V}_1$. In Chapter 11 we found that a $(1 - \alpha) \times 100\%$ credible interval for μ_i is given by $m_i \pm z_{\frac{\alpha}{2}} \sigma_i$. We could find the individual $(1 - \alpha) \times 100\%$ credible interval credible interval for every component, and their intersection forms a k-dimensional credible region for the mean vector $\boldsymbol{\mu}$. The mean vector $\boldsymbol{\mu}$ being in the credible region means that all components of the mean vector are simultaneously within their respective intervals.

However, when we combined the individual credible intervals like this we will lose control of the overall level of credibility. The posterior probability that all of the components are simultaneously contained in their respective individual credible intervals would be much less than the desired $(1-\alpha) \times 100\%$ level. So, we need to find a simultaneous credible region for all components of the mean vector. This credible region has the posterior probability $(1 - \alpha) \times 100\%$ that all components are simultaneously in this region.

We are assuming that the matrix $\mathbf{V}_1$ is full rank k, the same as the dimension of $\boldsymbol{\mu}$. Generalizing the bivariate normal case to the multivariate normal case, we find that the level surfaces of a multivariate normal distribution will be concentric ellipsoids. The $(1-\alpha) \times 100\%$ credible region will be the ellipsoid that contains posterior probability equal to α. The posterior distribution of the random variable $U = (\boldsymbol{\mu} - \mathbf{m}_1)' \mathbf{V}_1^{-1} (\boldsymbol{\mu} - \mathbf{m}_1)$ will be *chi-squared* with k degrees of freedom. If we find the upper α point on the chi-squared distribution with k degrees of freedom, $c_k(\alpha)$ in Table B.6, then

$$P(U \leq c_k(\alpha)) = 1 - \alpha.$$

Hence

$$P[(\boldsymbol{\mu} - \mathbf{m}_1)' \mathbf{V}_1^{-1} (\boldsymbol{\mu} - \mathbf{m}_1) \leq c_k(\alpha)] = 1 - \alpha.$$

Hence the $(1 - \alpha) \times 100\%$ confidence ellipsoid is the set of values $\boldsymbol{\mu}$ that lie within the ellipsoid determined by the equation

$$(\boldsymbol{\mu} - \mathbf{m}_1)' \mathbf{V}_1^{-1} (\boldsymbol{\mu} - \mathbf{m}_1) = c_k(\alpha).$$

Testing Point Hypothesis Using the Credible Region

We call the hypothesis $H_0 : \boldsymbol{\mu} = \boldsymbol{\mu}_0$ a point hypothesis because there is only one single point in the k-dimensional parameter space where it is true. Each component μ_i must equal its hypothesized value μ_{i0} for $i = 1, \ldots, k$. The point hypothesis will be false if at least one of the components is not equal to its hypothesized value regardless of whether or not the others are.

We can test the point hypothesis

$$H_0 : \boldsymbol{\mu} = \boldsymbol{\mu}_0 \quad \text{versus} \quad H_1 : \boldsymbol{\mu} \neq \boldsymbol{\mu}_0$$

at the level of significance α using the $(1 - \alpha) \times 100\%$ credible region in a similar way that we tested a single parameter point hypothesis against a two-sided alternative. Only in this case, we use the k-dimensional credible region in place of the single credible interval. When the point $\boldsymbol{\mu}_0$ lies outside the

$(1 - \alpha) \times 100\%$ credible region for $\boldsymbol{\mu}$, we can reject the null hypothesis and conclude that $\boldsymbol{\mu} \neq \boldsymbol{\mu}_0$ at the level[2]. However, if the point $\boldsymbol{\mu}_0$ lies in the credible region, then we conclude $\boldsymbol{\mu}_0$ still is a credible value, so we cannot reject the null hypothesis.

18.5   Multivariate Normal Distribution with Unknown Covariance Matrix

In this section we look at the $MVN(\boldsymbol{\mu}, \boldsymbol{\Sigma})$ where both the mean vector and covariance matrix are unknown. We will show how to do inference on the mean vector $\boldsymbol{\mu}$ where the covariance matrix $\boldsymbol{\Sigma}$ is considered to be a nuisance parameter. First we have to look at a distribution for the variances and covariances which we have arranged in matrix $\boldsymbol{\Sigma}$.

Inverse Wishart Distribution

The symmetric positive definite k by k matrix $\boldsymbol{\Sigma}$ has the k by k-dimensional *Inverse Wishart* $(\mathbf{S}^{-1})$ with ν degrees of freedom if its density has shape given by

$$f(\boldsymbol{\Sigma}|\mathbf{S}) \propto \frac{1}{|\boldsymbol{\Sigma}|^{\frac{\nu+k+1}{2}}} \times e^{-\frac{tr(\mathbf{S}\boldsymbol{\Sigma}^{-1})}{2}}, \tag{18.7}$$

where $\mathbf{S}$ is a symmetric positive definite k by k matrix, $tr(\mathbf{S}\boldsymbol{\Sigma}^{-1})$ is the trace (sum of the diagonal elements) of matrix $\mathbf{S}\boldsymbol{\Sigma}^{-1}$, and the degrees of freedom, ν, must be greater than k (O'Hagan and Forster, 2004). The *inverse Wishart* distribution is the multivariate generalization of the S times an *inverse chi-squared* distribution when the multivariate random variables are arranged in a symmetric positive definite matrix. That makes it the appropriate distribution for the covariance matrix of a multivariate normal distribution.

Likelihood for Multivariate Normal Distribution with Unknown Covariance Matrix

Suppose $\mathbf{y}$ is a single random vector drawn from the $MVN(\boldsymbol{\mu}, \boldsymbol{\Sigma})$ distribution where both the mean vector $\boldsymbol{\mu}$ and the covariance matrix $\boldsymbol{\Sigma}$ are unknown. The likelihood of the single draw is given by

$$f(\mathbf{y}|\boldsymbol{\mu}, \boldsymbol{\Sigma}) \propto \frac{1}{|\boldsymbol{\Sigma}|^{\frac{1}{2}}} e^{-\frac{1}{2}(\mathbf{y}-\boldsymbol{\mu})'\boldsymbol{\Sigma}^{-1}(\mathbf{y}-\boldsymbol{\mu})}.$$

Now suppose we have a random sample of vectors, $\mathbf{y}_1, \ldots, \mathbf{y}_n$ drawn from the $MVN(\boldsymbol{\mu}, \boldsymbol{\Sigma})$ distribution. The likelihood of the random sample will be the

---

[2]$\boldsymbol{\mu}_0$ being outside the credible region is equivalent to $(\boldsymbol{\mu}_0 - \mathbf{m}_1)'\mathbf{V}_1^{-1}(\boldsymbol{\mu}_0 - \mathbf{m}_1) > c_k(\alpha)$. We are comparing the distance the null hypothesis value $\boldsymbol{\mu}_0$ is from the posterior mean $\mathbf{m}_1$ using a distance based on the posterior covariance matrix $\mathbf{V}_1$.

product of the individual likelihoods and is given by

$$f(\mathbf{y}_1,\ldots,\mathbf{y}_n|\boldsymbol{\mu},\boldsymbol{\Sigma}) \propto \prod_{i=1}^{n} \frac{1}{|\boldsymbol{\Sigma}|^{\frac{1}{2}}} e^{-\frac{1}{2}(\mathbf{y}_i-\boldsymbol{\mu})'\boldsymbol{\Sigma}^{-1}(\mathbf{y}_i-\boldsymbol{\mu})}$$

$$\propto \frac{1}{|\boldsymbol{\Sigma}|^{\frac{n}{2}}} e^{-\frac{1}{2}\sum_{i=1}^{n}(\mathbf{y}_i-\boldsymbol{\mu})'\boldsymbol{\Sigma}^{-1}(\mathbf{y}_i-\boldsymbol{\mu})}.$$

We look at the exponent, subtract and add the vector of means $\bar{\mathbf{y}}$, and break it into four sums.

$$\sum_{i=1}^{n}(\mathbf{y}_i-\boldsymbol{\mu})'\boldsymbol{\Sigma}^{-1}(\mathbf{y}_i-\boldsymbol{\mu}) = \sum_{i=1}^{n}(\mathbf{y}_i-\bar{\mathbf{y}}+\bar{\mathbf{y}}-\boldsymbol{\mu})'\boldsymbol{\Sigma}^{-1}(\mathbf{y}_i-\bar{\mathbf{y}}+\bar{\mathbf{y}}-\boldsymbol{\mu})$$

$$= \sum_{i=1}^{n}(\mathbf{y}_i-\bar{\mathbf{y}})'\boldsymbol{\Sigma}^{-1}(\mathbf{y}_i-\bar{\mathbf{y}}) + \sum_{i=1}^{n}(\mathbf{y}_i-\bar{\mathbf{y}})'\boldsymbol{\Sigma}^{-1}(\bar{\mathbf{y}}-\boldsymbol{\mu})$$

$$+ \sum_{i=1}^{n}(\bar{\mathbf{y}}-\boldsymbol{\mu})'\boldsymbol{\Sigma}^{-1}(\mathbf{y}_i-\bar{\mathbf{y}}) + \sum_{i=1}^{n}(\bar{\mathbf{y}}-\boldsymbol{\mu})'\boldsymbol{\Sigma}^{-1}(\bar{\mathbf{y}}-\boldsymbol{\mu}).$$

The middle two sums each equal $\mathbf{0}$, so the likelihood equals

$$f(\mathbf{y}_1,\ldots,\mathbf{y}_n|\boldsymbol{\mu},\boldsymbol{\Sigma}) \propto \frac{1}{|\boldsymbol{\Sigma}|^{\frac{1}{2}}} e^{-\frac{n}{2}(\bar{\mathbf{y}}-\boldsymbol{\mu})'\boldsymbol{\Sigma}^{-1}(\bar{\mathbf{y}}-\boldsymbol{\mu})} \times \frac{1}{|\boldsymbol{\Sigma}|^{\frac{n-1}{2}}} e^{-\frac{1}{2}\sum_{i=1}^{n}(\mathbf{y}_i-\bar{\mathbf{y}})'\boldsymbol{\Sigma}^{-1}(\mathbf{y}_i-\bar{\mathbf{y}})}.$$

We note that

$$\sum_{i=1}^{n}(\mathbf{y}_i-\bar{\mathbf{y}})'\boldsymbol{\Sigma}^{-1}(\mathbf{y}_i-\bar{\mathbf{y}})$$

is the trace (sum of diagonal elements) of the matrix $\mathbf{Y}'\boldsymbol{\Sigma}^{-1}\mathbf{Y}$, where

$$\mathbf{Y}' = \begin{bmatrix} \mathbf{y}_1' \\ \vdots \\ \mathbf{y}_n' \end{bmatrix}$$

is the matrix where the row vector observations are all stacked up. The trace will be the same for any cyclical permutation of the order of the matrix factors. Thus, the likelihood

$$f(\mathbf{y}_1,\ldots,\mathbf{y}_n|\boldsymbol{\mu},\boldsymbol{\Sigma}) \propto \frac{1}{|\boldsymbol{\Sigma}|^{\frac{1}{2}}} e^{-\frac{n}{2}(\bar{\mathbf{y}}-\boldsymbol{\mu})'\boldsymbol{\Sigma}^{-1}(\bar{\mathbf{y}}-\boldsymbol{\mu})}$$

$$\times \frac{1}{|\boldsymbol{\Sigma}|^{\frac{n-1}{2}}} e^{-\frac{1}{2}tr(\mathbf{SS}_M\boldsymbol{\Sigma}^{-1})} \tag{18.8}$$

is a product of a $MVN(\bar{\mathbf{y}}, \frac{\boldsymbol{\Sigma}}{n})$ for $\boldsymbol{\mu}$ conditional on $\boldsymbol{\Sigma}$ times a *inverse Wishart*$(\mathbf{SS}_M)$ distribution with $n-k-2$ degrees of freedom for $\boldsymbol{\Sigma}$ where $\mathbf{SS}_M = (\mathbf{y}-\bar{\mathbf{y}})(\mathbf{y}-\bar{\mathbf{y}})'$.

Finding the Exact Posterior when the Joint Conjugate Prior is Used for all Parameters

The product of independent conjugate priors for each parameter will not be jointly conjugate for all the parameters together just as in the single sample case. In this case, we saw that the form of the joint likelihood is a product of a $MVN(\bar{\mathbf{y}}, \frac{\Sigma}{n})$ for $\boldsymbol{\mu}$ conditional on Σ times a *inverse Wishart*$(\mathbf{SS}_M)$ distribution with $n - k - 2$ degrees of freedom for Σ where $\mathbf{SS}_M = (\mathbf{y} - \bar{\mathbf{y}})(\mathbf{y} - \bar{\mathbf{y}})'$. The joint conjugate prior will have the same form as the joint likelihood. If we let the conditional distribution of $\boldsymbol{\mu}$ be $MVN(\mathbf{m}_0, \frac{\Sigma}{n_0})$ and the marginal distribution of Σ be *Inverse Wishart*$(\mathbf{S}, \nu)$, where $\nu > k - 1$, then the joint conjugate prior is given by

$$g(\boldsymbol{\mu}, \Sigma) \propto \frac{1}{|\Sigma|^{\frac{1}{2}}} e^{-\frac{n_0}{2}(\boldsymbol{\mu} - \mathbf{m}_0)'\Sigma^{-1}(\boldsymbol{\mu} - \mathbf{m}_0)} \times \frac{1}{|\Sigma|^{\frac{\nu+k+1}{2}}} e^{-\frac{1}{2}tr(S\Sigma^{-1})}.$$

Therefore, the joint posterior of $\boldsymbol{\mu}$ and Σ is given by

$$g(\boldsymbol{\mu}, \Sigma | \mathbf{y}_1, \dots, \mathbf{y}_n) \propto \frac{1}{|\Sigma|^{\frac{1}{2}}} e^{-\frac{n_0}{2}(\boldsymbol{\mu} - \mathbf{m}_0)'\Sigma^{-1}(\boldsymbol{\mu} - \mathbf{m}_0)} \times \frac{1}{|\Sigma|^{\frac{\nu+k+1}{2}}} e^{-\frac{1}{2}tr(S\Sigma^{-1})}$$

$$\times \frac{1}{|\Sigma|^{\frac{1}{2}}} e^{-\frac{n}{2}(\bar{\mathbf{y}} - \boldsymbol{\mu})'\Sigma^{-1}(\bar{\mathbf{y}} - \boldsymbol{\mu})} \times \frac{1}{|\Sigma|^{\frac{n-1}{2}}} e^{-\frac{1}{2}tr(\mathbf{SS}_M\Sigma^{-1})}.$$

It can be shown that

$$n_0(\boldsymbol{\mu} - \mathbf{m}_0)'\Sigma^{-1}(\boldsymbol{\mu} - \mathbf{m}_0) + n(\bar{\mathbf{y}} - \boldsymbol{\mu})'\Sigma^{-1}(\bar{\mathbf{y}} - \boldsymbol{\mu})$$

$$= (n_0 + n)(\boldsymbol{\mu} - \mathbf{m}_1)'\Sigma^{-1}(\boldsymbol{\mu} - \mathbf{m}_1) + \frac{n_0 n}{n_0 + n}(\mathbf{m}_0 - \bar{\mathbf{y}})'\Sigma^{-1}(\mathbf{m}_0 - \bar{\mathbf{y}}),$$

where

$$\mathbf{m}_1 = \frac{n_0 \mathbf{m}_0 + n\bar{\mathbf{y}}}{n_0 + n}.$$

See Abadir and Magnus (2005, p. 216–217) for a proof. It can also be shown that

$$\frac{n_0 n}{n_0 + n}(\mathbf{m}_0 - \bar{\mathbf{y}})'\Sigma^{-1}(\mathbf{m}_0 - \bar{\mathbf{y}}) = tr\left[\frac{n_0 n}{n_0 + n}(\mathbf{m}_0 - \bar{\mathbf{y}})(\mathbf{m}_0 - \bar{\mathbf{y}})'\Sigma^{-1}\right].$$

This lets us write the joint posterior as

$$g(\boldsymbol{\mu}, \Sigma | \mathbf{y}_1, \dots, \mathbf{y}_n) \propto \frac{1}{|\Sigma|^{\frac{1}{2}}} e^{-\frac{n_0+n}{2}(\boldsymbol{\mu} - \mathbf{m}_1)'\Sigma^{-1}(\boldsymbol{\mu} - \mathbf{m}_1)}$$

$$\times \frac{1}{|\Sigma|^{\frac{n+\nu+k+1}{2}}} e^{-\frac{1}{2}tr[\mathbf{S}_1 \Sigma^{-1}]},$$

where

$$\mathbf{S}_1 = \mathbf{S} + \mathbf{SS}_M + \frac{n_0 n}{n_0 + n}(\mathbf{m}_0 - \bar{\mathbf{y}})(\mathbf{m}_0 - \bar{\mathbf{y}})'.$$

This expression has the shape of a *MVN* times an *Inverse Wishart* distribution as desired. It also tells us that conditional distribution of $\boldsymbol{\mu}$ given the data and $\boldsymbol{\Sigma}$ is *MVN* with posterior mean $\mathbf{m}_1$ and variance–covariance $\frac{\boldsymbol{\Sigma}}{(n_0+n)}$. This is the exact same result we would get from our updating formulas if we regard $\boldsymbol{\Sigma}$ as known and assume a $MVN(\mathbf{m}_0, \frac{\boldsymbol{\Sigma}}{n_0})$ prior for $\boldsymbol{\mu}$. It also tells us that the marginal distribution of $\boldsymbol{\Sigma}$ given the data is an *Inverse Wishart* distribution with a scale matrix of $\mathbf{S}_1$ and with $n + \nu$ degrees of freedom.

Finding the Marginal Distribution of $\boldsymbol{\mu}$ Generally we are interested in making inferences about $\boldsymbol{\mu}$ and not joint inference on both $\boldsymbol{\mu}$ and $\boldsymbol{\Sigma}$. In this case $\boldsymbol{\Sigma}$ is a nuisance parameter which we need to integrate out. To do this we can use the same trick to collapse the exponents. That is, we can write

$$-\frac{n_0 + n}{2}(\boldsymbol{\mu} - \mathbf{m}_1)'\boldsymbol{\Sigma}^{-1}(\boldsymbol{\mu} - \mathbf{m}_1) = -\frac{1}{2}tr\left((n_0 + n)(\boldsymbol{\mu} - \mathbf{m}_1)(\boldsymbol{\mu} - \mathbf{m}_1)'\boldsymbol{\Sigma}^{-1}\right).$$

Then we can rewrite the joint posterior as

$$g(\boldsymbol{\mu}, \boldsymbol{\Sigma}|\mathbf{y}_1, \ldots, \mathbf{y}_n) \propto \frac{1}{|\boldsymbol{\Sigma}|^{1+\frac{n+\nu+k}{2}}} e^{-\frac{1}{2}tr\left([\mathbf{S}_1+(n_0+n)(\boldsymbol{\mu}-\mathbf{m}_1)(\boldsymbol{\mu}-\mathbf{m}_1)']\boldsymbol{\Sigma}^{-1}\right)}.$$

Following DeGroot (1970, p. 180), the integral of this expression with respect to the $\frac{k(k+1)}{2}$ distinct terms of $\boldsymbol{\Sigma}$ is

$$g(\boldsymbol{\mu}|\mathbf{y}_1, \ldots, \mathbf{y}_n) \propto |\mathbf{S}_1 + (n_0 + n)(\boldsymbol{\mu} - \mathbf{m}_1)(\boldsymbol{\mu} - \mathbf{m}_1)'|^{-\frac{(\nu+n+1)}{2}}.$$

We need to use a result from the theory of determinants to get our final result. Harville (1997, p. 419–420) showed that if $\mathbf{R}$ is an $n \times n$ nonsingular matrix, $\mathbf{S}$ is an $n \times m$ matrix, $\mathbf{T}$ is an $m \times m$ nonsingular matrix, and $\mathbf{U}$ is an $m \times n$ matrix, then

$$|\mathbf{R} + \mathbf{STU}| = |\mathbf{R}||\mathbf{T}||\mathbf{T}^{-1} + \mathbf{UR}^{-1}\mathbf{S}|.$$

A special case of this result occurs when $\mathbf{T}$ has the scalar (1×1 matrix) value 1, $\mathbf{S} = \mathbf{v}$ is an n-dimensional column vector, and $\mathbf{U} = \mathbf{S}' = \mathbf{v}'$, and thus

$$|\mathbf{R} + \mathbf{vv}'| = |\mathbf{R}|(1 + \mathbf{v}'\mathbf{R}^{-1}\mathbf{v})$$

and is sometimes called the *Matrix Determinant Lemma*. Therefore,

$$g(\boldsymbol{\mu}|\mathbf{y}_1, \ldots, \mathbf{y}_n) \propto (1 + (n_0 + n)(\boldsymbol{\mu} - \mathbf{m}_1)'\mathbf{S}_1^{-1}(\boldsymbol{\mu} - \mathbf{m}_1))^{-\frac{(\nu+n+1)}{2}},$$

which has the form of a *multivariate t* distribution with $\nu + n - k + 1$ degrees of freedom, mean $\mathbf{m}_1$, and variance–covariance matrix $\frac{\mathbf{S}_1}{(n_0+n)(n_0+n-k+1)}$.

Main Points

- Let $\mathbf{y}$ and $\boldsymbol{\mu}$ be vectors of length k and let $\boldsymbol{\Sigma}$ be a k by k matrix of full rank. The multivariate observation $\mathbf{y}$ has the multivariate normal distribution with mean vector $\boldsymbol{\mu}$ and known covariance matrix $\boldsymbol{\Sigma}$ when its joint density function is given by

$$f(y_1, y_2) = \frac{1}{(2\pi)^{\frac{k}{2}}|\boldsymbol{\Sigma}|^{\frac{1}{2}}} e^{-\frac{1}{2}(\mathbf{u}-\boldsymbol{\mu})'\boldsymbol{\Sigma}^{-1}(\mathbf{u}-\boldsymbol{\mu})}, \tag{18.9}$$

where $|\boldsymbol{\Sigma}|$ is the determinant of the covariance matrix and $\boldsymbol{\Sigma}^{-1}$ is the inverse of the covariance matrix.

- Each component y_i has the $Normal(\mu_i, \sigma_i^2)$ distribution where μ_i is the ith component of the mean vector and σ_i^2 is the ith diagonal element of the covariance matrix.

- Furthermore, each subset of components has the appropriate multivariate normal distribution where the mean vector is made up of the means of that subset of components, and the covariance matrix is made up of the covariances of that subset of components.

- When we have $\mathbf{y}_1 \ldots, \mathbf{y}_n$, a random sample of draws from the $MVN(\boldsymbol{\mu}, \boldsymbol{\Sigma})$ distribution, and we use a $MVN(\mathbf{m}_0, \mathbf{V}_0)$ prior distribution for $\boldsymbol{\mu}$, the posterior distribution will be $MVN(\mathbf{m}_1, \mathbf{V}_1)$, where

$$\mathbf{V}_1^{-1} = \mathbf{V}_0^{-1} + n\boldsymbol{\Sigma}^{-1}$$

and

$$\mathbf{m}_1 = \mathbf{V}_1\mathbf{V}_0^{-1}\mathbf{m}_0 + \mathbf{V}_1 n\boldsymbol{\Sigma}^{-1}\bar{\mathbf{y}}.$$

- The simple updating rules are:

  - The posterior precision matrix (inverse of posterior covariance matrix) is the sum of the prior precision matrix (inverse of prior covariance matrix) plus the precision matrix of the sample (inverse of the co-variance matrix of the data (represented by the sample mean vector $\bar{\mathbf{y}}$).

  - The posterior mean vector is the posterior covariance matrix (inverse of the posterior precision matrix) times the prior precision matrix times the prior mean vector plus the posterior covariance matrix (inverse of posterior precision matrix) times the precision matrix of the data times the sample mean vector.

These are similar to the rules for the univariate normal, but adapted to multivariate observations.

Computer Exercises

18.1. Curran (2010) gives the elemental concentrations of five elements (manganese, barium, strontium, zirconium, and titanium) measured in six different beer bottles. Each bottle has measurements taken from four different locations (shoulder, neck, body and base). The scientists who collected this evidence expected that there would be no detectable different between measurements made on the same bottle, but there might be differences between the bottles. The data is included in the R package dafs which can be downloaded from CRAN. It can also be downloaded as a comma separated value (CSV) file from the URL http://www.introbayes.ac.nz.

[**Minitab:**] After you have downloaded and saved the file on your computer, select *Open Worksheet...* from the *File* menu. Change the *Files of type* drop down box to Text (*.csv). Locate the file *bottle.csv* in your directory and click on *OK*.

[**R:**] Type

```
bottle.df = read.csv("https://www.introbayes.ac.nz/bottle.csv")
```

18.2. A matrix of scatterplots is a good way to examine data like this.

[**Minitab:**] Select *Matrix Plot...* from the *Graph* menu. Click on *Matrix of Plots, With Groups* which is the second option on the top row of the dialog box. Click on *OK*. Enter Mn-Ti or c3-c7 into the *Graph variables:* text box. Enter Part Number or c2 c1 into the *Categorical variables for grouping (0-3):* text box, and click on *OK*.

[**R:**] Type

```
pairs(bottle.df[, -c(1:2)], col = bottle.df$Number,
      pch = 15 + as.numeric(bottle.df$Part))
```

It should be very clear from this plot that one bottle seems quite different from the others. We can see from the legend in Minitab that this is bottle number 5. We can do this in R by changing the plotting symbol to the bottle number. This is done by altering the second line of code

```
pairs(bottle.df[, -c(1:2)], col = bottle.df$Number,
      pch = as.character(bottle.df$Number))
```

```
pairs(bottle.df[, -c(1:2)], col = bottle.df$Number,
    pch = as.character(bottle.df$Number))
```

18.3. In this exercise we will test the hypothesis that the mean vector for bottle number 5 is not contained in a credible interval centered around the mean of the remaining observations. We can think of this as a crude equivalent of a hypothesis test for a single mean from a normal distribution. It is crude because we will not take into account the uncertainty in the measurements in bottle 5, nor will we account for the fact that we do not know the true mean concentration. However, we can see from the plots that bottle number 5 is quite distinct from the other bottles.

   i. Firstly we need to separate the measurements from bottle number 5 out and calculate the mean concentration of each element

   [**R:**]

   ```
   no5 = subset(bottle.df, Number == 5)[,-c(1,2)]
   no5.mean = colMeans(no5)
   ```

   And we need to separate out the data for the remaining bottles.

   ```
   rest = subset(bottle.df, Number != 5)[,-c(1,2)]
   ```

   [**Minitab:**] We need to divide the data into two groups in Minitab. The simplest way do this is to select the 20 rows of data where the bottle number is equal to 5 and cut (**Ctrl-X**) and paste (**Ctrl-V**) them into columns c9–c15. To calculate the column means we click on the *Session* window and then select *Command Line Editor* from the *Edit* menu. Type the following into Minitab:

   ```
   statistics c11-15;
   mean c17-c21.
   stack c17-c21 c16
   ```

   This will calculate the column means for bottle number 5, initially store them in columns c17 to c21, and then transpose them and store them in column c16.

   ii. Next we use the Minitab macro *MVNorm* or R function mvnmvnp to calculate the posterior mean. We assume a prior mean of $(0, 0, 0, 0, 0)'$, and a prior variance of $10^6\mathbf{I}_5$ where $\mathbf{I}_5$ is the 5×5 identify matrix as our data is recorded on five elements. Note that to perform our calculations in Minitab we need to use Minitab's matrix commands. These can be quite verbose and tedious to deal with as Minitab cannot perform multiple matrix operations on the same line.

[R:]

```
result = mvnmvnp(rest, m0 = 0, V0 = 1e6 * diag(5))
```

[Minitab:]

```
name m1 'SIGMA'
covariance 'Mn'-'Ti' 'SIGMA'.
name c17 'm0'
set c17
5(0)
end
set c18
5(10000)
end
name m2 'V0'
diag c18 'V0'
name m3 'Y'
name m4 'V1'
name c19 'm1'
copy 'Mn'-'Ti' 'Y'
%<path here>MVNorm 'Y' 5;
CovMat 'SIGMA';
prior 'm0' 'V0';
posterior 'm1' 'V1'.
```

iii. Finally we calculate the test statistic and the the *P*-value.

    [R:]

```
m1 = result$mean
V1 = result$var
d = no5.mean - m1
X0 = t(d) %*% solve(V1) %*% d
p.value = 1 - pchisq(X0, 5)
p.value
```

[Minitab:] Note that the **name** commands are not necessary, but it makes the commands slightly more understandable.

```
mult 'm1' -1 'm1'
add c16 'm1' c20
name c20 'd'
name m6 'V1Inv*d'
mult 'V1Inv' 'd' 'V1Inv*d'
name c21 'X0'
name m7 't(d)'
```

```
transpose 'd' 't(d)'
mult 't(d)' 'V1Inv*d' 'X0'
name c22 'Pval'
cdf 'X0' 'Pval';
chisquare 5.
let 'Pval' = 1 - 'Pval'
```

CHAPTER 19

BAYESIAN INFERENCE FOR THE MULTIPLE LINEAR REGRESSION MODEL

In Chapter 14 we looked at fitting a linear regression for the response variable y on a single predictor variable x using data that consisted of ordered pairs of points $(x_1, y_1), \ldots, (x_n, y_n)$. We assumed an unknown linear relationship between the variables, and found how to estimate the intercept and slope parameters from the data using the method of least squares. The goal of regression modeling is to find a model that will use the predictor variable x to improve our predictions of the response variable y. In order to do inference on the parameters and predictions, we needed assumptions on the nature of the data. These include the mean assumption, the normal error assumption (including equal variance), and the independence assumption. These assumptions enable us to develop the likelihood for the data. Then, we used Bayes' theorem to find the posterior distribution of the parameters given the data. It combined the information in our prior belief summarized in the prior distribution and the information in the data as summarized by the likelihood.

In this chapter we develop the methods for fitting a linear regression model for the response variable y on a set of predictor variables $x_1, \ldots, x_p$ from data consisting of points $(x_{11}, \ldots, x_{1p}, y_1), \ldots, (x_{n1}, \ldots, x_{np}, y_n)$. We assume that the response is related to the p predictors by an unknown linear function.

In Section 19.1 we see how to estimate the intercept and the slopes using the principle of least squares in matrix form. In Section 19.2 we look at the assumptions for the multiple linear regression model. They are analogous to those for the simple linear regression model: the mean assumption, the normal error assumption, and the independence assumption. Again, we develop the likelihood from these assumptions. In Section 19.3 we use Bayes' theorem to find the posterior distribution of the intercept and slope parameters. In Section 19.4 we show how to do Bayesian inferences for the parameters of the multiple linear regression model. We find credible intervals for individual parameters, as well as credible regions for the vector of parameters. We use these to test point hypothesis for both cases. In Section 19.5 we find the predictive distribution for a future observation.

19.1   Least Squares Regression for Multiple Linear Regression Model

The linear function $y = \beta_0 + \beta_1 x_1 + \ldots + \beta_p x_p$ forms a hyperplane[1] in $(p+1)$-dimensional space. The ith residual is the vertical distance the observed value of the response variable y_i is from the hyperplane and is given by $y_i - (\beta_0 + \beta_1 x_{i1} + \cdots + \beta_p x_{ip})$. The sum of squares of the residuals from the hyperplane is given by

$$SS_{res} = \sum_{i=1}^{n} [y_i - (\beta_0 + x_{i1}\beta_1 + \cdots + x_{ip}\beta_p)]^2 .$$

According to the least squares principle, we should find the values of the parameters $\beta_0, \beta_1, \ldots, \beta_p$ that minimize the sum of squares of the residuals. We find them by setting the derivatives of the sum of squares of the residuals with respect to each of the parameters equal to zero and finding the simultaneous

---

[1]A hyperplane is a generalization of a plane into higher dimensions. It is flat like a plane, but since it is in a higher dimension, we cannot picture it.

solution. These give the equations

$$\frac{\partial SS_{res}}{\partial \beta_0} = \sum_{i=1}^{n} [y_i - (\beta_0 + x_{i1}\beta_1 + \cdots + x_{ip}\beta_p)]^{2-1} \times (-1) = 0$$

$$\frac{\partial SS_{res}}{\partial \beta_1} = \sum_{i=1}^{n} [y_i - (\beta_0 + x_{i1}\beta_1 + \cdots + x_{ip}\beta_p)]^{2-1} \times (-x_{i1}) = 0$$

$$\vdots \qquad\qquad \vdots \qquad\qquad\qquad \vdots$$

$$\frac{\partial SS_{res}}{\partial \beta_1} = \sum_{i=1}^{n} [y_i - (\beta_0 + x_{i1}\beta_1 + \cdots + x_{ip}\beta_p)]^{2-1} \times (-x_{ip}) = 0$$

We can write these as a single equation in matrix notation

$$\mathbf{X}'[\mathbf{y} - \mathbf{X}\boldsymbol{\beta}] = \mathbf{0},$$

where the response vector, the matrix of predictors, and the parameter vector are given by

$$\mathbf{y} = \begin{pmatrix} y_1 \\ \vdots \\ y_n \end{pmatrix}, \quad \mathbf{X} = \begin{bmatrix} 1 & x_{11} & \cdots & x_{1p} \\ 1 & x_{21} & \cdots & x_{2p} \\ \vdots & \vdots & \ddots & \vdots \\ 1 & x_{n1} & \cdots & x_{np} \end{bmatrix}, \quad \text{and} \quad \boldsymbol{\beta} = \begin{pmatrix} \beta_0 \\ \vdots \\ \beta_p \end{pmatrix},$$

respectively. We can rearrange the equation to give the normal equation.[2]

$$\mathbf{X}'\mathbf{X}\boldsymbol{\beta} = \mathbf{X}'\mathbf{y}. \tag{19.1}$$

We will assume that $\mathbf{X}'\mathbf{X}$ has full rank $p+1$ so that its inverse exists and is unique. (If its rank is less than $p+1$, the model is over parameterized and the least squares estimates are not unique. In that case, we would reduce the number of parameters in the model until we have a full rank model.) We multiply both sides of the normal equation by the inverse $\mathbf{X}'\mathbf{X}^{-1}$ and its solution is the least squares vector

$$\mathbf{b}_{LS} = (\mathbf{X}'\mathbf{X})^{-1}\mathbf{X}'\mathbf{y}. \tag{19.2}$$

---

[2]This is the equation the least squares estimates satisfy. The least squares estimates are the projection of the n dimensional observation vector onto the $(p+1)$-dimensional space spanned by the columns of $\mathbf{X}$. Normal refers to the right angles that the residuals make with the columns of $\mathbf{X}$, not to the normal distribution.

19.2   Assumptions of Normal Multiple Linear Regression Model

The method of least squares is only a data analysis tool. It depends only on the data, not the probability distribution of the data. We cannot make any inferences about the slopes or the intercept unless we have a probability model that underlies the data. Hence, we make the following assumptions for the multiple linear regression model:

1. *Mean assumption.* The conditional mean of the response variable y given the values of the predictor variables $x_1, \ldots, x_p$ is an unknown linear function

$$\mu_{y|x_1,\ldots,x_p} = \beta_0 + \beta_1 x_1 + \cdots + \beta_p x_p,$$

   where β_0 is the intercept and β_i is the slope in direction x_i for $i = 1, \ldots, p$. β_i is the direct effect of increasing x_1 by one unit on the mean of the response variable y.

2. *Error assumption.* Each observation y_i equals its mean, plus a random error e_i for $i = 1, \ldots, n$. The random errors all have the $normal(0, \sigma^2)$ distribution. They all have the same variance σ^2. We are assuming that the variance is a known constant. Under this assumption, the covariance matrix of the observation vector equals $\sigma^2 \mathbf{I}$, where $\mathbf{I}$ is the n by n identity matrix.

3. *Independence assumption.* The errors are all independent of each other.

We assume that the observed data comes from this model. Since the least squares vector $\mathbf{b}_{LS}$ is a linear function of the observation vector y, its covariance matrix under these assumptions is

$$\mathbf{V}_{LS} = (\mathbf{X}'\mathbf{X})^{-1}\mathbf{X}'(\sigma^2\mathbf{I})\mathbf{X}(\mathbf{X}'\mathbf{X})^{-1}$$
$$= \sigma^2(\mathbf{X}'\mathbf{X})^{-1}.$$

If σ^2 is unknown, then we can estimate it from the sum of squares of the residuals away from the least squares hyperplane. The vector of fitted values is given by

$$\hat{\mathbf{y}} = \mathbf{X}\mathbf{b}_{LS}$$

$$= \mathbf{X}(\mathbf{X}'\mathbf{X})^{-1}\mathbf{X}'\mathbf{y}$$

$$= \mathbf{H}\mathbf{y},$$

where the matrix $\mathbf{H} = \mathbf{X}(\mathbf{X}'\mathbf{X})^{-1}\mathbf{X}'$. We note $\mathbf{H}$ and $\mathbf{I} - \mathbf{H}$ are symmetric idempotent[3] matrices. The residuals from the least squares hyperplane are

---

[3]An idempotent matrix multiplied by itself yields itself. Both $\mathbf{HH} = \mathbf{H}$ and $(\mathbf{I}-\mathbf{H})(\mathbf{I}-\mathbf{H}) = (\mathbf{I} - \mathbf{H})$.

given by

$$\hat{\mathbf{e}} = \mathbf{y} - \hat{\mathbf{y}}$$

$$= (\mathbf{I} - \mathbf{H})\mathbf{y}$$

We estimate σ^2 by the sum of squares of the residuals divided by their degrees of freedom. This is given by

$$\hat{\sigma}^2 = \frac{\hat{\mathbf{e}}'\hat{\mathbf{e}}}{n - (p+1)}$$

$$= \frac{\mathbf{y}'(\mathbf{I} - \mathbf{H})\mathbf{y}}{n - p - 1}$$

19.3   Bayes' Theorem for Normal Multiple Linear Regression Model

We will use the assumptions of the multiple linear regression model to find the joint likelihood of the parameter vector $\boldsymbol{\beta}$. Then we will apply Bayes' theorem to find the joint posterior. In the general case, this requires the evaluation of a $(p+1)$-dimensional integral which is usually done numerically. However, we will look at two cases where we can find the exact posterior without having to do any numerical integration. In the first case, we use independent flat priors for all parameters. In the second case, we will use the conjugate prior for the parameter vector.

Likelihood of Single Observation

Under the assumptions, the single observation y_i given the values of the predictor variables $x_{i1}, \ldots, x_{ip}$ is $normal(\mu_{y_i|x_{i1},\ldots,x_{ip}}, \sigma^2)$ where its mean

$$\mu_{y_i|x_{i1},\ldots,x_{ip}} = \sum_{j=0}^{p} x_{ij}\beta_j$$

$$= \mathbf{x}_i\boldsymbol{\beta},$$

where $\mathbf{x}_i$ equals $(x_{i0}, \ldots, x_{ip})$, the row vector of predictor variable values for the ith observation. Note: $x_{i0} = 1$. Hence the likelihood equals

$$f(y_i|\boldsymbol{\beta}) \propto e^{-\frac{1}{2\sigma^2}(y_i - \mathbf{x}_i\boldsymbol{\beta})^2} .$$

Likelihood of a Random Sample of Observations

All the observations are independent of each other. Hence the likelihood of the random sample is the product of the likelihoods of the individual observations and is given by

$$f(\mathbf{y}|\boldsymbol{\beta}) \propto \prod_{i=1}^{n} f(y_i|\boldsymbol{\beta})$$

$$\propto e^{-\frac{1}{2\sigma^2}\sum_{i=1}^{n}(y_i - \mathbf{x}_i\boldsymbol{\beta})^2}.$$

We can put the likelihood of the random sample in matrix notation as

$$f(\mathbf{y}|\boldsymbol{\beta}) \propto e^{-\frac{1}{2\sigma^2}(\mathbf{y}-\mathbf{X}\boldsymbol{\beta})'(\mathbf{y}-\mathbf{X}\boldsymbol{\beta})}.$$

We add and subtract $\mathbf{Xb}_{LS}$ from each term in the exponent and multiply it out.

$$(\mathbf{y} - \mathbf{X}\boldsymbol{\beta})'(\mathbf{y} - \mathbf{X}\boldsymbol{\beta}) = (\mathbf{y} - \mathbf{Xb}_{LS} + \mathbf{Xb}_{LS} - \mathbf{X}\boldsymbol{\beta})'(\mathbf{y} - \mathbf{Xb}_{LS} + \mathbf{Xb}_{LS} - \mathbf{X}\boldsymbol{\beta})$$
$$= (\mathbf{y} - \mathbf{Xb}_{LS})'(\mathbf{y} - \mathbf{Xb}_{LS}) + (\mathbf{y} - \mathbf{Xb}_{LS})'(\mathbf{Xb}_{LS} - \mathbf{X}\boldsymbol{\beta})$$
$$+ (\mathbf{Xb}_{LS} - \mathbf{X}\boldsymbol{\beta})'(\mathbf{y} - \mathbf{Xb}_{LS}) + (\mathbf{Xb}_{LS} - \mathbf{X}\boldsymbol{\beta})'(\mathbf{Xb}_{LS} - \mathbf{X}\boldsymbol{\beta})$$

Look at the first middle term. (The other middle term is its transpose.)

$$(\mathbf{y} - \mathbf{Xb}_{LS})'(\mathbf{Xb}_{LS} - \mathbf{X}\boldsymbol{\beta}) = (\mathbf{y} - \mathbf{X}(\mathbf{X'X})^{-1}\mathbf{X'y})'(\mathbf{X}(\mathbf{b}_{LS} - \boldsymbol{\beta}))$$
$$= \mathbf{y}'(\mathbf{I} - \mathbf{X}(\mathbf{X'X})^{-1}\mathbf{X'})(\mathbf{X}(\mathbf{b}_{LS} - \boldsymbol{\beta}))$$
$$= 0.$$

Thus the two middle terms are 0 and the likelihood of the random sample

$$f(\mathbf{y}|\boldsymbol{\beta}) \propto e^{-\frac{1}{2\sigma^2}[(\mathbf{y}-\mathbf{Xb}_{LS})'(\mathbf{y}-\mathbf{Xb}_{LS})+(\mathbf{Xb}_{LS}-\mathbf{X}\boldsymbol{\beta})'(\mathbf{Xb}_{LS}-\mathbf{X}\boldsymbol{\beta})]}.$$

Since the first term does not contain the parameter, it can be absorbed into the constant and the likelihood can be simplified to

$$f(\mathbf{y}|\boldsymbol{\beta}) \propto e^{-\frac{1}{2\sigma^2}(\mathbf{b}_{LS}-\boldsymbol{\beta})'(\mathbf{X'X})(\mathbf{b}_{LS}-\boldsymbol{\beta})}. \tag{19.3}$$

Thus the likelihood has the form of a $MVN(\mathbf{b}_{LS}, \mathbf{V}_{LS})$ where $\mathbf{V}_{LS} = \frac{\sigma^2}{(\mathbf{X'X})}$.

Finding the Posterior when a Multivariate Continuous Prior is Used

Suppose we use a continuous multivariate prior $g(\boldsymbol{\beta}) = g(\beta_0, \ldots, \beta_p)$ for the parameter vector. In matrix form, the joint posterior will be proportional to the joint prior times the joint likelihood

$$g(\boldsymbol{\beta}|\mathbf{y}) \propto g(\boldsymbol{\beta}) \times f(\mathbf{y}|\boldsymbol{\beta})$$

We can write this out component-wise as

$$g(\beta_0, \ldots, \beta_p|y_1, \ldots, y_n) \propto g(\beta_0, \ldots, \beta_p) \times f(y_1, \ldots, y_n|\beta_0, \ldots, \beta_p).$$

To find the exact posterior we divide the proportional posterior by its integral over all parameter values. This gives

$$g(\boldsymbol{\beta}|\mathbf{y}) = \frac{g(\beta_0, \dots, \beta_p) \times f(y_1, \dots, y_n|\beta_0, \dots, \beta_p)}{\int \dots \int g(\beta_0, \dots, \beta_p) \times f(y_1, \dots, y_n|\beta_0, \dots, \beta_p) d\beta_0 \dots d\beta_p}.$$

For most prior distributions, this integral will have to be evaluated numerically, and this may be difficult. We will look at two cases where we can evaluate the exact posterior without having to do the numerical integration.

Finding the Posterior when a Multivariate Flat Prior is Used

If we use a multivariate flat prior

$$g(\beta_0, \dots, \beta_p) = 1 \qquad \text{for} \qquad \left\{ \begin{array}{c} -\infty < \beta_0 < \infty \\ \vdots \\ -\infty < \beta_p < \infty \end{array} \right. ,$$

then the joint posterior will be proportional to the joint likelihood.

$$g(\boldsymbol{\beta}|\mathbf{y}) \propto e^{-\frac{1}{2\sigma^2}(\mathbf{b}_{LS}-\boldsymbol{\beta})'(\mathbf{X}'\mathbf{X})(\mathbf{b}_{LS}-\boldsymbol{\beta})}.$$

We recognize this as a $MVN(\mathbf{b}_{LS}, \mathbf{V}_{LS})$ distribution. Therefore the posterior mean is equal the least squares vector

$$\mathbf{b}_1 = \hat{\boldsymbol{\beta}} = \mathbf{b}_{LS}.$$

The posterior covariance matrix is

$$\mathbf{V}_1 = \mathbf{V}_{LS} = \sigma^2 (\mathbf{X}'\mathbf{X})^{-1}.$$

Finding the Posterior when a Multivariate Normal Prior is Used

We observed that the likelihood has the form of a $MVN(\mathbf{b}_{LS}, \mathbf{V}_{LS})$ distribution. The conjugate prior will also be *multivariate normal* of the same dimension. We will find that when we use a $MVN(\mathbf{b}_0, \mathbf{V}_0)$ prior for $\boldsymbol{\beta}$ the posterior can be found using a simple updating rule, without the need for numerical integration. The joint posterior is proportional to the prior times the likelihood.

$$g(\beta|\mathbf{y}) \propto g(\beta) \times f(\mathbf{y}|\beta)$$

$$\propto e^{-\frac{1}{2}[(\beta-\mathbf{b_0})'\mathbf{V}_0^{-1}(\beta-\mathbf{b_0}))} \times e^{-\frac{1}{2}[(\beta-\mathbf{b}_{LS})'\mathbf{V}_{LS}^{-1}(\beta-\mathbf{b}_{LS})]}$$

$$\propto e^{-\frac{1}{2}[(\beta-\mathbf{b_0})'\mathbf{V}_0^{-1}(\beta-\mathbf{b_0})+(\beta-\mathbf{b}_{LS})'\mathbf{V}_{LS}^{-1}(\beta-\mathbf{b}_{LS})]}$$

$$\propto e^{-\frac{1}{2}[\beta'(\mathbf{V}_0^{-1}+\mathbf{V}_{LS}^{-1})\beta-\beta'(\mathbf{V}_{LS}^{-1}\mathbf{b}_{LS}+\mathbf{V}_0^{-1}\mathbf{b_0})-(\mathbf{b'_{LS}}\mathbf{V}_{LS}^{-1}+\mathbf{b'_0}\mathbf{V}_0^{-1})\beta}$$

$$+ (\mathbf{b'_{LS}}\mathbf{V}_{LS}^{-1}+\mathbf{b'_0}\mathbf{V}_0^{-1})(\mathbf{V}_{LS}^{-1}\mathbf{b}_{LS}+\mathbf{V}_0^{-1}\mathbf{b_0})] \, .$$

The last term does not contain β so it will not affect the shape of the posterior. It can be absorbed into the proportionality constant. We let $\mathbf{V}_1^{-1} = \mathbf{V}_0^{-1} + \mathbf{V}_{LS}^{-1}$. The posterior becomes

$$\propto e^{-\frac{1}{2}[\beta'\mathbf{V}_1^{-1}\beta-\beta'(\mathbf{V}_{LS}^{-1}\mathbf{b}_{LS}+\mathbf{V}_0^{-1}\mathbf{b_0})-(\hat{\beta}'_{LS}\mathbf{V}_{LS}^{-1}+\mathbf{b'_0}\mathbf{V}_0^{-1})\beta]} \, .$$

Let $\mathbf{U}'\mathbf{U} = \mathbf{V}_1^{-1}$, where $\mathbf{U}$ an orthogonal matrix. We are assuming $\mathbf{V}_1^{-1}$ is of full rank so both $\mathbf{U}$ and $\mathbf{U}'$ are also full rank, and their inverses exist. We complete the square by adding $(\mathbf{b}'_{LS}\mathbf{V}_{LS}^{-1} + \mathbf{b}'_0\mathbf{V}_0^{-1})\mathbf{U}(\mathbf{U}')^{-1}(\mathbf{V}_{LS}^{-1}\mathbf{b}_{LS} + \mathbf{V}_0^{-1}\mathbf{b_0})$. We subtract it as well, but since that does not contain the parameter β, that part gets absorbed into the constant. The posterior becomes

$$\propto e^{-\frac{1}{2}[\beta'\mathbf{U}'\mathbf{U}\beta-\beta'\mathbf{U}'(\mathbf{U}')^{-1}(\mathbf{V}_{LS}^{-1}\mathbf{b}_{LS}+\mathbf{V}_0^{-1}\mathbf{b_0})-(\mathbf{b}'_{LS}\mathbf{V}_{LS}^{-1}+\mathbf{b}'_0\mathbf{V}_0^{-1})\mathbf{U}^{-1}\mathbf{U}\beta+}$$

$$(\mathbf{b}'_0\mathbf{V}_0^{-1}+\mathbf{b}_{LS}\mathbf{V}_{LS}^{-1})\mathbf{U}^{-1}(\mathbf{U}')^{-1}(\mathbf{V}_0^{-1}\mathbf{b_0}+\mathbf{V}_{LS}^{-1}\mathbf{b}_{LS})]} \, .$$

When we factor the exponent the posterior becomes

$$\propto e^{-\frac{1}{2}(\beta'\mathbf{U}'-(\mathbf{V}_{LS}^{-1}\mathbf{b}_{LS}+\mathbf{V}_0^{-1}\mathbf{b_0})\mathbf{U}^{-1})(\mathbf{U}\beta-(\mathbf{U}')^{-1}(\mathbf{V}_0^{-1}\mathbf{b_0}+\mathbf{V}_{LS}^{-1}\mathbf{b}_{LS}))} \, .$$

When we factor $\mathbf{U}'$ out of the first factor and $\mathbf{U}$ out of the second factor in the product, we get

$$e^{-\frac{1}{2}[\beta-(\mathbf{b}'_0\mathbf{V}_0^{-1}+\mathbf{b}'_{LS}\mathbf{V}_{LS}^{-1})\mathbf{U}^{-1}(\mathbf{U}')^{-1}]'(\mathbf{U}'\mathbf{U})[\beta-\mathbf{U}^{-1}(\mathbf{U}')^{-1}(\mathbf{V}_0^{-1}\mathbf{b_0}+\mathbf{V}_{LS}^{-1}\mathbf{b}_{LS})]} \, .$$

Since $\mathbf{U}'\mathbf{U} = \mathbf{V}_1^{-1}$ and they are all of full rank we have $(\mathbf{U}')^{-1}\mathbf{U}^{-1} = \mathbf{V}_1$. When we substitute back into the posterior we get

$$g(\beta|\mathbf{y}) \propto e^{-\frac{1}{2}(\beta-\mathbf{b}_1)'\mathbf{V}_1^{-1}(\beta-\mathbf{b}_1)}.$$

where $\mathbf{b}_1 = \mathbf{V}_1\mathbf{V}_0^{-1}\mathbf{b_0} + \mathbf{V}_1\mathbf{V}_{LS}^{-1}\mathbf{b}_{LS}$. The posterior distribution of $\beta|\mathbf{y}$ will be $MVN(\mathbf{b}_1, \mathbf{V}_1)$.

The Updating Formulas. When we have the assumptions of the multivariate linear regression model satisfied, and we use a $MVN(\mathbf{b}_0, \mathbf{V}_0)$ prior density the posterior will be $MVN(\mathbf{b}_1, \mathbf{V}_1)$ where the constants are found by the updating formulas "the posterior precision matrix equals the sum of the prior precision matrix plus the precision matrix of the likelihood function

$$\mathbf{V}_1^{-1} = \mathbf{V}_0^{-1} + \mathbf{V}_{LS}^{-1} \tag{19.4}$$

and "the posterior mean vector is the weighted average of the prior mean vector and the least squares vector where their weights are the inverse of the posterior precision matrix (which is the posterior covariance matrix) multiplied by their respective precision matrices"

$$\mathbf{b}_1 = \mathbf{V}_1 \mathbf{V}_0^{-1} \mathbf{b}_0 + \mathbf{V}_1 \mathbf{V}_{LS}^{-1} \mathbf{b}_{LS}. \tag{19.5}$$

19.4   Inference in the Multivariate Normal Linear Regression Model

In this section we look at making inferences about the parameters in the multivariate normal linear regression model. First we will look at making inferences about a single slope parameter. Here we are trying to determine the effect of that single predictor on the response variable. Later on, we will look at making inference on all the slope parameters at the same time. Here we are trying to determine the effect of all the predictors simultaneously on the the response variable.

Inference on a Single Slope Parameter

In this section we consider making inferences about a single slope parameter in the multiple linear regression model. The other slopes and the intercept are considered to be nuisance parameters. We make the inference on the single parameter on the marginal posterior of that parameter. The posterior distribution of the parameter vector is $MVN(\mathbf{b}_1, \mathbf{V}_1)$. Suppose β_k is the parameter of interest. The marginal posterior distribution of β_k is *normal*$(m'_{\beta_k}, s_k^2(s'_{\beta_k})^2)$ where the mean m'_{β_k} is the kth component of posterior mean vector $\mathbf{b}_1$ and the variance $s_k^2(s'_{\beta_k})^2$ is the kth diagonal element of the posterior covariance matrix $\mathbf{V}_1$.

Credible Interval for a Single Slope   A $(1 - \alpha) \times 100\%$ credible interval for the slope β_k is any interval that has posterior probability equal to $(1 - \alpha)$. The equal tail area $(1 - \alpha) \times 100\%$ credible interval is given by $(m'_{\beta_k} - z_{\alpha_2} s_k s'_{\beta_k},\ m'_{\beta_k} + z_{\alpha_2} s_k s'_{\beta_k})$. If the true standard deviation is unknown and we are using the estimate from the sample, then we find the critical values using the *Student's t* with $n - p - 1$ degrees of freedom instead of the normal. This will give an approximate credible interval for β_k. This approximation is exactly correct when we are using independent Jeffrey's priors for all the parameters.

Testing a Two-Sided Hypothesis for a Single Slope Using the Credible Interval
We can test the credibility of the null hypothesis

$$H_0 : \beta_k = \beta_{k\,0} \quad \text{versus} \quad H_1 : \beta_k \neq \beta_{k\,0}$$

using the credible interval. If the null value $\beta_{k\,0}$ lies outside the equal tail area
$(1 - \alpha) \times 100\%$ credible interval for β_k, then we can reject the null hypothesis
at the α level of significance. However, if the null value lies inside the interval,
the null value remains credible and we cannot reject the null hypothesis.

Testing a One-Sided Hypothesis for a Single Slope   We test a one-sided hypothesis

$$H_0 : \beta_k \leq \beta_{k\,0} \quad \text{versus} \quad H_1 : \beta_k > \beta_{k\,0}$$

about the slope β_k by calculating the posterior probability of the null hypothesis using its marginal posterior distribution. If this probability is less than the level of significance α, then we reject the null hypothesis $H_0 : \beta_k \leq \beta_{k\,0}$ and conclude the alternative hypothesis $H_1 : \beta_k > \beta_{k\,0}$ is true.

Inference for the Vector of all Slopes

Here we are wanting to make our inferences on the vector of all slopes. In this case, the only nuisance parameter is the intercept β_0. We will use the marginal posterior of all the slope parameters to do our inferences. The vector of slope parameters

$$\boldsymbol{\beta} = \begin{pmatrix} \beta_1 \\ \vdots \\ \beta_p \end{pmatrix}$$

is $MVN(\mathbf{b}_\beta, \mathbf{V}_\beta)$, where the components of the mean vector and covariance matrix come from $\mathbf{b}_1$ and $\mathbf{V}_1$, the mean vector and covariance matrix of the posterior distribution of the whole parameter vector (including intercept). We will assume that the covariance matrix $\mathbf{V}_\beta$ is full rank. Otherwise, we will reduce the number of slope parameters until it is.

Credible Region for all the Slopes

We want to find a region in the p-dimensional space that has $(1 - \alpha) \times 100\%$ posterior probability. We know

$$U = (\boldsymbol{\beta} - \mathbf{b}_\beta)\mathbf{V}_\beta^{-1}(\boldsymbol{\beta} - \mathbf{b}_\beta)$$

will have the *chi-squared* distribution with p degrees of freedom. This means that the region made up of all the points $\boldsymbol{\beta}$ such that

$$(\boldsymbol{\beta} - \mathbf{b}_\beta)\mathbf{V}_\beta^{-1}(\boldsymbol{\beta} - \mathbf{b}_\beta) < U_\alpha,$$

where U_α is the upper α point in the *chi-squared* distribution with p degrees of freedom, will be a $(1-\alpha)\times 100\%$ credible interval for the parameter vector.[4]

Testing a Point Hypothesis about all the Slopes

We want to test the null hypothesis

$$H_0 : \boldsymbol{\beta} = \boldsymbol{\beta_0} \text{ versus } H_1 : \boldsymbol{\beta} \neq \boldsymbol{\beta_0}.$$

Under the null hypothesis each slope β_k equals its null value $\beta_{k\,0}$ for $k = 1, \ldots, p$. If any of the slopes are not equal its null value the alternative is true. Thus in the p-dimensional space, there is only a single point where the null hypothesis is true. We can test the credibility of the null hypothesis using the credible region. If $\boldsymbol{\beta_0}$ lies outside the credible region, then we can reject the null hypothesis at the α level of significance. On the other hand, if the null value $\boldsymbol{\beta_0}$ lies inside the credible interval, then we cannot reject the null hypothesis as it remains credible at the level α.

Most often, we want to know whether or not all the slopes are equal to zero. If they are, none of the predictor variables are of any use in modeling the response. We will be testing the null value $\boldsymbol{\beta_0} = \mathbf{0}$. Here we are testing whether all the slopes equal 0 versus the alternative where at least one of the slopes is not equal to 0.

Modeling Issues: Removing Unnecessary Variables

Often, the multiple linear regression model is run including all possible predictor variables that we have data for. Some of these variables may affect the response very little if at all. The true coefficient of such a variable β_j would be very close to zero. Leaving these unnecessary predictor variables in the model can complicate the determination of the effects of the remaining predictor variables if there is correlation among the predictors themselves in the data set. Removal of these unnecessary predictors will lead to an improved model for predictions. This is often referred to as the principal of parsimony.

We would like to remove all predictor variables x_j where the true coefficient β_j equals 0. This is not as easy as it sounds as we do not know which coefficients are truly equal to zero. We have a random sample from the joint posterior distribution of $\beta_1, \ldots, \beta_J$. When the predictor variables $x_1, \ldots, x_J$ are correlated, some of the predictor variables can be either enhancing or masking the effect of other predictors. This means that a coefficient value estimated from the posterior sample may look very close to zero, but the effect of its predictor variable actually may be larger. Other predictor variables are masking its effect. Sometimes a whole set of predictor variables can be

---

[4]This credible region contains all the points that are "close" to the posterior mean vector where the closeness is measured by the posterior distribution of the parameter vector.

masking each others' effect, making each individual predictor look unnecessary (not significant), yet the set as a whole is very significant.

We should not test the hypothesis for each slope individually, in sequence. The individual test for $H_0 : \beta_j = 0$ versus $H_1 : \beta_j \neq 0$ is based on the additional effect of predictor x_j given the other predictors are already in the model. Thus for each predictor, its effect may be hidden by other predictors already in the model.

Instead, we should examine the posterior distribution of all the slopes and identify all those with mean close to 0 (in standard deviation units.) Those give us the predictor variables that are candidates for removal. Let $x_{k1}, \ldots, x_{kq}$ be the set of q predictor variables that are candidates for removal. let

$$\boldsymbol{\beta} = \begin{pmatrix} \beta_{k1} \\ \vdots \\ \beta_{kq} \end{pmatrix}$$

be the vector of those slopes. It has the marginal posterior distribution $MVN(\mathbf{b}_{\boldsymbol{\beta}}, \mathbf{V}_{\boldsymbol{\beta}})$, where the component means and covariances are given by the corresponding components of the mean vector and covariance matrix $\mathbf{b}_1$ and $\mathbf{V}_1$, the mean vector and covariance matrix of the posterior distribution of the whole parameter vector (including intercept). We compute the $(1 - \alpha) \times 100\%$ credible region for the vector of those slopes, $\boldsymbol{\beta}$. It will be the region made up of all the points $\boldsymbol{\beta}$ such that

$$(\boldsymbol{\beta} - \mathbf{b}_{\boldsymbol{\beta}})' \mathbf{V}_{\boldsymbol{\beta}}^{-1} (\boldsymbol{\beta} - \mathbf{b}_{\boldsymbol{\beta}}) < U_{\alpha},$$

where U_{α} is the upper α point in the *chi-squared* distribution with q degrees of freedom. We test the null hypothesis

$$H_0 : \boldsymbol{\beta} = \mathbf{0} \text{ versus } H_1 : \boldsymbol{\beta} \neq \mathbf{0}$$

at the level of significance α using the credible region. If $\mathbf{0}$ lies inside the credible region, then we cannot reject the null hypothesis, and it is credible that the slopes of all those predictors are simultaneously 0.

If that is the case, we remove those predictors from the model and redo the analysis with the remaining predictors.

◼ EXAMPLE 19.1

The Bears data (which can be found in both the Minitab example folder and the `Bolstad` package) contains a set of morphometric measurements as well as sex on a number of bears of various ages — although the age data is incomplete. Ilze decides that she would like to build a regression model which uses these measurements to predict the weight (in pounds) of the bears. Some of the bears in this data set have been measured more than once, but not in any way that would let Ilze incorporate the

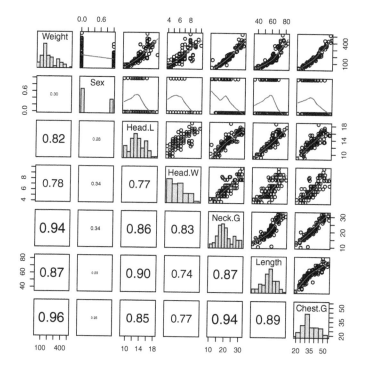

Figure 19.1   Scatterplot matrix of the variables in the Bears data.

correlation between successive measurements into her model. Therefore, Ilze discards all but the first measurement on each bear (Obs.No = 1), leaving a data set with 97 observations with which to build the model.

Ilze starts with some exploratory data analysis. Figure 19.1 shows a scatterplot matrix of each pair of variables she plans to use in the analysis. The figures in the lower triangle of the matrix are the linear correlation coefficients for the pairs of variables. Ilze can see that all of the continuous predictor variables have a moderately high correlation with weight, and she can also see that there is a moderate between the predictor variables themselves. The scatterplot matrices also reveal a slight increase in variability as the predictors increase, and they show that the relationship may be non-linear in some cases, and that the response variable Weight is right-skewed. All of these features suggest to Ilze that it may be better work to with the logarithm of Weight rather than Weight itself. This is often true for measurements like volume, concentration, time and income, that start at zero and (in theory) can increase indefinitely. The

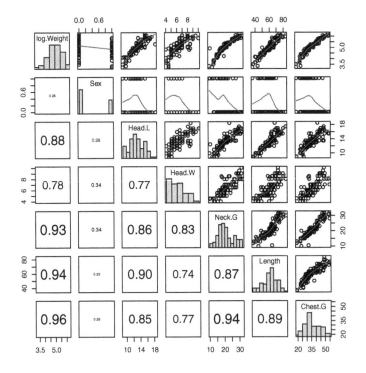

Figure 19.2   scatterplot matrix of the variables in the Bears data.

logarithmic transformation of the response is called a *variance stabilizing transformation* because, as well addressing issues of non-linearity, it often deals the issue of non-constant variance. The scatterplot matrix with a log transformed response is shown in Figure 19.2. Ilze decides to fit the model

$$\log(Weight_i) = \beta_0 + \beta_1 Sex_i + \beta_2 Head.L_i + \beta_3 Head.W_i$$
$$+ \beta_4 Neck.G_i + \beta_5 Length_i + \beta_6 Chest.G_i.$$

She chooses MVN prior with mean $\mathbf{b}_0 = \mathbf{0}$, and with prior variance of $\mathbf{V}_0 = 10^6 \times \mathbf{I}_7$. This is a very vague prior centered on zero. Ilze centers each of the covariates in the model by subtracting the mean from each variable. Centering can aid interpretation of the intercept, provide numerical stability, and in certain circumstances remove dependency between explanatory variables. The posterior estimates of the regression coefficients are shown in Table 19.1. Inspection of the coefficients and

Table 19.1   Regression coefficients

	Estimate	Std. Error	t value
Intercept	5.042562	0.014806	340.581
Sex	0.020919	0.033770	0.619
Head.L	0.001407	0.018223	0.077
Head.W	0.008746	0.018764	0.466
Neck.G	0.019491	0.009630	2.024
Length	0.024722	0.004130	5.986
Chest.G	0.034235	0.005568	6.148

their estimated standard deviations (standard errors) suggests that the variables Sex, Head.L and Head.W are not important, i.e. the coefficients are close to zero. Ilze decides to formally test this. If the point **0** is contained in the credible region, then it must satisfy the inequality $(\beta - \mathbf{b}_\beta)' \mathbf{V}_\beta^{-1} (\beta - \mathbf{b}_\beta) < U_\alpha$, where U_α in this case is the upper $\alpha = 0.05$ of the *chi-squared* distribution with three degrees of freedom. We have three degrees of freedom because we are considering removing three variables from the model. Therefore Ilze computes

$$(\mathbf{0} - \mathbf{b}_\beta)' \mathbf{V}_\beta^{-1} (\mathbf{0} - \mathbf{b}_\beta) = \mathbf{b}_\beta' \mathbf{V}_\beta^{-1} \mathbf{b}_\beta$$

and shows that it is less than $U_{0.05}^3 = 7.815$, where

$$\mathbf{b}_\beta = \begin{bmatrix} 0.02092 \\ 0.00141 \\ 0.00875 \end{bmatrix} \quad \text{and} \quad \mathbf{V}_\beta^{-1} = \begin{bmatrix} 892.95 & -113.26 & -203.02 \\ -113.26 & 3099.49 & 488.82 \\ -203.02 & 488.82 & 2955.83 \end{bmatrix}.$$

Using these numbers, Ilze shows that $\mathbf{b}_\beta' \mathbf{V}_\beta^{-1} \mathbf{b}_\beta = 0.554$ which is definitely less that 7.815; therefore these variables can be removed from the model. ∎

19.5   The Predictive Distribution for a Future Observation

In this section we consider Bayesian prediction using our linear regression model. As in the case of simple linear regression, we have a new observation, and we wish to predict the response, y_{n+1}. However, in this situation our new observation, $\mathbf{x}_{n+1}$, is a (row) vector of length $p + 1$, with the first element equal to 1, and the $(i + 1)^{\text{th}}$ element corresponding to a new value of the ith predictor. We will drop the subscripts from y and $\mathbf{x}$ in the following sections for mathematical simplicity.

If the coefficient vector β and the variance σ^2 of the residuals was known, and assuming the standard assumptions of independence, normality and equality of variance, then y would have a $normal(\mathbf{x}\beta, \sigma^2)$ distribution. However, we do not know β and σ^2. We only know their posterior distributions estimated from the data. Therefore, like simple linear regression, we need to find the joint density of the next observation and the model parameters and then integrate β and the variance σ^2 out of this expression. That is, the posterior predictive distribution for y is

$$f(y|\mathbf{x}, data) = \int f(y|\mathbf{x}, data, \beta, \sigma^2)g(\beta, \sigma^2|\mathbf{x}, data)\, d\beta d\sigma^2.$$

We start initially with the situation where σ^2 is known. The distribution does not depend on the data, and the distribution of β does not depend on $\mathbf{x}$. The predictive density for y is then given by

$$f(y|\mathbf{x}, data) = \int f(y|\mathbf{x}, \beta)g(\beta|data)\, d\beta.$$

We know that if β and σ^2 are known then y has a $normal(\mathbf{x}\beta, \sigma^2)$ distribution, and that the posterior distribution of β is $MVN(\mathbf{b}_1, \mathbf{V}_1)$, so

$$f(y|x, data) \propto \int e^{-\frac{1}{2\sigma^2}(y-x\beta)^2} \times e^{-\frac{1}{2}(\beta-\mathbf{b}_1)'\mathbf{V}_1^{-1}(\beta-\mathbf{b}_1)}\, d\beta$$

$$= \int e^{-\frac{1}{2\sigma^2}[y^2 - 2\mathbf{x}\beta y + (\mathbf{x}\beta)^2] - \frac{1}{2}[\beta'\mathbf{V}_1^{-1}\beta - 2\beta'\mathbf{V}_1^{-1}\mathbf{b}_1 + \mathbf{b}_1'\mathbf{V}_1^{-1}\mathbf{b}_1]}\, d\beta.$$

The term $\mathbf{b}_1'\mathbf{V}_1^{-1}\mathbf{b}_1$ does not depend on β and hence can be absorbed into the constant of integration. It is also convenient at this point to consider $\mathbf{x}$ as a column vector write $\beta'\mathbf{x}$ instead of $\mathbf{x}\beta$. Dealing solely with the exponent and letting $\tau = \frac{1}{\sigma^2}$, we have

$$-\frac{1}{2\sigma^2}\left[y^2 - 2\mathbf{x}\beta y + (\mathbf{x}\beta)^2\right] - \frac{1}{2}\left[\beta'\mathbf{V}_1^{-1}\beta - 2\beta'\mathbf{V}_1^{-1}\mathbf{b}_1\right]$$

$$= -\frac{1}{2\sigma^2}\left[y^2 - 2\beta'\mathbf{x}y + (\beta'\mathbf{x})^2\right] - \frac{1}{2}\left[\beta'\mathbf{V}_1^{-1}\beta - 2\beta'\mathbf{V}_1^{-1}\mathbf{b}_1\right]$$

$$= -\frac{1}{2}\left(\tau y^2 - 2\tau\beta'\mathbf{x}y + \tau\beta'\mathbf{x}\mathbf{x}'\beta + \beta'\mathbf{V}_1^{-1}\beta - 2\beta'\mathbf{V}_1^{-1}\mathbf{b}_1\right)$$

$$= -\frac{1}{2}\left(\beta'(\tau\mathbf{x}\mathbf{x}' + \mathbf{V}_1^{-1})\beta - 2\beta'(\tau y\mathbf{x} + \mathbf{V}_1^{-1}\mathbf{b}_1) + \tau y^2\right).$$

If we let $\mathbf{V} = \tau\mathbf{x}\mathbf{x}' + \mathbf{V}_1^{-1}$ and let $\mathbf{m} = \mathbf{V}^{-1}(\tau y\mathbf{x} + \mathbf{V}_1^{-1}\mathbf{b}_1)$ assuming $\mathbf{V}$ is invertible, then we can complete the square so that the form of the exponent is

$$(\beta - \mathbf{m})'\mathbf{V}(\beta - \mathbf{m}) - \mathbf{m}'\mathbf{V}\mathbf{m} + \tau y^2.$$

Substituting this back into our predictive posterior density, we have

$$f(y|\mathbf{x}, data) \propto \int e^{-\frac{1}{2}(\beta-\mathbf{m})'\mathbf{V}(\beta-\mathbf{m})} \times e^{\frac{1}{2}(\mathbf{m}'\mathbf{V}\mathbf{m} - \tau y^2)}\, d\beta.$$

The second term does not depend on $\boldsymbol{\beta}$ and therefore can be moved outside the integral

$$f(y|\mathbf{x}, data) \propto e^{\frac{1}{2}(\mathbf{m}'\mathbf{V}\mathbf{m}-\tau y^2)} \int e^{-\frac{1}{2}(\boldsymbol{\beta}-\mathbf{m})'\mathbf{V}(\boldsymbol{\beta}-\mathbf{m})} d\boldsymbol{\beta}.$$

The integral is proportional to a MVN density and hence integrates to a constant which can be absorbed into the constant of proportionality. It remains to rearrange $e^{\frac{1}{2}(\mathbf{m}'V\mathbf{m}-\tau y^2)}$ into a form we are familiar with. Working with the exponent again, we have

$$\tau y^2 - \mathbf{m}'\mathbf{V}\mathbf{m}.$$

This expression can be rewritten as a quadratic form, again by completing the square, so that

$$f(y|\mathbf{x}, data) \propto e^{-\frac{1}{2[\sigma^2+\mathbf{x}'\mathbf{V}_1\mathbf{x}]}(y-\mathbf{b}_1'\mathbf{x})^2}.$$

This last calculation is not trivial and requires the use of the Sherman–Morrison formula (Sherman and Morrison, 1949, 1950).

Theorem 19.1 *Suppose* $\mathbf{A}$ *is an invertible square matrix, and* $\mathbf{u}, \mathbf{v}$ *are column vectors. Furthermore, suppose that if* $1 + \mathbf{v}'\mathbf{A}^{-1}\mathbf{u} \neq 0$, *then*

$$(\mathbf{A} + \mathbf{u}\mathbf{v}')^{-1} = \mathbf{A}^{-1} - \frac{\mathbf{A}^{-1}\mathbf{u}\mathbf{v}'\mathbf{A}^{-1}}{1 + \mathbf{v}'\mathbf{A}^{-1}\mathbf{u}}.$$

This result is known as the Sherman–Morrison formula.

We know that the denominator condition holds because $\mathbf{V}_1$ is a variance–covariance matrix, hence it is invertible and positive–semidefinite which guarantees that the quadratic form $\mathbf{x}'\mathbf{V}_1^{-1}\mathbf{x}$ is always greater than or equal to zero.

This means that the posterior predictive distribution is proportional to a *normal* distribution with mean $\mathbf{b}_1'\mathbf{x}$ and variance $\sigma^2 + \mathbf{x}'\mathbf{V}_1\mathbf{x}$. The variance has two components: σ^2 represents sampling uncertainty, and $\mathbf{x}'\mathbf{V}_1\mathbf{x}$ represents uncertainty about $\boldsymbol{\beta}$. If we use flat priors for $\boldsymbol{\beta}$, then the posterior mean vector and covariance for matrix for $\boldsymbol{\beta}$ will be equal to the maximum likelihood estimates, which in this case are the least squares solutions. That is, if we use flat priors for $\boldsymbol{\beta}$, then the posterior distribution of $\boldsymbol{\beta}$ is MVN with parameters $\mathbf{b}_1 = \mathbf{b}_{LS}$ and $\mathbf{V}_1 = \mathbf{V}_{LS}$. The variance of the posterior predictive distribution then simplifies to $\sigma^2(1 + \mathbf{x}'(\mathbf{X}'\mathbf{X})^{-1}\mathbf{x})$.

So far we have assumed σ^2 that is known, which is unrealistic. In the case where σ^2 is unknown, it can be shown that the posterior predictive density has the shape of a *Student* t distribution with mean $\mathbf{b}_1'\mathbf{x}$, variance $s^2 + \mathbf{x}'(\mathbf{X}'\mathbf{X})^{-1}\mathbf{x}$, and $n - p$ degrees of freedom where s^2 is the *residual mean square*, i.e.,

$$s^2 = \frac{1}{n-p}(y - \mathbf{X}\mathbf{b}_{LS})'(y - \mathbf{X}\mathbf{b}_{LS}).$$

Note that this result holds exactly for a flat prior on $(\boldsymbol{\beta}, \sigma^2)$, and approximately if the prior is very uninformative. This derivation requires some lengthy algebraic manipulation and so is not shown here.

Main Points

- Multiple regression describes the situation where we are interested in relating a single vector of observed response values, $\mathbf{y}$, to a set of two or more possible explanatory (or predictor) variables, $\mathbf{x}_1, \mathbf{x}_2, \ldots, \mathbf{x}_p$.

- Bayesian multiple regression involves finding the posterior mean, $\mathbf{b}_1$, and covariance matrix $\mathbf{V}_1$ for the vector of regression coefficients $\boldsymbol{\beta}$. We are interested in making inferences about these parameters given our observed data.

- If we assume a multivariate flat prior for $\boldsymbol{\beta}$, then the posterior distribution of $\boldsymbol{\beta}$ is MVN with posterior mean and variance equal to the least squares estimates, i.e., $\mathbf{b}_1 = \mathbf{b}_{LS}$ and $\mathbf{V}_1 = \mathbf{V}_{LS}$.

- If we assume a $MVN(\mathbf{b}_0, \mathbf{V}_0)$ prior distribution for $\boldsymbol{\beta}$, then the posterior distribution is also MVN, and the parameters can be estimated by two simple updating formulas

$$\mathbf{V}_1^{-1} = \mathbf{V}_0^{-1} + \mathbf{V}_{LS}^{-1}$$

and

$$\mathbf{b}_1 = \mathbf{V}_1 \mathbf{V}_0^{-1} \mathbf{b}_0 + \mathbf{V}_1 \mathbf{V}_{LS}^{-1} \mathbf{b}_{LS}.$$

Computer Exercises

19.1. The data in this exercise and those that follow can be downloaded from http://www.stat.berkeley.edu/~statlabs/data/babies.data. The variables in the data set are:

Variable	Description
bwt	Birth weight in ounces (999 = unknown)
gestation	Length of pregnancy in days (999 = unknown)
parity	Order of birth (0 = first born, 9 = unknown)
age	Mother's age in years
height	Mother's height in inches (99 = unknown)
weight	Mother's pre-pregnancy weight in pounds (999 = unknown)
smoke	Smoking status of mother (0 = not now, 1 = yes now, 9 = unknown)

[**Minitab:**] The data on the webpage needs to be saved as as a text file (`*.txt`) import the data into Minitab. It is important to click on the `Options...` button and choose the *Free format* field definition before importing file; otherwise the data will not import correctly.

[**R:**] R can read the data directly from the URL. Simply type

```
url = "http://www.stat.berkeley.edu/~statlabs/data/babies.data"
bw.df = read.table(url, head = TRUE)
```

It is not necessary to do this in two steps, but it clarifies what R is doing.

If the data has been correctly imported then you should have 1,236 observations on seven variables.

19.2. It is important to make sure that we are working with only the complete data; otherwise we have to have a model for the missing values.

[**Minitab:**] Select *Copy > Columns to Columns...* from the *Data* menu. Enter `c1-c7` in the *Copy from columns:* text box. Click on the *Subset the data...* button. Click on the *Specify rows to include* and *Rows that match* radio buttons. Click on the *Condition...* button. Enter the following condition into the *Condition* text box:

```
bwt <> 999  And  gestation <> 999 And parity <> 9 And
height <> 99 And weight <> 999 And smoke <> 9
```

Finally click *OK* on each of the three dialogue boxes. This will produce a new worksheet with the 1,175 complete cases.

[**R:**] Type

```
bw.df = subset(bw.df, bwt != 999  &  gestation != 999
                & parity != 9 & height != 99
                & weight != 999 & smoke != 9)
nrow(bw.df)
```

The **nrow** function will tell you how many (complete) cases are left in the data after the subsetting operation.

19.3. It is always useful to plot the data before we contemplate models. This sometimes can reveal features which we may not have noticed in the data, and can warn us of potential issues. A scatterplot matrix is a good first choice for multiple regression.

[**Minitab:**] Select *Matrix Plot...* from the *Graph* menu, and then select *Matrix of plots* with the *With Smoother* option before clicking on *OK*. Enter either c1-c7 or bwt-smoke into the *Graph variables* text box and click on *OK*.

[**R:**] Type

```
pairs(bw.df, upper.panel = panel.smooth)
```

You should notice that there appears to be an unusual *age* value of 99, which we think can reasonably considered a missing value even though it is not mentioned in the data description. We should remove this point from our analysis.

[**Minitab:**] Hover (move the pointer with the mouse but do not click) over the the far right point in any of the plots in the *age* column. This should pop up a label telling you that the observation is in Row 401. Select *Delete...* from the *Data* menu. Enter 401 into the *Rows to delete* text box and enter bwt-smoke or c1-c7 into the *Columns from which to delete these rows* text box. Click on *OK*.

[**R:**] Type

```
bw.df = subset(bw.df, age != 99)
```

19.4. Use the Minitab macro *BayesMultReg* or the R function bayes.lm with a multivariate normal prior to fit a multiple linear regression model to this data set. An initial choice of prior might be $\mathbf{b}_0 = \mathbf{0}$ and $\mathbf{V}_0 = 10^6 \mathbf{I}_7$, where $\mathbf{I}_7$ is a 7×7 identity matrix. This is a very diffuse prior centered on zero.

19.5. Use the posterior mean and covariance of the regression coefficients to test the hypothesis

$$H_0 : \begin{pmatrix} \beta_{age} \\ \beta_{weight} \end{pmatrix} = \begin{pmatrix} 0 \\ 0 \end{pmatrix}.$$

CHAPTER 20

COMPUTATIONAL BAYESIAN STATISTICS INCLUDING MARKOV CHAIN MONTE CARLO

The posterior distribution itself is the essence of Bayesian inference. It summarizes all that we can believe about the parameter(s) after looking at the data. All further Bayesian inferences such as finding a point estimate of a parameter, finding a credible interval for a parameter, and testing a hypothesis about a parameter can be performed by calculations from the posterior distribution. However, finding the posterior itself by using Bayes' theorem is not always as easy as it seems. In earlier chapters we have shown that it is sometimes possible to find a formula for the exact posterior density. In other cases we have to calculate the posterior numerically. Even that may be difficult when there is a multivariate parameter. We need to find another way to do Bayesian inference.

In this chapter we show that there is another way we can make inferences about the parameter. They can be based on a random sample drawn from the posterior distribution. The histogram of the random sample from the posterior will approach the posterior density as the sample size increases towards infinity. Thus statistics calculated from the random sample will approach the parameters of the posterior distribution. This is the most basic idea of statistics; a random sample from a population becomes closer and closer to

Introduction to Bayesian Statistics, 3rd ed.
By Bolstad, W. M. and Curran, J. M. Copyright © 2016 John Wiley & Sons, Inc.

the population as the sample size increases. This is the basis for computational Bayesian statistics. It is the driving force behind the great resurgence of Bayesian statistics over the past quarter century. Consider the following example.

■ EXAMPLE 20.1

Suppose Aisha, Blair, and Chiara observe 5 successes out of 20 Bernoulli trials with success probability π. They decide to use a $beta(1,1)$ prior for π. Aisha says the posterior will be $beta(6,16)$. She calculates the posterior mean, median, and an equal tail area 95% credible interval for π using Minitab or R. Blair notes that the proportional posterior will be given by

$$g(\pi|y) \propto g(\pi)\, f(y|\pi)$$
$$\propto \pi^{1-1}(1-\pi)^{1-1}\, \pi^5(1-\pi)^{20-5}$$
$$\propto \pi^5\,(1-\pi)^{15}.$$

He integrates this proportional posterior over the whole range $0 \le \pi \le 1$ to find the scale factor needed to make this a probability density.

$$\int_0^1 \pi^5(1-\pi)^{15}\, d\pi = .000003071.$$

He finds the numerical posterior density is

$$g(\pi|y) = \frac{1}{.000003071}\pi^5(1-\pi)^{15}.$$

[**Minitab:**]   He calculates the posterior mean using the macro *tintegral* and the posterior median and the (equal tail area) 95% credible interval using the macro *CredIntNum*.

[**R:**]   He calculates the posterior density using `binogcp` and calculates the posterior mean using the R `mean` function, and he also calculates the posterior median and the (equal tail area) 95% credible interval using the `median` and `quantile` function respectively.

Chiara decides to take random samples from the posterior. The histograms of her samples are shown in Figure 20.1 together with the exact posterior. We see that the histogram of the random sample from the posterior is approaching the shape of the true posterior as the sample size is increasing. She calculates the sample mean and the sample median from her posterior sample, and she also calculates an equal tail area 95% credible interval from the posterior sample. Instead of calculating the tail area

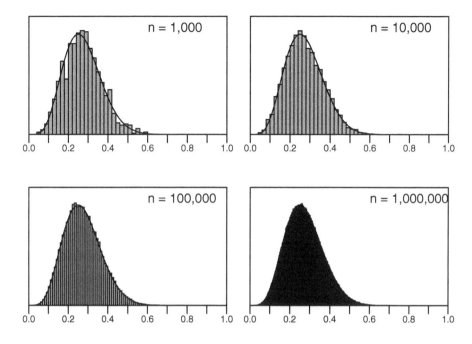

Figure 20.1   Chiara's samples from the posterior distribution for sample sizes 1,000,10,000,100,000, and 1,000,000 respectively.

based on probability, she calculates the tail area based on the proportion of the posterior sample. The exact, numerical, and sample results are shown in Table 20.1.   ∎

Table 20.1   The posterior mean, median, and equal tail area 95% credible interval

Person	Posterior	Mean	Median	95% Credible Interval	
				lower	upper
Aisha	exact	.27273	.26574	.11281	.47166
Blair	numerical	.27273	.26574	.11281	.47166
Chiara	sample (1,000)	.27314	.26730	.11110	.47316
Chiara	sample (10,000)	.27280	.26534	.11073	.47077
Chiara	sample (100,000)	.27303	.26605	.11312	.47290
Chiara	sample (1,000,000)	.27283	.26587	.11308	.47174

In the above example we see that the numerical posterior gives the same results as the exact posterior, as it should. Also, the statistics calculated from the posterior sample are approximations to the correct values, and the approximation improves as the sample size increases. This shows that we can base inferences on random samples drawn from the posterior distribution, provided the sample size is large enough. Of course, in this example we knew the exact posterior density and it was easily sampled from. We will see that it is not necessary to know the exact posterior density in order to draw samples from it. All we need to know is a formula that gives us the shape of the posterior. We do not need to know the scale factor needed to make it an exact density.

Bayesian Statistics: Easy in Theory, Difficult in Practice

Bayesian statistics is easy in theory: The *posterior is proportional to the prior times the likelihood*.

$$g(\theta|y) \propto g(\theta)\, f(y|\theta)\,.$$

Thus it is easy to find an equation that gives the shape of the posterior density. However this equation does not give the exact density as it does not give the scale factor needed to make it integrate to 1. Since it is not the exact density, neither probabilities nor moments can be calculated from it. It cannot be used for statistical inference. The exact posterior is found by dividing the proportional posterior by its integral over all parameter values

$$g(\theta|y) = \frac{g(\theta)\, f(y|\theta)}{\int g(\theta)\, f(y|\theta)\, d\theta}\,.$$

A closed form for the integral and hence for the posterior can only be found in a limited number of special cases. In other cases, it needs to be evaluated numerically. This numerical process quickly loses efficiency as the dimension of the parameter θ increases, since the number of points where the function has to be evaluated increases exponentially with the dimension. Also, the accuracy of the numerical integral depends on the placement of the evaluated points in the high dimension space. Thus, Bayesian statistics is often difficult in practice.

The difficulty of evaluating the posterior in the general case left Bayesian statistics out of mainstream applied statistical practice. Statisticians were aware from their studies in decision theory that Bayesian statistics offered real advantages in theory,[1] but in practice these advantages were not really available. Almost all applied statistics was done using frequentist methods.

Then, in the last quarter of the twentieth century, statisticians became aware of methods for drawing samples from the true posterior, even when we only know the unscaled version. Some of these methods had been developed

---

[1] Wald showed that admissible estimators were classified as Bayesian estimators.

much earlier, but until sufficient computing power became available to implement them, they were mostly unused. Computational Bayesian statistics is based on using these algorithms to draw samples from the posterior and then using the random sample from the posterior as the basis for inference. These methods work even when we do not know the exact posterior, only its unscaled version. They work for general distributions, not just for the exponential family with conjugate prior case. The statistician can focus on the statistical aspects of the model without worrying about calculability. This allows the applied statistician to use realistic models that are based on the underlying situation instead of being restricted to models that are mathematically easy to work with. These methods are not approximations as we are drawing a Monte Carlo random sample from the exact posterior. Estimates calculated from the sample can achieve any required accuracy by setting the sample size large enough. Existing exploratory data analysis (EDA) techniques can be used to explore the posterior. This essentially is the overall goal of Bayesian inference. Sensitivity analysis can be done on the model in a simple fashion.

In Section 20.1 we introduce acceptance–rejection sampling where we draw a random sample of candidates from an easily sampled density. Then we reshape this sample into a random sample from the posterior by only accepting some of the values into the final sample. This performs very satisfactorily as long as the candidate density dominates the target. However, it becomes inefficient as the number of parameters increases.

In Section 20.3, we introduce Markov chain Monte Carlo (MCMC) methods for drawing a sample from the posterior. Here we set up a Markov chain that has the posterior as its long-run distribution. We let the Markov chain run long enough so a random draw from the chain can be considered a random draw from the posterior. The Metropolis–Hastings algorithm and the Gibbs sampling algorithm are the two main Markov chain Monte Carlo methods. The Markov chain Monte Carlo samples will not be independent. There will be serial dependence in the Markov chain output due to the Markov property. Different chains have different mixing properties. That means they move around the parameter space at different rates. We show how to determine how much we must thin the sample to obtain a sample that well approximates a random sample from the posterior to be used for inference.

In Section 20.2, we look at performing the inferences from the posterior sample. The overall goal of Bayesian inference is knowing the posterior. The fundamental idea behind nearly all statistical methods is that as the sample size increases, the distribution of a random sample from a population approaches the distribution of the population. Thus, the histogram of the random sample from the posterior will approach the true posterior density. Other inferences such as point and interval estimates of the parameters can be constructed from the posterior sample. For example, if we had a random sample from the posterior, any parameter could be estimated by the corresponding statistic calculated from that random sample. We could achieve any required level of accuracy for our estimates by making sure our random sample from the posterior is large enough. Existing exploratory data analysis (EDA) techniques can be used on the sample from the posterior to explore the relationships between parameters in the posterior.

The computational approach to Bayesian statistics allows the posterior to be approached from a completely different direction. Instead of using the computer to calculate the posterior numerically, we use the computer to draw a Monte Carlo sample from the posterior. These methods have revolutionized Bayesian statistics. They have freed Bayesian statisticians from being restricted to those models where the posterior can be found analytically. Now, Bayesian statisticians can use observation models, choose prior distributions that are more realistic, and calculate estimates of the parameters from the Monte Carlo samples from the posterior. Computational Bayesian methods can easily deal with complicated models that have many parameters. This makes the advantages that the Bayesian approach offers accessible to a much wider class of useful models. Bayesian statisticians are no longer constrained by analytic or numerical tractability. Models that are based on the underlying situation can be used instead of models based on mathematical convenience. This allows the statistician to focus on the statistical aspects of the model without worrying about calculability.

20.1   Direct Methods for Sampling from the Posterior

In these direct methods we obtain our random sample from the posterior either by transforming a random sample drawn from another distribution, or by reshaping a random sample drawn from an easily sampled candidate distribution. We do this reshaping by only accepting some of the values into the final sample. The first method we will look at is called inverse proabality sampling. We will then look at acceptance–rejection sampling.

Inverse Probability Sampling

Inverse probability sampling relies on the *probability integral transform*.

Theorem 20.1 *If X is a continuous random variable with cumulative distribution function $F_X(x)$, then the random variable Y defined as*

$$Y = F_X(X)$$

has a uniform$(0, 1)$ *distribution.*

The inverse of this theorem, sometimes called the *inverse probability integral transform*, which comes from applying the inverse cumulative distribution function to Y says that the random variable defined by

$$X = F_X^{-1}(Y)$$

has the same distribution as X (because it is X). What it says in practice is that if we know the inverse cdf function for a continuous random variable X, then we can generate random variates from the distribution of X by generating $uniform(0, 1)$ random variates and transforming using the inverse cdf.

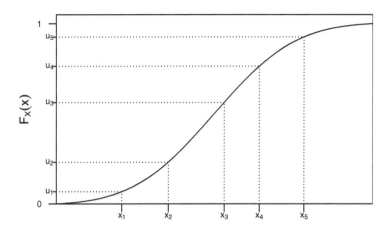

Figure 20.2   The cumulative distribution function maps values of a random variate X (which may take values in the interval (a, b)) to values in the interval $[0, 1]$, and the inverse cumulative distribution function maps values in the interval $[0, 1]$ to values in the interval (a, b).

This can be seen graphically in Figure 20.2. The cdf takes the values of the random variable X and *maps* them to values between zero and one. This means that if X is a random variable, then so is $Y = F_X(X)$ and that the values of Y are uniformly distributed between zero and one. If we switch the axes so that $Y = F_X(X)$ is on the x-axis and X is in the y-axis, then the curve is an inverse cdf for Y and maps the values of Y to the values of X.

■ EXAMPLE 20.2

Leah wants to use the inverse probability integral transform to sample from an *exponential* distribution with a rate parameter of $\lambda = 2$. We have not encountered the exponential distribution before. It is a special case of the *gamma* distribution with the shape parameter set to 1. That is, $exponential(\lambda) = gamma(r = 1, v = \lambda)$. As such, the pdf simplifies to

$$g(x; r = 1, v = \lambda) = \frac{\lambda^1 x^{1-1} e^{-\lambda x}}{\Gamma(1)}$$
$$= \lambda e^{-\lambda x}.$$

Therefore, the cdf is

$$G(x; \lambda) = \int_0^x \lambda e^{-\lambda t} dt$$
$$= \left[-e^{-\lambda t} \right]_0^x$$
$$= 1 - e^{-\lambda x}.$$

Leah can see that this function is easily invertible, so that

$$G^{-1}(p; \lambda) = -\lambda \log(1 - p).$$

She generates 10,000 $uniform(0, 1)$ random numbers and calculates $x_i = -2 \log(1 - u_i)$ for each uniform number u_i. Leah's sample can be seen in Figure 20.3.

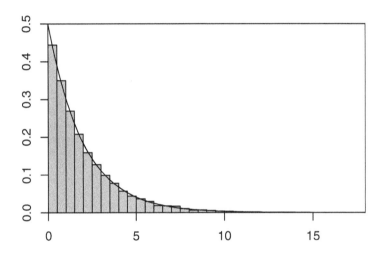

Figure 20.3   Leah's sample from an $exponential(\lambda = 2)$ distribution

Acceptance–Rejection Sampling

Acceptance–rejection sampling sampling, or more commonly *rejection sampling*, dates back to work by the famous mathematician John von Neumann (1951), and even further back to the 18th century in the specialized case of *Buffon's needle*. The idea, and its implementation, is very simple. We wish to sample from a distribution with probability density function (pdf) $f(x)$, which is difficult. However we can sample very easily from a distribution with pdf $g(x)$ which has the property that it *envelopes* $f(x)$. We express this mathematically as $f(x) \leq Mg(x)$ for all x where M is a constant. This is just a mathematical way of saying that the height of $g(x)$ or some scaled version of it $Mg(x)$ must be greater than $f(x)$. $f(x)$ is sometimes referred to as the *target density*, and $g(x)$ as the *candidate density* or *proposal density*. The word *distribution* is sometimes used instead of *density*. We can use our ability to sample from $g(x)$ to sample from $f(x)$ using the following algorithm:

1. Sample x from $g(x)$, and $u \sim U[0,1]$.

2. If $u < f(x)/Mg(x)$, then *accept* x.

3. Otherwise *reject* x.

4. Repeat steps 1–3 until the desired sample size is achieved.

The algorithm gets its name from steps 2 and 3. We actually do not even need to require $f(x)$ to be a proper probability density function, but simply that $f(x) \geq 0$. The reason is quite straightforward. If $f(x) \geq 0$ for all x, and

$$\int_{-\infty}^{\infty} f(x)\,dx = c$$

where c is some non-zero finite constant, then we can make $f(x)$ into a pdf by scaling it by $k = 1/c$. That is, if $h(x) = kf(x)$, then

$$\int_{-\infty}^{\infty} h(x)\,dx = \int_{-\infty}^{\infty} kf(x)\,dx$$
$$= k \int_{-\infty}^{\infty} f(x)\,dx$$
$$= \frac{c}{c}$$
$$= 1$$

If we knew k, then we could scale $f(x)$ appropriately. We would have to scale $g(x)$ by the same factor to ensure that $f(x) \leq Mg(x)$. Therefore the scaling factors would cancel out. That is, when we compute

$$\frac{kf(x)}{Mkg(x)},$$

k appears in both the numerator and denominator and hence cancels. This means we only have to be able to compute the pdf up to a constant which turns out to be very useful.

◼ EXAMPLE 20.3

Fiona wants to draw samples from a *beta*(2, 2) distribution. Her statistics package does not have a *beta* random number generator, but it does have a *uniform* random number generator. Fiona knows that the $beta(2,2)$ random variates have the same range as the uniform $uniform(0,1)$ random variates, and that the $beta(2,2)$ density is proportional to $\pi^{2-1}(1 - \pi)^{2-1} = \pi(1 - \pi)$. It is easy to show, either graphically or with calculus, that this function has a maximum value .25 when $\pi = .5$. Therefore, if Fiona chooses $M = .25$, then $M \times g(x) = M$ is greater than $f(x)$ for all $0 < x < 1$. This is shown in Figure 20.4. She draws approximately 15,000 pairs of uniform random numbers to get a sample of size 10,000. A histogram of Fiona's sample is shown in Figure 20.5.

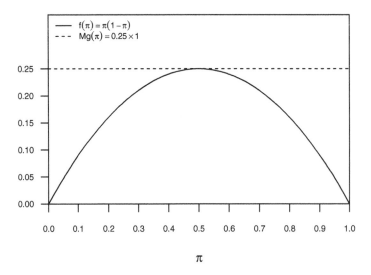

Figure 20.4   Fiona's target density and proposal density.

This example highlights one of the drawbacks of simple rejection sampling; namely, that it can be potentially quite inefficient. By inefficient we mean it requires taking a far larger sample from the candidate distribution than we need from the target distribution. The efficiency is governed, as you might have deduced, by the ratio of the area under the (scaled) target density (scaled) to the (scaled) candidate density. If the candidate density is very close to the target density, then the sampling will be very efficient. If the candidate density is quite far from the target density then the sampling will be inefficient. This problem is magnified when the target distribution is multivariate.

EXAMPLE 20.3   (continued)

The area under Fiona's scaled candidate density is

$$1 \times .25 = .25.$$

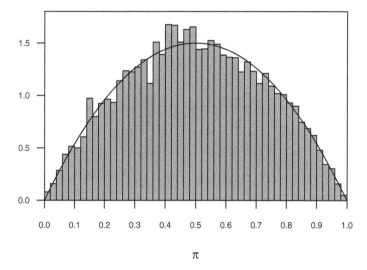

Figure 20.5   Fiona's sample of size 10,000

The area under Fiona's unscaled target density is

$$\int_0^1 \pi(1-\pi).d\pi = \left[\frac{\pi^2}{2} - \frac{\pi^3}{3}\right]_0^1$$
$$= \frac{1}{2} - \frac{1}{3}$$
$$= \frac{1}{6}.$$

The ratio between these two areas is

$$\frac{1}{6} \div \frac{1}{4} = \frac{4}{6} = \frac{2}{3}.$$

Therefore her sampling scheme is only approximately 66.7% efficient. This means on average she needs to draw three pairs of random uniforms to get two $beta(2,2)$ random variates. We can see that this theory closely matches reality as Fiona generated 14,989 ($\approx 15,000$) pairs of random variates to take a sample of size 10,000.

Daniel decides he can do a better job by using a trapezoidal candidate density. He proposes the function

$$g(\pi) = \begin{cases} \pi & \text{for} \quad 0 \le \pi < .25, \\ .25 & \text{for} \quad .25 \le \pi < .75, \\ 1 - \pi & \text{for} \quad .75 \le \pi \le 1. \end{cases}$$

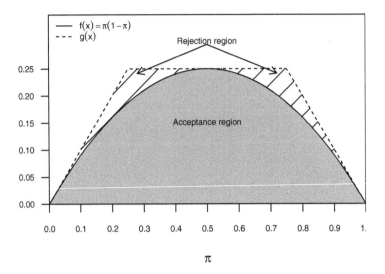

Figure 20.6   Daniel's target density and proposal density

Daniel's candidate density and the target density is shown in Figure 20.6. Daniel is going to exploit the *inverse probability transform* to sample from his candidate density.

In order to sample from Daniel's density he needs to find the associated cumulative distribution function. This will require the piecewise integration of $g(\pi)$ and some scaling so that it has area 1 under the curve. Daniel's function is a symmetric trapezium, so he needs to only work out the area the "box" and one of the triangles. In this case the area of the box is

$$
\begin{aligned}
Area &= \text{width} \times \text{height} \\
&= .5 \times .25 \\
&= .125.
\end{aligned}
$$

and the area of each of the triangles is

$$
\begin{aligned}
Area &= \frac{1}{2} \times \text{width} \times \text{height} \\
&= .5 \times .25 \times .25 \\
&= .03125.
\end{aligned}
$$

The total area is $k = .125 + .03125 + .03125 = .1875$. The cdf is

$$
G(\pi) = \frac{1}{k} \times \begin{cases} .5\pi^2 & \text{for} \quad 0 \leq \pi < .25, \\ .25\pi - .03125 & \text{for} \quad .25 \leq \pi < .75, \\ \pi - .5\pi^2 - .3125 & \text{for} \quad .75 \leq \pi \leq 1. \end{cases}
$$

With a bit work, Daniel finds the inverse cdf

$$
G^{-1}(p) = \begin{cases} \sqrt{2kp} & \text{for} \quad 0 \leq p < .25, \\ .125 + 4kp & \text{for} \quad .25 \leq p < .75, \\ 1 - .25\sqrt{6 - 32kp} & \text{,for} \quad .75 \leq p \leq 1. \end{cases}
$$

The inverse cdf allows Daniel to sample directly from his candidate density, and then use those proposals in his rejection sampling scheme. The candidate density is closer to the target density than the uniform distribution that Fiona used, but how close? Again we need to look at the ratio of the areas. The area under Fiona's target density is $1/6$. The area under Daniel's target density is .1875 or $6/32$, therefore the ratio of the two areas is

$$
\begin{aligned}
\frac{1}{6} \div \frac{6}{32} &= \frac{1}{6} \times \frac{32}{6} \\
&= \frac{32}{36} \\
&= \frac{8}{9}.
\end{aligned}
$$

This means that for every 9 pairs of uniform random numbers Daniel generates, he will get on average 8 $beta(2,2)$ random variates. Again, the theory closely resembles the practice. Daniel generated 11,255 pairs of uniform random numbers to get a sample of size 10,000 from the $beta(2,2)$ distribution.

Clearly, Daniel's sampling scheme is more efficient than Fiona's but it required a lot of work. Ideally, it would be good to have a way of doing this automatically. Automation of this process is one of the ideas behind *adaptive–rejection sampling* which is discussed in the next section.   ∎

Adaptive–Rejection Sampling

The essence of *adaptive–rejection sampling* is very easy to understand; we automatically update our candidate density with information based on the rejected proposals from a rejection sampling scheme. The implementation is slightly more complicated. It is important to note that in its simplest form, the adaptive–rejection sampling scheme only works for *log-concave* distribution functions. Formally, a function $f(x)$ is concave if

$$
f\left((1 - t)x + ty\right) \geq (1 - t)f(x) + tf(y)
$$

for all $t \in [0, 1]$. A function is log-concave if $h(x) = \log(f(x))$ obeys the same inequality. Proving that this inequality holds can be quite difficult and messy.

It is often simpler to show that $f(x)$ is concave if its second derivative $f''(x)$ is less than zero for all values of x where the function and its derivatives are defined. For example, the $normal(\mu, \sigma^2)$ distribution is log-concave because

$$h(x; \mu, \sigma) = \log \left[\frac{1}{\sqrt{2\pi}\sigma} e^{-\frac{1}{2}\left(\frac{x-\mu}{\sigma}\right)^2} \right]$$

$$= -\frac{1}{2} \left(\frac{x-\mu}{\sigma} \right)^2 - \log \left(\frac{1}{\sqrt{2\pi}\sigma} \right).$$

Therefore

$$h'(x; \mu, \sigma) = \frac{\partial h(x; \mu, \sigma)}{\partial x}$$

$$= -\frac{1}{2\sigma^2} 2(x - \mu)$$

and

$$h''(x; \mu, \sigma) = \frac{\partial^2 h(x; \mu, \sigma)}{\partial x^2}$$

$$= -\frac{1}{\sigma^2}.$$

We can see that $h''(x; \mu, \sigma) < 0$ for $\sigma > 0$. Some other examples of log-concave densities are

- $uniform(a, b)$

- $gamma(r, v)$ for $r \geq 1$

- $beta(a, b)$ for $a, b \geq 1$

Student's t distribution, on the other hand, is not log-concave. The method can be altered to cope with log-convex (the opposition of log-concave) distribution, but this is beyond the scope of this book.

Firstly we describe the adaptive–rejection sampling algorithm

1. Find $h(x) = h(x; \boldsymbol{\theta}) = \log f(x; \boldsymbol{\theta})$, where $\boldsymbol{\theta}$ is the vector of parameters that describe the distribution.

2. Find the first derivative of the log-density $h'(x)$, and solve $h'(x) = 0$ for x to find the maximum, x_{max} of $h(x)$. Note that it is not essential to do this exactly, and numerical methods will usually provide enough accuracy.

3. Choose two arbitrary points x_0 and x_1 such that $x_0 < x_{max}$ and $x_1 > x_{max}$.

4. Compute the tangent lines $t_0(x)$ and $t_1(x)$. The *tangent line* $t_i(x_i)$ is the line which passes through the point $(x_i, h(x_i))$ and has slope $h'(x_i)$. As such, the tangent line is defined by

$$t_i(x) = h'(x_i)x + (h(x_i) - h'(x_i)x).$$

Each tangent line can be described by a column vector containing its intercept and slope. That is

$$\begin{aligned} t_i(x) &= h'(x_i)x + (h(x_i) - h'(x_i)x) \\ &= \alpha_i + \beta_i x \\ &= [1, x] \begin{bmatrix} \alpha_i \\ \beta_i \end{bmatrix}. \end{aligned}$$

5. Compute the envelope density $g_0(x)$ by exponentiating $t_0(x_0)$ and $t_1(x_1)$:

$$g_0(x) = \begin{cases} e^{t_0(x)} = e^{\alpha_0 + \beta_0 x} & \text{for} \quad -\infty < x < x_{max}, \\ e^{t_1(x)} = e^{\alpha_1 + \beta_1 x} & \text{for} \quad x_{max} \le x < +\infty. \end{cases}$$

6. Compute the integrated envelope density $G_0(x)$

$$G_0(x) = \int_{-\infty}^{x} g_0(t)\, dt$$

and the constant $k_0 = 1/G_0(+\infty)$ that is required to scale the area under $G_0(x)$ to 1.

7. Compute the inverse cdf $G_0^{-1}(p)$ such that

$$G_0^{-1}(k_0 G_0(x)) = x.$$

8. Sample $(u, v) \sim uniform(0, 1)$.

9. Set $x = G_0^{-1}(v)$.

10. If $u \le f(x)/g_0(x)$, then accept x as a random variate from your target distribution. If you have achieved your target sample size, then stop; otherwise repeat steps 8–10.

11. Otherwise, add x to your set of tangent points, and repeat steps 4–11.

There are some implementation details we have skipped over in this description of the algorithm, but we will address those in Example 20.4.

◼ EXAMPLE 20.4

Lucy is interested in sampling from a $beta(2, 2)$ density using the adaptive-rejection sampling method. Her unscaled target density is the same as Fiona's and Daniel's, i.e.,

$$f(\pi) \propto \pi(1 - \pi), \quad 0 < \pi < 1.$$

The logarithm of this (unscaled) density is

$$h(\pi) = \log(\pi) + \log(1 - \pi).$$

Lucy knows that the $beta$ distribution is log-concave for $\alpha, \beta \geq 1$, but she will check anyway. She can show $h(\pi)$ is log-concave by checking that its second derivative is negative for all values of π. The second derivative of $h(\pi)$ is

$$\frac{\partial^2}{\partial \pi^2} h(\pi) = -\frac{1}{\pi^2} - \frac{1}{(1 - \pi)^2}$$

$$= -\left(\frac{1}{\pi^2} + \frac{1}{(1 - \pi)^2}\right),$$

which is clearly negative for any value of π such that $0 < \pi < 1$. Therefore this density is log-concave. **Note:** In general it is sufficient to only consider a candidate density up to a constant of proportionality. That is, we do not need to include terms that are constant given the parameters of the distribution as these do not effect the concavity of the function.

Lucy starts by finding the point that maximizes $h(\pi)$. In this example, she can do this by inspection since she knows that the function is symmetric around $\pi = 0.5$. In general, however, we can find the maximum by solving $h'(\pi) = \partial/\partial \pi \, h(\pi) = 0$. The first derivative is

$$h'(\pi) = \frac{1}{\pi} - \frac{1}{1 - \pi}$$

and so by setting this equal to zero and solving for π, Lucy finds

$$\frac{1}{\pi} - \frac{1}{1 - \pi} = 0$$

$$\frac{1 - \pi - \pi}{\pi(1 - \pi)} = 0$$

$$1 - 2\pi = 0$$

$$\pi = \frac{1}{2} = .5$$

as expected. She now chooses two arbitrary points, where one is below the maximum and the other is above the maximum, from the range of feasible values for π. Lucy chooses $\pi_1 = 0.2$, and $\pi_2 = 0.8$. Lucy needs to find the tangent line for each of these points. That is, she needs to find the equations of the lines that pass through the points $(\pi_i, h(\pi_i))$ and have slope $h'(\pi_i)$ for $i \in \{1, 2\}$. This simply involves solving

$$h(\pi_i) = h'(\pi_i)\pi_i + b$$

for the intercept b, which is given by rearrangement as

$$b = h(\pi_i) - h'(\pi_i)\pi_i.$$

Lucy's first two tangent lines are

$$\log g_{env}(\pi) = \begin{cases} 3.75\pi - 2.5825815 & \text{for} \quad 0 \le \pi < .5, \\ -3.75\pi + 1.1674185 & \text{for} \quad 0.5 \le \pi \le 1. \end{cases}$$

Lucy finds a piecewise exponential function that envelopes her target density by exponentiating $\log g_0(\pi)$. That is, she finds

$$g_0(\pi) = \begin{cases} e^{3.75\pi - 2.5825815} & \text{for} \quad 0 \le \pi < .5, \\ e^{-3.75\pi + 1.1674185} & \text{for} \quad 0.5 \le \pi \le 1 \end{cases}$$

which describes a function that consists of two exponential curves that envelope her unscaled target density. These curves can be seen in Figure 20.7.

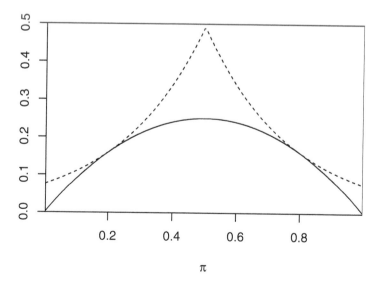

Figure 20.7   Lucy's first envelope function

The integral of $g_0(\pi)$ is also exponential

$$\int_0^\pi g_0(t)\,dt = \begin{cases} \frac{1}{3.75}\left[e^{3.75\pi-2.5825815} - e^{-2.5825815}\right] & \text{for} \quad 0 \le \pi < .5, \\ .1112683 + \frac{1}{-3.75}\left[e^{-3.75\pi+1.1674185}\right. \\ \qquad\qquad \left. -e^{-3.75\times.5+1.1674185}\right] & \text{for} \quad 0.5 \le \pi \le 1 \end{cases}$$

$$= \begin{cases} \frac{1}{3.75}\left[e^{3.75\pi-2.5825815} - e^{-2.5825815}\right] & \text{for} \quad 0 \le \pi < .5, \\ .1112683 + \frac{1}{-3.75}\left[e^{-3.75\pi+1.1674185}\right. \\ \qquad\qquad \left. -.4928347\right] & \text{, for} \quad 0.5 \le \pi \le 1. \end{cases}$$

We need to scale this function so that the area is one. We know that $g_0(\pi)$ is symmetric around .5, and that the area under the curve to the left of .5 is .1112683. Therefore the area under the whole function is twice this amount, i.e. .2225366. We set $k_0 = 1/.2225366 = 4.4936437$. The cdf for the envelope density is therefore given by

$$G_0(\pi) = k_0 \begin{cases} \frac{1}{\beta_0}\left[e^{\beta_0\pi+\alpha_0} - e^{\alpha_0}\right] & \text{for} \quad 0 \le \pi < .5, \\ \kappa_0 + \frac{1}{\beta_1}\left[e^{\beta_1\pi+\alpha_1} - e^{.5\beta_1+\alpha_1}\right] & \text{for} \quad 0.5 \le \pi \le 1, \end{cases}$$

where $\alpha_0 = -2.5825815$, $\beta_0 = 3.75$, $\alpha_1 = 1.1674185$, $\beta_1 = -3.75$ and $\kappa_0 = .1112683$. We can easily invert this cdf to find the inverse cdf,

$$G_0^{-1}(p) = \begin{cases} \frac{1}{\beta_0}\left(\log(\frac{\beta_0 p}{k_0} + e^{\alpha_0}) - \alpha_0\right) & \text{for} \quad 0 \le \pi < .5, \\ \frac{1}{\beta_1}\left(\log(\frac{\beta_1}{k_0}(p - \frac{\kappa_0}{k_0}) + e^{.5\beta_1+\alpha_1}) - \alpha_1\right) & \text{for} \quad .5 \le \pi \le 1. \end{cases}$$

Lucy generates a pair of $uniform(0, 1)$ random variates $(u, v) = (.2875775, .7883051)$. She then calculates

$$\pi = G_0^{-1}(u) = 0.3811180,$$
$$r = f(x)/g_0(x) = 0.7474424.$$

Lucy rejects $\pi = 0.3811180$ because $v > r$. A candidate value that is rejected can be thought of as being in an area where the envelope function does not match the target density very closely. Therefore, Lucy uses this information to adapt her envelope function. Firstly, she calculates a new tangent line at $\pi = 0.3811180$. This gives

$$t_2(\pi) = -1.8286698 + 1.0080420\pi.$$

The new envelope function $g_1(\pi)$ now includes $e^{t_2(\pi)}$. However, Lucy needs to decide which tangent line is closest to the log density (and hence which exponential function is closest to the original density) for any value of π. There are several ways to do this, and all of them are tedious. Lucy decides to exploit the fact that her new point lies between $\pi_0 = 0.2$

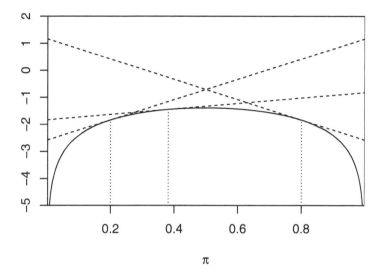

Figure 20.8   The new tangent line is closest to the log density over the range defined by between where it intersects with the other tangent lines

and $\pi_1 = 0.8$, therefore the new tangent line will be closest to the log density over the range where each of these lines intersect. This can be seen displayed graphically in Figure 20.8.

The lines intersect at the points $\pi = .2749537$ and $\pi = .2749537$. Therefore Lucy's new envelope function, $g_1(\pi)$ is

$$
g_1(\pi) = \begin{cases} e^{3.75\pi - 2.5825815} & \text{for} \quad 0 \le \pi < .2749537, \\ e^{-1.8286698 + 1.0080420\pi} & \text{for} \quad .2749537 \le \pi < .6296894, \\ e^{-3.75\pi + 1.1674185} & \text{, for} \quad .6296894 \le \pi \le 1. \end{cases}
$$

Lucy's updated envelope function is shown in Figure 20.9. The integration at this point becomes extremely tedious and error prone. However, because the components of the envelope are very smooth functions, they are can be numerically integrated extremely accurately using either `tintegral` in Minitab or `sintegral` in R. Lucy uses these functions to find that the area under the new envelope function is .1873906, so the sampling efficiency has gone from $100 \times \frac{1}{6} \div .2225366 \approx 74.9\%$ to $100 \times \frac{1}{6} \div .1873906 \approx 88.9\%$ in a single iteration. Equally, it is not necessary to find the intersections of all the tangent lines. If the column vectors describing each of the tangent lines are stored in a matrix $\boldsymbol{\beta}$, then we can define the envelope function after the nth update, $g_n(\pi)$, as

$$
g_n(\pi) = \min_{j=1,\dots,n} (1, \pi) \boldsymbol{\beta}.
$$

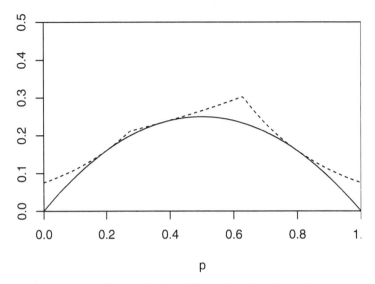

Figure 20.9 The updated envelope function

This "trick" works because we know that all of the tangent lines are upper bounds to the target density. That is, they lie above the target density. This is why we need the target to be log-concave. Therefore, the closest line is the one with the smallest value at a given value of π. This is a little wasteful in terms of the amount of computation required, but the time taken is trivial on a modern computer.

Using this algorithm, Lucy only had to generate 10,037 pairs of *uniform*$(0, 1)$ random numbers to get a sample size of 10,000 from a *beta*$(2, 2)$ density. The final envelope function (after 36 updates in total) was approximately 99.82% efficient; in fact the sampler was over 99% efficient after approximately 1,100 pairs of *uniform*$(0, 1)$ random numbers had been generated and after 14 updates. ∎

20.2 Sampling–Importance–Resampling

The topic of *importance sampling* often arises in situations where people wish to estimate the probability of a rare event. Importance sampling solves this problem by sampling from an *importance density* and reweighting the sampled observations accordingly.

If X is a random variable with probability density function $p(x)$, and $f(X)$ is some function of X, then the expected value of $f(X)$ is

$$E[f(X)] = \int_{-\infty}^{+\infty} f(x)p(x)\,dx.$$

If $h(x)$ is also a probability density whose support contains the support of $p(x)$ (i.e. $h(x) > 0 \quad \forall x$ such that $p(x) > 0$), then this integral can be rewritten as

$$E[f(X)] = \int_{-\infty}^{+\infty} f(x) \frac{p(x)}{h(x)} h(x) \, dx.$$

$h(x)$ is the *importance density*, and the ratio $p(x)/h(x)$ is called the *likelihood ratio*. Therefore, if we take a large sample from $h(x)$, then this integral can be approximated by

$$E[f(X)] = \frac{1}{N} \sum_{i=1}^{N} w_i f(x_i) dx,$$

where $w_i = p(x_i)/h(x_i)$ are the *importance weights*. The efficiency of this scheme is related to how closely the importance density follows the target density in the region of interest. A good importance density will be easy to sample from whilst still closely following the target density. It The process of choosing a good importance density is known as *tuning* and can often be very difficult.

■ **EXAMPLE 20.5**

Karin wants to use importance sampling to estimate the probability of observing a normal random variate greater than 5 standard deviations from the mean. That is, Karin wishes to estimate P such that

$$P = 1 - \Phi(5) = \int_{5}^{+\infty} \frac{1}{\sqrt{2\pi}} e^{-\frac{x}{2}} \, dx.$$

Karin knows that she can approximate p by taking a sample of size N from a standard normal distribution, and calculating

$$E[I(X > 5)] = \frac{1}{N} \sum_{i=1}^{N} I(x_i > 5).$$

However, this is incredibly inefficient as fewer than three random variates in 10 million will exceed 5 which means the vast majority of estimates will be zero unless N is extraordinarily large.

Karin knows, however, that the she can create a *shifted exponential distribution* which has a probability density function which dominates the standard normal density for all x greater than 5. The shifted exponential distribution arises when we consider the random variable $Y = X + \delta$, where X has an exponential distribution with mean 1 and $\delta > 0$. If $\delta = 5$, then the pdf of Y is

$$h(y) = e^{-(y-5)} = e^{(5-y)}.$$

If Karin uses $h(y)$ as her importance density, then her importance sampling scheme is as follows:

1. Take a random sample $x_1, x_2, \ldots, x_N$ such that $x_i \sim exp(1)$

2. Let $y_i = x_i + 5$

3. Calculate

$$\frac{1}{N} \sum_{i=1}^{N} \frac{p(y_i)}{h(y_i)} I(y_i > 5).$$

where p is the standard normal probability density function. Given that she knows all values of y_i are greater than 5, this simplifies to

$$\frac{1}{N} \sum_{i=1}^{N} \frac{p(y_i)}{h(y_i)}.$$

Karin chooses $N = 100,000$ and repeats this procedure 100 times to try and understand the variability in the importance sample estimates. Karin can see that the importance sampling method yields much better

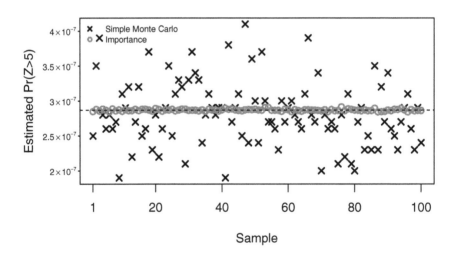

Figure 20.10 100 estimates of $Pr(Z > 5)$, $Z \sim normal(0, 1)$ using samples size 100,000 and importance sampling compared to samples of size 10^8 and naïve Monte Carlo methods.

estimates than simple Monte Carlo methods for far less effort. The true value (calculated using numerical integration) is 2.8665157×10^{-7}. The mean of Karin's 100 importance sampling estimates is 2.8675162×10^{-7}, and the mean of her 100 Monte Carlo estimates is 2.815×10^{-7}. The standard deviations are 4.7×10^{-8} for the Monte Carlo estimates and 1.4×10^{-9} for the importance sampling. The importance sampling method clearly gives much higher accuracy and precision in this example for considerably less computational effort. ∎

The importance sampling method is useful when we want to calculate a function of samples from an unscaled posterior density, such as a mean or a quantile. However, it does not help us draw a sample from the posterior density. We need to extend the importance sampling algorithm very slightly to do this.

1. Draw a large sample, $\boldsymbol{\theta} = \{\theta_1, \theta_2, \dots, \theta_N\}$, from the importance density.

2. Calculate the importance weights for each value sampled

$$w_i = \frac{p(\theta_i)}{h(\theta_i)}.$$

3. Calculate the *normalized weights*:

$$r_i = \frac{w_i}{\sum_{i=1}^{N} w_i}.$$

4. Draw a sample *with replacement* of size N' from $\boldsymbol{\theta}$ with probabilities given by r_i

The resampling combined with the importance weights gives this method its name — the *Sampling Importance Resampling* or *SIR* method. This method is sometimes referred to as the *Bayesian bootstrap*. If N is small, then N' should be smaller, otherwise the sample will contain too many repeated values. However, in general, if N is large ($N \geq 100,000$), the restrictions on N' can be relaxed.

■ EXAMPLE 20.6

Livia wants to draw a random sample, $\boldsymbol{\theta}$, from a *beta*$(2, 8)$ distribution. She knows that this density has a mean of $2/(2+8) = .2$ and a variance of

$$\frac{2 \times 8}{(2+8)^2 \times (2+8+1)} \approx .015;$$

therefore she decides to use a *normal*$(.2, .015)$ density as her importance density. The support of this density $(-\infty, +\infty)$ contains the support of the *beta*$(2, 8)$ density $[0, 1]$ which means it will function as an importance density. This choice may be inefficient as values outside of $[0, 1]$ will receive weights of zero, but Livia uses the properties the normal distribution to show that this will only happen just over 5% of the time on average – that is, if $X \sim normal(.2, .015)$, then $\Pr(X < 0) + \Pr(X > 1) \approx 0.05$.

Livia draws a sample of size $N = 100,000$, and calculates the importance weights

$$
w_i = \begin{cases} \dfrac{\theta_i^1(1-\theta_i)^7}{e^{-\frac{(\theta-.2)^2}{2\times.015}}}, & 0 < \theta_i < 1, \\ 0, & \text{otherwise.} \end{cases}
$$

Livia does not bother calculating the constants for each of her densities as they are the same for every θ_i, and hence only change the weights by a constant scale factor which cancels out when she calculates the normalized weights, r_i.

Livia then draws a sample of size $N' = 10,000$ from $\boldsymbol{\theta}$ with replacement. This can be seen in Figure 20.11.

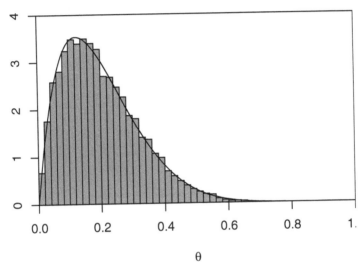

Figure 20.11   Sample of size 10,000 from a $beta(2,8)$ density using the SIR method with a $normal(.2, 0.015)$ importance density.

20.3   Markov Chain Monte Carlo Methods

The development of Markov chain Monte Carlo (MCMC) methods has been a huge step forward for Bayesian statistics. These methods allow users to draw a sample from the exact posterior $g(\theta|y)$, even though only the proportional form of the posterior $g(\theta) f(y|\theta)$ (prior times likelihood given by Bayes' theorem) is known. Inferences are based on this sample from the posterior rather than the exact posterior. MCMC methods can be used for quite complicated models having a large number of parameters. This has allowed applied statisticians to use Bayesian inference for many more models than was previously possible. First we give a brief summary of Markov chains.

Markov Chains

Markov chains are a model of a process that moves around a set of possible values called states where a random chance element is involved in the movement through the states. The future state will be randomly chosen using some probability of transitions. Markov chains have the "memoryless" property that, given the past and present states of the process, the future state only depends on the present state, not the past states. This is called the Markov property. The transition probabilities of a Markov chain will only depend on the current state, not the past states. Each state is only directly connected to the previous state and not to states further back. This way it is linked to the past like a chain, not like a rope. The set of states is called the state-space and can be either discrete or continuous.

We only use Markov chains where the transition probabilities stay the same at each step. This type of chain is called time-invariant.

Each state of a Markov chain can be classified as a transient state, a null recurrent state, or a positive recurrent state. A Markov chain will return to a transient state only a finite number of times. Eventually the chain will leave the state and never return. The Markov chain will return to a null recurrent state an infinite number of times. However, the mean time between returns to the state will also be infinite. The Markov chain will return to a positive recurrent state an infinite number of times, and the mean return time will be finite.

Markov chains where it is possible to reach every state from every other state are called irreducible Markov chains. All states in an irreducible Markov chain are the same type. An irreducible chain with all positive recurrent states will is called an ergodic Markov chain. We will only use ergodic Markov chains since they will have a unique long-run probability distribution (or probability density in the case of continuous state space). It can be found from the transition probabilities as the unique solution of the steady-state equation.

In Markov chain Monte Carlo, we need to find a Markov chain with long-run distribution that is the same as the posterior distribution $g(\theta|y)$. The set of states is the parameter space, the set of all possible parameter values.

There are two main methods for doing this: the Metropolis–Hastings algorithm and the Gibbs sampling algorithm. The Metropolis–Hastings algorithm is based on the idea of balancing the flow of the steady-state probabilities from every pair of states. The Metropolis–Hastings algorithm can either be applied to (a) all the parameters at once or (b) blockwise for each block of parameters given the values of the parameters in the other blocks. The Gibbs sampling algorithm cycles through the parameters, sampling each parameter in turn from the conditional distribution of that parameter, given the most recent values of the other parameters and the data. These conditional distributions may be hard to find in general. However, when the parameter model has a hierarchical structure, they can easily be found. Those are the cases where the Gibbs sampler is most useful. The Metropolis–Hastings algorithm is the most general and we shall see that the Gibbs sampler is a special case of the blockwise Metropolis–Hastings algorithm.

Metropolis–Hastings Algorithm for a Single Parameter

The Metropolis–Hastings, like a number of the techniques we have discussed previously, aims to sample from some *target* density by choosing values from a candidate density. The choice of whether to accept a candidate value, sometimes called a *proposal*, depends (only) on the previously accepted value. This means that the algorithm needs an initial value to start, and an acceptance, or transition, probability. If the transition probability is symmetric, then the sequence of values generated from this process form a Markov chain. By symmetric we mean that the probability of moving from state θ to state θ' is the same as the probability of moving from state θ' to state θ. If $g(\theta|y)$ is an unscaled posterior distribution, and $q(\theta, \theta')$ is a candidate density, then a transition probability defined by

$$\alpha(\theta, \theta') = \min\left[1, \frac{g(\theta'|y)q(\theta', \theta)}{g(\theta|y)q(\theta, \theta')}\right]$$

will satisfy the symmetric transition requirements. This acceptance probability was proposed by Metropolis et al. (1953). The steps of the Metropolis–Hastings algorithm are:

1. Start at an initial value $\theta^{(0)}$.

2. Do the following for $n = 1, \ldots, n$.

    (a) Draw θ' from $q(\theta^{(n-1)}, \theta')$.
    (b) Calculate the probability $\alpha(\theta^{(n-1)}, \theta')$.
    (c) Draw u from $U(0, 1)$.
    (d) If $u < \alpha(\theta^{(n-1)}, \theta')$, then let $\theta^{(n)} = \theta'$, else let $\theta^{(n)} = \theta^{(n-1)}$.

We should note that having the candidate density $q(\theta, \theta')$ close to the target $g(\theta|y)$ leads to more candidates being accepted. In fact, when the candidate density is exactly the same shape as the target

$$q(\theta, \theta') = k \times g(\theta'|y)$$

the acceptance probability is given by

$$\alpha(\theta, \theta') = \min\left[1, \frac{g(\theta'|y)\, q(\theta', \theta)}{g(\theta|y)\, q(\theta, \theta')}\right]$$

$$= \min\left[1, \frac{g(\theta'|y)g(\theta|y)}{g(\theta|y)g(\theta'|y)}\right]$$

$$= 1.$$

Thus, in that case, all candidates will be accepted.

There are two common variants of this algorithm. The first arises when the candidate density is a symmetric distribution centered on the current value. That is,

$$q(\theta, \theta') = q_1(\theta' - \theta),$$

where $q_1()$ is a function symmetric around zero. This is called a *random-walk candidate density*. The symmetry means that $q_1(\theta' - \theta) = q_1(\theta - \theta')$, and therefore the acceptance probability simplifies to

$$\alpha(\theta, \theta') = \min\left[1, \frac{g(\theta'|y)\, q(\theta', \theta)}{g(\theta|y)\, q(\theta, \theta')}\right]$$

$$= \min\left[1, \frac{g(\theta'|y)}{g(\theta|y)}\right].$$

This acceptance probability means that any proposal, θ', which has a higher value of the target density than the current value, θ, will always be accepted. That is, the chain will always move *uphill*. On the other hand, if the proposal is less probable than the current value, then the proposal will only be accepted with probability proportional to the ratio of the two target density values. That is, there is a non-zero probability that the chain will move *downhill*. This scheme allows the chain to explore the parameter over time, but in general the moves will be small and so it might take a long time to explore the whole parameter space.

■ EXAMPLE 20.7

Tamati has an unscaled target density given by

$$g(\theta|y) = .7 \times e^{-\frac{\theta^2}{2}} + 0.15 \times \frac{1}{.5} e^{-\frac{1}{2}\left(\frac{\theta-3}{.5}\right)^2} + 0.15 \times \frac{1}{.5} e^{-\frac{1}{2}\left(\frac{\theta+3}{.5}\right)^2}.$$

This is a mixture of a normal$(0, 1)$, a normal$(3, .5^2)$, and a normal$(-3, .5^2)$. Tamati decides to use a random-walk candidate density. Its shape is given by

$$q(\theta, \theta') = e^{-\frac{(\theta-\theta')^2}{2}}.$$

Let the starting value be $\theta = 2$. Figure 20.12 shows the first six consecutive draws from the Metropolis–Hastings chain with a random-walk candidate. Table 20.2 gives a summary of the first six draws from this chain. Tamati can see that the proposals in draws 1, 3, and 5 were simply more probably than the current state (and hence $\alpha = 1$), so the chain automatically moved to these states. The candidate values in draws 2 and 5 were slightly less probable ($0 < \alpha < 1$), but still had a fairly high chance of being accepted. However, the proposal on the sixth draw was very

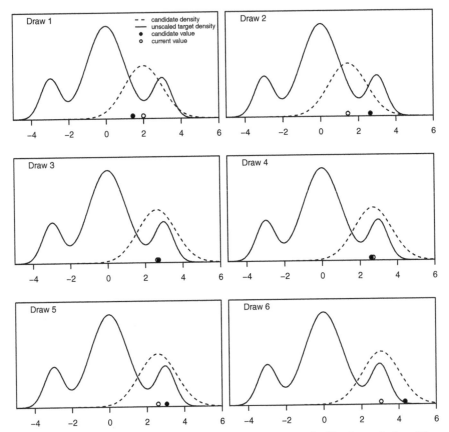

Figure 20.12 Six consecutive draws from a Metropolis–Hastings chain with a random walk candidate density. Note: the candidate density is centered around the current value.

poor ($\alpha = 0.028$) and consequently was not selected. Figure 20.13 shows the trace plot and histogram of the first 1,000 values in the Metropolis–Hastings chain with a random-walk candidate density. Tamati can see that the sampler is moving through the space fairly satisfactorily because the trace plot is changing start regularly. The trace plot would contain *flat* spots if the sampler was not moving well. This can happen when there are local maxima (or minima), or the likelihood surface is very flat. Tamati can also see that the sampler occasionally chooses extreme values from the tails but tends to *jump* back to the central region very quickly. The values sampled from the chain are starting to take the shape of the target density, but it is not quite there. Figure 20.14 shows the histograms for 5,000 and 20,000 draws from the Metropolis–Hastings chain. Tamati can see that the chain is getting closer to the true posterior density as the number of draws increases. ∎

Table 20.2   Summary of the first six draws from the chain using the random-walk candidate density

Draw	Current value	Candidate	α	u	Accept
1	2.000	1.440	1.000	0.409	Yes
2	1.440	2.630	0.998	0.046	Yes
3	2.630	2.700	1.000	0.551	Yes
4	2.700	2.591	0.889	0.453	Yes
5	2.591	3.052	1.000	0.103	Yes
6	3.052	4.333	0.028	0.042	No

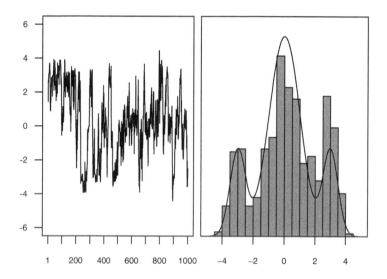

Figure 20.13   Trace plot and histogram of 1,000 Metropolis–Hastings values using the random-walk candidate density with a standard deviation of 1.

The second variant is called the *independent* candidate density. Hastings (1970) introduced Markov chains with candidate densities that did not depend on the current value of the chain. These are called *independent* candidate densities, and

$$q(\theta, \theta') = q_2(\theta'),$$

where $q_2(\theta)$ is some function that dominates the target density in the tails. This requirement is the same as that for the candidate density in acceptance–

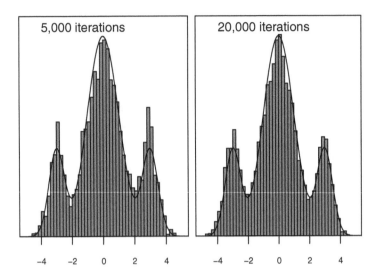

Figure 20.14   Histograms of 5,000 and 20,000 draws from the Metropolis–Hastings chain using the random-walk candidate density with a standard deviation of 1.

rejection sampling. The acceptance probability simplifies to

$$\alpha(\theta, \theta') = \min \left[1, \frac{g(\theta'|y)\, q(\theta', \theta)}{g(\theta|y)\, q(\theta, \theta')} \right]$$
$$= \min \left[1, \frac{g(\theta'|y) q_2(\theta)}{g(\theta|y) q_2(\theta')} \right].$$

for an independent candidate density.

■ EXAMPLE 20.7   (continued)

Tamati's friend Aroha thinks that she might be able to do a better job with an independent candidate density. Aroha chooses a $normal(0, 3^2)$ density as her independent candidate density because it *covers* the target density well. Table 20.3 gives a summary of the first six draws from this chain. Figure 20.15 shows the trace plot and a histogram for the first 1,000 draws from the Metropolis–Hastings chain using Aroha's independent candidate density with a mean of 0 and a standard deviation of 3. The independent candidate density allows larger jumps, but it may accept fewer proposals than the random-walk chain. However, the acceptances will be larger and so the chain will potentially explore the parameter space faster. Aroha can see that the chain is moving through the space very satisfactorily. In this example, Aroha's chain accepted approximately 2,400 fewer proposals than Tamati's chain over 20,000 iterations. The histogram

Table 20.3 Summary of the first six draws from the chain using the independent candidate density

Draw	Current value	Candidate	α	u	Accept
1	2.000	-4.031	0.526	0.733	No
2	2.000	3.137	1.000	0.332	Yes
3	3.137	-4.167	0.102	0.238	No
4	3.137	-0.875	0.980	0.218	Yes
5	-0.875	2.072	0.345	0.599	No
6	-0.875	1.164	0.770	0.453	Yes

shows that the chain has a little way to go before it is sampling from the true posterior. Figure 20.16 shows histograms for 5,000 and 20,000 draws from the chain. Aroha can see that the chain is getting closer to the true posterior as the number of draws increases. ∎

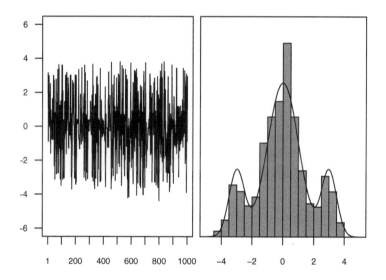

Figure 20.15 Trace plot and histogram of 1,000 Metropolis–Hastings values using the independent candidate density with a mean of 0 and a standard deviation of 3.

Gibbs Sampling

Gibbs sampling is more relevant in problems where we have multiple parameters in our problem. The Metropolis–Hastings algorithm is easily extended to problems with multiple parameters. However, as the number of parameters increases, the acceptance rate of the algorithm generally decreases. The

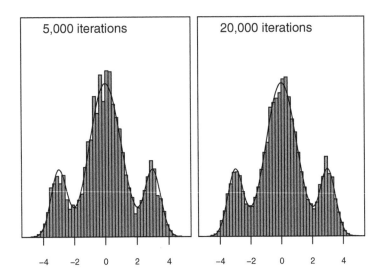

Figure 20.16   Histograms of 5,000 and 20,000 draws from the Metropolis–Hastings chain using the independent candidate density with a mean of 0 and a standard deviation of 3.

acceptance rate can be improved in Metropolis–Hastings by only updating a block of parameters at each iteration. This leads to the *blockwise Metropolis–Hastings* algorithm. The Gibbs sampling algorithm is a special case of the blockwise Metropolis–Hastings algorithm. It depends on being able to derive the true conditional density of one (block) of parameters given every other parameter value. The Gibbs sampling algorithm is particularly well suited to what are known as *hierarchical models*, because the dependencies between model parameters are well-defined.

Suppose we decide to use the true conditional density as the candidate density at each step for every parameter given all of the others. In that case

$$q(\theta_j, \theta'_j | \boldsymbol{\theta}_{-j}) = g(\theta_j | \boldsymbol{\theta}_{-j}, \mathbf{y}),$$

where $\boldsymbol{\theta}_{-j}$ is the set of all the parameters excluding the jth parameters. Therefore, the acceptance probability for θ_j at the nth step will be

$$\alpha\left(\theta_j^{(n-1)}, \theta'_j | \boldsymbol{\theta}_{-j}^{(n)}\right) = \min\left[1, \frac{g(\theta'_j | \boldsymbol{\theta}_{-j}, \mathbf{y}) \, q(\theta'_j, \theta_j | \boldsymbol{\theta}_{-j})}{g(\theta_j | \boldsymbol{\theta}_{-j}, \mathbf{y}) \, q(\theta_j, \theta_{-j} | \boldsymbol{\theta}_{-j})}\right]$$

$$= 1.$$

so the candidate will be accepted at each step. The case where we draw each candidate block from its true conditional density given all the other blocks at their most recently drawn values is known as Gibbs sampling. This algorithm was developed by Geman and Geman (1984) as a method for recreating images from a noisy signal. They named it after Josiah Willard Gibbs, who had determined a similar algorithm could be used to determine the energy states of gasses at equilibrium. He would cycle through the particles, drawing each one conditional on the energy levels of all other particles. His algorithm became the basis for the field of statistical mechanics.

■ EXAMPLE 20.8

Suppose there are two parameters, θ_1 and θ_2. It is useful to propose a target density that we know and could approach analytically, so we know what a random sample from the target should look like. We will use a bivariate normal$(\boldsymbol{\mu}, \mathbf{V})$ distribution with mean vector and covariance matrix equal to

$$\boldsymbol{\mu} = \begin{pmatrix} 0 \\ 0 \end{pmatrix} \quad \text{and} \quad \mathbf{V} = \begin{bmatrix} 1 & \rho \\ \rho & 1 \end{bmatrix}.$$

Suppose we let $\rho = .9$. Then the unscaled target (posterior) density has formula

$$g(\theta_1, \theta_2) \propto e^{-\frac{1}{2(1-.9^2)}(\theta_1^2 - 2 \times .9 \times \theta_1 \theta_2 + \theta_2^2)}.$$

$$g(\theta_1, \theta_2) \propto e^{-\frac{1}{2(1-.9^2)}(\theta_1^2 - 2 \times .9 \times \theta_1 \theta_2 + \theta_2^2)}.$$

The conditional density of θ_1 given θ_2 is normal(m_1, s_1^2), where

$$m_1 = \rho \theta_2 \quad \text{and} \quad s_1^2 = (1 - \rho^2).$$

Similarly, the conditional density of θ_2 given θ_1 is normal(m_2, s_2^2) where

$$m_2 = \rho \theta_1 \quad \text{and} \quad s_2^2 = (1 - \rho^2).$$

We will alternate back and forth, first drawing θ_1 from its density given the most recently drawn value of θ_2, then drawing θ_2 from its density given the most recently drawn value of θ_1. We don't have to calculate the acceptance probability since we know it will always be 1. Table 20.4 shows the first three steps of the algorithm. The initial value for θ_1 is 2. We then draw θ_2 from a $normal(.9 \times \theta_1 = .9 \times 2, 1 - .9^2)$ distribution. The value we draw is $\theta_2 = 1.5557$. This value is accepted because the candidate density is equal the target density, so the acceptance probability is 1. Next we draw θ_1 from a $normal(.9 \times \theta_2 = .9 \times 1.5556943, 1 - .9^2)$ distribution. The value we draw is $\theta_1 = 1.2998$, and so on.

Figure 20.17 shows traceplots for the first 1,000 steps of the Gibbs sampling chain.

Figure 20.18 shows the scatterplot of θ_2 versus θ_1 for 1,000 draws from the Gibbs sampling chain. Figure 20.19 shows histograms of θ_1 and θ_2

Table 20.4 Summary of the first three steps using Gibbs sampling

Step	Current value
1	(2.0000, 1.5557)
2	(1.2998, 1.8492)
3	(1.6950, 1.5819)

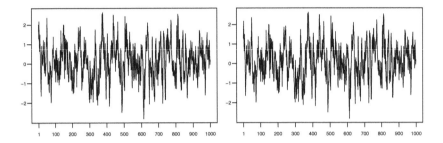

Figure 20.17 Trace plots of θ_1 and θ_2 for 1,000 steps of the Gibbs sampling chain.

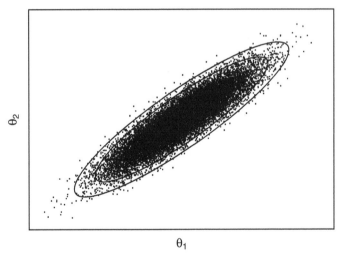

Figure 20.18 Scatterplot of θ_2 versus θ_1 for 1,000 draws from the Gibbs sampling chain with the contours from the exact posterior.

together with their exact marginal posteriors for 5,000 and 20,000 steps of the Gibbs sampler. ∎

There is no real challenge in sampling from the bivariate normal distribution, and in fact we can do it directly without using Gibbs sampling at all. To see the use of Gibbs sampling, we return to the problem of the difference in two

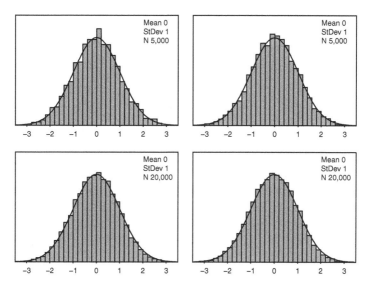

Figure 20.19   Histograms of θ_1 and θ_2 for 5,000 and 20,000 draws of the Gibbs sampling chain.

means when we do not make the assumption of equal variances. We discussed this problem at the end of Chapter 17.

Recall we have two independent samples of observations, $\mathbf{y}_1 = \{y_{1,1}, y_{1,2} \ldots, y_{1,n_1}\}$ and $\mathbf{y}_2 = \{y_{2,1}, y_{2,2} \ldots, y_{2,n_2}\}$ which come from normal distributions with unknown parameters (μ_1, σ_1) and (μ_2, σ_2) respectively. We are interested, primarily, in the posterior difference of the means $\delta = \mu_1 - \mu_2$ given the data. We will use independent conjugate priors for each of the parameters. We can consider the parameter sets independently for each distribution because the formulation of the problem specifies them as being independent. Therefore, we let the prior distribution for μ_j be $normal(m_j^2, s_j^2)$ and let the prior distribution for σ_j^2 be S_j times an inverse-chi-squared with κ_j degrees of freedom, where $j = 1, 2$.

We will draw samples from the incompletely known posterior using the Gibbs sampler. The conditional distributions for each parameter, given the other is known, are:

1. When we consider μ_j is known, the full conditional for σ_j^2 is

$$g_{\sigma_j^2}(\sigma_j^2 | \mu_j, y_{j,1}, \ldots, y_{j,n_j}) \propto g_{\sigma_j^2}(\sigma_j^2) f(y_{j,1}, \ldots, y_{j,n_j} | \mu_j, \sigma_j^2).$$

Since we are using S_j times an inverse chi-squared prior with κ_j degrees of freedom, this will be S_j' times an inverse chi-squared with κ_j' degrees of freedom where

$$S_j' = S_j + \sum_{i=1}^{n_k} (y_{j,i} - \mu_j)^2 \quad \text{and} \quad \kappa_j' = \kappa_j + n_j.$$

2. When we consider σ_j^2 known, the full conditional for μ_j is

$$g_{\mu_j}(\mu_j | \sigma_j^2, y_{j,1}, \ldots, y_{j,n_j}) \propto g_\mu(\mu) f(y_{j,1}, \ldots, y_{j,n_j} | \mu_j) . \qquad (20.1)$$

Since we are using a normal(m_j, s_j^2) prior, We know this will be normal$(m_j', (s_j')^2)$ where

$$\frac{1}{s_j^2} + \frac{n_j}{\sigma_j^2} = \frac{1}{(s_j')^2} \quad \text{and} \quad m_j' = \frac{\frac{1}{s_j^2}}{\frac{1}{(s_j')^2}} \times m_j + \frac{\frac{n_j}{\sigma_j^2}}{\frac{1}{(s_j')^2}} \times \bar{y}_j . \qquad (20.2)$$

To find an initial value to start the Gibbs sampler, we draw σ_j^2 at $t = 0$ for each population from the S_j times an inverse chi-squared with κ_j degree of freedom. Then we draw μ_j for each population from the $normal(m_j, s_j^2)$ distribution. This gives us the values to start the Gibbs sampler. We draw the Gibbs sample using the following steps:

- For $t = 1, \ldots, N$.

  - Calculate S_j' and κ_j' using Equation 20.1, where $\mu_j = \mu_j^{(t-1)}$.
  - Draw $(\sigma_j^{(t)})^2$ from S_j' times an inverse chi-squared distribution with κ_j' degrees of freedom.
  - Calculate $(s_j')^2$ and m_j' using Equation 20.2, where $\sigma_j^2 = (\sigma_j^{(t)})^2$.
  - Draw $\mu_j^{(t)}$ from normal$(m_j', (s_j')^2)$.
  - Calculate
    i. $\delta^{(t)} = \mu_2^{(t)} - \mu_1^{(t)}$,
    ii. $\sigma_\delta^{(t)} = \sqrt{\frac{(\sigma_1^{(t)})^2}{n_1} + \frac{(\sigma_2^{(t)})^2}{n_2}}$
    iii. and $T_0^{(t)} = \frac{\delta^{(t)}}{\sigma_\delta^{(t)}}$

◼ EXAMPLE 20.9

An ecological survey was determine the abundance of oysters recruiting from two different estuaries in New South Wales. The number of oysters observed in 10 cm by 10 cm panels (quadrats) was recorded at a number of different random locations within each site over a two-year period. The data are as follows:

Georges River	25	24	25	14	23	24	24	25	43	24
	30	21	33	27	18	38	30	35	23	30
	34	42	32	58	40	48	36	39	38	48
Port Stephens	72	118	48	103	81	107	80	91	94	104
	132	137	88	96	86	108	73	91	111	126
	74	67	65	103						

We can see from inspecting the data that there is clearly a difference in the average count at each estuary. However, the variance of count data often increases in proportion to the mean. This is a property of both the Poisson distribution and the binomial distribution which are often used to model counts. The sample standard deviations are 9.94 and 22.17 for Georges River and Port Stephens, respectively. Therefore, we have good reason to suspect that the variances are unequal. We can make inferences about the means using the normal distribution because both sets of data have large means, and hence the Poisson (and the binomial) can be approximated by the normal distribution.

We need to choose the parameters of the priors in order to carry out a Gibbs sampling procedure. If we believe, as we did in Chapter 17, a priori that there is no difference in the means of these two populations, then the choice of m_1 and m_2 is irrelevant, as long as $m_1 = m_2$. We say it is irrelevant, because we are interested in the difference in the means, hence if they are equal then this is the same as saying there is no difference. Therefore, we will choose $m_1 = m_2 = 0$. Given we do not know much about anything, we will choose a vague prior for μ_1 and μ_2. We can achieve this by setting $s_1 = s_2 = 10$. In previous problems, we have chosen S to be a median value. That is we have said something like "We are 50% sure that the true standard deviation is at least as large as c" or "The standard deviation is equally likely to be above or below c" where c is some arbitrarily chosen value. The reality it that in many cases the posterior scale factor, S', is heavily dominated by the total sums of squares, SS_T. In this example the sums are squares are 2864.30 and 11306.96 for Georges River and Port Stephens respectively.

Figure 20.20 shows the effect of letting S vary from 1 to 100. The scaling constants are so small compared to the total sum of squares for each site that the choice of S has very little effect on the medians and credible intervals for group the standard deviations. We will set $S_1 = S_2 = 10$ in this example; 95% of the prior probability is using this prior assigned to variances less than approximately 2,500. This seems reasonable given that the sample variances are 98.8 and 491.6 for Georges Rives and Port Stephens respectively.

To find an initial value to start the Gibbs sampler we draw σ_j^2 at $t = 0$ for each population from the $S_j = 10$ times an inverse chi-squared distribution with $\kappa_j = 1$ degree of freedom. We do this by first drawing two random variates from a chi-squared distribution with one degree of freedom and then dividing $S_j = 10$ by each of these numbers The values we draw are $(0.2318, 4.5614)$, so we calculate $(\sigma_1^2, \sigma_2^2) = (10/0.2318, 10/4.5614) = (43.13, 2.19)$. Then we draw μ_j for each population from a $normal(60, 10^2)$ distribution. As previously noted, it does not really matter what values we choose for the prior means, as we are interested in the difference. We have chosen 60 as being approximately half way between the two sample means. The values we draw are $(\mu_1, \mu_2) = (47.3494, 53.1315)$. These steps

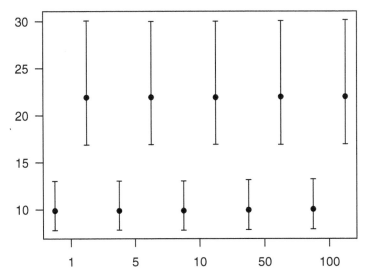

Figure 20.20   The effect of changing the prior value of the scaling constant, S, on the medians and credible intervals for the group standard deviations.

give us the values to start the Gibbs sampler. Table 20.5 shows the first

Table 20.5   First five draws and updated constants for a run of the Gibbs sampler using independent conjugate priors

t	S'	σ^2	s'	m'	μ
0		(43.1, 2.2)			(47.3, 53.1)
1	(10221.4, 51320.9)	(383.6, 1147.8)	(11.3, 32.4)	(34.9, 83.0)	(36.6, 86.1)
2	(3595.9, 12800.8)	(112.8, 378.6)	(3.6, 13.6)	(32.7, 89.3)	(30.7, 92.4)
3	(2902.6, 11376.9)	(98.2, 662.5)	(3.2, 21.6)	(32.6, 86.6)	(31.8, 85.1)
4	(2874.4, 13219.7)	(129.6, 962.4)	(4.1, 28.6)	(32.9, 84.2)	(31.6, 78.0)
5	(2874.6, 17426.5)	(119.8, 644.3)	(3.8, 21.2)	(32.8, 86.8)	(35.2, 85.4)

five steps from the Gibbs sampler. We can see that even in this small number of steps the means and variances are starting to converge. We take a sample of size $N = 100,000$, which should be more than sufficient for the inferences we wish to make. Recall that we are interested in the difference in the mean abundance between the two estuaries. We know that the Georges River counts are much lower than the Port Stephens counts, so we will phrase our inferences in terms of the differences between Port Stephens and Georges River. The first quantity of interest is the posterior probability that the difference is greater than zero. We can estimate this simply by counting the number of times the difference in the sampled means was greater than zero. More formally we calculate

$$\Pr(\delta > 0) \approx \frac{1}{N} \sum_{i=1}^{N} I(\delta_i > 0)$$

This happened zero times in our sample of 100,000, so our estimate of $\Pr(\delta)$ is 0, but it it is safer to say that it is simply less than $0.00001 = 10^{-5}$. Needless to say, we take this as very strong support for the hypothesis that there is a real difference in the mean abundance of oysters between the two estuaries. Sampling also makes it very easy for us to make inferential statements about functions of random variables which might be difficult to derive analytically. For example, we chose a model which allowed the variances for the two locations to be different. We might be interested in the posterior ratio of the variances. If the different variance model is well-justified, then we would expect to see the ratio of the two variances exceed 1 more often than not. To explore this hypothesis, we simply calculate and store σ_2^2/σ_1^2. We can see from Figure 20.21 that the ratio of

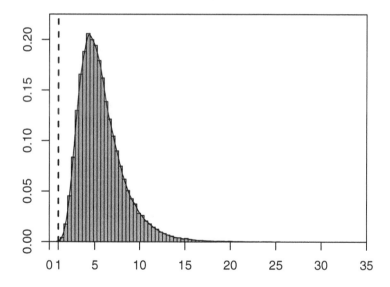

Figure 20.21   The posterior ratio of the variances for the two estuaries.

the variances is almost always greater than one. In fact more than 99% of the ratios are greater than two, thus providing very strong support that the different variance model was well justified.   ∎

20.4   Slice Sampling

Neal (1997, 2003) developed a MCMC method that can be used to draw a random point under a target density and called it *slice sampling*. First we note that when we draw a random point under the unscaled target density and then only look at the horizontal component θ, it will be a random draw from the target density. Thus we are concerned about the horizontal component θ. The vertical component g is an auxiliary variable.

▮ EXAMPLE 20.10

For example, suppose the unscaled target has density $g(\theta) \propto e^{-\frac{\theta^2}{2}}$. This unscaled target is actually an unscaled $normal(0,1)$, so we know almost all the probability is between -3.5 and 3.5. The unscaled target has maximum value at $\theta = 0$ and the maximum is 1. So we draw a random sample of 10,000 horizontal values uniformly distributed between -3.5 and 3.5. We draw a random sample of 10,000 vertical values uniformly distributed between 0 and 1. These are shown in Figure 20.22 along with the unscaled target

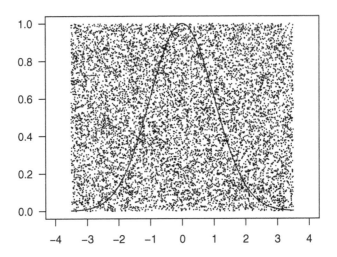

Figure 20.22   Uniformly distributed points and the unscaled target

Then we discard all the points that are above the unscaled target. The remaining points are shown in Figure 20.23 along with the unscaled target. We form a histogram of the horizontal values of the remaining points. This is shown in Figure 20.24 along with the target. This shows that it is a random sample from the target density.   ∎

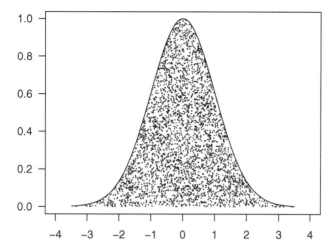

Figure 20.23   Uniformly distributed points under the unscaled target

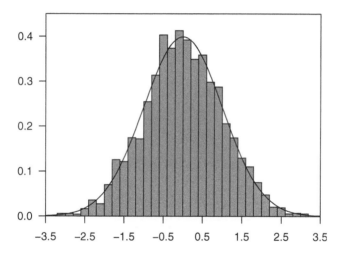

Figure 20.24   Histogram of the remaining points and the target density

Slice sampling is an MCMC method that has as its long-run distribution a random draw from under the target density. It is particularly effective for a one-dimension parameter with a unimodal density. Each step of the chain has two phases. At step i, we first draw the auxiliary variable g_i given the current value of the parameter θ_{i-1} from the $uniform(0,c)$ distribution where $c = g(\theta_{i-1})$. This is drawing uniformly from the vertical slice at the current value θ_{i-1}. Next, we draw θ_i given the current value of the auxiliary variable from the $uniform(a,b)$, where $g(a) = g(b) = g_i$, the current value of the auxiliary variable. The long-run distribution of this chain converges to a random draw of a point under the density of the parameter. Thus the horizontal component is a draw from the density of the parameter.

■ **EXAMPLE 20.8   (continued)**

We will start at time 0 with $\theta_0 = 0$ which is the mode of the unscaled target. The first four steps of a slice sampling chain are given in Figure 20.25.   ■

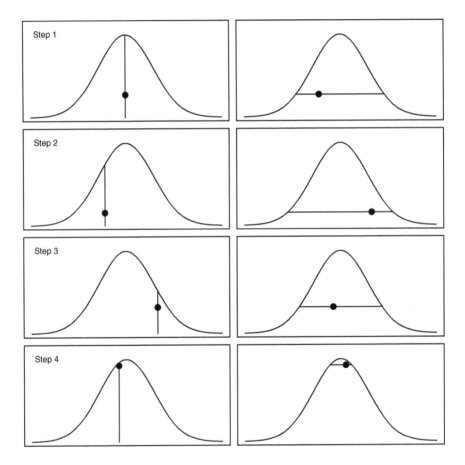

Figure 20.25   Four slice sampling steps. The left pane is sampling the auxiliary variable in the vertical dimension. The right pane is sampling θ in the horizontal dimension.

20.5 Inference from a Posterior Random Sample

The reason we have spent so long discussing sampling methods is that ultimately we want to use random samples from the posterior distribution to answer statistical questions. That is we want to use random samples from the posterior distribution of the statistic interest to make inferential statements about that statistic. The key idea to this process is that integration can be performed by sampling. If we are interested in finding the area under the curve defined by the function $f(x)$ between the points a and b, then we can approximate it by taking a sequence of (equally) spaced points between a and b and summing the areas of the rectangles defined by each pair of points and the (average) height of the curve at these points. That is, if

$$\Delta x = \frac{b-a}{N},$$

then

$$\int_a^b f(x)dx = \lim_{N \to \infty} \sum_{i=1}^N f(x_i)\Delta x,$$

where $x_i = a + (i - 0.5)\Delta x$. This is sometimes known as the *midpoint rule* and is a special case of a *Reimann sum* which leads to the most common type of integration (*Reimann integration*). In Monte Carlo integration we replace the sampling at regular intervals with a random sample of points in the interval $[a, b]$. The integral then becomes

$$\int_a^b f(x)\,dx = \lim_{N \to \infty} \frac{b-a}{N} \sum_{i=1}^N f(x_i),$$

which is exactly the same as the previous formula. This idea can be, and is, easily extended into multiple dimensions. This provides us with the basis for performing inference from a sample. We know, for example, that the expected value of a continuous random variable, X, is defined as

$$E[X] = \int_{-\infty}^{+\infty} x f(x)\,dx.$$

This is in the form of an integral, and if we are sampling with respect to the probability density function $f(x)$ (instead of uniformly), then we can approximate the expected value by

$$E[X] \approx \frac{1}{N} \sum_{i=1}^N x_i.$$

This tells us that we can estimate the posterior mean by taking a large sample from the posterior distribution and calculating the sample mean. There are some caveats to this method if the sample is generated from a Markov chain which we will discuss shortly. This method also works for any function of X. That is

$$E[g(X)] = \int_{-\infty}^{+\infty} g(x)f(x)dx \approx \frac{1}{N}\sum_{i=1}^{N} g(x_i).$$

This allows us to estimate the variance through sampling since

$$Var[X] = E[(x-\mu)^2] \int_{-\infty}^{+\infty} (x-\mu)^2 f(x)\, dx$$

$$\approx \frac{1}{N}\sum_{i=1}^{N} (x_i - \mu)^2$$

$$\approx \frac{1}{N}\sum_{i=1}^{N} (x_i - \bar{x})^2.$$

We note that the denominator should be $N-1$, but in practice this rarely makes any difference, and at any rate the concept of unbiased estimators does not exist in the Bayesian framework. If

$$g(x) = I(x < c) = \begin{cases} 0, & x < c, \\ 1, & \text{otherwise} \end{cases}$$

for some point $c \in (-\infty, +\infty)$, then we can see that

$$\int_{-\infty}^{+\infty} g(x)f(x)\, dx = \int_{-\infty}^{+\infty} I(x < c)f(x)\, dx = \int_{-\infty}^{c} f(x)\, dx.$$

It should be clear that defining $g(x)$ in this way leads to the definition of the cumulative distribution function

$$F(x) = \int_{-\infty}^{x} f(t)\, dt$$

and that this can be approximated by the *empirical distribution function*

$$F_N(x) = \frac{1}{N}\sum_{i=1}^{N} I(x_i < x).$$

The *Glivenko–Cantelli theorem*, which is way beyond the scope of this book, tells us that $F_N(x)$ converges almost surely to $F(x)$. This means we can use a large sample from the posterior to estimate posterior probabilities, and we can use the empirical quantiles to calculate credible intervals and posterior medians for example.

Posterior Inference from Samples Taken Using Markov Chains

One side effect of using a Markov chain-based sampling method is that the resulting sample is correlated. The net impact of correlation is that it reduces the amount of independent information available for estimation or inference, and therefore it is effectively like working with a smaller sample of observations. Bayesians employ a number of strategies to reduce the amount of correlation in their samples from the posterior. The two most common are a *burn-in* period and then *thinning* of the chain. Giving a sampler a burn-in period simply means we discard a certain number of the observations from the start of the Markov chain. For example, we might run a Gibbs sampler for 11,000 steps and discard the first 1,000 observations. This serves two related purposes. Firstly, it allows the sampler to move away from the initial values which may have been set manually. Secondly, it allows the sampler to move to a state where we might be more confident (but we will never know for sure) that the sampler is sampling from the desired target density. You might think about this in terms of the initial values being very unlikely. Therefore, the sampler might have to go through quite a few changes in state until it is sampling from more likely regions of the target density. Thinning the chain attempts to minimize autocorrelation, that is, correlation between successive samples, by only retaining every k^{th} value, where k is chosen to suit the problem. This can be a very useful strategy when the chain is not *mixing* well. If the chain is said to be *mixing poorly*, then it means generally that the proposals are not being accepted very often, or the proposals are not moving through the state space very efficiently. Some authors recommend the calculation of *effective sample size (ESS)* measures; however, the only advice we offer here is that careful checking is important when using MCMC.

20.6   Where to Next?

We have reached our journey's end in this text. You, the reader, might reasonably ask "Where to next?" We do have another book which takes that next step. *Understanding Computational Bayesian Statistics* (Bolstad, 2010) follows in the spirit of this book, by providing a hands-on approach to understanding the methods that are used in modern applications of Bayesian statistics. In particular, it covers in more detail the methods we discussed in this chapter. It also provides an introduction to the Bayesian treatment of count data through logistic and Poisson regression, and to survival with a Bayesian version of the Cox proportional hazards model.

A

INTRODUCTION TO CALCULUS

FUNCTIONS

A function $f(x)$ defined on a set of real numbers, A, is a rule that associates each real number x in the set A with one and only one other real number y. The number x is associated with the number y by the rule $y = f(x)$. The set A is called the *domain* of the function, and the set of all y that are associated with members of A is called the *range* of the function.

Often the rule is expressed as an equation. For example, the domain A might be all positive real numbers, and the function $f(x) = \log_e(x)$ associates each element of A with its natural logarithm. The range of this function is the set of all real numbers.

For a second example, the domain A might be the set of real numbers in the interval $[0, 1]$ and the function $f(x) = x^4 \times (1 - x)^6$. The range of this function is the set of real numbers in the interval $[0, .4^4 \times .6^6]$.

Note that the variable name is merely a cipher, or a place holder. $f(x) = x^2$ and $f(z) = z^2$ are the same function, where the rule of the function is associate each number with its square. The function is the rule by which the association is made. We could refer to the function as f without the variable name, but usually we will refer to it as $f(x)$. The notation $f(x)$ is used for two things. First, it represents the specific value associated by the function f to the point x. Second, it represents the function by giving the rule which it uses. Generally, there is no confusion as it is clear from the context which meaning we are using.

Combining Functions

We can combine two functions algebraically. Let f and g be functions having the same domain A, and let k_1 and k_2 be constants. The function $h = k_1 \times f$ associates a number x with $y = k_1 f(x)$. Similarly, the function $s = k_1 f \pm k_2 g$ associates the number x with $y = k_1 \times f(x) \pm k_2 \times g(x)$. The function $u = f \times g$ associates a number x with $y = f(x) \times g(x)$. Similarly, the function $v = \frac{f}{g}$ associates the number x with $y = \frac{f(x)}{g(x)}$.

If function g has domain A and function f has domain that is a subset of the range of the function g, then the composite function (function of a function) $w = f(g)$ associates a number x with $y = f(g(x))$.

Graph of a Function

The graph of the function f is the graph of the equation $y = f(x)$. The graph consists of all points $(x, f(x))$, where $x \in A$ plotted in the coordinate plane. The graph of the function f defined on the closed interval $A = [0, 1]$, where $f(x) = x^4 \times (1 - x)^6$, and is shown in Figure A.1. The graph of the function g defined on the open interval $A = (0, 1)$, where $g(x) = x^{-\frac{1}{2}} \times (1 - x)^{-\frac{1}{2}}$, is shown in Figure A.2.

Limit of a Function

The limit of a function at a point is one of the fundamental tools of calculus. We write

$$\lim_{x \to a} f(x) = b$$

to indicate that b is the limit of the function f when x approaches a. Intuitively, this means that as we take x values closer and closer to (but not equal to) a, their corresponding values of $f(x)$ are getting closer and closer to b. We note that the function $f(x)$ does not have to be defined at a to have a limit at a. For example, 0 is not in the domain A of the function $f(x) = \frac{\sin x}{x}$ because division by 0 is not allowed. Yet

$$\lim_{x \to 0} \frac{\sin(x)}{x} = 1$$

as seen in Figure A.3. We see that if we want to be within a specified closeness

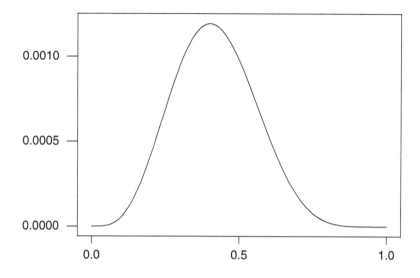

Figure A.1   Graph of function $f(x) = x^4 \times (1-x)^6$.

to $y = 1$, we can find a degree of closeness to $x = 0$ such that all points x that are within that degree of closeness to $x = 0$ and are in the domain A will have $f(x)$ values within that specified closeness to $y = 1$.

We should note that a function may not have a limit at a point a. For example, the function $f(x) = cos(1/x)$ does not have a limit at $x = 0$. This is shown in Figure A.4, which shows the function at three scales. No matter how close we get to $x = 0$, the possible $f(x)$ values always range from -1 to 1.

Theorem A.1 *Limit Theorems:*
Let $f(x)$ and $g(x)$ be functions that each have limit at a, and let k_1 and k_2 be scalars.

1 Limit of a sum (difference) of functions

$$\lim_{x \to a}[k_1 \times f(x) \pm k_2 \times g(x)] = k_1 \times \lim_{x \to a} f(x) \pm k_2 \times \lim_{x \to a} g(x).$$

2 Limit of a product of functions

$$\lim_{x \to a}[f(x) \times g(x)] = \lim_{x \to a} f(x) \times \lim_{x \to a} g(x).$$

3 Limit of a quotient of functions

$$\lim_{x \to a}\left[\frac{f(x)}{g(x)}\right] = \left[\frac{\lim_{x \to a} f(x)}{\lim_{x \to a} g(x)}\right].$$

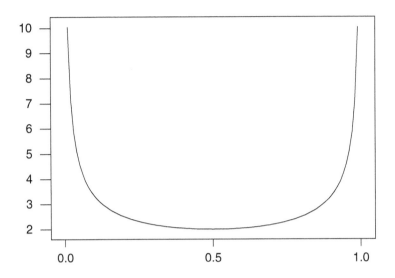

Figure A.2   Graph of function $f(x) = x^{-\frac{1}{2}} \times (1-x)^{-\frac{1}{2}}$.

4 Limit of a power of a function

$$\lim_{x \to a} [f^n(x)] = [\lim_{x \to a} f(x)]^n .$$

Let $g(x)$ be a function that has limit at a equal to b, and let $f(x)$ be a function that has a limit at b. Let $w(x) = f(g(x))$ be a composite function.

5. Limit of a composite function

$$\lim_{x \to a} w(x) = \lim_{x \to a} f(g(x)) = f(\lim_{x \to a} g(x)) = f(g(b)).$$

CONTINUOUS FUNCTIONS

A function $f(x)$ is *continuous* at point a if and only if

$$\lim_{x \to a} f(x) = f(a).$$

This says three things. First, the function has a limit at $x = a$. Second, a is in the domain of the function, so $f(a)$ is defined. Third, the limit of the function at $x = a$ is equal to the value of the function at $x = a$. If we want $f(x)$ to be some specified closeness to $f(a)$, we can find a degree of closeness so that for all x within that degree of closeness to a, $f(x)$ is within the specified closeness to $f(a)$.

A function that is continuous at all values in an interval is said to be continuous over the interval. Sometimes a continuous function is said to be

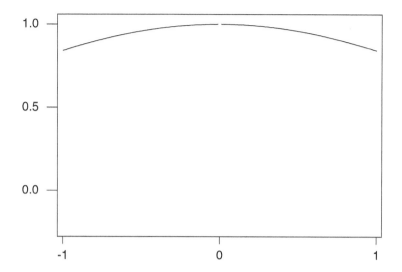

Figure A.3   Graph of $f(x) = \frac{\sin(x)}{x}$ on $A = (-1,0) \cup (0,1)$. Note that f is not defined at $x = 0$.

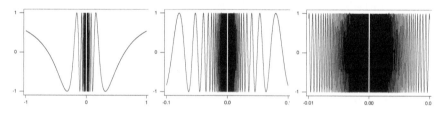

Figure A.4   Graph of $f(x) = \cos\left(\frac{1}{x}\right)$ at three scales. Note that f is defined at all real numbers except for $x = 0$.

a function that "can be graphed over the interval without lifting the pencil." Strictly speaking, this is not true for all continuous functions. However, it is true for all functions with formulas made from polynomial, exponential, or logarithmic terms.

Theorem A.2 *Let $f(x)$ and $g(x)$ be continuous functions, and let k_1 and k_2 be scalars. Then are all continuous functions on their range of definition:*

1. A linear function of continuous functions

$$s(x) = k_1 \times f(x) + k_2 \times g(x)$$

2. A product of continuous functions

$$u(x) = f(x) \times g(x)$$

3. *A quotient of continuous functions*

$$v(x) = \frac{f(x)}{g(x)}$$

4. *A composite function of continuous functions*

$$w(x) = f(g(x))$$

Minima and Maxima of Continuous Functions

One of the main achievements of calculus is that it gives us a method for finding where a continuous function will achieve minimum and/or maximum values.

Suppose $f(x)$ is a continuous function defined on a continuous domain A. The function achieves a local maximum at the point $x = c$ if and only if $f(x) \leq f(c)$ for *all* points $x \in A$ that are sufficiently close to c. Then $f(c)$ is called a local maximum of the function. The largest local maximum of a function in the domain A is called the global maximum of the function.

Similarly, the function achieves a local minimum at point $x = c$ if and only if $f(x) \geq f(c)$ for *all* points $x \in A$ that are sufficiently close to c, and $f(c)$ is called a local minimum of the function. The smallest local minimum of a function in the domain A is called the global minimum of the function.

A continuous function defined on a domain A that is a closed interval $[a, b]$ always achieves a global maximum (and minimum). It can occur either at one of the endpoints $x = a$ or $x = b$ or at an interior point $c \in (a, b)$. For example, the function $f(x) = x^4 \times (1 - x)^6$ defined on $A = [0, 1]$ achieves a global maximum at $x = \frac{4}{6}$ and a global minimum at $x = 0$ and $x = 1$ as can be seen in Figure A.1.

A continuous function defined on a domain A that is an open interval (a, b) may or may not achieve either a global maximum or minimum. For example, the function $f(x) = \frac{1}{x^{1/2} \times (x-1)^{1/2}}$ defined on the open interval $(0, 1)$ achieves a global minimum at $x = .5$, but it does not achieve a global maximum as can be seen from Figure A.2.

DIFFERENTIATION

The first important use of the concept of a limit is finding the derivative of a continuous function. The process of finding the derivative is known as *differentiation*, and it is extremely useful in finding values of x where the function takes a minimum or maximum.

We assume that $f(x)$ is a continuous function whose domain is an interval of the real line. The derivative of the function at $x = c$, a point in the interval is

$$f'(c) = \lim_{h \to 0} \left(\frac{f(c + h) - f(c)}{h} \right)$$

if this limit exists. When the derivative exists at $x = c$, we say the function $f(x)$ is *differentiable* at $x = c$. If this limit does not exist, the function $f(x)$ does not have a derivative at $x = c$. The limit is not easily evaluated, as plugging in $h = 0$ leaves the quotient $\frac{0}{0}$ which is undefined. We also use the notation for the derivative at point c

$$f'(c) = \frac{d}{dx}f(x)\bigg|_{x=c} .$$

We note that the derivative at point $x = c$ is the slope of the curve $y = f(x)$ evaluated at $x = c$. It gives the "instantaneous rate of change" in the curve at $x = c$. This is shown in Figure A.5, where $f(x)$, the line joining the point

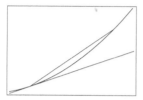  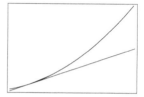

Figure A.5   The derivative at a point is the slope of the tangent to the curve at that point.

$(c, f(c))$ and point $(c + h, f(c + h))$ for decreasing values of h and its tangent at c, is graphed.

The Derivative Function

When the function $f(x)$ has a derivative at all points in an interval, the function

$$f'(x) = \lim_{h \to 0} \left(\frac{f(x + h) - f(x)}{h} \right)$$

is called the *derivative function*. In this case we say that $f(x)$ is a *differentiable function*. The derivative function is sometimes denoted $\frac{dy}{dx}$. The derivatives of some elementary functions are given in the following table:

$f(x)$	$f'(x)$
$a \times x$	a
x^b	$b \times x^{b-1}$
e^x	e^x
$\log_e(x)$	$\frac{1}{x}$
$\sin(x)$	$\cos(x)$
$\cos(x)$	$-\sin(x)$
$\tan(x)$	$-\sec^2(x)$

The derivatives of more complicated functions can be found from these using the following theorems:

Theorem A.3 *Let $f(x)$ and g be differentiable functions on an interval, and let k_1 and k_2 be constants.*

1. *The derivative of a constant times a function is the constant times the derivative of the function. Let $h(x) = k_1 \times f(x)$. Then $h(x)$ is also a differentiable function on the interval, and*

$$h'(x) = k_1 \times f'(x).$$

2. *The sum (difference) rule.*
   Let $s(x) = k_1 \times f(x) \pm k_2 \times g(x)$. Then $s(x)$ is also a differentiable function on the interval, and

$$s'(x) = k_1 \times f'(x) \pm k_2 \times g'(x).$$

3. *The product rule.*
   Let $u(x) = f(x) \times g(x)$. Then $u(x)$ is a differentiable function, and

$$u'(x) = f(x) \times g'(x) + f'(x) \times g(x).$$

4. *The quotient rule.*
   Let $v(x) = \frac{f(x)}{g(x)}$. Then $v(x)$ is also a differentiable function on the interval, and

$$v'(x) = \frac{g(x) \times f'(x) - f(x) \times g'(x)}{(g(x))^2}.$$

Theorem A.4 *The chain rule.*
Let $f(x)$ and $g(x)$ be differentiable functions (defined over appropriate intervals) and let $w(x) = f(g(x))$. Then $w(x)$ is a differentiable function and

$$w'(x) = f'(g(x)) \times g'(x).$$

Higher Derivatives

The second derivative of a differentiable function $f(x)$ at a point $x = c$ is the derivative of the derivative function $f'(x)$ at the point. The second derivative is given by

$$f''(c) = \lim_{h \to 0} \left(\frac{f'(c+h) - f'(c)}{h} \right)$$

if it exists. If the second derivative exists for all points x in an interval, then $f''(x)$ is the second derivative function over the interval. Other notation for the second derivative at point c and for the second derivative function are

$$f''(c) = f^{(2)}(c) = \frac{d}{dx} f'(x) \Big|_{x=c} \quad \text{and} \quad f^{(2)}(x) = \frac{d_2}{dx^2} f(x).$$

Similarly, the kth derivative is the derivative of the $k-1^{\text{th}}$ derivative function

$$f^{(k)}(c) = \lim_{h \to 0} \left(\frac{f^{(k-1)}(c+h) - f^{(k-1)}(c)}{h} \right)$$

if it exists.

Critical Points

For a function $f(x)$ that is differentiable over an open interval (a, b), the derivative function $f'(x)$ is the slope of the curve $y = f(x)$ at each x-value in the interval. This gives a method of finding where the minimum and maximum values of the function occur. The function will achieve its minimum and maximum at points where the derivative equals 0. When $x = c$ is a solution of the equation

$$f'(x) = 0,$$

c is called a critical point of the function $f(x)$. The critical points may lead to local maximum or minimum, or to global maximum or minimum, or they may be points of inflection. A point of inflection is where the function changes from being concave to convex, or vice versa.

Theorem A.5 *First derivative test: If $f(x)$ is a continuous differentiable function over an interval (a, b) having derivative function $f'(x)$, which is defined on the same interval. Suppose c is a critical point of the function. By definition, $f'(c) = 0$.*

1. *The function achieves a unique local maximum at $x = c$ if, for all points x that are sufficiently close to c,*
    when $x < c$,   then $f'(x) > 0$ and
    when $x > c$,   then $f'(x) < 0$.

2. *Similarly, the function achieves a unique local minimum at $x = c$ if, for all points x that are sufficiently close to c,*
    when $x < c$,   then $f'(x) < 0$ and
    when $x > c$,   then $f'(x) > 0$.

3. *The function has a point of inflection at critical point $x = c$ if, for all points x that are sufficiently close to c, either*
    when $x < c$,   then $f'(x) < 0$ and
    when $x > c$,   then $f'(x) < 0$
    or
    when $x < c$,   then $f'(x) > 0$ and
    when $x > c$,   then $f'(x) > 0$.
    At a point of inflection, either the function stops increasing and then resumes increasing, or it stops decreasing and then resumes decreasing.

For example, the function $f(x) = x^3$ and its derivative $f'(x) = 3 \times x^2$ are shown in Figure A.6. We see that the derivative function $f'(x) = 3x^2$ is positive for $x < 0$, so the function $f(x) = x^3$ is increasing for $x < 0$. The derivative function is positive for $x > 0$ so the function is also increasing for $x > 0$. However at $x = 0$, the derivative function equals 0, so the original function is not increasing at $x = 0$. Thus the function $f(x) = x^3$ has a point of inflection at $x = 0$.

Theorem A.6 *Second derivative test: If $f(x)$ is a continuous differentiable function over an interval (a, b) having first derivative function $f'(x)$ and second derivative function $f^{(2)}(x)$ both defined on the same interval. Suppose c is a critical point of the function. By definition, $f'(c) = 0$.*

1. The function achieves a maximum at $x = c$ if $f^{(2)}(c) < 0$.

2. The function achieves a minimum at $x = c$ if $f^{(2)}(c) > 0$.

INTEGRATION

The second main use of calculus is finding the area under a curve using *integration*. It turns out that *integration* is the inverse of *differentiation*. Suppose $f(x)$ is a function defined on an interval $[a, b]$. Let the function $F(x)$ be an *antiderivative* of $f(x)$. That means the derivative function $F'(x) = f(x)$. Note that the antiderivative of $f(x)$ is not unique. The function $F(x) + c$ will also be an antiderivative of $f(x)$. The antiderivative is also called the *indefinite integral*.

The Definite Integral: Finding the Area under a Curve

Suppose we have a nonnegative[1] continuous function $f(x)$ defined on a closed interval $[a, b]$. $f(x) \geq 0$ for all $x \in [a, b]$. Suppose we partition the the interval $[a, b]$ using the partition $x_0, x_1, \ldots, x_n$, where $x_0 = a$ and $x_n = b$ and $x_i < x_{i+1}$. Note that the partition does not have to have equal length intervals. Let the minimum and maximum value of $f(x)$ in each interval be

$$l_i = \sup_{x \in [x_{i-1}, x_i]} f(x) \quad \text{and} \quad m_i = \inf_{x \in [x_{i-1}, x_i]} f(x),$$

where sup is the least upper bound, and inf is the greatest lower bound. Then the area under the curve $y = f(x)$ between $x = a$ and $x = b$ lies between the lower sum

$$L_{x_0, \ldots, x_n} = \sum_{i=1}^{n} l_i \times (x_i - x_{i-1})$$

---

[1]The requirement that $f(x)$ be nonnegative is not strictly necessary. However, since we are using the definite integral to find the area under probability density functions that are nonnegative, we will impose the condition.

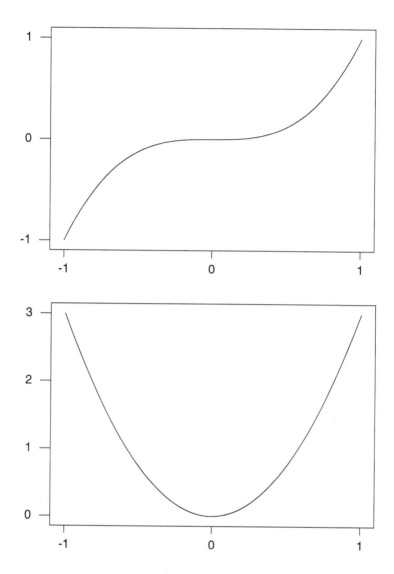

Figure A.6   Graph of $f(x) = x^3$ and its derivative. The derivative function is negative where the original function is increasing, and it is positive where the original function is increasing We see that the original function has a point of inflection at $x = 0$.

and the upper sum

$$M_{x_0,\dots,x_n} = \sum_{i=1}^{n} m_i \times (x_i - x_{i-1}).$$

We can refine the partition by adding one more x value to it. Let $x'_1, \ldots, x'_{n+1}$ be a refinement of the partition $x_1, \ldots, x_n$. Then $x'_0 = x_0$, $x'_{n+1} = x_n$, $x'_i = x_i$ for all $i < k$, and $x'_{i+i} = x_i$ for all $i > k$. x_k is the new value added to the partition. In the lower and upper sum, all the bars except for the kth are unchanged. The kth bar has been replaced by two bars in the refinement. Clearly,

$$M_{x'_0, \ldots, x'_{n+1}} \leq M_{x_0, \ldots, x_n}$$

and

$$L_{x'_0, \ldots, x'_{n+1}} \geq L_{x_0, \ldots, x_n}.$$

The lower and upper sums for a partition and its refinement are shown in Figure A.7. We see that refining a partition must make tighter bounds on the

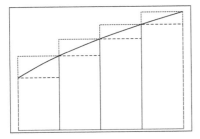 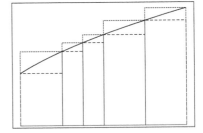

Figure A.7 Lower and upper sums over a partition and its refinement. The lower sum has increased and the upper sum has decreased in the refinement. The area under the curve is always between the lower and upper sums.

area under the curve.

Next we will show that for any continuous function defined on a closed interval $[a, b]$, we can find a partition $x_0, \ldots, x_n$ for some n that will make the difference between the upper sum and the lower sum as close to zero as we wish. Suppose $\epsilon > 0$ is the number we want the difference to be less than. We draw lines $\delta = \frac{\epsilon}{[b-a]}$ apart parallel to the horizontal (x) axis. (Since the function is defined on the closed interval, its maximum and minimum are both finite.) Thus a finite number of the horizontal lines will intercept the curve $y = f(x)$ over the interval $[a, b]$. Where one of the lines intercepts the curve, draw a vertical line down to the horizontal axis. The x values where these vertical lines hit the horizontal axis are the points for our partition. For example, the function $f(x) = 1 + \sqrt{4 - x^2}$ is defined on the interval $[0, 2]$. The difference between the upper sum and the lower sum for the partition for that ϵ is given by

$$M_{x_0, \ldots, x_n} - L_{x_0, \ldots, x_n} = \delta \times [(x_1 - x_0) + (x_2 - x_1) + \ldots + (x_n - x_{n-1})]$$
$$= \delta \times [b - a]$$
$$= \epsilon.$$

We can make this difference as small as we want to by choosing $\epsilon > 0$ small enough.

Let $\epsilon_k = \frac{1}{k}$ for $k = 1, \ldots, \infty$. This gives us a sequence of partitions such that $\lim_{k \to \infty} \epsilon_k = 0$. Hence

$$\lim_{k \to \infty} M_{x_0,\ldots,x_{n_k}} - L_{x_0,\ldots,x_{n_k}} = 0 \,.$$

The partitions for ϵ_1 and ϵ_2 are shown in Figure A.8. Note that $\delta_k = \frac{1}{2k}$.

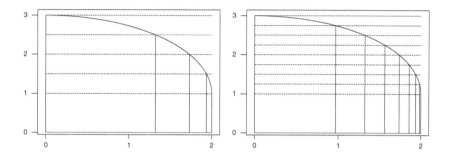

Figure A.8   The partition induced for the function $f(x) = 1 + \sqrt{4 - x^2}$ where $\epsilon_1 = 1$ and its refinement where $\epsilon_2 = \frac{1}{2}$.

That means that the area under the curve is the least upper bound for the lower sum, and the greatest lower bound for the upper sum. We call it the definite integral and denote it

$$\int_a^b f(x)\,dx \,.$$

Note that the variable x in the formula above is a dummy variable:

$$\int_a^b f(x)\,dx = \int_a^b f(y)\,dy \,.$$

Basic Properties of Definite Integrals

Theorem A.7 *Let $f(x)$ and $g(x)$ be functions defined on the interval $[a, b]$, and let c be a constant. Then the following properties hold.*

1. *The definite integral of a constant times a function is the constant times the definite integral of the function:*

$$\int_a^b cf(x)\,dx = c \int_a^b f(x)\,dx \,.$$

2. *The definite integral of a sum of two functions is a sum of the definite integrals of the two functions:*

$$\int_a^b (f(x) + g(x))\, dx = \int_a^b f(x)\, dx + \int_a^b g(x)\, dx.$$

Fundamental Theorem of Calculus

The methods of finding extreme values by differentiation and finding area under a curve by integration were known before the time of Newton and Liebniz. Newton and Liebniz independently discovered the fundamental theorem of calculus that connects differentiation and integration. Because each was unaware of the others work, they are both credited with the discovery of the calculus.

Theorem A.8 *Fundamental theorem of calculus. Let $f(x)$ be a continuous function defined on a closed interval. Then:*

1. *The function has antiderivative in the interval.*

2. *If a and b are two numbers in the closed interval such that $a < b$, and $F(x)$ is any antiderivative function of $f(x)$, then*

$$\int_a^b f(x)\, dx = F(b) - F(a).$$

Proof:
For $x \in (a, b)$, define the function

$$I(x) = \int_a^x f(x)\, dx.$$

This function shows the area under the curve $y = f(x)$ between a and x. Note that the area under the curve is additive over an extended region from a to $x + h$:

$$\int_a^{x+h} f(x)\, dx = \int_a^x f(x)\, dx + \int_x^{x+h} f(x)\, dx.$$

By definition, the derivative of the function $I(x)$ is

$$I'(x) = \lim_{h \to 0} \frac{I(x + h) - I(x)}{h} = \lim_{h \to 0} \frac{\int_x^{x+h} f(x)\, dx}{h}.$$

In the limit as h approaches 0,

$$\lim_{h \to 0} f(x') = f(x)$$

for all values $x' \in [x, x+h)$. Thus

$$I'(x) = \lim_{h \to 0} \frac{h \times f(x)}{h} = f(x).$$

In other words, $I(x)$ is an antiderivative of $f(x)$. Suppose $F(x)$ is any other antiderivative of $f(x)$. Then

$$F(x) = I(x) + c$$

for some constant c. Thus $F(b) - F(a) = I(b) - I(a) = \int_a^b f(x)\, dx$, and the theorem is proved.

For example, suppose $f(x) = e^{-2x}$ for $x \geq 0$. Then $F(x) = -\frac{1}{2} \times e^{-2x}$ is an antiderivative of $f(x)$. The area under the curve between 1 and 4 is given by

$$\int_1^4 f(x)\, dx = F(4) - F(1) = -\frac{1}{2} \times e^{-2 \times 4} + \frac{1}{2} \times e^{-2 \times 1}.$$

Definite Integral of a Function $f(x)$ Defined on an Open Interval

Let $f(x)$ be a function defined on the open interval (a, b). In this case, the antiderivative $F(x)$ is not defined at a and b. We define

$$F(a) = \lim_{x \to a} F(x) \quad \text{and} \quad F(b) = \lim_{x \to b} F(x).$$

provided that those limits exist. Then we define the definite integral with the same formula as before:

$$\int_a^b f(x) = F(b) - F(a).$$

For example, let $f(x) = x^{-1/2}$. This function is defined over the half-open interval $(0, 1]$. It is not defined over the closed interval $[0, 1]$ because it is not defined at the endpoint $x = 0$. This curve is shown in Figure A.9. We see that the curve has a vertical asymptote at $x = 0$. We will define

$$F(0) = \lim_{x \to 0} F(x)$$
$$= \lim_{x \to 0} 2x^{1/2}$$
$$= 0.$$

Then

$$\int_0^1 x^{-1/2} = 2x^{1/2} \Big|_0^1 = 2.$$

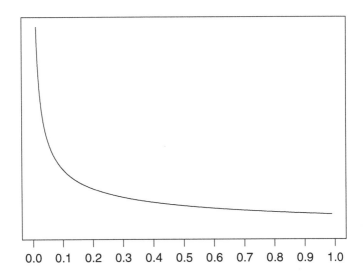

0.0   0.1   0.2   0.3   0.4   0.5   0.6   0.7   0.8   0.9   1.0

Figure A.9   The function $f(x) = x^{-1/2}$.

Theorem A.9 *Integration by parts.*   *Let $F(x)$ and $G(x)$ be differentiable functions defined on an interval $[a, b]$. Then*

$$\int_a^b F'(x) \times G(x)\, dx = F(x) \times G(x)\big|_a^b - \int_a^b F(x) \times G'(x)\, dx\,.$$

Proof: Integration by parts is the inverse of finding the derivative of the product $F(x) \times G(x)$:

$$\frac{d}{dx}[F(x) \times G(x)] = F(x) \times G'(x) + F(x) \times G'(x)\,.$$

Integrating both sides, we see that

$$F(b) \times G(b) - F(a) \times G(a) = \int_a^b F(x) \times G'(x)\, dx + \int_a^b F'(x) \times G(x)\, dx\,.$$

Theorem A.10 *Change of variable formula. Let $x = g(y)$ be a differentiable function on the interval $[a, b]$. Then*

$$\int_a^b f(g(y))g'(y)\, dy = \int_{g(a)}^{g(b)} f(y)\, dy\,.$$

The change of variable formula is the inverse of the chain rule for differentiation. The derivative of the function of a function $F(g(y))$ is

$$\frac{d}{dx}[F(g(y))] = F'(g(y)) \times g'(y)\,.$$

Integrating both sides from $y = a$ to $y = b$ gives

$$F(g(b)) - F(g(a)) = \int_a^b F'(g(y)) \times g'(y)\, dy\,.$$

The left-hand side equals $\int_{g(a)}^{g(b)} F'(y)\, dy$. Let $f(x) = F'(x)$, and the theorem is proved.

MULTIVARIATE CALCULUS

Partial Derivatives

In this section we consider the calculus of two or more variables. Suppose we have a function of two variables $f(x, y)$. The function is continuous at the point (a, b) if and only if

$$\lim_{(x,y) \to (a,b)} f(x, y) = f(a, b)\,.$$

The first *partial derivatives* at the point (a, b) are defined to be

$$\left.\frac{\partial f(x, y)}{\partial x}\right|_{(a,b)} = \lim_{h \to 0} \frac{f(a + h, b) - f(a, b)}{h}$$

and

$$\left.\frac{\partial f(x, y)}{\partial y}\right|_{(a,b)} = \lim_{h \to 0} \frac{f(a, b + h) - f(a, b)}{h}\,,$$

provided that these limits exist. In practice, the first partial derivative in the x-direction is found by treating y as a constant and differentiating the function with respect to x, and vice versa, to find the first partial derivative in the y-direction.

If the function $f(x, y)$ has first partial derivatives for all points (x, y) in a continuous two-dimensional region, then the first partial derivative function with respect to x is the function that has value at point (x, y) equal to the partial derivative of $f(x, y)$ with respect to x at that point. It is denoted

$$f_x(x, y) = \left.\frac{\partial f(x, y)}{\partial x}\right|_{(x,y)}\,.$$

The first partial derivative function with respect to y is defined similarly. The first derivative functions $f_x(x, y)$ and $f_y(x, y)$ give the instantaneous rate of change of the function in the x-direction and y-direction, respectively.

The second *partial derivatives* at the point (a, b) are defined to be

$$\left.\frac{\partial^2 f(x, y)}{\partial x^2}\right|_{(a,b)} = \lim_{h \to 0} \frac{f_x(x + h, y) - f_x(x, y)}{h}$$

and

$$\frac{\partial^2 f(x,y)}{\partial y^2}\bigg|_{(a,b)} = \lim_{h \to 0} \frac{f_y(x, y + h) - f_y(x,y)}{h}.$$

The second *cross partial derivatives* at (a, b) are

$$\frac{\partial^2 f(x,y)}{\partial x \partial y}\bigg|_{(a,b)} = \lim_{h \to 0} \frac{f_y(x + h, y) - f_y(x,y)}{h}$$

and

$$\frac{\partial^2 f(x,y)}{\partial y \partial x}\bigg|_{(a,b)} = \lim_{h \to 0} \frac{f_x(x, y + h) - f_x(x,y)}{h}.$$

For all the functions that we consider, the *cross partial derivatives* are equal, so it doesn't matter which order we differentiate.

If the function $f(x, y)$ has second partial derivatives (including cross partial derivatives) for all points (x, y) in a continuous two-dimensional region, then the second partial derivative function with respect to x is the function that has value at point (x, y) equal to the second partial derivative of $f(x, y)$ with respect to x at that point. It is denoted

$$f_{xx}(x,y) = \frac{\partial f_x(x,y)}{\partial x}\bigg|_{(x,y)}.$$

The second partial derivative function with respect to y is defined similarly. The second cross partial derivative functions are

$$f_{xy}(x,y) = \frac{\partial f_x(x,y)}{\partial y}\bigg|_{(x,y)}$$

and

$$f_{yx}(x,y) = \frac{\partial f_y(x,y)}{\partial x}\bigg|_{(x,y)}.$$

The two cross partial derivative functions are equal.

Partial derivatives of functions having more than 2 variables are defined in a similar manner.

Finding Minima and Maxima of a Multivariate Function

A univariate functions with a continuous derivative achieves minimum or maximum at an interior point x only at points where the derivative function $f'(x) = 0$. However, not all such points were minimum or maximum. We had to check either the first derivative test, or the second derivative test to see whether the critical point was minimum, maximum, or point of inflection.

The situation is more complicated in two dimensions. Suppose a continuous differentiable function $f(x, y)$ is defined on a two dimensional rectangle. It is not enough that both $f_x(x, y) = 0$ and $f_y(x, y) = 0$.

The directional derivative of the function $f(x,y)$ in direction θ at a point measures the rate of change of the function in the direction of the line through the point that has angle θ with the positive x-axis. It is given by

$$D_\theta f(x,y) = f_x(x,y)\cos(\theta) + f_y(x,y)\sin(\theta).$$

The function achieves a maximum or minimum value at points (x,y), where $D_\theta f(x,y) = 0$ for all θ.

Multiple Integrals

Let $f(x,y) > 0$ be a nonnegative function defined over a closed a rectangle $a_1 \leq x \leq b_1$ and $a_2 \leq y \leq b_2$. Let $x_0, \ldots, x_n$ partition the interval $[a_1, b_1]$, and let $y_1, \ldots, y_m$ partition the interval a_2, b_2. Together these partition the rectangle into $j = m \times n$ rectangles. The volume under the surface $f(x,y)$ over the rectangle A is between the upper sum

$$U = \sum_{j=1}^{mn} f(t_j, u_j)$$

and the lower sum

$$U = \sum_{j=1}^{mn} f(v_j, w_j),$$

where (t_j, u_j) is the point where the function is maximized in the jth rectangle, and (v_j, w_j) is the point where the function is minimized in the jth rectangle. Refining the partition always lowers the upper sum and raises the lower sum. We can always find a partition that makes the upper sum arbitrarily close to the lower sum. Hence the total volume under the surface denoted

$$\int_{a_1}^{b_1} \int_{a_2}^{b_2} f(x,y)\, dx\, dy$$

is the least upper bound of the lower sum and the greatest lower bound of the upper sum.

B

USE OF STATISTICAL TABLES

Tables or Computers?

In this appendix we will learn how to use statistical tables in order to answer various probability questions. In many ways this skill is largely redundant as computers and statistical software have replaced the tables, giving users the ability to obtain lower and upper tail probabilities, or the quantiles for an associated probability, for nearly any choice of distribution. Some of this functionality is available in high-school students' calculators, and is certainly in any number of smartphone apps.

However, the associated skills that one learns along with learning how to use the tables are important and therefore there is still value in this information. In this appendix we retain much of the original information from the first and second editions of this text. In this edition we have also added instructions on how to obtain the same (or in some cases more accurate) results from Minitab and R.

Introduction to Bayesian Statistics, 3rd ed. **497**
By Bolstad, W. M. and Curran, J. M. Copyright © 2016 John Wiley & Sons, Inc.

Binomial Distribution

Table B.1 contains values of the *binomial* (n, π) probability distribution for $n = 2, 3, 4, 5, 6, 7, 8, 9, 10, 11, 12, 15,$ and 20 and for $\pi = .05, .10, \ldots, .95$. Given the parameter π, the *binomial* probability is obtained by the formula

$$P(Y = y | \pi) = \binom{n}{y} \pi^y (1 - \pi)^{n-y}. \tag{B.1}$$

When $\pi \leq .5$, use the π value along the top row to find the correct column of probabilities. Go down to the correct n. The probabilities correspond to the y values found in the left-hand column. For example, to find $P(Y = 6)$ when Y has the *binomial* $(n = 10, \pi = .3)$ distribution, go down the table to $n = 10$ and find the row $y = 6$ on the *left* side. Look across the top to find the column labeled .30. The value in the table at the intersection of that row and column is $P(Y = 6) = .0368$ in this example. When $\pi > .5$ use the π value along the bottom row to find the correct column of probabilities. Go down to the correct n. The probabilities correspond to the y values found in the right-hand column. For example, to find $P(Y = 3)$ when y has the *binomial* $(n = 8, \pi = .65)$ distribution, go down the table to $n = 8$ and find the row $y = 3$ on the *right* side. Look across the bottom to find the column labeled .65. The value in the table at the intersection of that row and column is $P(Y = 3) = .0808$ in this example.

[**Minitab:**]: Enter the value 6 into the first cell of column c1. Select *Probability Distributions* from the *Calc* menu, and then *Binomial....* Click the *Probability* radio button. Enter 10 into the *Number of trials* text box, .3 into the *Event Probability* text box, and c1 into the *Input Column* text box. Finally, click on *OK*. Alternatively, if you have enabled command line input by selecting *Enable Commands* from the *Editor* menu, or you are using the *Command Line Editor* from the *Edit* menu, you can type

```
pdf c1;
binomial 10 .3.
```

This should return a value of 0.0367569. It is also useful to be able to answer questions of the form $Pr(Y \leq y)$ — for example, what is the probability we will see six or fewer successes in 10 trials where the probability of success is .3? This is the value can be obtained from the table by adding all of the values in the same column for a given value of n. Using Minitab, we can answer this question in second by simply clicking the *Cumulative probability* radio button, or by replacing the command **pdf** with **cdf**.

[**R:**]: All R distribution functions have the same basic naming structure dxxx, pxxx, and qxxx. These three functions return the probability (density) function, the cumulative distribution function, and the inverse-cdf or quantile function, respectively, for distribution xxx. In this section we want the binomial pdf and cdf, which are provided by the functions dbinom and pbinom. To find the probability $\Pr(Y = 6)$ in the example, we type

```
dbinom(6, 10, 0.3)
[1] 0.03675691
```

To find the probability $Pr(Y \leq 6)$, we type

```
pbinom(6, 10, 0.3)
[1] 0.9894079
```

If we are interested in the upper tail probability — i.e. $\Pr(Y > 6)$ — then we can obtain this by noting that

$$\Pr(Y \leq y) + \Pr(Y > y) = 1, \text{ so } \Pr(Y > y) = 1 - Pr(Y \leq y)$$

or by setting the lower.tail argument of the R functions to FALSE, e.g.

```
1 - pbinom(6, 10, 0.3)
[1] 0.01059208
pbinom(6, 10, 0.3, FALSE)
[1] 0.01059208
```

Standard Normal Distribution

This section contains two tables. Table B.2 contains the area under the standard normal density. Table B.3 contains the ordinates (height) of the standard normal density. The standard normal density has mean equal to 0 and variance equal to 1. Its density is given by the formula

$$f(z) = \frac{1}{\sqrt{2\pi}} e^{-\frac{1}{2}z^2}. \tag{B.2}$$

We see that the standard normal density is symmetric about 0. The graph of the standard normal density is shown in Figure B.1.

Area Under Standard Normal Density

Table B.2 tabulates the area under the standard normal density function between 0 and z for nonnegative values of z from 0.0 to 3.99 in steps of .01. We read down the z column until we come to the value that has the correct *units* and tenths digits of z. This is the correct row. We look across the top row to find the *hundredth* digit of z. This is the correct column. The

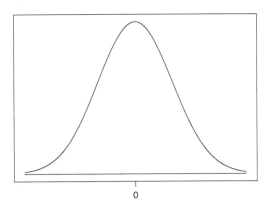

Figure B.1   Standard normal density.

tabulated value at the intersection of the correct row and correct column is $P(0 \le Z \le z)$, where Z has the *normal* $(0, 1)$ distribution. For example, to find $P(0 \le Z \le 1.23)$ we go down the z column to 1.2 for the correct row and across top to 3 for correct column. We find the tabulated value at the intersection of this row and column. For this example, $P(0 \le Z \le 1.23) =$ .3907. Because the standard normal density is symmetric about 0,

$$P(-z \le Z \le 0) = P(0 \le Z \le z).$$

Also, since it is a density function, the total area underneath it equals 1.0000, so the total area to the right of 0 must equal .5000. We can proceed to find

$$P(Z > z) = .5000 - P(Z \le z).$$

Finding Any Normal Probability

We can standardize any normal random variable to a standard normal random variable having mean 0 and variance 1. For instance, if W is a normal random variable having mean m and variance s^2, we standardize by subtracting the mean and dividing by the standard deviation.

$$Z = \frac{W - m}{s}.$$

This lets us find any normal probability by using the standard normal tables.

EXAMPLE B.1

Suppose W has the normal distribution with mean 120 and variance 225. (The standard deviation of W is 15.) Suppose we wanted to find the probability

$$P(W \leq 129).$$

We can subtract the mean from both sides of an inequality without changing the inequality:

$$P(W - 120 \leq 129 - 120).$$

We can divide both sides of an inequality by the standard deviation (which is positive) without changing the inequality:

$$P\left(\frac{W - 120}{15} \leq \frac{9}{15}\right).$$

On the left-hand side we have the standard normal Z, and on the right-hand side we have the number .60. Therefore

$$P(W \leq 129) = P(Z \leq .60) = .5000 + .2258 = .7258.$$

[**Minitab:**] We can answer this question in Minitab, as with the Binomial

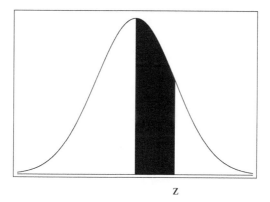

Figure B.2   Shaded area under standard normal density. These values are shown in Table B.2.

distribution, either using the menus or by entering Minitab commands. To use the menus, we first enter the value of interest (129) in column c1. We then select *Probability Distributions* from the *Calc* and then *Normal....* We select the *Cumulative probability* radio button, enter 120 into the *Mean* text box, 15 into the *Standard deviation* text box, and c1 into the *Input column*, and finally click on *OK*. Alternatively, we enter the following commands into Minitab:

```
cdf c1;
normal 120 15.
```

[R:] The R function pnorm returns values from the normal cumulative distribution function. To answer the question we would type

```
pnorm(129, 120, 15)
```

■

Ordinates of the Standard Normal Density

Figure B.3 shows the ordinate of the standard normal table at z. We see the ordinate is the height of the curve at z. Table B.3 contains the ordinates of the standard normal density for nonnegative z values from 0.00 to 3.99 in steps of .01. Since the standard normal density is symmetric about 0, $f(-z) = f(z)$, we can find ordinates of negative z values. This table is used to find values of the likelihood when we have a discrete prior distribution for μ. We go down the z column until we find the value that has the *units* and *tenths* digits. This gives us the correct row. We go across the top until we find the *hundredth* digit. This gives us the correct column. The value at the intersection of this row and column is the ordinate of the standard normal density at the value z. For instance, if we want to find the height of the standard normal density at $z = 1.23$ we go down z column to 1.2 to find the correct row and go across the top to 3 to find the correct column. The ordinate of the standard normal at $z = 1.23$ is equal to .1872. (Note: You can verify this is correct by plugging $z = 1.23$ into Equation B.2.)

▣ EXAMPLE B.2

Suppose the distribution of Y given μ is normal$(\mu, \sigma^2 = 1)$. Also suppose there are four possible values of μ. They are 3, 4, 5, and 6. We observe $y = 5.6$. We calculate

$$z_i = \left(\frac{5.6 - \mu_i}{1} \right).$$

The likelihood is found by looking up the ordinates of the normal distribution for the z_i values. We can put them in the following table:

3	2.60	.136
4	1.6	.1109
5	.6	.3332
6	-.4	.3683

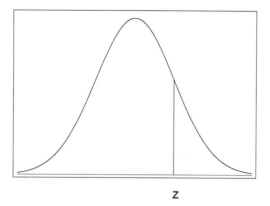

z

Figure B.3   Ordinates of standard normal density function. These values are shown in Table B.3.

[**Minitab:**] In this example we are interested in finding the height of the normal density at the point $y = 5.6$ for four possible values of the mean $\mu \in \{3, 4, 5, 6\}$. There are several approaches we might take. We might calculate the standardized values of y for each value of μ, i.e.,

$$z_i = \frac{y - \mu_i}{\sigma} = y - \mu_i,$$

and then compute the height of the standard normal at each of these values. Alternatively, we might exploit the fact that because the normal density computes the squared difference between the observation and the mean, $(y - \mu)^2$, it does not matter if the order of these values is reversed. That is $f(y|\mu, \sigma^2) = f(\mu|y, \sigma^2)$ for the normal distribution. We will take this second approach as it requires less calculation. Firstly we enter values for μ into column c1. We then select *Probability Distributions* from the *Calc* and then *Normal. . . .* We select the *Probability density* radio button, enter 5.6 into the *Mean* text box, 1 into the *Standard deviation* text box, and c1 into the *Input column*, and finally click on *OK*. Alternatively, we enter the following commands into Minitab:

```
pdf c1;
normal 5.6 1.
```

[**R:**] The R function **dnorm** returns values from the normal probability density function. To answer the question, we would type

```
dnorm(5.6, 3:6, 1)
```

■

Student's t Distribution

Figure B.4 shows the *Student's* t distribution for several different degrees
of freedom, along with the standard *normal*$(0, 1)$ distribution. We see the
Student's t family of distributions are similar to the standard *normal* in that
they are symmetric bell shaped curves; however, they have more weight in
the tails. The heaviness of the tails of the *Student's* t decreases as the degrees
of freedom increase.[1] The *Student's* t distribution is used when we use the
unbiased estimate of the standard deviation $\hat{\sigma}$ instead of the true unknown
standard deviation σ in the standardizing formula

$$z = \frac{y - \mu}{\sigma_y}$$

and y is a normally distributed random variable. We know that z will have
the *normal*$(0, 1)$ distribution. The similar formula

$$t = \frac{y - \mu}{\hat{\sigma}_y}$$

will have the *Student's* t distribution with k degrees of freedom. The degrees

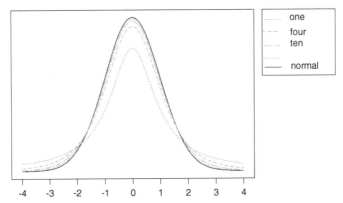

Figure B.4   *Student's* t densities for selected degrees of freedom, together with the
standard *normal* $(0, 1)$ density which corresponds to *Student's* t with ∞ degrees of
freedom.

of freedom k will equal the sample size minus the number of parameters estimated in the equation for $\hat{\sigma}$. For instance, if we are using $\bar{y}$ the sample mean,
its estimated standard deviation is given by $\hat{\sigma}_{\bar{y}} = \frac{\hat{\sigma}}{n}$, where

$$\hat{\sigma} = \sum_{i=1}^{n}(y_i - \bar{y})^2$$

---

[1]The *normal*$(0, 1)$ distribution corresponds to the *Student's* t distribution with ∞ degrees
of freedom.

and we observe that to use the above formula we have to first estimate $\bar{y}$. Hence, in the single sample case we will have $k = n - 1$ degrees of freedom. Table B.4 contains the tail areas for the *Student's t* distribution family. The *degrees of freedom* are down the left column, and the tabulated tail areas are across the rows for the specified tail probabilities.

[**Minitab:**] We can use Minitab to find $\Pr(T \le t)$ for a given value of t and a fixed number of degrees of freedom ν. As an illustration we will choose $t = 1.943$ and with $\nu = 6$ degrees of freedom. We can see from Table B.4 that the upper tail probability, $\Pr(T \ge 1.943)$, is approximately 0.05. This means the lower tail probability, which is what Minitab will calculate, is approximately 0.95. We say approximately because the values in Tables B.4 are rounded. Firstly we enter value for t (1.943) into column c1. We then select *Probability Distributions* from the *Calc* and then t.... We select the *Cumulative probability* radio button, enter 6 into the *Degrees of freedom* text box, c1 into the *Input column*, and finally click on *OK*. Alternatively, we enter the following commands into Minitab:

```
pdf c1;
t 6.
```

[**R:**] The R function **pt** returns values from the Student's t cumulative distribution function. To answer the same question we used to demonstrate Minitab, we would type

```
pt(1.943, 5, lower.tail = FALSE)
```

pt allows the user to choose whether they want the upper or lower tail probability.

Poisson Distribution

Table B.5 contains values of the *Poisson*(μ) distribution for some selected values of μ going from .1 to 4 in increments of .1, from 4.2 to 10 in increments of .2, and from 10.5 to 15 in increments of .5. Given the parameter μ, the *Poisson* probability is obtained from the formula

$$P(Y = y|\mu) = \frac{\mu^y e^{-\mu}}{y!} \tag{B.3}$$

for $y = 0, 1, \ldots$. Theoretically y can take on all non-negative integer values. In Table B.5 we include all possible values of y until the probability becomes less than .0001.

☐ EXAMPLE B.3

Suppose the distribution of Y given μ is Poisson(μ) and there are 3 possible values of μ, namely, .5, .75, and 1.00. We observed $y = 2$. The likelihood is found by looking up values in row $y = 2$ for the possible values of μ. Note that the value for $\mu = .75$ is not in the table. It is found by linearly interpolating between the values for $\mu = .70$ and $\mu = .80$.

μ_i	(Interpolation if necessary)	*Likelihood*
.50		.0758
.75	$(.5 \times .1217 + .5 \times .1438)$	.1327
1.00		.1839

[**Minitab:**] We cannot use the same trick we used for the normal distribution to repeat these calculations in Minitab this time, as $f(y|\mu) \neq f(mu|y)$. This means we have to repeat the steps we are about to describe for each value of μ. That is a little laborious, but not too painful as Minitab remembers your inputs to dialog boxes. Firstly we enter the values for y into column c1. We then select *Probability Distributions* from the *Calc* and then *Poisson....* We select the *Probability* radio button, enter 0.5 into the *Mean* text box, enter c1 into the *Input column*, and finally click on *OK*. Alternatively we enter the following commands into Minitab

```
pdf c1;
poisson 0.5.
```

These steps can be repeated using the subsequent values of $\mu = 0.75$ and $\mu = 1$.

[**R:**] The R function `dpois` returns values from the Poisson probability function. To answer the question, we would type

```
dpois(2, c(0.5, 0.75, 1))
```

■

Chi-Squared Distribution

Table B.6 contains $P(U > \alpha)$, the upper tail area when U has the *chi-squared* distribution. The values in the table correspond to the shaded area in Figure B.5. The posterior distribution of the variance σ^2 is $S' \times$ an *inverse chi-squared* distribution with κ' degrees of freedom. This means that $\frac{S'}{\sigma^2}$ has the *chi-squared* distribution with κ' degrees of freedom, so we can use the *chi-squared* table to find credible intervals for σ and test hypotheses about σ.

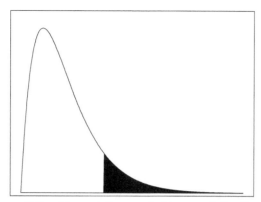

Figure B.5   Upper tail area of *chi-squared* distribution.

◼ EXAMPLE B.4

Suppose that the posterior distribution of σ^2 is 110× an inverse chi-squared distribution with 12 degrees of freedom. Then $\frac{110}{\sigma^2}$ has the chi-squared distribution with 12 degrees of freedom. So a 95% Bayesian credible interval is found by

$$.95 = P(4.404 < \frac{110}{\sigma^2} < 23.337)$$

$$= P\left((\sqrt{\frac{110}{23.337}} < \sigma < \sqrt{\frac{110}{4.404}} \right)$$

$$= P(2.17107 < \sigma < 4.99773).$$

We would test the two-sided hypothesis

$$H_0 : \sigma = 2.5 \quad \text{versus} \quad H_1 : \sigma \neq 2.5$$

at the 5% level of significance by observing that 2.5 lies within the credible interval, so we must accept the null hypothesis that $\sigma = 2.5$ is still a credible value. On the other hand, if we wanted to test the one-sided hypothesis

$$H_0 : \sigma \leq 2.5 \quad \text{versus} \quad H_1 : \sigma > 2.5$$

at the 5% level, we would calculate the posterior probability of the null hypothesis.

$$P(\sigma \leq 2.5) = P\left((\frac{110}{\sigma^2} \geq \frac{110}{2.5^2}\right)$$

$$= P\left((\frac{110}{\sigma^2} \geq 17.600\right).$$

The value 17.60 lies between 11.340 and 18.549, so the posterior probability of the null hypothesis is between .50 and .10. This is larger than the level of significance of 5%, so we would not reject the null hypothesis.

[**Minitab:**] The example states that σ^2 has a posterior density of 110 times in inverse chi-squared distribtion with 12 degrees of freedom. To find a 95% posterior credible interval for σ, we need to carry out two steps

1. Find the points $q_{0.025}$ and $q_{0.975}$ of a chi-squared distribution with 12 degrees of freedom such that

$$\Pr(X < q_{0.025}) = 0.025, \quad \text{and} \quad \Pr(X < q_{0.975}) = 0.975$$

2. Calculate
$$l = \sqrt{\frac{110}{q_{0.975}}} \quad \text{and} \quad u = \sqrt{\frac{110}{q_{0.025}}}$$

which are the lower and upper bounds, respectively, of our credible interval. To do this, we use the inverse cdf facilities of Minitab. Firstly, we enter the probabilities 0.025 and 0.975 into column c1. We then select *Probability Distributions* from the *Calc* and then *Chi-Square. . . .* We select the *Inverse cumulative probability* radio button, enter 12 into the *Degrees of freedom* text box, c1 into the *Input column*, c2 into the *Optional storage* and finally click on *OK*. We then select *Calculator* from the *Calc*. We enter c3 into the *Store result in variable* text box, sqrt(110/c2) in the *Expression* text box, and click on *OK*. Alternatively, we enter the following commands into Minitab:

```
invcdf c1 c2;
chisquare 12.
let c3 = sqrt(110/c2)
```

To calculate the posterior probability of the null hypothesis, we follow the same steps to calculate the critical value of 17.6 which we enter into column c4. We then select *Probability Distributions* from the *Calc* and then *Chi-Square. . . .* We select the *Cumulative probability* radio button, enter 12 into the *Degrees of freedom* text box, c4 into the *Input column*, and c5 into the *Optional storage*, and finally click on *OK*. We then select *Calculator* from the *Calc*. We enter c6 into the *Store result in variable* text box, 1-c5 in the *Expression* text box, and click on *OK*. Alternatively, we enter the following commands into Minitab:

```
cdf c4 c5;
chisquare 12.
let c6 = 1 - c5
```

Both of these methods return a value of $\Pr(H_0|data) = 0.1284$ which is indeed between 0.1 and 0.5.

[**R:**] We can repeat the steps described in the Minitab section above using the R functions `qchisq` and `pchisq` which return values of the chi-squared inverse cdf and cdf respectively. To compute a 95% posterior credible interval for σ, we type

```
sqrt(110 / qchisq(c(0.975, 0.025), 12))
```

The probabilities are provided to `qchisq` in reverse order so that the credible interval bounds are in the correct order. To compute the one-sided probability of the null hypothesis we type

```
pchisq(17.6, 12, lower.tail = FALSE)
```

■

Table B.1: Binomial probability table

n	y	.05	.10	.15	.20	.25	.30	.35	.40	.45	.50	
2	0	.9025	.81	.7225	.64	.5625	.49	.4225	.36	.3025	.25	2
	1	.0950	.18	.2550	.32	.3750	.42	.4550	.48	.4950	.50	1
	2	.0025	.01	.0225	.04	.0625	.09	.1225	.16	.2025	.25	0
3	0	.8574	.729	.6141	.512	.4219	.343	.2746	.216	.1664	.125	3
	1	.1354	.243	.3251	.384	.4219	.441	.4436	.432	.4084	.375	2
	2	.0071	.027	.0574	.096	.1406	.189	.2389	.288	.3341	.375	1
	3	.0001	.001	.0034	.008	.0156	.027	.0429	.064	.0911	.125	0
4	0	.8145	.6561	.5220	.4096	.3164	.2401	.1785	.1296	.0915	.0625	4
	1	.1715	.2916	.3685	.4096	.4219	.4116	.3845	.3456	.2995	.2500	3
	2	.0135	.0486	.0975	.1536	.2109	.2646	.3105	.3456	.3675	.3750	2
	3	.0005	.0036	.0115	.0256	.0469	.0756	.1115	.1536	.2005	.2500	1
	4	.0000	.0001	.0005	.0016	.0039	.0081	.0150	.0256	.0410	.0625	0
5	0	.7738	.5905	.4437	.3277	.2373	.1681	.1160	.0778	.0503	.0313	5
	1	.2036	.3281	.3915	.4096	.3955	.3601	.3124	.2592	.2059	.1563	4
	2	.0214	.0729	.1382	.2048	.2637	.3087	.3364	.3456	.3369	.3125	3
	3	.0011	.0081	.0244	.0512	.0879	.1323	.1811	.2304	.2757	.3125	2
	4	.0000	.0005	.0022	.0064	.0146	.0284	.0488	.0768	.1128	.1563	1
	5	.0000	.0000	.0001	.0003	.0010	.0024	.0053	.0102	.0185	.0313	0
6	0	.7351	.5314	.3771	.2621	.1780	.1176	.0754	.0467	.0277	.0156	6
	1	.2321	.3543	.3993	.3932	.3560	.3025	.2437	.1866	.1359	.0937	5
	2	.0305	.0984	.1762	.2458	.2966	.3241	.3280	.3110	.2780	.2344	4
	3	.0021	.0146	.0415	.0819	.1318	.1852	.2355	.2765	.3032	.3125	3
	4	.0001	.0012	.0055	.0154	.0330	.0595	.0951	.1382	.1861	.2344	2
	5	.0000	.0001	.0004	.0015	.0044	.0102	.0205	.0369	.0609	.0937	1
	6	.0000	.0000	.0000	.0001	.0002	.0007	.0018	.0041	.0083	.0156	0
7	0	.6983	.4783	.3206	.2097	.1335	.0824	.0490	.0280	.0152	.0078	7
	1	.2573	.3720	.3960	.3670	.3115	.2471	.1848	.1306	.0872	.0547	6
	2	.0406	.1240	.2097	.2753	.3115	.3177	.2985	.2613	.2140	.1641	5
	3	.0036	.0230	.0617	.1147	.1730	.2269	.2679	.2903	.2918	.2734	4
	4	.0002	.0026	.0109	.0287	.0577	.0972	.1442	.1935	.2388	.2734	3
	5	.0000	.0002	.0012	.0043	.0115	.0250	.0466	.0774	.1172	.1641	2
	6	.0000	.0000	.0001	.0004	.0013	.0036	.0084	.0172	.0320	.0547	1
	7	.0000	.0000	.0000	.0000	.0001	.0002	.0006	.0016	.0037	.0078	0
8	0	.6634	.4305	.2725	.1678	.1001	.0576	.0319	.0168	.0084	.0039	8
	1	.2793	.3826	.3847	.3355	.2670	.1977	.1373	.0896	.0548	.0313	7
	2	.0515	.1488	.2376	.2936	.3115	.2965	.2587	.2090	.1569	.1094	6
		.95	.90	.85	.80	.75	.70	.65	.60	.55	.50	
						π						y

Table B.1 – continued from previous page

n	y	.05	.10	.15	.20	.25	.30	.35	.40	.45	.50	
							π					
8	3	.0054	.0331	.0839	.1468	.2076	.2541	.2786	.2787	.2568	.2188	5
	4	.0004	.0046	.0185	.0459	.0865	.1361	.1875	.2322	.2627	.2734	4
	5	.0000	.0004	.0026	.0092	.0231	.0467	.0808	.1239	.1719	.2188	3
	6	.0000	.0000	.0002	.0011	.0038	.0100	.0217	.0413	.0703	.1094	2
	7	.0000	.0000	.0000	.0001	.0004	.0012	.0033	.0079	.0164	.0313	1
	8	.0000	.0000	.0000	.0000	.0000	.0001	.0002	.0007	.0017	.0039	0
9	0	.6302	.3874	.2316	.1342	.0751	.0404	.0207	.0101	.0046	.0020	9
	1	.2985	.3874	.3679	.3020	.2253	.1556	.1004	.0605	.0339	.0176	8
	2	.0629	.1722	.2597	.3020	.3003	.2668	.2162	.1612	.1110	.0703	7
	3	.0077	.0446	.1069	.1762	.2336	.2668	.2716	.2508	.2119	.1641	6
	4	.0006	.0074	.0283	.0661	.1168	.1715	.2194	.2508	.2600	.2461	5
	5	.0000	.0008	.0050	.0165	.0389	.0735	.1181	.1672	.2128	.2461	4
	6	.0000	.0001	.0006	.0028	.0087	.0210	.0424	.0743	.1160	.1641	3
	7	.0000	.0000	.0000	.0003	.0012	.0039	.0098	.0212	.0407	.0703	2
	8	.0000	.0000	.0000	.0000	.0001	.0004	.0013	.0035	.0083	.0176	1
	9	.0000	.0000	.0000	.0000	.0000	.0000	.0001	.0003	.0008	.0020	0
10	0	.5987	.3487	.1969	.1074	.0563	.0282	.0135	.0060	.0025	.0010	10
	1	.3151	.3874	.3474	.2684	.1877	.1211	.0725	.0403	.0207	.0098	9
	2	.0746	.1937	.2759	.3020	.2816	.2335	.1757	.1209	.0763	.0439	8
	3	.0105	.0574	.1298	.2013	.2503	.2668	.2522	.2150	.1665	.1172	7
	4	.0010	.0112	.0401	.0881	.1460	.2001	.2377	.2508	.2384	.2051	6
	5	.0001	.0015	.0085	.0264	.0584	.1029	.1536	.2007	.2340	.2461	5
	6	.0000	.0001	.0012	.0055	.0162	.0368	.0689	.1115	.1596	.2051	4
	7	.0000	.0000	.0001	.0008	.0031	.0090	.0212	.0425	.0746	.1172	3
	8	.0000	.0000	.0000	.0001	.0004	.0014	.0043	.0106	.0229	.0439	2
	9	.0000	.0000	.0000	.0000	.0000	.0001	.0005	.0016	.0042	.0098	1
	10	.0000	.0000	.0000	.0000	.0000	.0000	.0000	.0001	.0003	.0010	0
11	1	.3293	.3835	.3248	.2362	.1549	.0932	.0518	.0266	.0125	.0054	10
	2	.0867	.2131	.2866	.2953	.2581	.1998	.1395	.0887	.0513	.0269	9
	3	.0137	.0710	.1517	.2215	.2581	.2568	.2254	.1774	.1259	.0806	8
	4	.0014	.0158	.0536	.1107	.1721	.2201	.2428	.2365	.2060	.1611	7
	5	.0001	.0025	.0132	.0388	.0803	.1321	.1830	.2207	.2360	.2256	6
	6	.0000	.0003	.0023	.0097	.0268	.0566	.0985	.1471	.1931	.2256	5
	7	.0000	.0000	.0003	.0017	.0064	.0173	.0379	.0701	.1128	.1611	4
	8	.0000	.0000	.0000	.0002	.0011	.0037	.0102	.0234	.0462	.0806	3
	9	.0000	.0000	.0000	.0000	.0001	.0005	.0018	.0052	.0126	.0269	2
	10	.0000	.0000	.0000	.0000	.0000	.0000	.0002	.0007	.0021	.0054	1
	11	.0000	.0000	.0000	.0000	.0000	.0000	.0000	.0000	.0002	.0005	0
		.95	.90	.85	.80	.75	.70	.65	.60	.55	.50	
							π					y

Table B.1 – continued from previous page

n	y	π										
		.05	.10	.15	.20	.25	.30	.35	.40	.45	.50	
12	0	.5404	.2824	.1422	.0687	.0317	.0138	.0057	.0022	.0008	.0002	12
	1	.3413	.3766	.3012	.2062	.1267	.0712	.0368	.0174	.0075	.0029	11
	2	.0988	.2301	.2924	.2835	.2323	.1678	.1088	.0639	.0339	.0161	10
	3	.0173	.0852	.1720	.2362	.2581	.2397	.1954	.1419	.0923	.0537	9
	4	.0021	.0213	.0683	.1329	.1936	.2311	.2367	.2128	.1700	.1208	8
	5	.0002	.0038	.0193	.0532	.1032	.1585	.2039	.2270	.2225	.1934	7
	6	.0000	.0005	.0040	.0155	.0401	.0792	.1281	.1766	.2124	.2256	6
	7	.0000	.0000	.0006	.0033	.0115	.0291	.0591	.1009	.1489	.1934	5
	8	.0000	.0000	.0001	.0005	.0024	.0078	.0199	.0420	.0762	.1208	4
	9	.0000	.0000	.0000	.0001	.0004	.0015	.0048	.0125	.0277	.0537	3
	10	.0000	.0000	.0000	.0000	.0000	.0002	.0008	.0025	.0068	.0161	2
	11	.0000	.0000	.0000	.0000	.0000	.0000	.0001	.0003	.0010	.0029	1
	12	.0000	.0000	.0000	.0000	.0000	.0000	.0000	.0000	.0001	.0002	0
15	0	.4633	.2059	.0874	.0352	.0134	.0047	.0016	.0005	.0001	.0000	15
	1	.3658	.3432	.2312	.1319	.0668	.0305	.0126	.0047	.0016	.0005	14
	2	.1348	.2669	.2856	.2309	.1559	.0916	.0476	.0219	.0090	.0032	13
	3	.0307	.1285	.2184	.2501	.2252	.1700	.1110	.0634	.0318	.0139	12
	4	.0049	.0428	.1156	.1876	.2252	.2186	.1792	.1268	.0780	.0417	11
	5	.0006	.0105	.0449	.1032	.1651	.2061	.2123	.1859	.1404	.0916	10
	6	.0000	.0019	.0132	.0430	.0917	.1472	.1906	.2066	.1914	.1527	9
	7	.0000	.0003	.0030	.0138	.0393	.0811	.1319	.1771	.2013	.1964	8
	8	.0000	.0000	.0005	.0035	.0131	.0348	.0710	.1181	.1647	.1964	7
	9	.0000	.0000	.0001	.0007	.0034	.0116	.0298	.0612	.1048	.1527	6
	10	.0000	.0000	.0000	.0001	.0007	.0030	.0096	.0245	.0515	.0916	5
	11	.0000	.0000	.0000	.0000	.0001	.0006	.0024	.0074	.0191	.0417	4
	12	.0000	.0000	.0000	.0000	.0000	.0001	.0004	.0016	.0052	.0139	3
	13	.0000	.0000	.0000	.0000	.0000	.0000	.0001	.0003	.0010	.0032	2
	14	.0000	.0000	.0000	.0000	.0000	.0000	.0000	.0000	.0001	.0005	1
	15	.0000	.0000	.0000	.0000	.0000	.0000	.0000	.0000	.0000	.0000	0
20	0	.3585	.1216	.0388	.0115	.0032	.0008	.0002	.0000	.0000	.0000	20
	1	.3774	.2702	.1368	.0576	.0211	.0068	.0020	.0005	.0001	.0000	19
	2	.1887	.2852	.2293	.1369	.0669	.0278	.0100	.0031	.0008	.0002	18
	3	.0596	.1901	.2428	.2054	.1339	.0716	.0323	.0123	.0040	.0011	17
	4	.0133	.0898	.1821	.2182	.1897	.1304	.0738	.0350	.0139	.0046	16
	5	.0022	.0319	.1028	.1746	.2023	.1789	.1272	.0746	.0365	.0148	15
	6	.0003	.0089	.0454	.1091	.1686	.1916	.1712	.1244	.0746	.0370	14
	7	.0000	.0020	.0160	.0545	.1124	.1643	.1844	.1659	.1221	.0739	13
	8	.0000	.0004	.0046	.0222	.0609	.1144	.1614	.1797	.1623	.1201	12
		.95	.90	.85	.80	.75	.70	.65	.60	.55	.50	
						π						y

Table B.1 – continued from previous page

n	y	.05	.10	.15	.20	.25	π .30	.35	.40	.45	.50	
20	9	.0000	.0001	.0011	.0074	.0271	.0654	.1158	.1597	.1771	.1602	11
	10	.0000	.0000	.0002	.0020	.0099	.0308	.0686	.1171	.1593	.1762	10
	11	.0000	.0000	.0000	.0005	.0030	.0120	.0336	.0710	.1185	.1602	9
	12	.0000	.0000	.0000	.0001	.0008	.0039	.0136	.0355	.0727	.1201	8
	13	.0000	.0000	.0000	.0000	.0002	.0010	.0045	.0146	.0366	.0739	7
	14	.0000	.0000	.0000	.0000	.0000	.0002	.0012	.0049	.0150	.0370	6
	15	.0000	.0000	.0000	.0000	.0000	.0000	.0003	.0013	.0049	.0148	5
	16	.0000	.0000	.0000	.0000	.0000	.0000	.0000	.0003	.0013	.0046	4
	17	.0000	.0000	.0000	.0000	.0000	.0000	.0000	.0000	.0002	.0011	3
	18	.0000	.0000	.0000	.0000	.0000	.0000	.0000	.0000	.0000	.0002	2
	19	.0000	.0000	.0000	.0000	.0000	.0000	.0000	.0000	.0000	.0000	1
	20	.0000	.0000	.0000	.0000	.0000	.0000	.0000	.0000	.0000	.0000	0
		.95	.90	.85	.80	.75	.70	.65	.60	.55	.50	
							π					y

Table B.2: Area under standard normal density

z	.00	.01	.02	.03	.04	.05	.06	.07	.08	.09
0.0	.0000	.0040	.0080	.0120	.0160	.0199	.0239	.0279	.0319	.0359
0.1	.0398	.0438	.0478	.0517	.0557	.0596	.0636	.0675	.0714	.0753
0.2	.0793	.0832	.0871	.0910	.0948	.0987	.1026	.1064	.1103	.1141
0.3	.1179	.1217	.1255	.1293	.1331	.1368	.1406	.1443	.1480	.1517
0.4	.1554	.1591	.1628	.1664	.1700	.1736	.1772	.1808	.1844	.1879
0.5	.1915	.1950	.1985	.2019	.2054	.2088	.2123	.2157	.2190	.2224
0.6	.2257	.2291	.2324	.2357	.2389	.2422	.2454	.2486	.2517	.2549
0.7	.2580	.2611	.2642	.2673	.2703	.2734	.2764	.2794	.2823	.2852
0.8	.2881	.2910	.2939	.2967	.2995	.3023	.3051	.3078	.3106	.3133
0.9	.3159	.3186	.3212	.3238	.3264	.3289	.3315	.3340	.3365	.3389
1.0	.3413	.3438	.3461	.3485	.3508	.3531	.3554	.3577	.3599	.3621
1.1	.3643	.3665	.3686	.3708	.3729	.3749	.3770	.3790	.3810	.3830
1.2	.3849	.3869	.3888	.3907	.3925	.3944	.3962	.3980	.3997	.4015
1.3	.4032	.4049	.4066	.4082	.4099	.4115	.4131	.4147	.4162	.4177
1.4	.4192	.4207	.4222	.4236	.4251	.4265	.4279	.4292	.4306	.4319
1.5	.4332	.4345	.4357	.4370	.4382	.4394	.4406	.4418	.4429	.4441
1.6	.4452	.4463	.4474	.4484	.4495	.4505	.4515	.4525	.4535	.4545
1.7	.4554	.4564	.4573	.4582	.4591	.4599	.4608	.4616	.4625	.4633
1.8	.4641	.4649	.4656	.4664	.4671	.4678	.4686	.4693	.4699	.4706
1.9	.4713	.4719	.4726	.4732	.4738	.4744	.4750	.4756	.4761	.4767
2.0	.4772	.4778	.4783	.4788	.4793	.4798	.4803	.4808	.4812	.4817
2.1	.4821	.4826	.4830	.4834	.4838	.4842	.4846	.4850	.4854	.4857
2.2	.4861	.4864	.4868	.4871	.4875	.4878	.4881	.4884	.4887	.4890
2.3	.4893	.4896	.4898	.4901	.4904	.4906	.4909	.4911	.4913	.4916
2.4	.4918	.4920	.4922	.4925	.4927	.4929	.4931	.4932	.4934	.4936
2.5	.4938	.4940	.4941	.4943	.4945	.4946	.4948	.4949	.4951	.4952
2.6	.4953	.4955	.4956	.4957	.4959	.4960	.4961	.4962	.4963	.4964
2.7	.4965	.4966	.4967	.4968	.4969	.4970	.4971	.4972	.4973	.4974
2.8	.4974	.4975	.4976	.4977	.4977	.4978	.4979	.4979	.4980	.4981
2.9	.4981	.4982	.4982	.4983	.4984	.4984	.4985	.4985	.4986	.4986
3.0	.4987	.4987	.4987	.4988	.4988	.4989	.4989	.4989	.4990	.4990
3.1	.4990	.4991	.4991	.4991	.4992	.4992	.4992	.4992	.4993	.4993
3.2	.4993	.4993	.4994	.4994	.4994	.4994	.4994	.4995	.4995	.4995
3.3	.4995	.4995	.4995	.4996	.4996	.4996	.4996	.4996	.4996	.4997
3.4	.4997	.4997	.4997	.4997	.4997	.4997	.4997	.4997	.4997	.4998
3.5	.4998	.4998	.4998	.4998	.4998	.4998	.4998	.4998	.4998	.4998
3.6	.4998	.4998	.4999	.4999	.4999	.4999	.4999	.4999	.4999	.4999

Table B.3: Ordinates of standard normal density

z	.00	.01	.02	.03	.04	.05	.06	.07	.08	.09
0.0	.3989	.3989	.3989	.3988	.3986	.3984	.3982	.3980	.3977	.3973
0.1	.3970	.3965	.3961	.3956	.3951	.3945	.3939	.3932	.3925	.3918
0.2	.3910	.3902	.3894	.3885	.3876	.3867	.3857	.3847	.3836	.3825
0.3	.3814	.3802	.3790	.3778	.3765	.3752	.3739	.3725	.3712	.3697
0.4	.3683	.3668	.3653	.3637	.3621	.3605	.3589	.3572	.3555	.3538
0.5	.3521	.3503	.3485	.3467	.3448	.3429	.3410	.3391	.3372	.3352
0.6	.3332	.3312	.3292	.3271	.3251	.3230	.3209	.3187	.3166	.3144
0.7	.3123	.3101	.3079	.3056	.3034	.3011	.2989	.2966	.2943	.2920
0.8	.2897	.2874	.2850	.2827	.2803	.2780	.2756	.2732	.2709	.2685
0.9	.2661	.2637	.2613	.2589	.2565	.2541	.2516	.2492	.2468	.2444
1.0	.2420	.2396	.2371	.2347	.2323	.2299	.2275	.2251	.2227	.2203
1.1	.2179	.2155	.2131	.2107	.2083	.2059	.2036	.2012	.1989	.1965
1.2	.1942	.1919	.1895	.1872	.1849	.1826	.1804	.1781	.1758	.1736
1.3	.1714	.1691	.1669	.1647	.1626	.1604	.1582	.1561	.1539	.1518
1.4	.1497	.1476	.1456	.1435	.1415	.1394	.1374	.1354	.1334	.1315
1.5	.1295	.1276	.1257	.1238	.1219	.1200	.1182	.1163	.1145	.1127
1.6	.1109	.1092	.1074	.1057	.1040	.1023	.1006	.0989	.0973	.0957
1.7	.0940	.0925	.0909	.0893	.0878	.0863	.0848	.0833	.0818	.0804
1.8	.0790	.0775	.0761	.0748	.0734	.0721	.0707	.0694	.0681	.0669
1.9	.0656	.0644	.0632	.0620	.0608	.0596	.0584	.0573	.0562	.0551
2.0	.0540	.0529	.0519	.0508	.0498	.0488	.0478	.0468	.0459	.0449
2.1	.0440	.0431	.0422	.0413	.0404	.0396	.0387	.0379	.0371	.0363
2.2	.0355	.0347	.0339	.0332	.0325	.0317	.0310	.0303	.0297	.0290
2.3	.0283	.0277	.0270	.0264	.0258	.0252	.0246	.0241	.0235	.0229
2.4	.0224	.0219	.0213	.0208	.0203	.0198	.0194	.0189	.0184	.0180
2.5	.0175	.0171	.0167	.0163	.0158	.0154	.0151	.0147	.0143	.0139
2.6	.0136	.0132	.0129	.0126	.0122	.0119	.0116	.0113	.0110	.0107
2.7	.0104	.0101	.0099	.0096	.0093	.0091	.0088	.0086	.0084	.0081
2.8	.0079	.0077	.0075	.0073	.0071	.0069	.0067	.0065	.0063	.0061
2.9	.0060	.0058	.0056	.0055	.0053	.0051	.0050	.0048	.0047	.0046
3.0	.0044	.0043	.0042	.0040	.0039	.0038	.0037	.0036	.0035	.0034
3.1	.0033	.0032	.0031	.0030	.0029	.0028	.0027	.0026	.0025	.0025
3.2	.0024	.0023	.0022	.0022	.0021	.0020	.0020	.0019	.0018	.0018
3.3	.0017	.0017	.0016	.0016	.0015	.0015	.0014	.0014	.0013	.0013
3.4	.0012	.0012	.0012	.0011	.0011	.0010	.0010	.0010	.0009	.0009
3.5	.0009	.0008	.0008	.0008	.0008	.0007	.0007	.0007	.0007	.0006
3.6	.0006	.0006	.0006	.0005	.0005	.0005	.0005	.0005	.0005	.0004
3.7	.0004	.0004	.0004	.0004	.0004	.0004	.0003	.0003	.0003	.0003
3.8	.0003	.0003	.0003	.0003	.0003	.0002	.0002	.0002	.0002	.0002
3.9	.0002	.0002	.0002	.0002	.0002	.0002	.0002	.0002	.0001	.0001

Table B.4: Critical values of the *Student's t* distribution

Degrees of freedom (df)	Upper Tail Area							
	.20	.10	.05	.025	.01	.005	.001	.0005
1	1.376	3.078	6.314	12.71	31.82	63.66	318.3	636.6
2	1.061	1.886	2.920	4.303	6.965	9.925	22.33	31.60
3	.979	1.638	2.353	3.182	4.541	5.841	10.21	12.92
4	.941	1.533	2.132	2.776	3.747	4.604	7.173	8.610
5	.920	1.476	2.015	2.571	3.365	4.032	5.893	6.868
6	.906	1.440	1.943	2.447	3.143	3.707	5.208	5.959
7	.896	1.415	1.895	2.365	2.998	3.499	4.785	5.408
8	.889	1.397	1.860	2.306	2.896	3.355	4.501	5.041
9	.883	1.383	1.833	2.262	2.821	3.250	4.297	4.781
10	.879	1.372	1.812	2.228	2.764	3.169	4.144	4.587
11	.876	1.363	1.796	2.201	2.718	3.106	4.025	4.437
12	.873	1.356	1.782	2.179	2.681	3.055	3.930	4.318
13	.870	1.350	1.771	2.160	2.650	3.012	3.852	4.221
14	.868	1.345	1.761	2.145	2.624	2.977	3.787	4.140
15	.866	1.341	1.753	2.131	2.602	2.947	3.733	4.073
16	.865	1.337	1.746	2.120	2.583	2.921	3.686	4.015
17	.863	1.333	1.740	2.110	2.567	2.898	3.646	3.965
18	.862	1.330	1.734	2.101	2.552	2.878	3.610	3.922
19	.861	1.328	1.729	2.093	2.539	2.861	3.579	3.883
20	.860	1.325	1.725	2.086	2.528	2.845	3.552	3.850
21	.859	1.323	1.721	2.080	2.518	2.831	3.527	3.819
22	.858	1.321	1.717	2.074	2.508	2.819	3.505	3.792
23	.858	1.319	1.714	2.069	2.500	2.807	3.485	3.768
24	.857	1.318	1.711	2.064	2.492	2.797	3.467	3.745
25	.856	1.316	1.708	2.060	2.485	2.787	3.450	3.725
26	.856	1.315	1.706	2.056	2.479	2.779	3.435	3.707
27	.855	1.314	1.703	2.052	2.473	2.771	3.421	3.690
28	.855	1.313	1.701	2.048	2.467	2.763	3.408	3.674
29	.854	1.311	1.699	2.045	2.462	2.756	3.396	3.659
30	.854	1.310	1.697	2.042	2.457	2.750	3.385	3.646
40	.851	1.303	1.684	2.021	2.423	2.704	3.307	3.551
60	.848	1.296	1.671	2.000	2.390	2.660	3.232	3.460
80	.846	1.292	1.664	1.990	2.374	2.639	3.195	3.416
100	.845	1.290	1.660	1.984	2.364	2.626	3.174	3.390
∞	.842	1.282	1.645	1.960	2.326	2.576	3.090	3.291

Table B.5: Poisson probability table

y	μ									
	.1	.2	.3	.4	.5	.6	.7	.8	.9	1.0
0	.9048	.8187	.7408	.6703	.6065	.5488	.4966	.4493	.4066	.3679
1	.0905	.1637	.2222	.2681	.3033	.3293	.3476	.3595	.3659	.3679
2	.0045	.0164	.0333	.0536	.0758	.0988	.1217	.1438	.1647	.1839
3	.0002	.0011	.0033	.0072	.0126	.0198	.0284	.0383	.0494	.0613
4	.0000	.0001	.0003	.0007	.0016	.0030	.0050	.0077	.0111	.0153
5	.0000	.0000	.0000	.0001	.0002	.0004	.0007	.0012	.0020	.0031
6	.0000	.0000	.0000	.0000	.0000	.0000	.0001	.0002	.0003	.0005
7	.0000	.0000	.0000	.0000	.0000	.0000	.0000	.0000	.0000	.0001

y	μ									
	1.1	1.2	1.3	1.4	1.5	1.6	1.7	1.8	1.9	2.0
0	.3329	.3012	.2725	.2466	.2231	.2019	.1827	.1653	.1496	.1353
1	.3662	.3614	.3543	.3452	.3347	.3230	.3106	.2975	.2842	.2707
2	.2014	.2169	.2303	.2417	.2510	.2584	.2640	.2678	.2700	.2707
3	.0738	.0867	.0998	.1128	.1255	.1378	.1496	.1607	.1710	.1804
4	.0203	.0260	.0324	.0395	.0471	.0551	.0636	.0723	.0812	.0902
5	.0045	.0062	.0084	.0111	.0141	.0176	.0216	.0260	.0309	.0361
6	.0008	.0012	.0018	.0026	.0035	.0047	.0061	.0078	.0098	.0120
7	.0001	.0002	.0003	.0005	.0008	.0011	.0015	.0020	.0027	.0034
8	.0000	.0000	.0001	.0001	.0001	.0002	.0003	.0005	.0006	.0009
9	.0000	.0000	.0000	.0000	.0000	.0000	.0001	.0001	.0001	.0002

y	μ									
	2.1	2.2	2.3	2.4	2.5	2.6	2.7	2.8	2.9	3.0
0	.1225	.1108	.1003	.0907	.0821	.0743	.0672	.0608	.0550	.0498
1	.2572	.2438	.2306	.2177	.2052	.1931	.1815	.1703	.1596	.1494
2	.2700	.2681	.2652	.2613	.2565	.2510	.2450	.2384	.2314	.2240
3	.1890	.1966	.2033	.2090	.2138	.2176	.2205	.2225	.2237	.2240
4	.0992	.1082	.1169	.1254	.1336	.1414	.1488	.1557	.1622	.1680
5	.0417	.0476	.0538	.0602	.0668	.0735	.0804	.0872	.0940	.1008
6	.0146	.0174	.0206	.0241	.0278	.0319	.0362	.0407	.0455	.0504
7	.0044	.0055	.0068	.0083	.0099	.0118	.0139	.0163	.0188	.0216
8	.0011	.0015	.0019	.0025	.0031	.0038	.0047	.0057	.0068	.0081
9	.0003	.0004	.0005	.0007	.0009	.0011	.0014	.0018	.0022	.0027
10	.0001	.0001	.0001	.0002	.0002	.0003	.0004	.0005	.0006	.0008
11	.0000	.0000	.0000	.0000	.0000	.0001	.0001	.0001	.0002	.0002
12	.0000	.0000	.0000	.0000	.0000	.0000	.0000	.0000	.0000	.0001

y	μ									
	3.1	3.2	3.3	3.4	3.5	3.6	3.7	3.8	3.9	4.0
0	.0450	.0408	.0369	.0334	.0302	.0273	.0247	.0224	.0202	.0183
1	.1397	.1304	.1217	.1135	.1057	.0984	.0915	.0850	.0789	.0733
2	.2165	.2087	.2008	.1929	.1850	.1771	.1692	.1615	.1539	.1465
3	.2237	.2226	.2209	.2186	.2158	.2125	.2087	.2046	.2001	.1954
4	.1733	.1781	.1823	.1858	.1888	.1912	.1931	.1944	.1951	.1954
5	.1075	.1140	.1203	.1264	.1322	.1377	.1429	.1477	.1522	.1563

Table B.5 – continued from previous page

y	μ									
	3.1	3.2	3.3	3.4	3.5	3.6	3.7	3.8	3.9	4.0
6	.0555	.0608	.0662	.0716	.0771	.0826	.0881	.0936	.0989	.1042
7	.0246	.0278	.0312	.0348	.0385	.0425	.0466	.0508	.0551	.0595
8	.0095	.0111	.0129	.0148	.0169	.0191	.0215	.0241	.0269	.0298
9	.0033	.0040	.0047	.0056	.0066	.0076	.0089	.0102	.0116	.0132
10	.0010	.0013	.0016	.0019	.0023	.0028	.0033	.0039	.0045	.0053
11	.0003	.0004	.0005	.0006	.0007	.0009	.0011	.0013	.0016	.0019
12	.0001	.0001	.0001	.0002	.0002	.0003	.0003	.0004	.0005	.0006
13	.0000	.0000	.0000	.0000	.0001	.0001	.0001	.0001	.0002	.0002
14	.0000	.0000	.0000	.0000	.0000	.0000	.0000	.0000	.0000	.0001

y	μ									
	4.2	4.4	4.6	4.8	5.0	5.2	5.4	5.6	5.8	6.0
0	.0150	.0123	.0101	.0082	.0067	.0055	.0045	.0037	.0030	.0025
1	.0630	.0540	.0462	.0395	.0337	.0287	.0244	.0207	.0176	.0149
2	.1323	.1188	.1063	.0948	.0842	.0746	.0659	.0580	.0509	.0446
3	.1852	.1743	.1631	.1517	.1404	.1293	.1185	.1082	.0985	.0892
4	.1944	.1917	.1875	.1820	.1755	.1681	.1600	.1515	.1428	.1339
5	.1633	.1687	.1725	.1747	.1755	.1748	.1728	.1697	.1656	.1606
6	.1143	.1237	.1323	.1398	.1462	.1515	.1555	.1584	.1601	.1606
7	.0686	.0778	.0869	.0959	.1044	.1125	.1200	.1267	.1326	.1377
8	.0360	.0428	.0500	.0575	.0653	.0731	.0810	.0887	.0962	.1033
9	.0168	.0209	.0255	.0307	.0363	.0423	.0486	.0552	.0620	.0688
10	.0071	.0092	.0118	.0147	.0181	.0220	.0262	.0309	.0359	.0413
11	.0027	.0037	.0049	.0064	.0082	.0104	.0129	.0157	.0190	.0225
12	.0009	.0013	.0019	.0026	.0034	.0045	.0058	.0073	.0092	.0113
13	.0003	.0005	.0007	.0009	.0013	.0018	.0024	.0032	.0041	.0052
14	.0001	.0001	.0002	.0003	.0005	.0007	.0009	.0013	.0017	.0022
15	.0000	.0000	.0001	.0001	.0002	.0002	.0003	.0005	.0007	.0009
16	.0000	.0000	.0000	.0000	.0000	.0001	.0001	.0002	.0002	.0003
17	.0000	.0000	.0000	.0000	.0000	.0000	.0000	.0001	.0001	.0001

y	μ									
	6.2	6.4	6.6	6.8	7.0	7.2	7.4	7.6	7.8	8.0
0	.0020	.0017	.0014	.0011	.0009	.0007	.0006	.0005	.0004	.0003
1	.0126	.0106	.0090	.0076	.0064	.0054	.0045	.0038	.0032	.0027
2	.0390	.0340	.0296	.0258	.0223	.0194	.0167	.0145	.0125	.0107
3	.0806	.0726	.0652	.0584	.0521	.0464	.0413	.0366	.0324	.0286
4	.1249	.1162	.1076	.0992	.0912	.0836	.0764	.0696	.0632	.0573
5	.1549	.1487	.1420	.1349	.1277	.1204	.1130	.1057	.0986	.0916
6	.1601	.1586	.1562	.1529	.1490	.1445	.1394	.1339	.1282	.1221
7	.1418	.1450	.1472	.1486	.1490	.1486	.1474	.1454	.1428	.1396
8	.1099	.1160	.1215	.1263	.1304	.1337	.1363	.1381	.1392	.1396
9	.0757	.0825	.0891	.0954	.1014	.1070	.1121	.1167	.1207	.1241
10	.0469	.0528	.0588	.0649	.0710	.0770	.0829	.0887	.0941	.0993

Table B.5 – continued from previous page

y	μ									
	6.2	6.4	6.6	6.8	7.0	7.2	7.4	7.6	7.8	8.0
11	.0265	.0307	.0353	.0401	.0452	.0504	.0558	.0613	.0667	.0722
12	.0137	.0164	.0194	.0227	.0263	.0303	.0344	.0388	.0434	.0481
13	.0065	.0081	.0099	.0119	.0142	.0168	.0196	.0227	.0260	.0296
14	.0029	.0037	.0046	.0058	.0071	.0086	.0104	.0123	.0145	.0169
15	.0012	.0016	.0020	.0026	.0033	.0041	.0051	.0062	.0075	.0090
16	.0005	.0006	.0008	.0011	.0014	.0019	.0024	.0030	.0037	.0045
17	.0002	.0002	.0003	.0004	.0006	.0008	.0010	.0013	.0017	.0021
18	.0001	.0001	.0001	.0002	.0002	.0003	.0004	.0006	.0007	.0009
19	.0000	.0000	.0000	.0001	.0001	.0001	.0002	.0002	.0003	.0004
20	.0000	.0000	.0000	.0000	.0000	.0000	.0001	.0001	.0001	.0002
21	.0000	.0000	.0000	.0000	.0000	.0000	.0000	.0000	.0000	.0001

y	μ									
	8.2	8.4	8.6	8.8	9.0	9.2	9.4	9.6	9.8	10.0
0	.0003	.0002	.0002	.0002	.0001	.0001	.0001	.0001	.0001	.0000
1	.0023	.0019	.0016	.0013	.0011	.0009	.0008	.0007	.0005	.0005
2	.0092	.0079	.0068	.0058	.0050	.0043	.0037	.0031	.0027	.0023
3	.0252	.0222	.0195	.0171	.0150	.0131	.0115	.0100	.0087	.0076
4	.0517	.0466	.0420	.0377	.0337	.0302	.0269	.0240	.0213	.0189
5	.0849	.0784	.0722	.0663	.0607	.0555	.0506	.0460	.0418	.0378
6	.1160	.1097	.1034	.0972	.0911	.0851	.0793	.0736	.0682	.0631
7	.1358	.1317	.1271	.1222	.1171	.1118	.1064	.1010	.0955	.0901
8	.1392	.1382	.1366	.1344	.1318	.1286	.1251	.1212	.1170	.1126
9	.1269	.1290	.1306	.1315	.1318	.1315	.1306	.1293	.1274	.1251
10	.1040	.1084	.1123	.1157	.1186	.1210	.1228	.1241	.1249	.1251
11	.0776	.0828	.0878	.0925	.0970	.1012	.1049	.1083	.1112	.1137
12	.0530	.0579	.0629	.0679	.0728	.0776	.0822	.0866	.0908	.0948
13	.0334	.0374	.0416	.0459	.0504	.0549	.0594	.0640	.0685	.0729
14	.0196	.0225	.0256	.0289	.0324	.0361	.0399	.0439	.0479	.0521
15	.0107	.0126	.0147	.0169	.0194	.0221	.0250	.0281	.0313	.0347
16	.0055	.0066	.0079	.0093	.0109	.0127	.0147	.0168	.0192	.0217
17	.0026	.0033	.0040	.0048	.0058	.0069	.0081	.0095	.0111	.0128
18	.0012	.0015	.0019	.0024	.0029	.0035	.0042	.0051	.0060	.0071
19	.0005	.0007	.0009	.0011	.0014	.0017	.0021	.0026	.0031	.0037
20	.0002	.0003	.0004	.0005	.0006	.0008	.0010	.0012	.0015	.0019
21	.0001	.0001	.0002	.0002	.0003	.0003	.0004	.0006	.0007	.0009
22	.0000	.0000	.0001	.0001	.0001	.0001	.0002	.0002	.0003	.0004
23	.0000	.0000	.0000	.0000	.0000	.0001	.0001	.0001	.0001	.0002
24	.0000	.0000	.0000	.0000	.0000	.0000	.0000	.0000	.0001	.0001

Table B.5 – continued from previous page

y	μ									
	10.5	11.0	11.5	12	12.5	13.0	13.5	14.0	14.5	15.0
0	.0000	.0000	.0000	.0000	.0000	.0000	.0000	.0000	.0000	.0000
1	.0003	.0002	.0001	.0001	.0000	.0000	.0000	.0000	.0000	.0000
2	.0015	.0010	.0007	.0004	.0003	.0002	.0001	.0001	.0001	.0000
3	.0053	.0037	.0026	.0018	.0012	.0008	.0006	.0004	.0003	.0002
4	.0139	.0102	.0074	.0053	.0038	.0027	.0019	.0013	.0009	.0006
5	.0293	.0224	.0170	.0127	.0095	.0070	.0051	.0037	.0027	.0019
6	.0513	.0411	.0325	.0255	.0197	.0152	.0115	.0087	.0065	.0048
7	.0769	.0646	.0535	.0437	.0353	.0281	.0222	.0174	.0135	.0104
8	.1009	.0888	.0769	.0655	.0551	.0457	.0375	.0304	.0244	.0194
9	.1177	.1085	.0982	.0874	.0765	.0661	.0563	.0473	.0394	.0324
10	.1236	.1194	.1129	.1048	.0956	.0859	.0760	.0663	.0571	.0486
11	.1180	.1194	.1181	.1144	.1087	.1015	.0932	.0844	.0753	.0663
12	.1032	.1094	.1131	.1144	.1132	.1099	.1049	.0984	.0910	.0829
13	.0834	.0926	.1001	.1056	.1089	.1099	.1089	.1060	.1014	.0956
14	.0625	.0728	.0822	.0905	.0972	.1021	.1050	.1060	.1051	.1024
15	.0438	.0534	.0630	.0724	.0810	.0885	.0945	.0989	.1016	.1024
16	.0287	.0367	.0453	.0543	.0633	.0719	.0798	.0866	.0920	.0960
17	.0177	.0237	.0306	.0383	.0465	.0550	.0633	.0713	.0785	.0847
18	.0104	.0145	.0196	.0255	.0323	.0397	.0475	.0554	.0632	.0706
19	.0057	.0084	.0119	.0161	.0213	.0272	.0337	.0409	.0483	.0557
20	.0030	.0046	.0068	.0097	.0133	.0177	.0228	.0286	.0350	.0418
21	.0015	.0024	.0037	.0055	.0079	.0109	.0146	.0191	.0242	.0299
22	.0007	.0012	.0020	.0030	.0045	.0065	.0090	.0121	.0159	.0204
23	.0003	.0006	.0010	.0016	.0024	.0037	.0053	.0074	.0100	.0133
24	.0001	.0003	.0005	.0008	.0013	.0020	.0030	.0043	.0061	.0083
25	.0001	.0001	.0002	.0004	.0006	.0010	.0016	.0024	.0035	.0050
26	.0000	.0000	.0001	.0002	.0003	.0005	.0008	.0013	.0020	.0029
27	.0000	.0000	.0000	.0001	.0001	.0002	.0004	.0007	.0011	.0016
28	.0000	.0000	.0000	.0000	.0001	.0001	.0002	.0003	.0005	.0009
29	.0000	.0000	.0000	.0000	.0000	.0001	.0001	.0002	.0003	.0004
30	.0000	.0000	.0000	.0000	.0000	.0000	.0000	.0001	.0001	.0002
31	.0000	.0000	.0000	.0000	.0000	.0000	.0000	.0000	.0001	.0001
32	.0000	.0000	.0000	.0000	.0000	.0000	.0000	.0000	.0000	.0001

Table B.6: Chi-squared distribution

df	Upper Tail Area										
	.995	.99	.975	.95	.90	.50	.10	.05	.025	.01	.005
1	.0000	.0002	.0010	.0039	.0158	.4549	2.706	3.842	5.024	6.635	7.879
2	.0100	.0201	.0506	.1026	.2107	1.386	4.605	5.992	7.378	9.210	10.597
3	.0717	.1148	.2158	.3518	.5844	2.366	6.251	7.815	9.349	11.345	12.838
4	.2070	.2971	.4844	.7107	1.064	3.357	7.779	9.488	11.143	13.277	14.860
5	.4117	.5543	.8312	1.146	1.610	4.352	9.236	11.071	12.833	15.086	16.750
6	.6757	.8721	1.237	1.635	2.204	5.348	10.645	12.592	14.449	16.812	18.548
7	.9893	1.239	1.690	2.167	2.833	6.346	12.017	14.067	16.013	18.475	20.2
8	1.344	1.647	2.180	2.733	3.490	7.344	13.362	15.507	17.535	20.090	21.955
9	1.735	2.088	2.700	3.325	4.168	8.343	14.684	16.919	19.023	21.666	23.589
10	2.156	2.558	3.247	3.940	4.865	9.342	15.987	18.307	20.483	23.209	25.188
11	2.603	3.054	3.816	4.575	5.578	10.341	17.275	19.675	21.920	24.725	26.757
12	3.074	3.571	4.404	5.226	6.304	11.340	18.549	21.026	23.337	26.217	28.300
13	3.565	4.107	5.009	5.892	7.042	12.340	19.812	22.362	24.736	27.688	29.820
14	4.075	4.660	5.629	6.571	7.790	13.339	21.064	23.685	26.119	29.141	31.319
15	4.601	5.229	6.262	7.261	8.547	14.339	22.307	24.996	27.488	30.578	32.801
16	5.142	5.812	6.908	7.962	9.312	15.339	23.542	26.296	28.845	32.000	34.26
17	5.697	6.408	7.564	8.672	10.085	16.338	24.769	27.587	30.191	33.409	35.719
18	6.265	7.015	8.231	9.391	10.865	17.338	25.989	28.869	31.526	34.805	37.15
19	6.844	7.633	8.907	10.117	11.651	18.338	27.204	30.144	32.852	36.191	38.582
20	7.434	8.260	9.591	10.851	12.443	19.337	28.412	31.410	34.170	37.566	39.997
21	8.034	8.897	10.283	11.591	13.240	20.337	29.615	32.671	35.479	38.932	41.401
22	8.643	9.543	10.982	12.338	14.042	21.337	30.813	33.924	36.781	40.289	42.79
23	9.260	10.196	11.689	13.091	14.848	22.337	32.007	35.173	38.076	41.638	44.181
24	9.886	10.856	12.401	13.848	15.659	23.337	33.196	36.415	39.364	42.980	45.559
25	10.520	11.524	13.120	14.611	16.473	24.337	34.382	37.652	40.647	44.314	46.928
26	11.160	12.198	13.844	15.379	17.292	25.337	35.563	38.885	41.923	45.642	48.290
27	11.808	12.879	14.573	16.151	18.114	26.336	36.741	40.113	43.195	46.963	49.64
28	12.461	13.565	15.308	16.928	18.939	27.336	37.916	41.337	44.461	48.278	50.993
29	13.121	14.257	16.047	17.708	19.768	28.336	39.088	42.557	45.722	49.588	52.33
30	13.787	14.954	16.791	18.493	20.599	29.336	40.256	43.773	46.979	50.892	53.672
31	14.458	15.656	17.539	19.281	21.434	30.336	41.422	44.985	48.232	52.191	55.003
32	15.134	16.362	18.291	20.072	22.271	31.336	42.585	46.194	49.480	53.486	56.328
33	15.815	17.074	19.047	20.867	23.110	32.336	43.745	47.400	50.725	54.776	57.648
34	16.501	17.789	19.806	21.664	23.952	33.336	44.903	48.602	51.966	56.061	58.964
35	17.192	18.509	20.569	22.465	24.797	34.336	46.059	49.802	53.203	57.342	60.275
36	17.887	19.233	21.336	23.269	25.643	35.336	47.212	50.999	54.437	58.619	61.581
37	18.586	19.960	22.106	24.075	26.492	36.336	48.363	52.192	55.668	59.893	62.883
38	19.289	20.691	22.879	24.884	27.343	37.336	49.513	53.384	56.896	61.162	64.181
39	19.996	21.426	23.654	25.695	28.196	38.335	50.660	54.572	58.120	62.428	65.476
40	20.707	22.164	24.433	26.509	29.051	39.335	51.805	55.759	59.342	63.691	66.766

C

USING THE INCLUDED MINITAB MACROS

Minitab macros for performing Bayesian analysis and for doing Monte Carlo simulations are included. The macros may be downloaded from the web page for this text on the site

http://www.introbayes.ac.nz.

The macros are in a compressed ZIP file called *BayesMacros_YYYYMMDD.zip*, where *YYYYMMDD* indicates the year, month and day the macros were uploaded. You should make sure you get the latest version. Some Minitab worksheets are also included at that site.

In order to run the macros it is necessary to know the fully qualified file name. That means we need to know the drive and the directory name as the macro name. The simplest way to find the full directory name is to unzip the files to a commonly used location and then single click to select on one of the macros. Once the macro file is highlighted, right click on it to bring up the context menu and select *Properties*. This will bring up a dialog box with a whole lot of information about the file. The *Location* file

Introduction to Bayesian Statistics, 3rd ed.
By Bolstad, W. M. and Curran, J. M. Copyright © 2016 John Wiley & Sons, Inc.

directory name. For example, on my computer I unzipped the macros in the the *My Documents* folder. The location of the macros is

```
C:\Users\jcur002\Documents\BayesMacros
```

That is, every macro is stored on the C: drive in a directory called Users\jcur002\Documents\BayesMacros. This means that every time I see *<insert path>* in the set of Minitab commands below, I will type C:\Users\jcur002\Documents\BayesMacros. For example, if I was using the set of commands in Table C.1, I would type

```
%C:\Users\jcur002\Documents\BayesMacros\sscsample c1 100;
```

for the first line of commands.

Note: This chapter has been updated to work with Minitab 17 (version 17.3.10). Readers with earlier versions of Minitab should still be able to use the macros; however, some of the menus or menu commands may be in different locations. You may also find that you have to put the fully qualified file names (including the .mac file extension) inside a set of single quotes, e.g.,

```
%'C:\Users\jcur002\Documents\BayesMacros\sscsample.mac' c1 100;
```

in order to make them work.

Chapter 2: Scientific Data Gathering

Sampling Methods

We use the Minitab macro *sscsample* to perform a small-scale Monte Carlo study on the efficiency of simple, stratified, and cluster random sampling on the population data contained in *sscsample.mtw*. In the *File* menu select *Open Worksheet...* command. When the dialog box opens, find the directory BAYESMTW and type in *sscsample.mtw* in the filename box and click on "open". In the *Edit* menu select *Command Line Editor* and type in the commands from Table C.1 into the command line editor:

Experimental Design

We use the Minitab macro *Xdesign* to perform a small-scale Monte Carlo study, comparing *completely randomized design* and *randomized block design* in their effectiveness for assigning experimental units into treatment groups. In the *Edit* menu select *Command Line Editor* and type in the commands from Table C.2.

Table C.1   Sampling Monte Carlo study

Minitab Commands	Meaning
%<*insert path*>sscsample c1 100;	data are in c1, $N = 100$
strata c2 3;	there are 3 strata stored in c2
cluster c3 20;	there are 20 clusters stored in c3
type 1;	1 = simple, 2 = stratified, 3 = cluster
size 20;	sample size $n = 20$
mcarlo 200;	Monte Carlo sample size 200
output c6 c7 c8 c9.	c6 contains sample means, c7–c9 contain numbers in each strata

Table C.2   Experimental design Monte Carlo study

Minitab Commands	Meaning
let k1=.8	correlation between other and response variables
random 80 c1 c2;	generate 80 other and response variables
normal 0 1.	in c1 and c2, respectively
let c2=sqrt(1-k1**2)*c2+k1*c11	give them correlation k1
desc c1 c2	summary statistics
corr c1 c2	
plot c2*c1	shows relationship
%<*insert path*>Xdesign c1 c2;	other variable in c1, response in c2
size 20;	treatment groups of 20 units
treatments 4;	4 treatment groups
mcarlo 500;;	Monte Carlo sample size 500
output c3 c4 c5.	c3 contains other means,
	c4 contains response means,
—	c5 contains treatment groups
	1–4 from completely randomized design
	5–8 from randomized block design
code (1:4) 1 (5:8) 2 c5 c6	
desc c4;	summary statistics
by c6.	

Table C.3   Discrete prior distribution for binomial proportion π

π	$g(\pi)$
.3	.2
.4	.3
.5	.5

Table C.4   Finding the posterior distribution of binomial proportion with a discrete prior for π

Minitab Commands	Meaning
set c1	puts π in c1
.3 .4 .5	
end	
set c2	puts $g(\pi)$ in c2
.2 .3 .5	
end	
%<*insert path*>BinoDP 6 5;	$n = 6$ trials, $y = 5$ successes observed
prior c1 c2;	π in c1, prior $g(\pi)$ in c2
likelihood c3;	store likelihood in c3
posterior c4.	store posterior $g(\pi|y=5)$ in c4

Chapter 6: Bayesian Inference for Discrete Random Variables

Binomial Proportion with Discrete Prior

BinoDP is used to find the posterior when we have *binomial* (n, π) observation, and we have a discrete prior for π. For example, suppose π has the discrete distribution with three possible values, .3, .4, and .5. Suppose the prior distribution is given in Table C.3, and we want to find the posterior distribution after $n = 6$ trials and observing $y = 5$ successes. In the *Edit* menu pull down *Command Line Editor* and type in the commands from Table C.4.

Poisson Parameter with Discrete Prior

PoisDP is used to find the posterior when we have a *Poisson*(μ) observation, and a discrete prior for μ. For example, suppose μ has three possible values $\mu = 1, 2$, or 3 where the prior probabilities are given in Table C.5, and we want to find the posterior distribution after observing $y = 4$. In the *Edit* menu go down to *Command Line Editor* and type in the commands from Table C.6.

Table C.5   Discrete prior distribution for Poisson parameter μ

μ	$g(\mu)$
1	.3
2	.4
3	.3

Table C.6   Finding the posterior distribution of Poisson parameter with a discrete prior for μ

Minitab Commands	Meaning
`set c5`	puts observation(s) y in c5
`44`	
`end`	
`set c1`	puts μ in c1
`1 2 33`	
`end`	
`set c2`	puts $g(\mu)$ in c2
`.3 .4 .3`	
`end`	
`%<insert path>PoisDP c5;`	observations in c5
`prior c1 c2;`	μ in c1, prior $g(\mu)$ in c2
`likelihood c3;`	store likelihood in c3
`posterior c4.`	store posterior $g(\pi \mid y = 5)$ in c4

Chapter 8: Bayesian Inference for Binomial Proportion

$Beta(a, b)$ Prior for π

BinoBP is used to find the posterior when we have *binomial* (n, π) observation, and we have a *beta(a, b)* prior for π. The *beta* family of priors is conjugate for *binomial* (n, π) observations, so the posterior will be another member of the family, *beta(a', b')* where $a' = a + y$ and $b' = b + n - y$. For example, suppose we have $n = 12$ trials, and observe $y = 4$ successes, and we use a *beta(3, 3)* prior for π. In the *Edit* menu select *Command Line Editor* and type in the commands from Table C.7. We can find the posterior mean and standard deviation from the output. We can determine a Bayesian credible interval for π by looking at the values of π by pulling down the *Calc* menu to *Probability Distributions* and over to "beta" and selecting "inverse cumulative probability". We can test $H_0 : \pi \leq \pi_0$ vs. $H_1 : \pi > \pi_0$ by pulling down

Table C.7   Finding the posterior distribution of binomial proportion with a *beta* prior for π

Minitab Commands	Meaning
%<*insert path*>BinoBP 12 4;	$n = 12$ trials, $y = 4$ was observed
beta 3 3;	the beta prior
prior c1 c2;	stores π and the prior $g(\pi)$
likelihood c3;	store likelihood in c3
posterior c4.	store posterior $g(\pi|y = 4)$ in c4

Table C.8   Finding the posterior distribution of binomial proportion with a continuous prior for π

Minitab Commands	Meaning
%<*insert path*>BinoGCP 12 4;	$n = 12$ trials, $y = 4$ successes observed
prior c1 c2;	inputs π in c1, prior $g(\pi)$ in c2
likelihood c3;	store likelihood in c3
posterior c4.	store posterior $g(\pi|y = 4)$ in c4

the *Calc* menu to *Probability Distributions* and over to "beta" and selecting cumulative probability" and inputting the value of π_0.

General Continuous Prior for π

BinoGCP is used to find the posterior when we have *binomial* (n, π) observation, and we have a general continuous prior for π. Note that π must go from 0 to 1 in equal steps, and $g(\pi)$ must be defined at each of the π values. For example, suppose we have $n = 12$ trials, and observe $y = 4$ successes, where π is stored in c1 and a general continuous prior $g(\pi)$ is stored in c2. In the *Edit* menu select *Command Line Editor* and type in the commands from Table C.8. The output of *BinoGCP* does not print out the posterior mean and standard deviation. Nor does it print out the values that give the tail areas of the integrated density function that we need to determine credible interval for π. Instead we use the macro *tintegral* which numerically integrates a function over its range to determine these things. We can find the integral of the posterior density $g(\pi|y)$ using this macro. We can also use *tintegral* to find the posterior mean and variance by numerically evaluating

$$m' = \int_0^1 \pi g(\pi|y) \, d\pi$$

Table C.9 Bayesian inference using posterior density of binomial proportion π

Minitab Commands	Meaning
%<*insert path*>tintegral c1 c4;	integrates posterior density
output k1 c6.	stores definite integral over range in k1
	stores definite integral function in c6
let c7=c1*c4	$\pi \times g(\pi\|y)$
%<*insert path*>tintegral c1 c7;	finds posterior mean
output k1 c8.	
let c9=(c1-k1)**2 * c4	
%<*insert path*>tintegral c1 c9;	finds posterior variance
output k2 c10.	
let k3=sqrt(k2)	finds posterior st. deviation
print k1-k3	

and

$$(s')^2 = \int_0^1 (\pi - m')^2 g(\pi|y)\, d\pi \,.$$

In the *Edit* menu select *Command Line Editor* and type in the commands from Table C.9 . A 95% Bayesian credible interval for π is found by taking the values in c1 that correspond to .025 and .975 in c6. To test the hypothesis $H_0 : \pi \le \pi_0$ vs. $H_1 : \pi > \pi_0$, we find the value in c6 that corresponds to the value π_0 in c1. If it is less than the desired level of significance α, then we can reject the null hypothesis.

Chapter 10: Bayesian Inference for Poisson

$Gamma(r, v)$ Prior for μ

PoisGamP is used to find the posterior when we have a random sample from a *Poisson*(μ) distribution, and we have a *gamma*(r, v) prior for μ. The *gamma* family of priors is the conjugate family for *Poisson* observations, so the posterior will be another member of the family, *gamma*(r', v') where $r' = r + \sum y$ and $v' = v + n$. The simple rules are "add sum of observations to r" and "add number of observations to v". For example, suppose in column 5 there is a sample five observations from a *Poisson*(μ) distribution. Suppose we want to use a *gamma*(6, 3) prior for μ. select the *Edit* menu to the *Command Line Editor* command and type in the commands from Table C.10. We can determine a Bayesian credible interval for μ by looking at the values of μ by pulling down the *Calc* menu to *Probability Distributions* and over to *Gamma...* and

Table C.10   Finding the posterior distribution of a *Poisson* parameter with a *gamma* prior for μ

Minitab Commands	Meaning
`set c5`	Put observations in c5
`3 4 3 0 1`	
`end`	
`let k1=6`	r
`let k2=3`	v
`%<insert path>PoisGamP c5 ;`	observations in c5
`gamma k1 k2;`	the gamma prior
`prior c1 c2;`	stores μ and the prior $g(\mu)$
`likelihood c3;`	store likelihood in c3
`posterior c4.`	store posterior $g(\mu|y)$ in c4

selecting *Inverse cumulative probability*. Note: Minitab uses the parameter $1/v$ instead of v. We can test $H_0 : \mu \leq \mu_0$ vs. $H_1 : \mu > \mu_0$ by pulling down the *Calc* menu to *Probability Distributions* and over to "gamma" and selecting cumulative probability" and inputting the value of μ_0.

General Continuous Prior for Poisson Parameter μ

PoisGCP is used to find the posterior when we have a random sample from a *Poisson*(μ) distribution and we have a continuous prior for μ. Suppose we have a random sample of five observations in column c5. The prior density of μ is found by linearly interpolating the values in Table C.11. In the *Edit*

Table C.11   Continuous prior distribution for Poisson parameter μ has shape given by interpolating between these values

μ	$g(\mu)$
0	0
2	2
4	2
8	0

menu select *Command Line Editor* and type in the commands in Table C.12. The output of *PoisGCP* does not include the posterior mean and standard deviation. Nor does it print out the cumulative distribution function that allows us to find credible intervals. Instead we use the macro *tintegral* which numerically integrates the posterior to do these things. In the *Edit* menu select *Command Line Editor* and type in the commands from Table C.13. A

Table C.12   Finding the posterior distribution of a *Poisson* parameter with a continuous parameter for μ

Minitab Commands	Meaning
set c5	Put observations in c5
3 4 3 0 1	
end	
set c1	set μ
0:8/ .001	
end	
set c2	set $g(\mu)$
0:2 / .001 1999(2) 2:0 /-.0005	
end	
%<*insert path*>PoisGCP c5 ;	observations in c5
prior c1 c2;	μ and the prior $g(\mu)$ in c1 and c2
likelihood c3;	store likelihood in c3
posterior c4.	store posterior $g(\mu|y)$ in c4

Table C.13   Bayesian inference using posterior distribution of Poisson parameter μ

Minitab Commands	Meaning
%<*insert path*>tintegral c1 c4;	integrates posterior density
output k1 c6.	stores definite integral over range in k1
	stores definite integral function in c6
let c7=c1*c4	$\mu \times g(\mu|y_1, \ldots, y_n)$
%<*insert path*>tintegral c1 c7;	finds posterior mean
output k1 c8.	
let c9=(c1-k1)**2 * c4	
%<*insert path*>tintegral c1 c9;	finds posterior variance
output k2 c10.	
let k3=sqrt(k2)	finds posterior st. deviation
print k1-k3	

95% Bayesian credible interval for μ is found by taking the values in c1 that correspond to .025 and .975 in c6. To test the null hypothesis $H_0 : \mu \leq \mu_0$ vs. $H_1 : \mu > \mu_0$, find the value in c6 that corresponds to μ_0 in c1. If it is less than the desired level of significance, then we can reject the null hypothesis at that level.

Table C.14   Discrete prior distribution for normal mean μ

μ	$f(\mu)$
2	.1
2.5	.2
3	.4
3.5	.2
4	.1

Chapter 11: Bayesian Inference for Normal Mean

Discrete Prior for μ

NormDP is used to find the posterior when we have a column of $normal(\mu, \sigma^2)$ observations and σ^2 is known, and we have a discrete prior for μ. If the standard deviation σ is not entered, then the estimate from the observations is used, and the approximation to the posterior is found. For example, suppose μ has the discrete distribution with 5 possible values, 2, 2.5, 3, 3.5, and ,4. Suppose the prior distribution is given in Table C.14. and we want to find the posterior distribution after a random sample of $n = 5$ observations from a $normal(\mu, 1^2)$ that are 1.52, 0.02, 3.35, 3.49, 1.82. In the *Edit* menu select *Command Line Editor* and type in the commands from Table C.15.

Normal(m, s^2) Prior for μ

NormNP is used when we have a column c5 containing a random sample of n observations from a $normal(\mu, \sigma^2)$ distribution (with σ^2 known) and we use a $normal(m, s^2)$ prior distribution. If the observation standard deviation σ is not input, the estimate calculated from the observations is used, and the approximation to the posterior is found. If the *normal* prior is not input, a flat prior is used. The normal family of priors is conjugate for *normal* (μ, σ^2) observations, so the posterior will be another member of the family, $normal[m', (s')^2]$ where the new constants are given by

$$\frac{1}{(s')^2} = \frac{1}{s^2} + \frac{n}{\sigma^2}$$

and

$$m' = \frac{\frac{1}{b^2}}{\frac{1}{(b')^2}} \times m + \frac{\frac{n}{\sigma^2}}{\frac{1}{(b')^2}} \times \bar{y}.$$

For example, suppose we have a normal random sample of 4 observations from $normal(\mu, 1^2)$ which are 2.99, 5.56, 2.83, and 3.47. Suppose we use a $normal(3, 2^2)$ prior for μ. In the *Edit* menu select *Command Line Editor* and type in the commands from Table C.16 . We can determine a Bayesian credible

Table C.15   Finding the posterior distribution of a normal mean with discrete prior for μ

Minitab Commands	Meaning
set c1	puts μ in c1
2:4/.5	
end	
set c2	puts $g(\mu)$ in c2
.1 .2 .4 .2 .11	
end	
set c5	puts *data* in c5
1.52, 0.02, 3.35, 3.49 1.82	
end	
%<*insert path*>NormDP c5 ;	observed data in c5
sigma 1;	known $\sigma = 1$ is used
prior c1 c2;	μ in c1, prior $g(\mu)$ in c2
likelihood c3;	store likelihood in c3
posterior c4.	store posterior $g(\mu|data)$ in c4

Table C.16   Finding the posterior distribution of a normal mean with a normal prior for μ

Minitab Commands	Meaning
set c5	puts *data* in c5
2.99, 5.56, 2.83, 3.47	
end	
%<*insert path*>NormNP c5 ;	observed data in c5
sigma 1;	known $\sigma = 1$ is used
norm 3 2;	prior mean 3, prior std 2
prior c1 c2;	store μ in c1, prior $g(\mu)$ in c2
likelihood c3;	store likelihood in c3
posterior c4.	store posterior $g(\mu|data)$ in c4

Table C.17   Finding the posterior distribution of a normal mean with a continuous prior for μ

Minitab Commands	Meaning
set c5	puts *data* in c5
2.99, 5.56, 2.83, 3.47	
end	
%<*insert path*>NormGCP c5 ;	observed data in c5
sigma 1;;	known $\sigma = 1$ is used
prior c1 c2;	μ in c1, prior $g(\mu)$ in c2
likelihood c3;	store likelihood in c3
posterior c4.	store posterior $g(\mu|data)$ in c4

interval for μ by looking at the values of μ by pulling down the *Calc* menu to *Probability Distributions* and over to *Normal...* and selecting "inverse cumulative probability." We can test $H_0 : \mu \leq \mu_0$ vs. $H_1 : \mu > \mu_0$ by pulling down the *Calc* menu to *Probability Distributions* and over to *Normal...* and selecting cumulative probability" and inputting the value of μ_0.

General Continuous Prior for μ

NormGCP is used when we have (a) a column c5 containing a random sample of n observations from a *normal* (μ, σ^2) distribution (with σ^2 known), (b) a column c1 containing values of μ, and (c) a column c2 containing values from a continuous prior $g(\mu)$. If the standard deviation σ is not input, the estimate calculated from the data is used, and the approximation to the posterior is found.

   For example, suppose we have a normal random sample of 4 observations from *normal* $(\mu, \sigma^2 = 1)$ which are 2.99, 5.56, 2.83, and 3.47. In the *Edit* menu select *Command Line Editor* and type the following commands from Table C.17. The output of *NormGCP* does not print out the posterior mean and standard deviation. Nor does it print out the values that give the tail areas of the integrated density function that we need to determine credible interval for μ. Instead we use the macro *tintegral* which numerically integrates a function over its range to determine these things. In the *Edit* menu select *Command Line Editor* and type in the commands from Table C.18. To find a 95% Bayesian credible interval we find the values in c1 that correspond to .025 and .975 in c6. To test a hypothesis $H_0 : \mu \leq \mu_0$ versus $H_1 : \mu > \mu_0$, we find the value in c6 that corresponds to μ_0 in c1. If this is less than the chosen level of significance, then we can reject the null hypothesis at that level.

Table C.18   Bayesian inference using posterior distribution of normal mean μ

Minitab Commands	Meaning
%<*insert path*>tintegral c1 c4;	integrates posterior density
output k1 c6.	stores definite integral over range in k1
	stores definite integral function in c6
print c1 c6	
let c7=c1*c4	$\mu \times g(\mu \vert data)$
%<*insert path*>tintegral c1 c7;	finds posterior mean
output k1 c8.	
let c8=(c1-k1)**2 * c4	
%<*insert path*>tintegral c1 c8;	finds posterior variance
output k2 c9.	
let k3=sqrt(k2)	finds posterior std. deviation
print k1-k3	

Chapter 14: Bayesian Inference for Simple Linear Regression

BayesLinReg is used to find the posterior distribution of the simple linear regression slope β when we have a random sample of ordered pairs (x_i, y_i) from the simple linear regression model

$$y_i = \alpha_0 + \beta \times x_i + e_i,$$

where the observation errors e_i are independent $normal(0, \sigma^2)$ with known variance. If the variance is not known, then the posterior is found using the variance estimate calculated from the least squares residuals. We use independent priors for the slope β and the intercept $\alpha_{\bar{x}}$. These can be either flat priors or *normal* priors. (The default is flat priors for both slope and intercept of $x = \bar{x}$.) This parameterization yields independent posterior distribution for slope and intercept with simple updating rules "posterior precision equals prior precision plus precision of least squares estimate" and "posterior mean is weighted sum of prior mean and the least squares estimate where the weights are the proportions of the precisions to the posterior precision. Suppose we have y and x in columns c5 and c6, respectively, and we know the standard deviation $\sigma = 2$. We wish to use a $normal(0, 3^2)$ prior for β and a $normal(30, 10^2)$ prior for $\alpha_{\bar{x}}$. In the *Edit* menu select *Command Line Editor* and type in the commands from Table C.19. If we want to find a credible interval for the slope, use Equation 14.9 or Equation 14.10 depending on whether we knew the standard deviation or used the value calculated from the residuals. To find the credible interval for the predictions, use Equation 14.13 when we

Table C.19   Bayesian inference for simple linear regression model

Minitab Commands	Meaning
%<*insert path*>BayesLinReg c5 c6;	y (response) in c5, x (predictor) in c6
Sigma 2;	known standard deviation $\sigma = 2$
PriSlope 0 3;;	$normal(m_\beta = 0, s_\beta = 3)$ prior
PriIntcpt 30 10;;	$normal(m_{\alpha_{\bar{x}}} = 30, s_{\alpha_{\bar{x}}} = 10)$ prior
predict c7 c8 c9.	predict for x-values in c7, prediction in c8, standard deviations in c9
invcdf .975 k10;	Find critical value. Use *normal* when
norm 0 1.	variance is known, use *student's t* with $n - 2$ df when variance not known
let c10=c8-k10*c9	Lower credible bound for predictions
let c11=c8+k10*c9	Upper credible bound for predictions

know the variance or use Equation 14.14 when we use the estimate calculated from the residuals.

Chapter 15: Bayesian Inference for Standard Deviation

$S\times$ an *Inverse Chi-Squared*(κ) Prior for σ^2

NVarICP is used when we have a column c5 containing a random sample of n observations from a $normal(\mu, \sigma^2)$ distribution where the mean μ is known. The $S\times$ an *inverse chi-squared*(κ) family of priors is the conjugate family for normal observations with known mean. The posterior will be another member of the family where the constants are given by the simple updating rules "add the sum of squares around the mean to S" and "add the sample size to the degrees of freedom." For example, suppose we have five observations from a $normal(\mu, \sigma^2)$ where $\mu = 200$ which are 206.4, 197.4, 212.7, 208.5, and 203.4. We want to use a prior that has prior median equal to 8. In the *Edit* menu select *Command Line Editor* and type in the commands from Table C.20. Note: The graphs that are printed out are the prior distributions of the standard deviation σ even though we are doing the calculations on the variance.

If we want to make inferences on the standard deviation σ using the posterior distribution we found, pull down the *Edit* menu, select *Command Line Editor* and type in the commands given in Table C.21. To find an equal tail area 95% Bayesian credible interval for σ, we find the values in c1 that correspond to .025 and .975 in c6.

Table C.20   Finding posterior distribution of normal standard deviation σ using $S\times$ an *inverse chi-squared*(κ) prior for σ^2

Minitab Commands	Meaning
set c5	puts *data* in c5
206.4, 197.4, 212.7, 208.5, 203.4	
end	
%<*insert path*>NVarICP c5 200;	observed data in c5, known $\mu = 200$
IChiSq 29.11 1;	29.11$\times$ *inverse chi-squared*(1) has
	prior median 8
prior c1 c2;	σ in c1, prior $g(\sigma)$ in c2
likelihood c3;	store likelihood in c3
posterior c4;	store posterior $g(\sigma|data)$ in c4
constants k1 k2.	store S' in k1, κ' in k2

Table C.21   Bayesian inference using posterior distribution of normal standard deviation σ

Minitab Commands	Meaning
let k3=sqrt(k1/(k2-2))	The estimator for σ using posterior mean,
Print k3	Note: k2 must be greater than 2
InvCDF .5 k4;	store median of chi-squared (k2)
ChiSquare k2.	in k4
let k5=sqrt(k1/k4)	The estimator for σ using posterior median
Print k5	
%<*insert path*>tintegral c1 c4;	integrates posterior density
output k6 c6.	stores definite integral over range in k6,
	stores definite integral function in c6

Table C.22 Finding the posterior distribution of binomial proportion with a mixture prior for π

Minitab Commands	Meaning
%<*insert path*>BinoMixP 60 15 ;	$n = 60$ trials, $y = 15$ successes observed
bet0 10 6;	The precise beta prior
bet1 1 1;	The fall-back beta prior
prob .95;	prior probability of first component
Output c1-c4.	store π, prior, likelihood, and posterior in c1–c4

Chapter 16: Robust Bayesian Methods

BinoMixP is used to find the posterior when we have a *binomial*(n, π) observations and use a mixture of a *beta*(a_0, b_0) and a *beta*(a_1, b_1) for the prior distribution for π. Generally, the first component summarizes our prior belief, so that we give it a high prior probability. The second component has more spread to allow for our prior belief being mistaken, and we give it a low prior probability. For example, suppose our first component is *beta*$(10, 6)$, and the second component is *beta*$(1, 1)$ and we give a prior probability of .95 to the first component. We have taken 60 trials and observed $y = 15$ successes. In the *Edit* menu select *Command Line Editor* and type in the commands from Table C.22.

NormMixP is used to find the posterior when we have normal(μ, σ^2) observations with known variance σ^2 and our prior for μ is a mixture of two normal distributions, a *normal*(m_0, s_0^2) and a *normal*(m_1, s_1^2). Generally, the first component summarizes our prior belief, so we give it a high prior probability. The second component is a fall-back prior that has a much larger standard deviation to allow for our prior belief being wrong and has a much smaller prior probability. For example, suppose we have a random sample of observations from a *normal*(μ, σ^2) in column c5, where $\sigma^2 = .2^2$. Suppose we use a mixture of a *normal*$(10, .1^2)$ and a *normal*$(10, .4^2)$ prior where the prior probability of the first component is .95. In the *Edit* menu select *Command Line Editor* and type in the commands from Table C.23.

Table C.23   Finding the posterior distribution of normal mean with the mixture prior for μ

Minitab Commands	Meaning
%<*insert path*>NormMixP c5 ;	c5 contains observations of $normal(\mu, \sigma^2)$
sigma .2 ;	known value $\sigma = .2$ is used
np0 10 .1;	The precise $normal(10, .1^2)$ prior
np1 10 .4;	The fall-back $normal(10, .4^2)$ prior
prob .95;	prior probability of first component
Output c1-c4..	store μ, prior, likelihood, and posterior in c1–c4

Chapter 18: Bayesian Inference for Multivariate Normal Mean Vector

MVNorm is used to find the posterior mean and variance–covariance matrix for a set of multivariate normal data with known variance–covariance Σ assuming a multivariate normal prior. If the prior density is *MVN* with mean vector $\mathbf{m}_0$ and variance–covariance matrix $\mathbf{V}_0$, and $\mathbf{Y}$ is a sample of size n from a $MVN(\boldsymbol{\mu}, \boldsymbol{\Sigma})$ matrix (where $\boldsymbol{\mu}$ is unknown), then the posterior density of $\boldsymbol{\mu}$ is $MVN(\mathbf{m}_1, \mathbf{V}_1)$ where

$$\mathbf{V}_1 = (\mathbf{V}_0^{-1} + n\boldsymbol{\Sigma}^{-1})^{-1}$$

and

$$\mathbf{m}_1 = \mathbf{V}_1\mathbf{V}_0^{-1}\mathbf{m}_0 + n\mathbf{V}_1\boldsymbol{\Sigma}^{-1}\bar{\mathbf{y}}$$

If $\boldsymbol{\Sigma}$ is not specified, then the sample variance–covariance matrix can be used. In this case the posterior distribution is *multivariate t* so using the results with a *MVN* is only (approximately) valid for large samples.

In the *Edit* menu go down to *Command Line Editor* and type in the commands from Table C.24.

Chapter 19: Bayesian Inference for the Multiple Linear Regression Model

BayesMultReg is is used to find the posterior distribution of vector of regression coefficients $\boldsymbol{\beta}$ when we have a random sample of ordered pairs $(\mathbf{x}_i, y_i)$ from the multiple linear regression model

$$y_i = \beta_0 + \beta_1 x_{i1} + \beta_2 x_{i2} + \cdots + \beta_p x_{ip} + e_i = \mathbf{x}_i\boldsymbol{\beta} + e_i,$$

where (usually) the observation errors e_i are independent $normal(0, \sigma^2)$ with known variance. If σ^2 is not known, then we estimate it from the variance of the residuals. Our prior for β is either a flat prior or a $MVN(\mathbf{b}_0, \mathbf{V}_0)$ prior. It is not necessary to use the macro if we assume a flat prior as the posterior mean of β is equal to the least squares estimate β_{LS}, which is also the maximum likelihood estimate. This means that the Minitab regression procedure found in the *Stat* menu gives us the posterior mean and variance for a flat prior. The posterior mean of β when we use a $MVN(\mathbf{b}_0, \mathbf{V}_0)$ prior is found through the simple updating rules given Equations 19.5 and 19.4.

In the *Edit* menu go down to *Command Line Editor* and type in the commands from Table C.25.

Table C.24   Finding the posterior distribution of multivariate normal mean with the MVN prior for μ

Minitab Commands	Meaning
set c1	Set the true mean $\mu = (0,2)'$
0 2	and store it in c1
end	
set c2	Set the true variance–covariance
1 0.9	matrix to $\Sigma = \begin{pmatrix} 1 & 0.9 \\ 0.9 & 1 \end{pmatrix}$
end	
set c3	
0.9 1	
end	
copy c2-c3 m1	Store Σ in matrix m1
set c4	Set the prior mean to $\mathbf{m}_0 = (0,0)'$
2(0)	and store it in c4
end	
set c5	Set the prior variance–covariance
2(10000)	to $\mathbf{V}_0 = 10^4\mathbf{I}_2$ where $\mathbf{I}_2$ is the
end	2×2 identity matrix and store
diag c5 m2	it in matrix m2
Random 50 c6-c7;	Generate 50 observations from
Mnormal c1 m1.	$MVN(\mu, \Sigma)$ and store them
copy c6-c7 m3	in columns c6–c7. Assign the
	observations to matrix m3
%<insert path>MVNorm m3 2;	2 is the number of rows for μ
covmat m1;	Σ is stored in m1
prior c4 m2;	$\mathbf{m}_0$ and $\mathbf{V}_0$ are in c4 and m1
	respectively
posterior c9 m4.	Store the posterior mean $\mathbf{m}_1$
	in column c9 and the posterior
	variance in matrix m4

Table C.25   Finding the posterior distribution of regression coefficient vector $\boldsymbol{\beta}$

Minitab Commands	Meaning
`Random 100 c2-c5;`	Generate three random covariates and
`Normal 0.0 1.0.`	some errors and store them in c2–c5
`let c1 = 10 + 3*c2 + 1*c3 -5*c4 + c5`	Let the response be
	$y_i = 10 + 3x_{1i} + x_{2i} - 5x_{3i} + \epsilon_i$
`let c2 = c2 - mean(c2)`	Center the explanatory variables on
`let c3 = c3 - mean(c3)`	on their means
`let c4 = c4 - mean(c4)`	
`copy c2-c4 m1`	Copy the explanatory variables into
	matrix m1
`set c10`	Set the prior for $\boldsymbol{\beta}$ to be $(0,0,0,0)'$
`4(0)`	and store it in c10. **Note:** The prior
`end`	includes β_0
`set c11`	Set the prior variance to be $10^4 \times \mathbf{I}_4$
`4(10000)`	and store the diagonal in c11
`end`	
`diag c11 m10;`	Assign the diagonal to matrix m10
`%<insert path>BayesMultReg 4 c1 m1;`	4 is the number of coefficients
	including the intercept β_0
`sigma 1;`	In this case we know $\sigma = 1$
`prior c10 m10;`	The prior mean for $\boldsymbol{\beta}$ is in c10,
	and the covariance is in m10
`posterior c12 m12;`	Store the posterior mean for $\boldsymbol{\beta}$ in c12,
	and store the posterior variance–covariance
	in matrix m12

D

USING THE INCLUDED
R FUNCTIONS

A Note Regarding the Previous Edition

A considerable effort has been made to make the R functions easier to use since the previous edition. In particular, calculation of the mean, median, variance, standard deviation, interquartile range, quantiles, and the cumulative distribution function for the posterior density can now be achieved by using the natural choice of function you would expect to use for each of these operations (`mean`, `mean`, `var`, `sd`, `IQR`, `quantile`, `cdf`). The numerical integration and interpolation often associated with these calculations is handled seamlessly under the hood for you. It is also possible to plot the results for any posterior density using the `plot` function.

Introduction to Bayesian Statistics, 3rd ed.   **543**
By Bolstad, W. M. and Curran, J. M. Copyright © 2016 John Wiley & Sons, Inc.

Obtaining and using R

R is a free software environment for statistical computing and graphics. It compiles and runs on a wide variety of UNIX platforms, Windows, and Mac OS X. The latest version of R (currently 3.3.0) may always be found at http://cran.r-project.org. There is also a large number of mirror sites https://cran.r-project.org/mirrors.html which may be closer to you and faster. Compiled versions of R for Linux, Mac OS X and Windows, and the source code (for those who wish to compile R themselves) may also be found at this address.

Installation of R for Windows and Mac OS X requires no more effort than downloading the latest installer and running it. On Windows this will be an executable file. On Mac OS X it will be a package.

R Studio

If you plan to use R with this text, then we highly recommend that you also download and install R Studio (http://rstudio.com). R Studio is a free integrated development environment (IDE) for R which offers many features that make R easier and more pleasant to use. R Studio is developed by a commercial company which also offers a paid version with support for those who work in a commercial environment. If you choose to use R Studio, then make sure that you install it after R has been installed.

Obtaining the R Functions

The R functions used in conjunction with this book have been collated into an R package. R packages provide a simple mechanism to increase the functionality of R. The simplicity and attractiveness of this mechanism is extremely obvious with the more than 5,000 R packages available from the Comprehensive R Archive Network (CRAN). The name of the R package that contains the functions for this book is `Bolstad`. The latest version can be downloaded from CRAN using either the `install.packages` function or using the pulldown menus in R or R Studio. We give instructions below for both of these methods.

Installing the `Bolstad` Package

We assume, in the following instructions, that you have a functioning internet connection. If you do not, then you will at least need some way to download the package to your computer.

Installation from the Console

1. Start R or R Studio.

2. Type the following into the console window and hit Enter.

```
install.packages("Bolstad")
```

The capitalization is important. It does not matter whether you use double or single quotation marks.

Installation using R

1. Pull down the *Packages* menu on Windows, or the *Packages and Data* menu on Mac OS X.

2. [**Windows:**] Select *Install package(s)...*, and then select the CRAN mirror that is closest to you and click on *OK*.

   [**Mac OS X:**] Select *Package Installer*, click on the *Get List* button, and select the CRAN mirror that is closest to you.

3. [**Windows:**] Scroll down until you find the `Bolstad` package, select it by clicking on it, and click on *OK*.

   [**Mac OS X:**] Type `Bolstad` into the search text box (opposite the *Get List* button) and hit Enter. Select the `Bolstad` package and click on `Install Selected`. Click the red *Close Window* icon at the top-left of the dialog box.

Installation using R Studio

1. Select *Install Packages...* from the *Tools* menu. Type `Bolstad` into *Packages* text box and click on *Install*.

Installation from a Local File   Both R and R Studio offer options to install the `Bolstad` package from a local file. Again this can be achieved using the `install.package` function, or by using the menus. To install the package using the console, first either use *Change dir...* from the *File* menu on Windows, or *Change Working Directory...* from the *Misc* menu on Mac OS X, or use the `settled` function in the console, to set the working directory to the location where you have stored the `Bolstad` package you downloaded. Type

```
install.packages("Bolstad_X.X-XX.EXT")
```

where `X.X-XX` is the version number of the file you downloaded (e.g. `0.2-33`) and `EXT` is either `zip` on Windows or `tar.gz` on Mac OS X.

Loading the R Package

R will now recognize the package `Bolstad` as a package it can load. To use the functions in the package `Bolstad`, type

```
library(Bolstad)
```

into the console.

To see the list of functions contained within the package, type

```
library(help = Bolstad)
```

Help on each of the R functions is available once you have loaded the `Bolstad` package. There are a number of ways to access help files under R. The traditional way is to use the `help` or `?` function. For example, to see the help file on the `binodp` function, type

```
help(binodp)
```

or

```
?binodp
```

All of the examples listed in the help file may be executed by using the `example` command. For example, to run the examples listed in the `binodp` help file type

```
example(binodp)
```

Each help file has a standard layout, which is as follows:

Title: a brief title that gives some idea of what the function is supposed to do or show

Description: a fuller description of the what the function is supposed to do or show

Usage: the formal calling syntax of the function

Arguments: a description of each of the arguments of the function

Values: a description of the values (if any) returned by the function

See also: a reference to related functions

Examples: some examples of how the function may be used. These examples may be run either by using the `example` command (see above) or copied and pasted into the R console window

The R language has two special features that may make it confusing to users of other programming and statistical languages: default or optional arguments, and variable ordering of arguments. An R function may have arguments for which the author has specified a default value. Let us take the function `binobp` as an example. The syntax of `binobp` is

```
binobp(x, n, a = 1, b = 1, pi = seq(0.01, 0.999, by = 0.001),
        plot = TRUE)
```

The function takes six arguments: `x`, `n`, `a`, `b`, `pi`, and `plot`. However, the author has specified default values for `a`, `b`, `pi`, and `plot` namely a = 1, b = 1, pi = = seq(0.01, 0.999, by = 0.001), and plot = TRUE. This means that the user only has to supply the arguments x and n. Therefore the arguments `a`, `b`, `pi`, and `plot` are said to be optional or default. In this example, by default, a $beta(a = 1, b = 1)$ prior is used and a plot is drawn (`plot` = TRUE). Hence the simplest example for `binobp` is given as `binobp(6,8)`. If the user wanted to change the prior used, say to $beta(5,6)$, then they would type `binobp(6, 8, 5, 6)`. There is a slight catch here, which leads into the next feature. Assume that the user wanted to use a $beta(1,1)$ prior, but did not want the plot to be produced. One might be tempted to type `binobp(6, 8, FALSE)`. This is incorrect. R will think that the value `FALSE` is the value being assigned to the parameter `a`, and convert it from a logical value, `FALSE`, to the numerical equivalent, 0, which will of course give an error because the parameters of the beta distribution must be greater than zero. The correct way to make such a call is to use named arguments, such as `binobp(6, 8, plot = FALSE)`. This specifically tells R which argument is to be assigned the value `FALSE`. This feature also makes the calling syntax more flexible because it means that the order of the arguments does not need to be adhered to. For example, `binobp(n = 8, x = 6, plot = FALSE, a = 1, b = 3)` would be a perfectly legitimate function call.

Chapter 2: Scientific Data Gathering

In this chapter we use the function `sscsample` to perform a small-scale Monte Carlo study on the efficiency of simple, stratified, and cluster random sampling on the population data contained in `sscsample.data`. Make sure the `Bolstad` package is loaded by typing

```
library(Bolstad)
```

first. Type the following commands into the R console:

```
sscsample(20, 200)
```

This calls the `sscsample` function and asks for 200 samples of size 20 to be drawn from the data set `sscsample.data`. To return the means and the samples themselves, type

```
results = sscsample(20, 200)
```

This will store all 200 samples and their means in an R list structure called `results`. The means of the sample may be accessed by typing

```
results$means
```

The samples themselves are stored in the columns of a 20 × 200 matrix called `results$samples`. To access the ith sample, where $i = 1, \ldots, 200$, type

```
results$samples[, i]
```

For example, to access the 50th sample, type

```
results$samples[, 50]
```

Experimental Design

We use the function `xdesign` to perform a small-scale Monte Carlo study comparing *completely randomized design* and *randomized block design* in their effectiveness for assigning experimental units into treatment groups. Suppose we want to carry out our study with four treatment groups, each of size 20, and with a correlation of 0.8 between the response and the blocking variable. Type the following commands into the R console:

```
xdesign()
```

Suppose we want to carry out our study with five treatment groups, each of size 25, and with a correlation of −0.6 between the response and the blocking variable. We also want to store the results of the simulation in a variable called `results`. Type the following commands into the R console:

```
results = xdesign(corr = -0.6, size = 25, n.treatments = 5)
```

`results` is a list containing three member vectors of length 2×n.treatments×n.rep. Each block of `n.rep` elements contains the simulated means for each Monte Carlo replicate with in a specific treatment group. The first `n.treatments` blocks correspond to the *completely randomized design*, and the second `n.treatments` blocks correspond to *randomized block design*

- `block.means`: a vector of the means of the blocking variable

- `treat.means`: a vector of the means of the response variable

- `ind`: a vector indicating which means belong to which treatment group

An example of using these results might be

```
boxplot(block.means ~ ind, data = results)
boxplot(treat.means ~ ind, data = results)
```

Chapter 6: Bayesian Inference for Discrete Random Variables

Binomial Proportion with Discrete Prior

The function `binodp` is used to find the posterior when we have a *binomial* (n, π) observation, and we have a discrete prior for π. For example, suppose π has the discrete distribution with three possible values, .3, .4, and .5. Suppose the prior distribution is as given in Table D.1

Table D.1   An example discrete prior for a binomial proportion π

π	$f(\pi)$
.3	.2
.4	.3
.5	.5

and we want to find the posterior distribution after $n = 6$ trials and observing $y = 5$ successes. Type the following commands into the R console:

```
pi = c(0.3, 0.4, 0.5)
pi.prior = c(0.2, 0.3, 0.5)
results = binodp(5, 6, pi = pi, pi.prior = pi.prior)
```

Poisson Parameter with Discrete Prior

`poisdp` is used to find the posterior when we have a *Poisson*(μ) observation, and a discrete prior for μ. For example, suppose μ has three possible values

Table D.2 Discrete prior distribution for Poisson parameter μ

μ	$g(\mu)$
1	.3
2	.4
3	.3

$\mu = 1, 2$, or 3 where the prior probabilities are given in Table D, and we want to find the posterior distribution after observing $y = 4$.
Type the following commands into the R console:

```
mu = 1:3
mu.prior = c(0.3, 0.4, 0.3)
poisdp(4, mu, mu.prior)
```

Chapter 8: Bayesian Inference for Binomial Proportion

beta(a, b) Prior for π

binobp is used to find the posterior when we have a $binomial(n, \pi)$ observation, and we have a $beta(a, b)$ prior for π. The $beta$ family of priors is conjugate for $binomial(n, \pi)$ observations, so the posterior will be another member of the family, $beta(a', b')$ where $a' = a + y$ and $b' = b + n - y$. For example, suppose we have $n = 12$ trials, and observe $y = 4$ successes, and use a $beta(3, 3)$ prior for π. Type the following command into the R console:

```
binobp(4, 12, 3, 3)
```

We can find the posterior mean and standard deviation from the output. We can determine an equal tail area credible interval for π by taking the appropriate quantiles that correspond to the desired tail area values of the interval. For example, for 95% credible interval we take the quantiles with probability 0.025 and 0.975, respectively. These are 0.184 and 0.617. Alternatively, we can store the results and use the **mean**, **sd**, and **quantile** functions to find the posterior mean, standard deviation, and credible interval. Type the following commands into the R console:

```
results = binobp(4, 12, 3, 3)
mean(results)
sd(results)
quantile(results, probs = c(0.025, 0.975))
```

We can test $H_0 : \pi \leq \pi_0$ versus $H_1 : \pi > \pi_0$ by using the qbeta function in conjunction with the parameters of the posterior beta distribution. For example, assume that $\pi_0 = 0.1$, and that $y = 4$ successes were observed in $n = 12$ trials. If we use a $beta(3,3)$ prior, then the posterior distribution of π is $beta(3 + 4 = 7, 3 + 12 - 4 = 11)$. Therefore we can test $H_0 : \pi \leq \pi_0 = 0.1$ versus $H_1 : \pi > \pi_0 = 0.1$ by typing

```
pbeta(0.1, 7, 11)

## or alternatively use the cdf

results = binobp(4, 12, 3, 3)
Fpi = cdf(results)
Fpi(0.1)
```

General Continuous Prior for π

binogcp is used to find the posterior when we have a *binomial* (n, π) observation, and we have a general continuous prior for π. Note that π must go from 0 to 1 in equal steps of at least 0.01, and $g(\pi)$ must be defined at each of the π values. For example, suppose we have $n = 12$ trials and observe $y = 4$ successes. In this example our continuous prior for π is a *normal*$(\mu = 0.5, \sigma = 0.25)$. Type the following commands into the R console:

```
binogcp(4, 12, density = "normal", params = c(0.5, 0.25))
```

This example is perhaps not quite general as it uses some of the built in functionality of binogcp. In this second example we use a "user-defined" general continuous prior. Let the probability density function be a triangular distribution defined by

$$g(\pi) = \begin{cases} 4\pi & \text{for } 0 \leq \pi \leq 0.5, \\ 4 - 4\pi & \text{for } 0.5 < \pi \leq 1. \end{cases}$$

Type the following commands into the R console:

```
pi = seq(0, 1, by = 0.001)
prior = createPrior(c(0, 0.5, 1), c(0, 1, 0))
pi.prior = prior(pi)
results = binogcp(4, 12, "user", pi = pi, pi.prior = pi.prior)
```

The createPrior function is good for creating piecewise priors where the user can give the weights each point. The result is a function which uses linear interpolation to provide the value of the prior. The output of `binogcp` does not print out the posterior mean and standard deviation. Nor does it print out the values that give the tail areas of the integrated density function that we need to determine credible interval for π. Instead, we use the functions `mean`, `sd`, and `cdf` which numerically integrate a function over its range to determine these quantities. We can find the cumulative distribution function for posterior density $g(\pi|y)$ using the R function `cdf`. Type the following commands into the R console:

```
Fpi = cdf(results)
curve(Fpi, from = pi[1], to = pi[length(pi)],
      xlab=expression(pi[0]),
      ylab=expression(Pr(pi<=pi[0])))
```

These commands created a new function `Fpi`, which returns $\Pr(Y \leq x)$, for a given value x. To find a 95% credible interval (with equal tail areas) we use the `quantile` function.

```
ci = quantile(results, probs = c(0.025, 0.975))
ci = round(ci, 4)
cat(paste0("Approximate 95% credible interval : [", paste0(ci,
    collapse = ", "), "]\n"))
```

To test the hypothesis $H_0 : \pi \leq \pi_0$ versus $H_1 : \pi > \pi_0$, we calculate the value of the `cdf` at π_0. If the value is less than the desired level of significance α, then we can reject the null hypothesis. For example, if $\alpha = 0.05$ in our previous example, and $\pi_0 = 0.1$, then we would type

```
Fpi = cdf(results)
Fpi(0.1)
```

This should give the following output:

```
[1] 0.001593768
```

Given that 0.0016 is substantially less than our significance value of 0.05, then we would reject H_0. We can also find the posterior mean and variance by numerically evaluating

$$m' = \int_0^1 \pi g(\pi|y)\, d\pi$$

and

$$(s')^2 = \int_0^1 (\pi - m')^2 g(\pi|y)\, d\pi$$

This integration is handled for us by the R functions `mean` and `sd`. Type the following commands into the R console:

```
post.mean = mean(results)
post.sd = sd(results)
```

Of course we can use these values to calculate an approximate 95% credible interval using standard theory:

```
ci = post.mean + c(-1, 1) * qnorm(0.975) * post.sd
ci = round(ci, 4)
cat(paste0("Approximate 95% credible interval : [", paste0(ci,
    collapse = ", "), "]\n"))
```

Chapter 10: Bayesian Inference for Poisson

$gamma(R, v)$ Prior for μ

The function `poisgamp` is used to find the posterior when we have a random sample from a $Poisson(\mu)$ distribution, and we have a $gamma(r, v)$ prior for μ. The *gamma* family of priors is the conjugate family for *Poisson* observations, so the posterior will be another member of the family, $gamma(r', v')$ where $r' = r + \sum y$ and $v' = v + n$. The simple rules are "add sum of observations to r" and "add number of observations to v". For example, suppose we have a sample five observations from a $Poisson(\mu)$ distribution, 3, 4, 3, 0, 1. Suppose we want to use a $gamma(6, 3)$ prior for μ. Type the following commands into the R console:

```
y = c(3, 4, 3, 0, 1)
poisgamp(y, 6, 3)
```

By default `poisgamp` returns a 99% Bayesian credible interval for μ. If we want a credible interval of different width, then we can use the R functions relating to the posterior *gamma* distribution function. For example, if we wanted a 95% credible interval using the data above, then we would type

```
y = c(3, 4, 3, 0, 1)
results = poisgamp(y, 6, 3)
ci = quantile(results, probs = c(0.025, 0.975))
```

We can test $H_0 : \mu \leq \mu_0$ versus $H_1 : \mu > \mu_0$ using the **pgamma** function. For example, if in the example above we hypothesize $\mu_0 = 3$ and $\alpha = 0.05$, then we type

```
Fmu = cdf(results)
Fmu(3)
```

General Continuous Prior for Poisson Parameter μ

The function poisgcp is used to find the posterior when we have a random sample from a $Poisson(\mu)$ distribution and we have a continuous prior for μ. Suppose we have a sample five observations from a $Poisson(\mu)$ distribution, 3, 4, 3, 0, 1. The prior density of μ is found by linearly interpolating the values in Table D.3. To find the posterior density for μ with this prior, type the following commands into the R console:

```
y = c(3, 4, 3, 0, 1)
mu = seq(0, 8, by = 0.001)
prior = createPrior(c(0, 2, 4, 8), c(0, 2, 2, 0))
poisgcp(y, "user", mu = mu, mu.prior = prior(mu))
```

The output of poisgcp does not include the posterior mean and standard de-

Table D.3   Continuous prior distribution for Poisson parameter μ has shape given by interpolating between these values.

μ	$g(\mu)$
0	0
2	2
4	2
8	0

viation by default. Nor does it print out the cumulative distribution function that allows us to find credible intervals. Instead we use the functions mean, sd. cdf and quantile which numerically integrate the posterior in order to compute the desired quantities. Type the following commands into the R console to obtain the posterior cumulative distribution function:

```
results = poisgcp(y, "user", mu = mu, mu.prior = prior(mu))
Fmu = cdf(results)
```

We can use the inverse cumulative distribution function to find 95% Bayesian credible interval for μ. This is done by finding the values of μ that correspond to the probabilities .025 and .975. Type the following into the R console:

```
quantile(results, probs = c(0.025, 0.975))
```

We can use the cumulative distribution function to test the null hypothesis $H_0 : \mu \leq \mu_0$ versus $H_1 : \mu > \mu_0$. For example, if we hypothesis $\mu_0 = 1.8$ and our significance level is $\alpha = 0.05$, then

```
Fmu(1.8)
```

returns 0.1579979. Given that this is greater than the desired level of significance, we fail to reject the null hypothesis at that level.

Chapter 11: Bayesian Inference for Normal Mean

Discrete Prior for μ

Table D.4 A discrete prior for the normal mean μ

μ	$f(\pi)$
2	.1
2.5	.2
3	.4
3.5	.2
4	.1

The function **normdp** is used to find the posterior when we have a vector of $normal(\mu, \sigma^2)$ observations and σ^2 is known, and we have a discrete prior for μ. If $sigma^2$ is not known, then is it is estimated from the observations. For example, suppose μ has the discrete distribution with five possible values: 2, 2.5, 3, 3.5, and 4. Suppose the prior distribution is given in Table D.4 and we want to find the posterior distribution after we've observed a random sample of $n = 5$ observations from a *normal* $(\mu, \sigma^2 = 1)$ that are 1.52, 0.02, 3.35, 3.49, and 1.82. Type the following commands into the R console:

```
mu = seq(2, 4, by = 0.5)
mu.prior = c(0.1, 0.2, 0.4, 0.2, 0.1)
y = c(1.52, 0.02, 3.35, 3.49, 1.82)
normdp(y, 1, mu, mu.prior)
```

$normal(M, s^2)$ Prior for μ

The function **normnp** is used when we have a vector containing a random sample of n observations from a $normal(\mu, \sigma^2)$ distribution (with σ^2 known) and we use a *normal*(m, s^2) prior distribution. If the observation standard deviation σ is not entered, the estimate calculated from the observations is used, and the approximation to the posterior is found. If the *normal* prior is not entered, a flat prior is used. The normal family of priors is conjugate for *normal*(μ, σ^2) observations, so the posterior will be another member of the family, *normal*$[m', (s')^2]$ where the new constants are given by

$$\frac{1}{(s')^2} = \frac{1}{s^2} + \frac{n}{\sigma^2}$$

and

$$m' = \frac{\frac{1}{b^2}}{\frac{1}{(b')^2}} \times m + \frac{\frac{n}{\sigma^2}}{\frac{1}{(b')^2}} \times \bar{y}.$$

For example, suppose we have a normal random sample of four observations from *normal*$(\mu, \sigma^2 = 1)$ that are 2.99, 5.56, 2.83, and 3.47. Suppose we use a *normal* $(3, 2^2)$ prior for μ. Type the following commands into the R console:

```
y = c(2.99, 5.56, 2.83, 3.47)
normnp(y, 3, 2, 1)
```

This gives the following output:

```
Known standard deviation :1
Posterior mean           : 3.6705882
Posterior std. deviation : 0.4850713


Prob. Quantile
----------------
0.005  2.4211275
0.010  2.5421438
0.025  2.7198661
0.050  2.8727170
0.500  3.6705882
0.950  4.4684594
0.975  4.6213104
0.990  4.7990327
0.995  4.9200490
```

We can find the posterior mean and standard deviation from the output. We can determine an (equal tail area) credible interval for μ by taking the appropriate quantiles that correspond to the desired tail area values of the interval. For example, for 99% credible interval we take the quantiles with probability 0.005 and 0.995, respectively. These are 2.42 and and 4.92. Alternatively, we can determine a Bayesian credible interval for μ by using the posterior mean and standard deviation in the normal inverse cumulative distribution function qnorm. Type the following commands into the R console:

```
y = c(2.99, 5.56, 2.83, 3.47)
results = normnp(y, 3, 2, 1)
ci = quantile(results, probs = c(0.025, 0.975))
```

We can test $H_0 : \mu \leq \mu_0$ versus $H_1 : \mu > \mu_0$ by using the posterior mean and standard deviation in the normal cumulative distribution function pnorm. For example, if $H_0 : \mu_0 = 2$ and our desired level of significance is $\alpha = 0.05$, then

```
Fmu = cdf(results)
Fmu(2)

## Alternatively

pnorm(2, mean(results), sd(results))
```

returns 2.87×10^{-4} which would lead us to reject H_0.

General Continuous Prior for μ

The function normgcp is used when we have a vector containing a random sample of n observations from a *normal* (μ, σ^2) distribution (with σ^2 known) and we have a vector containing values of μ, and a vector containing values from a continuous prior $g(\mu)$. If the standard deviation σ is not entered, the estimate calculated from the data is used, and the approximation to the posterior is found.

For example, suppose we have a random sample of four observations from a *normal* $(\mu, \sigma^2 = 1)$ distribution. The values are 2.99, 5.56, 2.83, and 3.47. Suppose we have a triangular prior defined over the range -3 to 3 by

$$g(\mu) = \begin{cases} \frac{1}{3} + \frac{\mu}{9} & \text{for} \quad -3 \leq \mu \leq 0, \\ \frac{1}{3} - \frac{\mu}{9} & \text{for} \quad 0 < \mu \leq 3. \end{cases}$$

Type the following commands into the R console:

```
y = c(2.99, 5.56, 2.83, 3.47)
mu = seq(-3, 3, by = 0.1)
prior = createPrior(c(-3, 0, 3), c(0, 1, 0))
results = normgcp(y, 1,density = "user", mu = mu,
                  mu.prior = prior(mu))
```

The output of normgcp does not print out the posterior mean and standard deviation. Nor does it print out the values that give the tail areas of the integrated density function that we need to determine credible interval for μ. Instead we use the function `cdf` which numerically integrates a function over its range to determine these things. We can find the integral of the posterior density $g(\mu|data)$ using this function. Type the following commands into the R console:

```
Fmu = cdf(results)
curve(Fmu,from = mu[1], to = mu[length(mu)],
      xlab = expression(mu[0]),
      ylab=expression(Pr(mu<=mu[0])))
```

These commands created a new function `Fmu`, which returns $\Pr(Y \leq x)$ for a given value of x, i.e. the cumulative distribution function (cdf). To find a 95% credible interval (with equal tail areas), we use the quantile function

```
ci = quantile(results, probs = c(0.025, 0.975))
ci = round(ci, 4)

cat(paste0("Approximate 95% credible interval : [", paste(ci,
    collapse = ", "), "]\n"))
```

To test a hypothesis $H_0 : \mu \leq \mu_0$ versus $H_1 : \mu > \mu_0$, we can use our cdf `Fmu` at μ_0. If this is less than the chosen level of significance, we can reject the null hypothesis at that level.

We can also find the posterior mean and variance by numerically evaluating

$$m' = \int \mu g(\mu|data)\, d\mu$$

and

$$(s')^2 = \int (\mu - m')^2 g(\mu|data)\, d\mu.$$

using the functions `mean` and `var` which handle the numerical integration for us. Type the following commands into the R console:

```
post.mean = mean(results)
post.var = var(results)
post.sd = sd(results)
```

Of course, we can use these values to calculate an approximate 95% credible interval using standard theory:

```
z = qnorm(0.975)
ci = post.mean + c(-1, 1) * z * post.sd
ci = round(ci, 4)
```

```
cat(paste0("Approximate 95% credible interval : [", paste0(ci,
    collapse = ", "), "]\n"))
```

Chapter 14: Bayesian Inference for Simple Linear Regression

The function `bayes.lin.reg` is used to find the posterior distribution of the simple linear regression slope β when we have a random sample of ordered pairs (x_i, y_i) from the simple linear regression model

$$y_i = \alpha_0 + \beta \times x_i + e_i,$$

where the observation errors e_i are independent $normal(0, \sigma^2)$ with known variance. If the variance is not known, then the posterior is found using the variance estimate calculated from the least squares residuals. We use independent priors for the slope β and the intercept $\alpha_{\bar{x}}$. These can be either flat priors, or *normal* priors. (The default is flat priors for both slope and intercept of $x = \bar{x}$.) This parameterization yields independent posterior distribution for slope and intercept with simple updating rules "posterior precision equals prior precision plus precision of least squares estimate" and "the posterior mean is weighted sum of prior mean and the least squares estimate where the weights are the proportions of the precisions to the posterior precision." Suppose we have vectors y and x, respectively, and we know the standard deviation $\sigma = 2$. We wish to use a $normal(0, 3^2)$ prior for β and a $normal(30, 10^2)$ prior for $\alpha_{\bar{x}}$. First we create some data for this example.

```
set.seed(100)
x = rnorm(100)
y = 3 * x + 22 + rnorm(100, 0, 2)
```

Now we can use `bayes.lin.reg`

```
bayes.lin.reg(y, x, "n", "n", 0, 3, 30, 10, 2)
```

If we want to find a credible interval for the slope, then we use Equation 14.9 or Equation 14.10 depending on whether we knew the standard deviation or used the value calculated from the residuals. In the example above, we know the standard deviation, therefore we would type the following into R to find a 95% credible interval for the slope:

```
results = bayes.lin.reg(y, x, "n", "n", 0, 3, 30, 10, 2)
ci = quantile(results$slope, probs = c(0.025, 0.975))
```

To find the credible interval for the predictions use Equation 14.13 when we know the variance or Equation 14.14 when we use the estimate calculated from the residuals. In the example above we can ask for predicted values for $x = 1, 2, 3$ by typing:

```
results = bayes.lin.reg(y, x, "n", "n", 0, 3, 30, 10, 2,
                pred.x = c(1, 2, 3))
```

The list `results` will contain three extra vectors `pred.x`, `pred.y` and `pred.se`. We can use these to get a 95% credible interval on each of the predicted values. To do this type the following into the R console:

```
z = qnorm(0.975)
ci = cbind(results$pred.y -  z * results$pred.se,
           results$pred.y +  z * results$pred.se)
```

Chapter 15: Bayesian Inference for Standard Deviation

$S \times$ an *Inverse Chi-Squared*(κ) Prior for σ^2

The function `nvaricp` is used when we have a vector containing a random sample of n observations from a $normal(\mu, \sigma^2)$ distribution where the mean μ is known. The $S \times$ an *inverse chi-squared*(κ) family of priors is the conjugate family for normal observations with known mean. The posterior will be another member of the family where the constants are given by the simple updating rules "add the sum of squares around the mean to S" and "add the sample size to the degrees of freedom." For example, suppose we have five observations from a $normal(\mu, \sigma^2)$ where $\mu = 200$, which are 206.4, 197.4, 212.7, 208.5, and 203.4. We want to use a prior that has prior median equal to 8. It turns out that $29.11 \times$ *inverse chi-squared*$(\kappa = 1)$ distribution has prior median equal to 8. Type the following into the R console:

```
y = c(206.4, 197.4, 212.7, 208.5, 203.4)
results = nvaricp(y, 200, 29.11, 1)
```

Note: The graphs that are printed out are the prior distributions of the standard deviation σ even though we are doing the calculations on the variance.

If we want to make inferences on the standard deviation σ using the posterior distribution we found, such as finding an equal tail area 95% Bayesian credible interval for σ, type the following commands into the R console:

```
quantile(results, probs = c(0.025, 0.975))
```

We can also estimate σ using the posterior mean of κ and S if $\kappa' > 2$ or posterior median.

```
post.mean = mean(results)
post.median = median(results)
```

Chapter 16: Robust Bayesian Methods

The function `binomixp` is used to find the posterior when we have a *binomial*(n, π) observations and use a mixture of a *beta*(a_0, b_0) and a *beta*(a_1, b_1) for the prior distribution for π. Generally, the first component summarizes our prior belief so that we give it a high prior probability. The second component has more spread to allow for our prior belief being mistaken so we give the the second component a low prior probability. For example, suppose our first component is *beta*$(10, 6)$ and the second component is *beta*$(1, 1)$ and we give a prior probability of .95 to the first component. We have performed 60 trials and observed $y = 15$ successes. To find the posterior distribution of the binomial proportion with a mixture prior for π, type the following commands into the R console:

```
binomixp(15, 60, c(10, 6), p = 0.95)
```

The function `normmixp` is used to find the posterior when we have normal(μ, σ^2) observations with known variance σ^2 and our prior for μ is a mixture of two normal distributions, a *normal*(m_0, s_0^2) and a *normal*(m_1, s_1^2). Generally, the first component summarizes our prior belief so we give it a high prior probability. The second component is a fall-back prior that has a much larger standard deviation to allow for our prior belief being wrong and has a much smaller prior probability. For example, suppose we have a random sample of observations from a *normal*(μ, σ^2) in a vector x where $\sigma^2 = .2^2$. Suppose we use a mixture of a *normal*$(10, .1^2)$ and a *normal*$(10, .4^2)$ prior where the prior probability of the first component is .95. To find the posterior distribution of normal mean with the mixture prior for μ, type the following commands into the R console:

```
x = c(9.88, 9.78, 10.05, 10.29, 9.77)
normmixp(x, 0.2, c(10, 0.01), c(10, 1e-04), 0.95)
```

Chapter 17 Bayesian Inference for Normal with Unknown Mean and Variance

The function `bayes.t.test` is used for performing Bayesian inference with data from a *normal* with unknown mean and variance. The function can also be used in the two sample case. This function has been designed to work as much like `t.test` as possible. In fact, large portions of the code are taken from `t.test`. If the user is carrying out a two-sample test with the assumption of unequal variances, then no exact solution is possible. However, a numerical solution is provided through a Gibbs sampling routine. This should be fairly fast and stable; however, because it involves a sampling scheme, it will take a few seconds to finish computation.

Chapter 18: Bayesian Inference for Multivariate Normal Mean Vector

`mvnmvnp` is used to find the posterior mean and variance–covariance matrix for a set of multivariate normal data with known variance–covariance Σ assuming a multivariate normal prior. If the prior density is MVN with mean vector $\mathbf{m}_0$ and variance–covariance matrix $\mathbf{V}_0$, and $\mathbf{Y}$ is a sample of size n from a $MVN(\boldsymbol{\mu}, \Sigma)$ matrix (where $\boldsymbol{\mu}$ is unknown), then the posterior density of $\boldsymbol{\mu}$ is $MVN(\mathbf{m}_1, \mathbf{V}_1)$ where

$$\mathbf{V}_1 = (\mathbf{V}_0^{-1} + n\Sigma^{-1})^{-1}$$

and

$$\mathbf{m}_1 = \mathbf{V}_1 \mathbf{V}_0^{-1} \mathbf{m}_0 + n\mathbf{V}_1 \Sigma^{-1} \bar{\mathbf{y}}$$

If Σ is not specified, then the sample variance–covariance matrix is used. In this case the posterior distribution is *multivariate t* so using the results with a MVN is only (approximately) valid for large samples.

We demonstrate the use of the function with some simulated data. We will sample 50 observations from a MVN with a true mean of $\boldsymbol{\mu} = (0, 2)'$ and a variance–covariance matrix of

$$\Sigma = \begin{pmatrix} 1 & 0.9 \\ 0.9 & 2 \end{pmatrix}.$$

This requires the use of the **mvtnorm** package which you should have been asked to install when you installed the **Bolstad** package.

```
set.seed(100)
mu = c(0, 2)
Sigma = matrix(c(1, 0.9, 0.9, 1), nc = 2, byrow = TRUE)

library(mvtnorm)
Y = rmvnorm(50, mu, Sigma)
```

Once we have our random data we can use the `mvnmvnp` function. We will choose a MVN prior with a mean of $\mathbf{m}_0 = (0,0)'$ and a variance–covariance matrix $\mathbf{V}_0 = 10^4 \times \mathbf{I}_2$. This is a very diffuse prior centred on $\mathbf{0}$.

```
m0 = c(0, 0)
V0 = 10000 * diag(c(1, 1))
results = mvnmvnp(Y, m0, V0, Sigma)
```

The posterior mean and variance–covariance matrix can be obtained with the `mean` and `var` functions. The cdf and inverse-cdf can be obtained with the `cdf` and `quantile` functions. These latter two functions make calls to the `pmvnorm` and `qmvnorm` functions from the `mvtnorm` package.

Chapter 19: Bayesian Inference for the Multiple Linear Regression Model

`bayes.lm` is is used to find the posterior distribution of vector of regression coefficients $\boldsymbol{\beta}$ when we have a random sample of ordered pairs $(\mathbf{x}_i, y_i)$ from the multiple linear regression model

$$y_i = \beta_0 + \beta_1 x_{i1} + \beta_2 x_{i2} + \cdots + \beta_p x_{ip} + e_i = \mathbf{x}_i \boldsymbol{\beta} + e_i \,,$$

where (usually) the observation errors e_i are independent $normal(0, \sigma^2)$ with known variance. If σ^2 is not known, then we estimate it from the variance of the residuals. Our prior for $\boldsymbol{\beta}$ is either a flat prior or a $MVN(\mathbf{b}_0, \mathbf{V}_0)$ prior. It is not necessary to use the function if we assume a flat prior as the posterior mean of $\boldsymbol{\beta}$ is equal to the least squares estimate $\boldsymbol{\beta}_{LS}$, which is also the maximum likelihood estimate. This means that the R linear model function `lm` gives us the posterior mean and variance for a flat prior. The posterior mean of $\boldsymbol{\beta}$ when we use a $MVN(\mathbf{b}_0, \mathbf{V}_0)$ prior is found through the simple updating rules given Equations 19.5 and 19.4. We will generate some random data from the model

$$y_i = 10 + 3x_{1i} + x_{2i} - 5x_{3i} + \epsilon_i, \quad \epsilon_i iidN(0, \sigma^2 = 1)$$

to demonstrate the use of the function.

```
set.seed(100)
example.df = data.frame(x1 = rnorm(100),
                        x2 = rnorm(100),
                        x3 = rnorm(100))

example.df = within(example.df,
                    {y = 10 + 3 * x1 + x2 - 5 * x3 + rnorm(100)})
```

The bayes.lm function has been designed to work as much like lm as possible. If a flat prior is used (default), then bayes.lm calls lm. However, if we specify $\mathbf{b}_0$ and $\mathbf{V}_0$, then it uses the estimates from lm in conjunction with the simple updating rules. **Note:** bayes.lm centers each of the covariates before fitting the model. That is, $\mathbf{X}_i$ is replaced with $\mathbf{X}_i - \bar{\mathbf{X}}_i$. This will make the regression more stable, and alter the estimate of β_0. We will use a MVN prior with a mean of $\mathbf{b}_0 = (0,0,0,0)'$ and a variance–covariance matrix $\mathbf{V}_0 = 10^4 \times \mathbf{I}_4$.

```
b0 = rep(0, 4)
V0 = 1e4 * diag(rep(1, 4))
fit = bayes.lm(y ~ x1 + x2 + x3, data = example.df,
                prior = list(b0 = b0, V0 = V0))
```

A modified regression table can be obtained by using the **summary** function.

```
summary(fit)
```

The **mean** and **var** functions are not implemented for the fitted object in this case because they are not implemented for lm. However, the posterior mean and posterior variance–covariance matrix of the coefficients can be obtained using the $ notation. That is,

```
b1 = fit$post.mean
V1 = fit$post.var
```

These quantities may then, in turn, be used to carry out inference on β.

E

ANSWERS TO

SELECTED EXERCISES

Chapter 3: Displaying and Summarizing Data

3.1. (a) Stem-and-leaf plot for sulfur dioxide (SO_2) data

```
                     leaf unit    1
        0  ║  33
        0  ║  5799
        1  ║  1334
        1  ║  6789
        2  ║  33
        2  ║  56789
        3  ║
        3  ║  5
        4  ║  34
        4  ║  6
```

Introduction to Bayesian Statistics, 3rd ed.
By Bolstad, W. M. and Curran, J. M. Copyright © 2016 John Wiley & Sons, Inc.

(b) Median $Q_2 = X_{[13]} = 18$,

Lower quartile $Q_1 = X_{[\frac{26}{4}]} = \frac{X_6 + X_7}{2} = 10$, and

Upper quartile $Q_3 = X_{[\frac{78}{4}]} = \frac{X_{19} + X_{20}}{2} = 27.5$

(c) Boxplot of SO_2 data

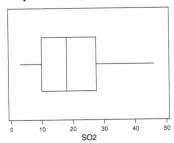

3.3. (a) Stem-and-leaf plot for distance measurements data

leaf unit   .01

299.4	0
299.5	0
299.6	0
299.7	00
299.8	000
299.9	000000
300.0	0000000
300.1	00000000
300.2	0000000
300.3	00
300.4	00000
300.5	000
300.6	00
300.7	00

(b) Median $= 300.1$ $Q_1 = 299.9$ $Q_3 = 300.35$

(c) Boxplot of distance measurement data

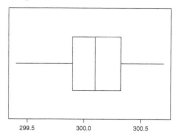

(d) Histogram of distance measurement data

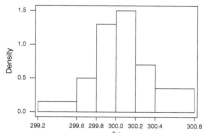

(e) Cumulative frequency polygon of distance measurement data

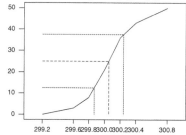

3.5. (a) Histogram of liquid cash reserve

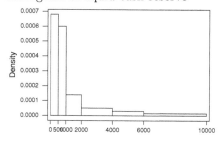

(b) Cumulative frequency polygon of liquid cash reserve

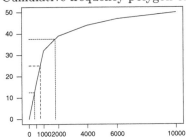

(c) Grouped mean = 1600

3.7. (a) Plot of weight versus length (slug data)

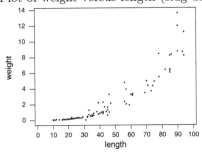

(b) Plot of log(weight) versus log(length)

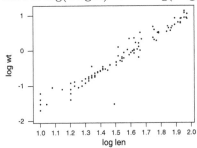

(c) The point $(1.5, -1.5)$ does not seem to fit the pattern. This corresponds to observation 90. Dr. Harold Henderson at AgResearch New Zealand has told me that there are two possible explanations for this point. Either the digits of length were transposed at recording or the decimal place for weight was misplaced.

Chapter 4: Logic, Probability, and Uncertainty

4.1. (a) $P(\tilde{A}) = .6$

(b) $P(A \cap B) = .2$

(c) $P(A \cup B) = .7$

4.3. (a) $P(\tilde{A} \cap B) = .24$ and $P(B) = .4$, therefore $P(A \cap B) = .16$. $P(A \cap B) = P(A) \times P(B)$, therefore they are independent.

(b) $P(A \cup B) = .4 + .4 - .16 = .64$

4.5. (a) $\Omega = \{1, 2, 3, 4, 5, 6\}$

(b) $A = \{2, 4, 6\}, \quad P(A) = \frac{3}{6}$

(c) $B = \{3, 6\}, \quad P(B) = \frac{2}{6}$

(d) $A \cap B = \{6\}, \quad P(A \cap B) = \frac{1}{6}$

(e) $P(A \cap B) = P(A) \times P(B)$, therefore they are independent.

4.7. (a)

$$A = \begin{cases} (1,1)\ (1,3)\ (1,5) \\ (2,2)\ (2,4)\ (2,6) \\ (3,1)\ (3,3)\ (3,5) \\ (4,2)\ (4,4)\ (4,6) \\ (5,1)\ (5,3)\ (5,5) \\ (6,2)\ (6,4)\ (6,6) \end{cases}$$

$P(A) = \frac{18}{36}$

(b)

$$B = \begin{cases} (1,2)\ (1,5)\ (2,1)\ (2,4)\ (3,3)\ (3,6) \\ (4,2)\ (4,5)\ (5,1)\ (5,4)\ (6,3)\ (6,6) \end{cases}$$

$P(B) = \frac{12}{36}$

(c) $A \cap B = \{(1,5)(2,4)(3,3)(4,2)(5,1)(6,6)\}$
$P(A \cap B) = \frac{6}{36}$

(d) $P(A \cap B) = P(A) \times P(B)$, yes they are independent.

4.9. Let D be "the person has the disease" and let T be "The test result was positive."

$$P(D|T) = \frac{P(D \cap T)}{P(T)} = .0875 \,.$$

4.10. Let A be ace drawn, and let F be face card or ten drawn.

$$P(\text{"Blackjack"}) = P(A) \times P(F|A) + P(F) \times P(A|F)$$

(they are disjoint ways of getting "Blackjack")

$$P(\text{"Blackjack"}) = \frac{16}{208} \times \frac{64}{207} + \frac{64}{208} \times \frac{16}{207} = 0.047566 \,.$$

Chapter 5: Discrete Random Variables

5.1. (a) $P(1 < Y \le 3) = .4$

(b) $E[Y] = 1.6$

(c) $\text{Var}[Y] = 1.44$

(d) $E[W] = 6.2$

(e) $\text{Var}[W] = 5.76$

5.3. (a) The filled-in table:

y_i	$f(y_i)$	$y_i \times f(y_i)$	$y_i^2 \times f(y_i)$
0	.0102	.0000	.0000
1	.0768	.0768	.0768
2	.2304	.4608	.9216
3	.3456	1.0368	3.1104
4	.2592	1.0368	4.1472
5	.0778	.3890	1.9450
Sum	1.0000	3.0000	10.2000

      i. $E[Y] = 3$

     ii. $\text{Var}[Y] = 10.2 - 3^2 = 1.2$

(b) Using formulas

      i. $E[Y] = 5 \times .6 = 3$

     ii. $\text{Var}[Y] = 5 \times .6 \times .4 = 1.2$

5.5. (a)

Outcome	Probability	Outcome	Probability
RRRR	$\frac{30}{50} \times \frac{30}{50} \times \frac{30}{50} \times \frac{30}{50}$	RRRG	$\frac{30}{50} \times \frac{30}{50} \times \frac{30}{50} \times \frac{20}{50}$
RRGR	$\frac{30}{50} \times \frac{30}{50} \times \frac{20}{50} \times \frac{30}{50}$	RGRR	$\frac{30}{50} \times \frac{20}{50} \times \frac{30}{50} \times \frac{30}{50}$
GRRR	$\frac{20}{50} \times \frac{30}{50} \times \frac{30}{50} \times \frac{30}{50}$	GRRG	$\frac{20}{50} \times \frac{30}{50} \times \frac{30}{50} \times \frac{20}{50}$
GRGR	$\frac{20}{50} \times \frac{30}{50} \times \frac{20}{50} \times \frac{30}{50}$	GGRR	$\frac{20}{50} \times \frac{20}{50} \times \frac{30}{50} \times \frac{30}{50}$
RRGG	$\frac{30}{50} \times \frac{30}{50} \times \frac{20}{50} \times \frac{20}{50}$	RGGR	$\frac{30}{50} \times \frac{20}{50} \times \frac{20}{50} \times \frac{30}{50}$
RGRG	$\frac{30}{50} \times \frac{20}{50} \times \frac{30}{50} \times \frac{20}{50}$	GGGR	$\frac{20}{50} \times \frac{20}{50} \times \frac{20}{50} \times \frac{30}{50}$
GGRG	$\frac{20}{50} \times \frac{20}{50} \times \frac{30}{50} \times \frac{20}{50}$	GRGG	$\frac{20}{50} \times \frac{30}{50} \times \frac{20}{50} \times \frac{20}{50}$
RGGG	$\frac{30}{50} \times \frac{20}{50} \times \frac{20}{50} \times \frac{20}{50}$	GGGG	$\frac{20}{50} \times \frac{20}{50} \times \frac{20}{50} \times \frac{20}{50}$

The outcomes having same number of green balls have the same probability.

(b)

Y = 0	Y = 1	Y = 2	Y = 3	Y = 4
RRRR	RRRG	RRGG	RGGG	GGGG
	RRGR	RGRG	GRGG	
	RGRR	RGGR	GGRG	
	GRRR	GRRG	GGGR	
		GRGR		
		GGRR		

(c) $P(Y = y)$ equals the "number of sequences having $Y = y$" times "the probability of any individual sequence having $Y = y$."

(d) The number of sequences having $Y = y$ is $\binom{n}{y}$ and the probability of any sequence having $Y = y$ successes is $\pi^y (1 - \pi)^{n-y}$ where in this case $n = 4$ and $\pi = \frac{20}{50}$. This gives the *binomial*(n, π) probability distribution.

5.7. (a) $P(Y = 2) = \frac{2^2 e^{-2}}{2!} = .2707$

(b) $P(Y \le 2) = \frac{2^0 e^{-2}}{0!} + \frac{2^1 e^{-2}}{1!} + \frac{2^2 e^{-2}}{2!} = .1353 + .2707 + .2707 = .6767$

(c) $P(1 \le Y < 4) = \frac{2^1 e^{-2}}{1!} + \frac{2^2 e^{-2}}{2!} + \frac{2^3 e^{-2}}{3!} = .2707 + .2707 + .1804 = .7218$

5.9.   The filled-in table:

X	Y					f(x)
	1	2	3	4	5	
1	.02	.04	.06	.08	.05	.25
2	.08	.02	.10	.02	.03	.25
3	.05	.05	.03	.02	.10	.25
4	.10	.04	.05	.03	.03	.25
f(y)	.25	.15	.24	.15	.21	

(a) The marginal distribution of X is found by summing across rows.

(b) The marginal distribution of Y is found by summing down columns.

(c) No they are not. The entries in the joint probability table aren't all equal to the products of the marginal probabilities.

(d) $P(X = 3 | Y = 1) = \frac{.05}{.25} = .2$

Chapter 6: Bayesian Inference for Discrete Random Variables

6.1. (a) Bayesian universe:

$$
\left\{
\begin{array}{ll}
(0,0) & (0,1) \\
(1,0) & (1,1) \\
(2,0) & (2,1) \\
(3,0) & (3,1) \\
(4,0) & (4,1) \\
(5,0) & (5,1) \\
(6,0) & (6,1) \\
(7,0) & (7,1) \\
(8,0) & (8,1) \\
(9,0) & (9,1)
\end{array}
\right\}
$$

(b) The filled-in table:

X	$prior$	$Y = 0$	$Y = 1$
0	$\frac{1}{10}$	$\frac{1}{10} \times \frac{9}{9}$	$\frac{1}{10} \times \frac{0}{9}$
1	$\frac{1}{10}$	$\frac{1}{10} \times \frac{8}{9}$	$\frac{1}{10} \times \frac{1}{9}$
2	$\frac{1}{10}$	$\frac{1}{10} \times \frac{7}{9}$	$\frac{1}{10} \times \frac{2}{9}$
3	$\frac{1}{10}$	$\frac{1}{10} \times \frac{6}{9}$	$\frac{1}{10} \times \frac{3}{9}$
4	$\frac{1}{10}$	$\frac{1}{10} \times \frac{5}{9}$	$\frac{1}{10} \times \frac{4}{9}$
5	$\frac{1}{10}$	$\frac{1}{10} \times \frac{4}{9}$	$\frac{1}{10} \times \frac{5}{9}$
6	$\frac{1}{10}$	$\frac{1}{10} \times \frac{3}{9}$	$\frac{1}{10} \times \frac{6}{9}$
7	$\frac{1}{10}$	$\frac{1}{10} \times \frac{2}{9}$	$\frac{1}{10} \times \frac{7}{9}$
8	$\frac{1}{10}$	$\frac{1}{10} \times \frac{1}{9}$	$\frac{1}{10} \times \frac{8}{9}$
9	$\frac{1}{10}$	$\frac{1}{10} \times \frac{0}{9}$	$\frac{1}{10} \times \frac{9}{9}$

which simplifies to

X	$prior$	$Y = 0$	$Y = 1$
0	$\frac{1}{10}$	$\frac{9}{90}$	$\frac{0}{90}$
1	$\frac{1}{10}$	$\frac{8}{90}$	$\frac{1}{90}$
2	$\frac{1}{10}$	$\frac{7}{90}$	$\frac{2}{90}$
3	$\frac{1}{10}$	$\frac{6}{90}$	$\frac{3}{90}$
4	$\frac{1}{10}$	$\frac{5}{90}$	$\frac{4}{90}$
5	$\frac{1}{10}$	$\frac{4}{90}$	$\frac{5}{90}$
6	$\frac{1}{10}$	$\frac{3}{90}$	$\frac{6}{90}$
7	$\frac{1}{10}$	$\frac{2}{90}$	$\frac{7}{90}$
8	$\frac{1}{10}$	$\frac{1}{90}$	$\frac{8}{90}$
9	$\frac{1}{10}$	$\frac{0}{90}$	$\frac{9}{90}$
		$\frac{45}{90}$	$\frac{45}{90}$

(c) The marginal distribution was found by summing down the columns.

(d) The reduced Bayesian universe is

$$\left\{ \begin{array}{c} (0,1) \\ (1,1) \\ (2,1) \\ (3,1) \\ (4,1) \\ (5,1) \\ (6,1) \\ (7,1) \\ (8,1) \\ (9,1) \end{array} \right\}.$$

(e) The posterior probability distribution is found by dividing the joint probabilities on the reduced Bayesian universe, by the sum of the joint probabilities over the reduced Bayesian universe.

(f) The simplified table is

X	prior	likelihood	prior × likelihood	posterior
0	$\frac{1}{10}$	$\frac{0}{9}$	$\frac{0}{90}$	$\frac{0}{45}$
1	$\frac{1}{10}$	$\frac{1}{9}$	$\frac{1}{90}$	$\frac{1}{45}$
2	$\frac{1}{10}$	$\frac{2}{9}$	$\frac{2}{90}$	$\frac{2}{45}$
3	$\frac{1}{10}$	$\frac{3}{9}$	$\frac{3}{90}$	$\frac{3}{45}$
4	$\frac{1}{10}$	$\frac{4}{9}$	$\frac{4}{90}$	$\frac{4}{45}$
5	$\frac{1}{10}$	$\frac{5}{9}$	$\frac{5}{90}$	$\frac{5}{45}$
6	$\frac{1}{10}$	$\frac{6}{9}$	$\frac{6}{90}$	$\frac{6}{45}$
7	$\frac{1}{10}$	$\frac{7}{9}$	$\frac{7}{90}$	$\frac{7}{45}$
8	$\frac{1}{10}$	$\frac{8}{9}$	$\frac{8}{90}$	$\frac{8}{45}$
9	$\frac{1}{10}$	$\frac{9}{9}$	$\frac{9}{90}$	$\frac{9}{45}$
	Sum		$\frac{45}{90}$	1

6.3. Looking at the two draws together, the simplified table is

X	prior	likelihood	prior × likelihood	posterior
0	$\frac{1}{10}$	$\frac{0}{9} \times 1$	$\frac{0}{90}$	$\frac{0}{120}$
1	$\frac{1}{10}$	$\frac{1}{9} \times \frac{8}{8}$	$\frac{8}{720}$	$\frac{8}{120}$
2	$\frac{1}{10}$	$\frac{2}{9} \times \frac{7}{8}$	$\frac{14}{720}$	$\frac{14}{120}$
3	$\frac{1}{10}$	$\frac{3}{9} \times \frac{6}{8}$	$\frac{18}{720}$	$\frac{18}{120}$
4	$\frac{1}{10}$	$\frac{4}{9} \times \frac{5}{8}$	$\frac{20}{720}$	$\frac{20}{120}$
5	$\frac{1}{10}$	$\frac{5}{9} \times \frac{4}{8}$	$\frac{20}{720}$	$\frac{20}{120}$
6	$\frac{1}{10}$	$\frac{6}{9} \times \frac{3}{8}$	$\frac{18}{720}$	$\frac{18}{120}$
7	$\frac{1}{10}$	$\frac{7}{9} \times \frac{2}{8}$	$\frac{14}{720}$	$\frac{14}{120}$
8	$\frac{1}{10}$	$\frac{8}{9} \times \frac{1}{8}$	$\frac{8}{720}$	$\frac{8}{120}$
9	$\frac{1}{10}$	$\frac{9}{9} \times \frac{0}{8}$	$\frac{0}{720}$	$\frac{0}{120}$
	Sum		$\frac{120}{720}$	1

6.4. The filled-in table

π	prior	likelihood	prior × likelihood	posterior
.2	.0017	.2048	.0004	.0022
.4	.0924	.3456	.0319	.1965
.6	.4678	.2304	.1078	.6633
.8	.4381	.0512	.0224	.1380
	marginal $P(Y_2 = 2)$		.1625	1.000

6.5. The filled in table

μ	prior	likelihood	prior $\times$ likelihood	posterior
1	.2	.1839	.0368	.2023
2	.2	.2707	.0541	.2976
3	.2	.2240	.0448	.2464
4	.2	.1465	.0293	.1611
5	.2	.0842	.0168	.0926
marginal $P(Y = 2)$			.1819	1.000

Chapter 7: Continuous Random Variables

7.1. (a) $E[X] = \frac{3}{8} = .375$

(b) $Var[X] = \frac{15}{8^2 \times 9} = 0.0260417$

7.3. The uniform distribution is also the $beta(1,1)$ distribution.

(a) $E[X] = \frac{1}{2} = .5$

(b) $Var[X] = \frac{1}{2^2 \times 3} = .08333$

(c) $P(X \le .25) = \int_0^{.25} 1\ dx = .25$

(d) $P(.33 < X < .75) = \int_{.33}^{.75} 1\ dx = .42$

7.5. (a) $P(0 \le Z < .65) = .2422$

(b) $P(Z \ge .54) = .2946$

(c) $P(-.35 \le Z \le 1.34) = .5467$

7.7. (a) $P(Y \le 130) = .8944$

(b) $P(Y \ge 135) = .0304$

(c) $P(114 \le Y \le 127) = .5826$

7.9. (a) $E[Y] = \frac{10}{10+12} = .4545$

(b) $Var[Y] = \frac{10 \times 12}{(22)^2 \times (23)} = .0107797$

(c) $P(Y > .5) = .3308$

7.10. (a) $E[Y] = \frac{12}{4} = 3$

(b) $Var[Y] = \frac{12}{4^2} = .75$

(c) $P(Y \le 4) = .873$

Chapter 8: Bayesian Inference for Binomial Proportion

8.1. (a) $binomial(n = 150, \pi)$ distribution

(b) $beta(30, 122)$

8.3. (a) a and b are the simultaneous solutions of

$$\frac{a}{a+b} = .5 \quad \text{and} \quad \frac{a \times b}{(a+b)^2 \times (a+b+1)} = .15^2$$

Solution is $a = 5.05$ and $b = 5.05$

(b) The equivalent sample size of her prior is 11.11

(c) $beta(26.05, 52.05)$

8.5. (a) $binomial(n = 116, \pi)$

(b) $beta(18, 103)$

(c)

$$E[\pi|y] = \frac{18}{18 + 103} \quad \text{and} \quad Var[\pi|y] = \frac{18 \times 103}{(121)^2 \times (122)}$$

(d) $normal(.149, .0322^2)$

(e) $(.086, .212)$

8.7. (a) $binomial(n = 174, \pi)$

(b) $beta(11, 168)$

(c)

$$E[\pi|y] = \frac{11}{11 + 168} = .0614$$

and

$$Var[\pi|y] = \frac{11 \times 168}{(179)^2 \times (180)} = .0003204$$

(d) $normal(.061, .0179^2)$

(e) $(.026, .097)$

Chapter 9: Comparing Bayesian and Frequentist Inferences for Proportion

9.1. (a) $binomial(n = 30, \pi)$

(b) $\hat{\pi}_f = \frac{8}{30} = .267$

(c) $beta(9, 23)$

(d) $\hat{\pi}_B = \frac{9}{32} = .281$

9.3. (a) $\hat{\pi}_f = \frac{11}{116} = .095$

(b) $beta(12, 115)$

(c) $E[\pi|y] = .094$ and $Var[\pi|y] = .0006684$
The Bayesian estimator $\hat{\pi}_B = .094$.

(d) $(.044, .145)$

(e) The null value $\pi = .10$ lies in the credible interval, so it remains a credible value at the 5% level

9.5. (a) $\hat{\pi}_f = \frac{24}{176} = .136$

(b) $beta(25, 162)$

(c) $E[\pi|y] = .134$ and $Var[\pi|y] = .0006160$
The Bayesian estimator $\hat{\pi}_B = .134$.

(d)
$$P(\pi \geq .15) = .255 \,.$$

This is greater than level of significance .05, so we can't reject the null hypothesis $H_0 : \pi \geq .15$.

Chapter 10: Bayesian Inference for Poisson

10.1. (a) Using positive uniform prior $g(\mu) = 1$ for $\mu > 0$:
   i. The posterior is $gamma(13, 5)$.
   ii. The posterior mean, median, and variance are

$$E[\mu|y_1, \ldots, y_5] = \frac{13}{5}, \quad median = 2.534,$$
$$Var[\mu|y_1, \ldots, y_5] = \frac{13}{5^2} \,.$$

(b) Using Jeffreys prior $g(\mu) = \mu^{-\frac{1}{2}}$:
   i. The posterior is $gamma(12.5, 5)$.
   ii. The posterior mean, median , and variance are

$$E[\mu|y_1, \ldots, y_5] = \frac{12.5}{5}, \quad median = 2.434,$$
$$Var[\mu|y_1, \ldots, y_5] = \frac{12.5}{5^2} \,.$$

10.2. (a) Using positive uniform prior $g(\mu) = 1$ for $\mu > 0$:
   i. The posterior is $gamma(123, 200)$.
   ii. The posterior mean, median, and variance are

$$E[\mu|y_1, \ldots, y_{200}] = \frac{123}{200}, \quad median = .6133,$$
$$Var[\mu|y_1, \ldots, y_{200}] = \frac{123}{200^2} \,.$$

(b) Using Jeffreys prior $g(\mu) = \mu^{-\frac{1}{2}}$:

     i. The posterior is $gamma(122.5, 200)$

     ii. The posterior mean, median, and variance are

$$E[\mu|y_1, \ldots, y_{200}] = \frac{122.5}{200}, \quad median = .6108,$$

$$\mathrm{Var}[\mu|y_1, \ldots, y_{200}] = \frac{122.5}{200^2}.$$

Chapter 11: Bayesian Inference for Normal Mean

11.1. (a) posterior distribution

Value	Posterior Probability
991	.0000
992	.0000
993	.0000
994	.0000
995	.0000
996	.0021
997	.1048
998	.5548
999	.3183
1000	.0198
1001	.0001
1002	.0000
1003	.0000
1004	.0000
1005	.0000
1006	.0000
1007	.0000
1008	.0000
1009	.0000
1010	.0000

(b) $P(\mu < 1000) = .9801.$

11.3. (a) The posterior precision equals

$$\frac{1}{(s')^2} = \frac{1}{10^2} + \frac{10}{3^2} = 1.1211.$$

The posterior variance equals $(s')^2 = \frac{1}{1.1211} = .89197$. The posterior standard deviation equals $s' = \sqrt{.89197} = .9444$. The posterior mean equals

$$m' = \frac{\frac{1}{10^2}}{1.1211} \times 30 + \frac{\frac{10}{3^2}}{1.1211} \times 36.93 = 36.87.$$

The posterior distribution of μ is $normal(36.87, .9444^2)$.

(b) Test

$$H_0 : \mu \leq 35 \qquad \text{versus} \qquad H_1 : \mu > 35.$$

Note that the alternative hypothesis is what we are trying to determine. The null hypothesis is that mean yield is unchanged from that of the standard process.

(c)

$$P(\mu \leq .35) = P\left(\frac{\mu - 36.87}{.944} \leq \frac{35 - 36.87}{.944}\right)$$
$$= P(Z \leq -1.9739) = .024$$

This is less than the level of significance $\alpha = .05\%$, so we reject the null hypothesis and conclude the yield of the revised process is greater than .35.

11.5. (a) The posterior precision equals

$$\frac{1}{(s')^2} = \frac{1}{200^2} + \frac{4}{40^2} = .002525.$$

The posterior variance equals $(s')^2 = \frac{1}{002525} = 396.0$ The posterior standard deviation equals $s' = \sqrt{396.0} = 19.9$. The posterior mean equals

$$m' = \frac{\frac{1}{200^2}}{.002525} \times 1000 + \frac{\frac{4}{40^2}}{.002525} \times 970 = 970.3.$$

The posterior distribution of μ is $normal(970.3, .19.9^2)$.

(b) The 95% credible interval for μ is is (931.3, 1009.3).

(c) The posterior distribution of θ is $normal(1392.8, 16.6^2)$.

(d) The 95% credible interval for θ is (1360,1425).

Chapter 12: Comparing Bayesian and Frequentist Inferences for Mean

12.1. (a) Posterior precision is given by

$$\frac{1}{(s')^2} = \frac{1}{10^2} + \frac{10}{2^2} = 2.51.$$

The posterior variance is $(s')^2 = \frac{1}{2.51} = .3984$ and the posterior standard deviation is $s' = \sqrt{.3984} = .63119$. The posterior mean is given by

$$m' = \frac{\frac{1}{10^2}}{2.51} \times 75 + \frac{\frac{10}{2^2}}{2.51} \times 79.430 = 79.4124.$$

The posterior distribution is $normal(79.4124, .63119^2)$.

(b) The 95% Bayesian credible interval is $(78.18, 80.65)$.

(c) To test

$$H_0 : \mu \geq 80 \qquad \text{versus} \qquad \mu < 80,$$

calculate the posterior probability of the null hypothesis.

$$P(\mu \geq 80) = P\left(\frac{\mu - 79.4124}{.63119} \geq \frac{80 - 79.4124}{.63119}\right)$$
$$= P(Z \geq .931) = .176.$$

This is greater than the level of significance, so we cannot reject the null hypothesis.

12.3. (a) Posterior precision

$$\frac{1}{(s')^2} = \frac{1}{80^2} + \frac{25}{80^2} = .0040625.$$

The posterior variance is $(s')^2 = \frac{1}{.0040625} = 246.154$ and the posterior standard deviation is $s' = \sqrt{246.154} = 15.69$. The posterior mean is

$$m' = \frac{\frac{1}{80^2}}{.0040625} \times 325 + \frac{\frac{25}{80^2}}{.0040625} \times 401.44 = 398.5.$$

The posterior distribution is $normal(398.5, 15.69^2)$.

(b) The 95% Bayesian credible interval is $(368, 429)$.

(c) To test

$$H_0 : \mu = 350 \qquad \text{versus} \qquad \mu \neq 350,$$

we observe that the null value (350) lies outside the credible interval, so we reject the null hypothesis $H_0 : \mu = 350$ at the 5% level of significance.

(d) To test

$$H_0 : \mu \leq 350 \qquad \text{versus} \qquad \mu > 350$$

we calculate the posterior probability of the null hypothesis.

$$P(\mu \leq 350) = P\left(\frac{\mu - 399}{15.69} \leq \frac{350 - 399}{15.69}\right)$$
$$= P(Z \leq -3.12) = .0009.$$

This is less than the level of significance, so we reject the null hypothesis and conclude $\mu > 350$.

Chapter 13: Bayesian Inference for Difference Between Means

13.1. (a) The posterior distribution of μ_A is $normal(119.4, 1.888^2)$, the posterior distribution of μ_B is $normal(122.7, 1.888^2)$, and they are independent.

   (b) The posterior distribution of $\mu_d = \mu_A - \mu_B$ is $normal(-3.271, 2.671^2)$.

   (c) The 95% credible interval for $\mu_A - \mu_B$ is $(-8.506, 1.965)$.

   (d) We note that the null value 0 lies inside the credible interval. Hence we cannot reject the null hypothesis.

13.3. (a) The posterior distribution of μ_1 is $normal(14.96, .3778^2)$, the posterior distribution of μ_2 is $normal(15.55, .3778^2)$, and they are independent.

   (b) The posterior distribution of $\mu_d = \mu_1 - \mu_1$ is $normal(-.5847, .5343^2)$.

   (c) The 95% credible interval for $\mu_1 - \mu_1$ is $(-1.632, .462)$.

   (d) We note that the null value 0 lies inside the credible interval. Hence we cannot reject the null hypothesis.

13.5. (a) The posterior distribution of μ_1 is $normal(10.283, .816^2)$, the posterior distribution of μ_2 is $normal(9.186, .756^2)$, and they are independent.

   (b) The posterior distribution of $\mu_d = \mu_1 - \mu_2$ is $normal(1.097, 1.113^2)$.

   (c) The 95% credible interval for $\mu_1 - \mu_2$ is $(-1.08, 3.28)$.

   (d) We calculate the posterior probability of the null hypothesis

$$P(\mu_1 - \mu_2 \leq 0) = .162 .$$

   This is greater than the level of significance, so we cannot reject the null hypothesis.

13.7. (a) The posterior distribution of μ_1 is $normal(1.51999, .000009444^2)$.

   (b) The posterior distribution of μ_2 is $normal(1.52001, .000009444^2)$.

   (c) The posterior distribution of $\mu_d = \mu_1 - \mu_2$ is $normal(-.00002, .000013^2)$.

   (d) A 95% credible interval for μ_d is $(-.000046, .000006)$.

   (e) We observe that the null value 0 lies inside the credible interval, so we cannot reject the null hypothesis.

13.9. (a) The posterior distribution of π_1 is $beta(172, 144)$.

   (b) The posterior distribution of π_2 is $beta(138, 83)$.

   (c) The approximate posterior distribution of $\pi_1 - \pi_2$ is $normal(-.080, .0429^2)$.

   (d) The 99% Bayesian credible interval for $\pi_1 - \pi_2$ is $(-.190, .031)$.

   (e) We observe that the null value 0 lies inside the credible interval, so we cannot reject the null hypothesis that the proportions of New Zealand women who are in paid employment are equal for the two age groups.

13.11. (a) The posterior distribution of π_1 is $beta(70, 246)$.

(b) The posterior distribution of π_2 is $beta(115, 106)$.

(c) The approximate posterior distribution of $\pi_1 - \pi_2$ is $normal(-.299, .0408^2)$.

(d) We calculate the posterior probability of the null hypothesis:

$$P(\pi_1 - \pi_2 \geq 0) = P(Z \geq 7.31) = .0000.$$

We reject the null hypothesis and conclude that the proportion of New Zealand women in the younger group who have been married before age 22 is less than the proportion of New Zealand women in the older group who have been married before age 22.

13.13. (a) The posterior distribution of π_1 is $beta(137, 179)$.

(b) The posterior distribution of π_2 is $beta(136, 85)$.

(c) The approximate posterior distribution of $\pi_1 - \pi_2$ is $normal(-.182, .0429^2)$.

(d) The 99% Bayesian credible interval for $\pi_1 - \pi_2$ is $(-.292, -.071)$.

(e) We calculate the posterior probability of the null hypothesis:

$$P(\pi_1 - \pi_2 \geq 0) = P(Z \geq 4.238) = .0000.$$

We reject the null hypothesis and conclude that the proportion of New Zealand women in the younger group who have given birth before age 25 is less than the proportion of New Zealand women in the older group who have given birth before age 25.

13.15. (a) The measurements on the same cow form a pair.

(c) The posterior precision equals

$$\frac{1}{3^2} + \frac{7}{1^2} = .703704.$$

The posterior variance equals $\frac{1}{.703704} = .142105$ and the posterior mean equals

$$\frac{\frac{1}{3^2}}{.703704} \times 0 + \frac{7}{1^2} .703704 \times -3.9143 = -3.89368 .$$

The posterior distribution of μ_d is $normal(-3.89, .377^2)$.

(d) The 95% Bayesian credible interval is $(-4.63, -3.15)$.

(e) To test the hypothesis

$$H_0 : \mu_d = 0 \qquad \text{versus} \qquad H_1 : \mu_d \neq 0 ,$$

we observe that the null value 0 lies outside the credible interval, so we reject the null hypothesis.

Chapter 14: Bayesian Inference for Simple Linear Regression

14.1. (a) and (c) The scatterplot of oxygen uptake on heart rate with least squares line

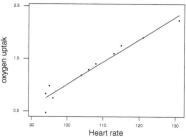

(b) The least squares slope

$$B = \frac{145.610 - 107 \times 1.30727}{11584.1 - 107^2} = 0.0426514 \,.$$

The least squares y-intercept equals

$$A_0 = 1.30727 - .0426514 \times 107 = -3.25643 \,.$$

(d) The estimated variance about the least squares line is found by taking the sum of squares of residuals and dividing by $n - 2$ and equals $\hat{\sigma}^2 = .1303^2$.

(e) The *likelihood* of β is proportional to a $normal(B, \frac{\sigma^2}{SS_x})$, where B is the least squares slope and $SS_x = n \times (\overline{x^2} - \bar{x}^2) = 1486$ and $\sigma^2 = .13^2$. The prior for β is $normal(0, 1^2)$. The posterior precision will be

$$\frac{1}{(s')^2} = \frac{1}{1^2} + \frac{SS_x}{.13^2} = 87930 \,,$$

the posterior variance will be $(s')^2 = \frac{1}{87930} = .000011373$, and the posterior mean is

$$m' = \frac{\frac{1}{1^2}}{87930} \times 0 + \frac{\frac{SS_x}{.13^2}}{87930} \times .0426514 = .0426509 \,.$$

The posterior distribution of β is $normal(.0426, .00337^2)$

(f) A 95% Bayesian credible interval for β is $(.036, .049)$.

(g) We observe that the null value 0 lies outside the credible interval, so we reject the null hypothesis.

14.3. (a) and (c) The scatterplot of distance on speed with least squares line

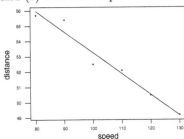

(b) The least squares slope

$$B = \frac{5479.83 - 105 \times 52.5667}{11316.7 - 105^2} = -0.136000 \,.$$

The least squares y-intercept equals

$$A_0 = 52.5667 - -0.136000 \times 105 = 66.8467 \,.$$

(d) The estimated variance about the least squares line is found by taking the sum of squares of residuals and dividing by $n - 2$ and equals $\hat{\sigma}^2 = .571256^2$.

(e) The *likelihood* of β is proportional to a *normal*$(B, \frac{\sigma^2}{SS_x})$ where B is the least squares slope and $SS_x = n \times (\overline{x^2} - \bar{x}^2) = 1750$ and $\sigma^2 = .57^2$. The prior for β is *normal*$(0, 1^2)$. The posterior precision will be

$$\frac{1}{(s')^2} = \frac{1}{1^2} + \frac{SS_x}{.57^2} = 5387.27 \,,$$

the posterior variance $(s')^2 = \frac{1}{5387.27} = .000185623$, and the posterior mean is

$$m' = \frac{\frac{1}{1^2}}{5387.27} \times 0 + \frac{\frac{SS_x}{.57^2}}{5387.27} \times (-0.136000) = -.135975 \,.$$

The posterior distribution of β is *normal*$(-.136, .0136^2)$.

(f) A 95% Bayesian credible interval for β is $(-.163, -0.109)$.

(g) We calculate the posterior probability of the null hypothesis.

$$P(\beta \geq 0) = P(Z \geq 9.98) = .0000 \,.$$

This is less than the level of significance, so we reject the null hypothesis and conclude that $\beta < 0$.

14.5. (a) and (c) Scatterplot of strength on fiber length with least squares line

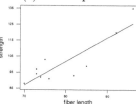

(b) The least squares slope

$$B = \frac{8159.3 - 79.6 \times 101.2}{6406.4 - 79.6^2} = 1.47751.$$

The least squares y-intercept equals

$$A_0 = 101.2 - 1.47751 \times 79.6 = -16.4095.$$

(d) The estimated variance about the least squares line is found by taking the sum of squares of residuals and dividing by $n - 2$ and equals $\hat{\sigma}^2 = 7.667^2$.

(e) The *likelihood* of β is proportional to a $normal(B, \frac{\sigma^2}{SS_x})$, where B is the least squares slope and $SS_x = n \times (\overline{x^2} - \bar{x}^2) = 702.400$ and $\sigma^2 = 7.7^2$. The prior for β is $normal(0, 10^2)$. The posterior precision will be

$$\frac{1}{10^2} + \frac{SS_x}{7.7^2} = 11.8569,$$

the posterior variance $= \frac{1}{11.8569} = .0843394$, and the posterior mean is

$$\frac{\frac{1}{10^2}}{11.8569} \times 0 + \frac{\frac{SS_x}{7.7^2}}{11.8569} \times 1.47751 = 1.47626.$$

The posterior distribution of β is $normal(1.48, .29^2)$.

(f) A 95% Bayesian credible interval for β is $(.91, 2.05)$.

(g) To test the hypothesis

$$H_0 : \beta \leq 0 \qquad \text{versus} \qquad H_1 : \beta > 0,$$

we calculate the posterior probability of the null hypothesis.

$$P(\beta \leq 0) = P\left(\frac{\beta - 1.48}{.29} \leq \frac{0 - 1.48}{.29}\right)$$
$$= P(Z \leq -5.08) = .0000.$$

This is less than the level of significance, so we reject the null hypothesis and conclude $\beta > 0$.

(h) The predictive distribution for the next observation y_{11} taken for a yarn with fiber length $x_{11} = 90$ is $normal(116.553, 8.622^2)$.

(i) A 95% credible interval for the prediction is

$$116.553 \pm 1.96 \times 8.622 = (99.654, 133.452).$$

14.7. (a) The scatterplot of number of ryegrass plants on the weevil infestation rate where the ryegrass was infected with endophyte. The does not look linear. It has a dip at infestation rate of 10.

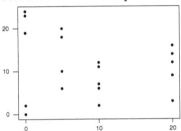

(c) The least squares slope is given by

$$B = \frac{19.9517 - 8.75 \times 2.23694}{131.250 - 8.75^2} = .00691966 .$$

The least squares y-intercept equals

$$A_0 = 2.23694 - .00691966 \times 8.75 = 2.17640 .$$

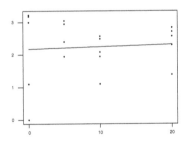

(d) $\hat{\sigma}^2 = .850111^2$.

(e) The *likelihood* of β is proportional to a $normal(B, \frac{\sigma^2}{SS_x})$, where B is the least squares slope and $SS_x = n \times (\overline{x^2} - \bar{x}^2) = 1093.75$ and $\sigma^2 = .850111^2$. The prior for β is $normal(0, 1^2)$. The posterior precision is

$$\frac{1}{(s')^2} = \frac{1}{1^2} + \frac{SS_x}{.850111^2} = 1514.45 .$$

the posterior variance is $(s')^2 = \frac{1}{1514.45} = .000660307$, and the posterior mean is

$$m' = \frac{\frac{1}{1^2}}{1514.45} \times 0 + \frac{\frac{SS_x}{.850111^2}}{1514.45} \times .00691966 = .00691509 .$$

The posterior distribution of β is $normal(.0069, .0257^2)$.

14.9. (a) To find the posterior distribution of $\beta_1 - \beta_2$, we take the difference between the posterior means, and add the posterior variances since they are independent. The posterior distribution of $\beta_1 - \beta_2$ is $normal(1.012, .032^2)$.

(b) The 95% credible interval for $\beta_1 - \beta_2$ is $(.948, 1.075)$.

(c) We calculate the posterior probability of the null hypothesis:

$$P(\beta_1 - \beta_2 \leq 0) = P(Z \leq -31) = .0000 \,.$$

This is less than the level of significance, so we reject the null hypothesis and conclude $\beta_1 - \beta_2 > 0$. This means that infection by endophyte offers ryegrass some protection against weevils.

Chapter 15: Bayesian Inference for Standard Deviation

15.1. (a) The shape of the likelihood function for the variance σ^2 is

$$f(y_1, \ldots, y_n | \sigma^2) \propto (\sigma^2)^{-\frac{n}{2}} e^{-\frac{\sum(y_i - \mu)^2}{2\sigma^2}}$$

$$\propto (\sigma^2)^{-\frac{10}{2}} e^{-\frac{1428}{2\sigma^2}} \,.$$

(b) The prior distribution for the variance is positive uniform $g(\sigma^2) = 1$ for $\sigma^2 > 1$. (This improper prior can be represented as $S \times$ an *inverse chi-squared* distribution with -2 degrees of freedom where $S = 0$.) The shape of the prior distribution for the standard deviation σ is found by applying the change of variable formula. It is

$$g_\sigma(\sigma) \propto g_{\sigma^2}(\sigma^2) \times \sigma$$

$$\propto \sigma.$$

(c) The posterior distribution of the variance is $1428 \times$ an *inverse chi-squared* with 8 degrees of freedom. Its formula is

$$g_{\sigma^2}(\sigma^2 | y_1, \ldots, y_{10}) = \frac{1428^{\frac{8}{2}}}{2^{\frac{8}{2}} \Gamma(\frac{8}{2})} \times \frac{1}{(\sigma^2)^{\frac{8}{2}+1}} e^{-\frac{1428}{2\sigma^2}} \,.$$

(d) The posterior distribution of the standard deviation is found by using the change of variable formula. It has shape given by

$$g_{\sigma^2}(\sigma^2 | y_1, \ldots, y_{10}) = \frac{1428^{\frac{8}{2}}}{2^{\frac{8}{2}} \Gamma(\frac{8}{2})} \times \frac{1}{(\sigma)^{8+1}} e^{-\frac{1428}{2\sigma^2}} \,.$$

(e) A 95% Bayesian credible interval for the standard deviation is

$$\left(\sqrt{\frac{1428}{17.5345}}, \sqrt{\frac{1428}{2.17997}} \right) = (9.024, 25.596) \,.$$

(f) To test $H_0 : \sigma \leq 8$ versus $H_1 : \sigma > 8$, we calculate the posterior probability of the null hypothesis.

$$P(\sigma \leq 8) = P(W \geq \frac{1428}{8^2})$$
$$= P(W \geq 22.3125),$$

where W has the *chi-squared* distribution with 8 degrees of freedom. From Table B.5 we see that this lies between the upper tail values for .005 and .001. The exact probability of the null hypothesis found using Minitab is .0044. Hence we would reject the null hypothesis and conclude $\sigma > 8$ at the 5% level of significance.

15.2. (a) The shape of the likelihood function for the variance σ^2 is

$$f(y_1, \ldots, y_n | \sigma^2) \propto (\sigma^2)^{-\frac{n}{2}} e^{-\frac{\sum(y_i - \mu)^2}{2\sigma^2}}$$
$$\propto (\sigma^2)^{-\frac{10}{2}} e^{-\frac{9.4714}{2\sigma^2}}.$$

(b) The prior distribution for the variance is Jeffreys' prior $g(\sigma^2) = (\sigma^2)^{-1}$ for $\sigma^2 > 1$. (This improper prior can be represented as $S\times$ an *inverse chi-squared* distribution with 0 degrees of freedom where $S = 0$.) The shape of the prior distribution for the standard deviation σ is found by applying the change of variable formula. It is

$$g_\sigma(\sigma) \propto g_{\sigma^2}(\sigma^2) \times \sigma$$
$$\propto \sigma^{-1}.$$

(c) The posterior distribution of the variance is $9.4714\times$ an *inverse chi-squared* with 10 degrees of freedom. Its formula is

$$g_{\sigma^2}(\sigma^2 | y_1, \ldots, y_{10}) = \frac{9.4714^{\frac{10}{2}}}{2^{\frac{10}{2}}\Gamma(\frac{10}{2})} \times \frac{1}{(\sigma^2)^{\frac{10}{2}+1}} e^{-\frac{9.4714}{2\sigma^2}}.$$

(d) The posterior distribution of the standard deviation is found by using the change of variable formula. It has shape given by

$$g_{\sigma^2}(\sigma^2 | y_1, \ldots, y_{10}) = \frac{9.4714^{\frac{10}{2}}}{2^{\frac{10}{2}}\Gamma(\frac{10}{2})} \times \frac{1}{(\sigma)^{10+1}} e^{-\frac{9.4714}{2\sigma^2}}.$$

(e) A 95% Bayesian credible interval for the standard deviation is

$$\left(\sqrt{\frac{9.4714}{20.483}}, \sqrt{\frac{9.4714}{3.247}} \right) = (.680, 1.708).$$

(f) To test $H_0 : \sigma \leq 1.0$ versus $H_1 : \sigma > 1.0$, we calculate the posterior probability of the null hypothesis.

$$P(\sigma \leq 1.0) = P(W \geq \frac{9.4714}{1^2})$$
$$= P(W \geq 9.4714),$$

where W has the *chi-squared* distribution with 10 degrees of freedom. From Table B.5 we see that this lies between the upper tail values for .50 and .10. (The exact probability of the null hypothesis found using Minitab is .4880.) Hence we can not reject the null hypothesis and must conclude $\sigma \leq 1.0$ at the 5% level of significance.

15.3. (a) The shape of the likelihood function for the variance σ^2 is

$$f(y_1, \ldots, y_n | \sigma^2) \propto (\sigma^2)^{-\frac{n}{2}} e^{-\frac{\sum(y_i - \mu)^2}{2\sigma^2}}$$
$$\propto (\sigma^2)^{-\frac{5}{2}} e^{-\frac{26.119}{2\sigma^2}}.$$

(b) The prior distribution is $S \times$ an *inverse chi-squared* distribution with 1 degree of freedom where $S = .4549 \times 4^2 = 7.278$. Its formula is

$$g_{\sigma^2}(\sigma^2) = \frac{7.278^{\frac{1}{2}}}{2^{\frac{1}{2}}\Gamma(\frac{1}{2})} \times \frac{1}{(\sigma^2)^{\frac{1}{2}+1}} e^{-\frac{7.278}{2\sigma^2}}.$$

(c) The shape of the prior distribution for the standard deviation is found by applying the change of variable formula. It is

$$g_\sigma(\sigma) \propto g_{\sigma^2}(\sigma^2) \times \sigma$$
$$\propto \frac{1}{(\sigma)^2} e^{-\frac{7.278}{2\sigma^2}}.$$

(d) The posterior distribution of the variance is $33.40 \times$ an *inverse chi-squared* with 6 degrees of freedom. Its formula is

$$g_{\sigma^2}(\sigma^2 | y_1, \ldots, y_5) = \frac{33.40^{\frac{6}{2}}}{2^{\frac{6}{2}}\Gamma(\frac{6}{2})} \times \frac{1}{(\sigma^2)^{\frac{6}{2}+1}} e^{-\frac{33.40}{2\sigma^2}}.$$

(e) The posterior distribution of the standard deviation is found by using the change of variable formula. It has shape given by

$$g_\sigma(\sigma | y_1, \ldots, y_5) \propto g_{\sigma^2}(\sigma^2 | y_1, \ldots, y_5) \times \sigma$$
$$\propto \frac{1}{(\sigma^2)^{\frac{5}{2}+1}} e^{-\frac{33.40}{2\sigma^2}}.$$

(f) A 95% Bayesian credible interval for the standard deviation is

$$\left(\sqrt{\frac{33.40}{14.449}}, \sqrt{\frac{33.40}{1.237}}\right) = (1.520, 5.195).$$

(g) To test $H_0 : \sigma \le 5$ versus $H_1 : \sigma > 5$ we calculate the posterior probability of the null hypothesis.

$$P(\sigma \le 5) = P\left(W \ge \frac{33.40}{5^2}\right)$$
$$= P(W \ge 1.336),$$

where W has the *chi-squared* distribution with 6 degrees of freedom. From Table B.5 we see that this lies between the upper tail values for .975 and .95. (The exact probability of the null hypothesis found using Minitab is .9696.) Hence we would accept the null hypothesis and conclude $\sigma \le 5$ at the 5% level of significance.

Chapter 16: Robust Bayesian Methods

16.1. (a) The posterior $g_0(\pi|y = 10)$ is *beta*$(7 + 10, 13 + 190)$.

(b) The posterior $g_1(\pi|y = 10)$ is *beta*$(1 + 10, 1 + 190)$.

(c) The posterior probability $P(I = 0|y = 10) = .163$.

(d) The marginal posterior $g(\pi|y = 10) = .163 \times g_0(\pi|y = 10) + .837 \times g_1(\pi|y = 10)$. This is a mixture of the two beta posteriors where the proportions are the posterior probabilities of I.

16.3. (a) The posterior $g_0(\mu|y_1, \ldots, y_6)$ is *normal*$(1.10061, .000898^2)$.

(b) The posterior $g_1(\mu|y_1, \ldots, y_6)$ is *normal*$(1.10302, .002^2)$.

(c) The posterior probability $P(I = 0|y_1, \ldots, y_6) = .972$.

(d) The marginal posterior

$$g(\mu|y_1, \ldots, y_6) = .972 \times g_0(\mu|y_1, \ldots, y_6) + .028 \times g_1(\mu|y_1, \ldots, y_6).$$

This is a mixture of the two normal posteriors where the proportions are the posterior probabilities of I.

References

K. M. Abadir and J. R. Magnus. *Matrix Algebra*. Cambridge University Press, Cambridge, 2005.

G. Barker and R. McGhie. The biology of introduced slugs (Pulmonata) in New Zealand: Introduction and notes on *limax maximus*. *New Zealand Entomologist*, 8:106–111, 1984.

T. Bayes. An essay towards solving a problem in the doctrine of chances. *Philosophical Transactions of the Royal Society*, 53:370–418, 1763.

R. L. Bennett, J. M. Curran, N. D. Kim, S. A. Coulson, and A. W. N. Newton. Spatial variation of refractive index in a pane of float glass. *Science and Justice*, 43(2):71–76, 2003.

D. Berry. *Statistics: A Bayesian Perspective*. Duxbury, Belmont, CA, 1996.

W. M. Bolstad. *Understanding Computational Bayesian Statistics*. John Wiley & Sons, New York, NY, 2010.

W. M. Bolstad, L. A. Hunt, and J. L. McWhirter. Sex, drugs, and rock & roll survey in a first-year service course in statistics. *The American Statistician*, 55:145–149, 2001.

J. M. Curran. *Introduction to Data Analysis with R for Forensic Scientists.* CRC Press, Boca Raton, FL, 2010.

B. de Finetti. *Theory of Probability, Volume 1 and Volume 2.* John Wiley & Sons, New York, NY, 1991.

M. H. DeGroot. *Optimal Statistical Decisions.* McGraw–Hill, New York, NY, 1970.

D. M Fergusson, J. M. Boden, and L. J. Horwood. Circumcision status and risk of sexually transmitted infection in young adult males: an analysis of a longitudinal birth cohort. *Pediatrics,* 118:1971–1977, 2006.

R. A. Fisher. The fiducial argument in statistical inference. *Annals of Eugenics,* 8:391–398, 1935.

A. Gelman. Prior distributions for variance parameters in hierarchical models. *Bayesian Analysis,* 1(3):515–533, 2006.

A. Gelman, J. B. Carlin, H. S. Stern, and D. B. Rubin. *Bayesian Data Analysis.* Chapman and Hall, London, 2nd edition, 2003.

S. Geman and D. Geman. Stochastic relaxation, Gibbs distributions, and the Bayesian restoration of images. *IEEE Transactions on Pattern Analysis and Machine Intelligence,* 6(6):721–741, 1984.

D. A. Harville. *Matrix Algebra from a Statistician's Perspective.* Springer–Verlag, New York, NY, 1997.

W. K. Hastings. Monte Carlo sampling methods using Markov chains and their applications. *Biometrika,* 57(1):97–109, 1970.

P. G. Hoel. *Introduction to Mathematical Statistics.* John Wiley & Sons, New York, 5th edition, 1984.

E. T. Jaynes and G. L. Bretthorst (Editor). *Probability Theory: The Logic of Science.* Cambridge University Press, Cambridge, 2003.

H. Jeffreys. *Theory of Probability.* Oxford University Press, Oxford, 1961.

N. Johnson, S. Kotz, and N. Balakrishnan. *Continuous Univariate Distributions, Volume 1.* John Wiley & Sons, New York, NY, 1970.

K. Johnstone, S. Baxendine, A. Dharmalingam, S. Hillcoat-Nallétamby, I. Pool, and N. Paki Paki. Fertility and family surveys in countries of the ECE region: Standard Country Report: New Zealand, 2001.

P. Lee. *Bayesian Statistics: An Introduction.* Edward Arnold, London, 1989.

A. Marsault, I. Poole, A. Dharmalingam, S. Hillcoat-Nallétamby, K. Johnstone, C. Smith, and M. George. Technical and methodological report: New Zealand women: Family, employment and education survey, 1997.

G. McBride, D. Till, T. Ryan, A. Ball, G. Lewis, S. Palmer, and P. Weinstein. Freshwater microbiology research programme pathogen occurrence and human risk assessment analysis, 2002.

L. M. McLeay, V. R. Carruthers, and P. G. Neil. Use of a breath test to determine the fate of swallowed fluids in cattle. *American Journal of Veterinary Research*, 58:1314–1319, 1997.

N. Metropolis, A. W. Rosenbluth, M. N. Rosenbluth, A. H. Teller, and E. Teller. Equations of state calculations by fast computing machines. *Journal of Chemical Physics*, 21(6):1087–1092, 1953.

A. M. Mood, F. A. Graybill, and D. C. Boes. *Introduction to the Theory of Statistics*. McGraw–Hill, New York, NY, 1974.

R. M. Neal. Technical Report 97222: Markov chain Monte Carlo methods based on "slicing" the density function, Department of Statistics, University of Toronto, 1997.

R. M. Neal. Slice sampling. *Annals of Statistics*, 31(3):705–767, 2003.

A. O'Hagan. *Kendall's Advanced Theory of Statistics, Volume 2B, Bayesian Inference*. Edward Arnold, London, 1994.

A. O'Hagan and J. J. Forster. *Kendall's Advanced Theory of Statistics, Volume 2B: Bayesian Inference*. Edward Arnold, London, 2nd edition, 2004.

F. Petchey. Radiocarbon dating fish bone from the Houhora archeological site, New Zealand. *Archeology Oceania*, 35:104–115, 2000.

F. Petchey and T. Higham. Bone diagenesis and radiocarbon dating of fish bones at the Shag River mouth site, New Zealand. *Journal of Archeological Science*, 27:135–150, 2000.

S. J. Press. *Bayesian Statistics: Principles, Models, and Applications*. John Wiley & Sons, New York, NY, 1989.

J. Sherman and W. J. Morrison. Adjustment of an inverse matrix corresponding to changes in the elements of a given column or a given row of the original matrix (abstract). *Annals of Mathematical Statistics*, 20:621, 1949.

J. Sherman and W. J. Morrison. Adjustment of an inverse matrix corresponding to a change in one element of a given matrix. *Annals of Mathematical Statistics*, 21:124–127, 1950.

S. M. Stigler. Do robust estimators work with real data? *The Annals of Statistics*, 5:1055–1098, 1977.

M. Stuiver, P. J. Reimer, and S. Braziunas. High precision radiocarbon age calibration for terrestial and marine samples. *Radiocarbon*, 40:1127–1151, 1998.

E. Thorp. *Beat the Dealer*. Blaisdell Publishing Company, New York, NY, 1st edition, 1962.

J. von Neumann. Various techniques used in connection with random digits. *Monte Carlo Methods: National Bureau of Standards Applied Mathematics Series*, 12:36–38, 1951.

A. Wald. *Statistical Decision Functions*. John Wiley & Sons, New York, NY, 1950.

B. L. Welch. The significance of the difference between two means when the population variances are unequal. *Biometrika*, 29:350–362, 1938.

C. J. Wild and G. A. F. Seber. *Chance Encounters*. John Wiley & Sons, New York, NY, 1999.

Index

Bayes factor, 75
Bayes' theorem, 71, 79
    analyzing the observations all to-
        gether, 114, 215
    analyzing the observations sequen-
        tially, 114, 215
    binomial observation
      beta prior, 151
      continuous prior, 150
      discrete prior, 116
      mixture prior, 342
      uniform prior, 150
    discrete random variables, 109
    events, 68, 71, 73
    linear regression model, 292
    mixture prior, 340
    normal observations known mean
      inverse gamma prior for σ^2, 324
      inverse-chi-squared prior for σ^2,
        321
      Jeffreys' prior for σ^2, 320
      positive uniform prior for σ^2,
        319
    normal observations with known
      variance
      continuous prior for μ, 218
      discrete prior for μ, 211
      flat prior for μ, 219
      mixture prior, 345
      normal prior for μ, 219
    Poisson
      Jeffreys' prior, 195
    Poisson observation
      continuous prior, 193
      gamma prior, 195
      positive uniform prior, 194
Bayes' theorem using table

binomial observation with discrete prior, 118

discrete observation with discrete prior, 113

normal observation with discrete prior, 212

Poisson observation with discrete prior, 120

Bayesian approach to statistics, 6, 12

Bayesian bootstrap, 453

Bayesian credible interval, 162

binomial proportion π, 162

difference between normal means $\mu_1 - \mu_2$

equal variances, 257

unequal variances, 262

difference between proportions $\pi_1 - \pi_2$, 265

normal mean μ, 224, 241

normal standard deviation σ, 328

Poisson parameter μ, 203

regression slope β, 297

used for Bayesian two-sided hypothesis test, 185

Bayesian estimator

normal mean μ, 238

binomial proportion π, 161

normal σ, 326

Bayesian hypothesis test

one-sided

binomial proportion π, 182

normal mean μ, 244

normal standard deviation σ, 329

Poisson parameter μ, 203

regression slope β, 297

two-sided

binomial proportion π, 185

normal mean μ, 249

Poisson parameter μ, 205

Bayesian inference for standard deviation, 315

Bayesian universe, 71, 109, 121

parameter space dimension, 74, 79, 110, 121

reduced, 72, 110, 121

sample space dimension, 74, 79, 109, 121

beta distribution, 135

mean, 136

normal approximation, 142

probability density function, 135

shape, 135

variance, 137

bias

response, 16

sampling, 15

binomial distribution, 90, 103, 149, 498

characteristics of, 90

mean, 91

probability function, 91

variance, 92

blackjack, 76, 82

boxplot, 32, 51

stacked, 39

Buffon's needle, 438

burn-in, 475

candidate density, 439

central limit theorem, 140, 211

chi-squared distribution, 506

conditional probability, 78

conditional random variable

continuous

conditional density, 144

confounded, 3

conjugate family of priors

binomial observation, 153, 163

Poisson observation, 195, 196

continuous random variable, 129

pdf, 131

probability density function, 131, 145

probability is area under density, 133, 145

correlation

bivariate data set, 49, 52

covariance

bivariate data set, 49

cumulative frequency polygon, 37, 51

deductive logic, 60
degrees of freedom, 45
    simple linear regression, 297
    two samples unknown equal variances, 260
    two samples unknown unequal variances
        Satterthwaite's adjustment, 263
    unknown variance, 226
derivative, 483
    higher, 484
    partial, 493
designed experiment, 19, 23
    completely randomized design, 19, 23, 26, 27
    randomized block design, 19, 23, 26, 27
differentiation, 482
discrete random variable, 83, 84, 102
    expected value, 86
    probability distribution, 83, 86, 102
    variance, 87
distribution
    exponential, 437
    shifted exponential, 451
dotplot, 32
    stacked, 39

ecdf, 474
effective sample size, 475
ESS, 475
empirical distribution function, 474
envelope, 438
equivalent sample size
    beta prior, 155
    gamma prior, 197
    normal prior, 222
estimator
    frequentist, 171, 238
    mean squared error, 173
    minimum variance unbiased, 172, 238

    sampling distribution, 171
    unbiased, 172, 238
event, 62
events
    complement, 62, 78
    independent, 65, 66
    intersection, 62, 78
    mutually exclusive (disjoint), 63, 66, 78
    partitioning universe, 69
    union, 62, 78
expected value
    continuous random variable, 133
    discrete random variable, 86, 103
experimental units, 18–20, 26

finite population correction factor, 93
five number summary, 33
frequency table, 35
frequentist
    interpretation of probability and parameters, 170
frequentist approach to statistics, 5, 11
frequentist confidence interval, 175
    normal mean μ, 241
    regression slope β, 297
frequentist confidence intervals
    relationship to frequentist hypothesis tests, 185
frequentist hypothesis test
    P-value, 181
    level of significance, 179
    null distribution, 180
    one-sided
        binomial proportion π, 179
        normal mean μ, 244
    rejection region, 181
    two-sided
        binomial proportion π, 182
        normal mean μ, 247
function, 477
    antiderivative, 486
    continuous, 480
        maximum and minimum, 482

differentiable, 483
  critical points, 485
graph, 478
limit at a point, 478
fundamental theorem of calculus, 490

gamma distribution, 137
  mean, 138
  probability density function, 138
  shape, 137
  variance, 139
Gibbs sampling
  special case of Metropolis–Hastings
    algorithm, 461
Glivenko–Cantelli theorem, 474

hierarchical models, 462
histogram, 36, 37, 51
hypergeometric distribution, 92
  mean, 93
  probability function, 93
  variance, 93

integration, 486
  definite integral, 486, 489, 491
  midpoint rule, 473
  multiple integral, 495
interquartile range
  data set, 45, 52
  posterior distribution, 160
inverse chi-squared distribution, 316,
    331
  density, 316
inverse probability transform, 442

Jeffreys' prior
  binomial, 153
  normal mean, 219
  normal variance, 320
  Poisson, 195
joint likelihood
  linear regression sample, 292
joint random variables
  conditional probability, 100
  conditional probability distribution, 101

continuous, 143
  joint density, 143
  marginal density, 143
continuous and discrete, 144
discrete, 96
  joint probability distribution, 96
  marginal probability distribution, 96
independent, 98
joint probability distribution, 103
marginal probability distribution, 103

likelihood
  binomial, 117
  binomial proportional, 120
  discrete parameter, 111, 112
  events partitioning universe, 71, 72
  mean
    single normal observation, 212
  multiplying by constant, 73, 120
  normal
    sample mean $\bar{y}$, 217
  normal mean
    random sample of size n, 217
    using density function, 213
    using ordinates table, 213
  normal variance, 318
  Poisson, 194
  regression
    intercept $\alpha_{\bar{x}}$, 294
    slope β, 294
  sample mean from normal distribution, 221
log-concave, 443
logic
  deductive, 77
  inductive, 78
lurking variable, 2, 11, 19, 21, 26, 27

marginalization, 227, 299
marginalizing out the mixture parameter, 341
matrix determinant lemma, 405

MCMC
    mixing, 475
mean
    continuous random variable, 133
    data set, 42, 51
    difference between random variables, 99, 104
    discrete random variable, 87
    grouped data, 43
    of a linear function, 88, 103
    sum of random variables, 97, 104
    trimmed, 44, 52
mean squared error, 239
measures of location, 41
measures of spread, 44
median
    data set, 43, 50, 51
Metropolis–Hastings algorithm
    steps, 456
mixture prior, 338
Monte Carlo study, 7, 12, 23, 26, 77
multiple linear regression model, 411
    removing unnecessary variables, 421
multivariate normal distribution
    known covariance matrix
        posterior, 399

non-sampling errors, 16
normal distribution, 139
    area under standard normal density, 499
    computing the normal density, 502
    find normal probabilities, 500
    mean, 140
    ordinates of standard normal density, 502
    probability density function, 140
    shape, 140
    standard normal probabilities, 141
    variance, 140
normal linear regression model
    posterior, 418
nuisance parameter, 7, 315
    marginalization, 227, 299

observational study, 18, 22
Ockham's razor, 4, 178
odds, 75
order statistics, 33, 35, 50
outcome, 62
outlier, 43

parameter, 5, 6, 14, 21, 74
parameter space, 74
pdf, 131
    continuous random variable, 131
plausible reasoning, 60, 78
point estimation, 171
Poisson distribution, 93, 193, 505
    characteristics of, 94
    mean, 95
    probability function, 94
    variance, 95
population, 5, 14, 21
posterior distribution, 7
    discrete parameter, 112
    normal
        discrete prior, 212
    regression
        slope β, 295
posterior mean
    as an estimate for π, 161
    beta distribution, 159
    gamma distribution, 200
posterior mean square
    of an estimator, 161
posterior median
    as an estimate for π, 161
    beta distribution, 159
    gamma distribution, 200
posterior mode
    beta distribution, 158
    gamma distribution, 200
posterior probability
    of an unobservable event, 71
posterior probability distribution
    binomial with discrete prior, 117
posterior standard deviation, 160
posterior variance
    beta distribution, 160

pre-posterior analysis, 12
precision
    normal
        $\bar{y}$, 221
        observation, 221
        posterior, 221
        prior, 221
    regression
        likelihood, 295
        posterior, 295
        prior, 295
predictive distribution
    normal, 227
    regression model, 298
pre-posterior analysis, 8
prior distribution, 6
    choosing beta prior for π
        matching location and scale, 154, 163
        vague prior knowledge, 154
    choosing inverse chi-squared prior for σ^2, 322
    choosing normal prior for μ, 222
    choosing normal priors for regression, 294
    constructing continuous prior for μ, 222
    constructing continuous prior for π, 155, 164
    discrete parameter, 111
    multiplying by constant, 73, 119
    uniform prior for π, 163
prior probability
    for an unobservable event, 71
probability, 62
    addition rule, 64
    axioms, 64, 78
    conditional, 66
        independent events, 67
    degree of belief, 74
    joint, 65
    law of total probability, 69, 79
    long-run relative frequency, 74
    marginal, 66
    multiplication rule, 67, 79, 102

probability density function
    continuous random variable, 131
probability distribution
    conditional, 101
probability integral transform, 436
proposal, 456
proposal density, 439

quartiles
    data set, 33, 50
    from cumulative frequency polygon, 38
    posterior distribution, 160

random experiment, 62, 78
random sampling
    cluster, 16, 22
    simple, 15, 22
    stratified, 16, 22
randomization, 5, 11
randomized response methods, 17, 22
range
    data set, 44, 52
regression
    Bayes' theorem, 292
    least squares, 284
    normal equations, 285
    simple linear regression assumptions, 290
robust Bayesian methods, 337

sample, 5, 14, 21
sample space, 74, 78
    of a random experiment, 62
sampling
    acceptance–rejection, 438
    adaptive–rejection, 443
    importance, 450
        importance density, 450
        importance weights, 451
        likelihood ratio, 451
        normalized weights, 453
        tuning, 451
    inverse probability, 436
    Metropolis–Hastings
        blockwise, 462

independent candidate, 459
    random-walk candidate, 457
  rejection, 438
  sampling-importance-resampling, 453
  SIR, 453
  slice, 470
sampling distribution, 6, 7, 12, 23–26, 170
sampling frame, 15
scatterplot, 47, 52, 283
scatterplot matrix, 47, 52
scientific method, 3, 11
  role of statistics, 11
  role of statistics, 5
Sherman–Morrison formula, 427
sign test, 385
standard deviation
  data set, 46, 52
statistic, 14, 22
statistical inference, 1, 14, 78
statistics, 5
stem-and-leaf diagram, 34, 51
  back-to-back, 39
strata, 16
stratum, 16
Student's t distribution, 225, 315, 504

t-test
  paired, 384
tangent line, 445
target density, 439
thinning, 475
traceplot
  Gibbs sampling chain, 463
transformation
  variance stabilizing, 424

uniform distribution, 135
universe, 62
  of a joint experiment, 96
  reduced, 66, 69, 100
updating rule
  binomial proportion π, 153
  normal mean μ, 221

normal variance σ^2, 322
Poisson parameter μ, 196

variance
  continuous random variable, 134
  data set, 45, 52
  difference between independent random variables, 99, 104
  discrete random variable, 87, 103
  grouped data, 45
  linear function, 103
  of linear function, 88
  sum of independent random variables, 99, 104
Venn diagram, 62, 65

Printed in the USA/Agawam, MA
September 23, 2021

781412.010